T0135299

Springer Series in Bio-/Neuroinformatics

Volume 6

Series editor

Nikola Kasabov, Auckland, New Zealand

More information about this series at http://www.springer.com/series/10088

Péter Érdi · Basabdatta Sen Bhattacharya
Amy L. Cochran

Editors

Computational Neurology and Psychiatry

 Springer

Editors
Péter Érdi
Department of Physics
Kalamazoo College
Kalamazoo, MI
USA

Amy L. Cochran
Department of Mathematics
University of Michigan
Ann Arbor, MI
USA

Basabdatta Sen Bhattacharya
School of Engineering
University of Lincoln
Lincoln
UK

ISSN 2193-9349   ISSN 2193-9357   (electronic)
Springer Series in Bio-/Neuroinformatics
ISBN 978-3-319-84284-4   ISBN 978-3-319-49959-8   (eBook)
DOI 10.1007/978-3-319-49959-8

Printed on acid-free paper

This Springer imprint is published by Springer Nature
The registered company is Springer International Publishing AG
The registered company address is: Gewerbestrasse 11, 6330 Cham, Switzerland

Contents

Introduction

Basabdatta Sen Bhattacharya, Amy L. Cochran and Péter Erdi

Computational Neurology and Psychiatry

The mysterious ways of the brain and mind have eluded great thinkers and scientists for centuries now [1]. Indications of early, albeit vague, understanding of the intricate connections between the brain and the mind can be traced back to the Indo-Aryan settlers in the central Asian region, who would depict a third eye on the mid-forehead of figurines [2], an area that is now known to be the seat of the medial pre-frontal cortex (a brain structure that "reaches its greatest complexity in the primate brain" and is thought to play a central role in cognitive behaviour [3]). Such depictions convey a deep reverence and fear of the unknown realms of the mind, perhaps implying a belief in the superiority of 'beings' who can control the brain and the mind *to foresee or predict*. Indeed, early history is littered with evidence of many a battle won by the virtue of (superior) foresight and prediction capabilities of kings, queens and warlords.

Fast forward to the 20th Century: Jeff Hawkins in "On Intelligence" argues that prediction is the basis of human intelligence [4]. The term 'intelligence' is used here in context to the neocortical structure that is much larger in humans, making them 'superior' to all other creatures. The neocortex, in association with the sub-cortical structures, stores environmental and experiential information, termed as 'memory', which allows humans to learn and recall, and thus make predictions in time and

B. Sen Bhattacharya (✉)
School of Engineering, University of Lincoln, Lincoln, UK
e-mail: basab.sen.b@gmail.com

P. Erdi
Department of Physics, Kalamazoo College, Michigan, USA
e-mail: perdi@kzoo.edu

A.L. Cochran
Department of Mathematics, University of Michigan, Michigan, USA
e-mail: alcochra@gmail.com

© Springer International Publishing AG 2017
P. Érdi et al. (eds.), *Computational Neurology and Psychiatry*,
Springer Series in Bio-/Neuroinformatics 6,
DOI 10.1007/978-3-319-49959-8_1

space; this in turn facilitates fine skills such as language and art. Hawkin's theory provides an integral mutual dependency between the evolved advancement in brain structure (large neocortex) and the advancement of memory and prediction, both of which are representations of the mind (intelligence). Note the uncanny resemblance between the concept of superiority of the humans as the possessor of the predictive powers of the mind, and the concept of the third eye in times BC (around 3000 years ago). Karl Friston lends further credence to this concept with his Free Energy Principle, hypothesising that the brain acts most efficiently by 'minimising surprises'; in other words, by making better predictions! [5].

So then: *where are we in our understanding of the brain and mind?* Notwithstanding human's superior foresight and prediction, we have yet to put forth a coherent understanding of the brain (neurology) and how it functions as a mind (psychiatry). The 125th Anniversary Special issue of *Science* posed '125 problems for scientists to solve', of which the second problem was posed as: "What Is the Biological Basis of Consciousness?" [6]. Suffice to say that understanding the structure and functional ways of the brain is considered as one of the 'last frontiers in science' [7]. As if in a race to conquer this frontier, many countries have set socio-political agenda around brain research with huge monetary investments. The sudden surge of interest can be attributed to the near 50 % increase in mental health problems worldwide, and within a span of around a decade as reported by World Health Organisation during April 2016 [8]:

> Common mental disorders are increasing worldwide. Between 1990 and 2013, the number of people suffering from depression and/or anxiety increased by nearly 50 %, from 416 million to 615 million. Close to 10 % of the world's population is affected, and mental disorders account for 30 % of the global non-fatal disease burden.

The report is based on recent research arguing the need for investment in tackling the growing menace, and suggesting a quadruple increase in returns [9, 10].

Inability to address the growing world burden of brain disorders is not more painfully apparent than when examining current treatment options. The fundamental drawback of existing pharmacological and therapeutic interventions in several brain disorders are some serious secondary effects that hinder a return to normalcy, even if the symptoms of the primary disease condition are often alleviated successfully. For example, depression is treated with drugs that increase the levels of serotonin; however, the quality of life in patients is often impaired seriously. Similarly, Deep Brain Stimulation (DBS) successfully alleviates Parkinsonian tremor and restores a large degree of autonomy and dignity in patients, and yet, it has not been possible to formalise a 'closed-loop' treatment, where the neural conditions can be monitored, and based on this 'feedback' data, stimulation can be provided on a need-based manner. Furthermore, the neuronal mechanisms that make DBS effective in the disease are understood poorly, which is essential for treating the underlying disease condition. On the other hand, several brain disorders such as tinnitus and dementia do not have any pharmacological relief, while insomnia and other sleeping disorders are treated currently with pills for symptomatic relief only and do not address the root problem.

An essential pre-requisite for enhanced treatment of brain disorders is physiological data that can facilitate the proposition of testable hypotheses and the design of testable models of the brain and the mind [11, 12]. Thus, it is no wonder that millions of rodents, mammals and primates are sacrificed in the quest for data on brain physiology and its complex electro-chemical structure. In this context, the reader may refer to Alex Boese's concise (and tragically humorous) account of some 'bizarre' experiments that were undertaken on animals and humans during the 20th century [13]. However, with increased awareness of both animal and human rights, and formulated regulations and standards, it is getting increasingly difficult to use animals for collecting experimental data. More importantly, where animal models are used successfully for drug discovery, many a times, the developed drugs have proved ineffective in humans [14].

Offering a pioneering direction to brain research during 1940s and 50s, Sir Alan Hodgkin and Sir Andrew Huxley, for the first time, used mathematics to capture the underlying physiology of the basic information transfer mechanism used by neurons viz. the action potential (or *spike*) [15]. Commonly referred to as the Hodgkin-Huxley neuronal model, this is the first neuronal model to be firmly based on physiological data and continues to inspire novel directions in computational neuroscience research [16]. It will indeed not be far-fetched to attribute the popularity of mathematical models in current times to the advent of affordable desktop computers with user-friendly software, so much so that the term 'computational model' is now generally accepted as referring to computer-based modelling and simulation in neuroscience research. The primary advantage of computational models is that hypotheses can be designed and explored in a manner that is quick, easy, and inexpensive (the reader can find evidence of this in the following chapters of this book); this is in stark contrast to expensive and time-consuming experimental studies in vivo and in–vitro. An initial exploration of various hypotheses in silico (i.e. modelling and simulation using computers) can potentially narrow the targets in future experiments, thus saving on time and cost. Historically, the importance of in silico studies is perhaps most apparent in the study of epilepsy, and even early models used biophysically detailed features underlying epileptogenesis [17]. The field is sufficiently mature for an excellent book nearly a decade ago [18]; current state-of-the-art in epilepsy research is sampled in several chapters in this book (see section "Scientific Problems Addressed in the Book" for details) as well as in a recent compilation [19].

However, constraints on computational time and memory requirements have been one of the primary deterrents to a more rapid progression in the field. Moreover, during the early 1970s [20–22], there was a growing notion that to emulate higher order dynamics observed via brain imaging methods (e.g. electroencephalography, functional magnetic resonance imaging), building massive networks of single neurons may lead to a case of 'not seeing the wood for the trees'; rather, modelling the population dynamics underlying such higher level brain behaviour would be desirable. Thus, the concept of neural field and neural mass models were proposed in a series of pioneering works, which enabled the simulation of population behaviour in silico, therefore bypassing the need for building computationally intensive networks of single neuron models. Subsequently, another novel computational framework viz. dynamic causal modelling was proposed and developed based on the statistical technique of Bayesian inference [23]. Collectively, these models play a major role in advancing the understanding of neurological disorders in current times, and allow the exploration of the fascinating world of brain oscillatory dynamics, also referred to as "brain rhythms" [24].

More recently, novel perspectives to computational models have emerged in the fledgling area of computational psychiatry, where both theory-driven and data-driven approaches have proved to be useful [25–27]; this research area is further consolidated by the first online journal in the field that is only just established (see http://computationalpsychiatry.org/). It is speculated that computational psychiatry "might enable the field to move from a symptom-based description of mental illness to descriptors based on objective computational multidimensional functional variables…" [28] (the term 'field' within the quote refers to Psychiatry). Indeed, in silico psychiatry has the potential to provide a holistic understanding of how brain oscillations and neurotransmitters are fundamental to the brain-body-mind function and to the evolution of species [11].

In section "Scientific Problems Addressed in the Book", we present an overview of the computational approaches addressed in this book and against the thematic backdrop of neurological and psychiatric disorders. Some concluding remarks and thoughts are presented in section "Closing Remarks".

Scientific Problems Addressed in the Book

This book brings together diverse computational approaches to study disorders related to the brain and mind, i.e. disorders that fall into the medical fields of neurology, psychiatry, or both.

The book starts off in chapter "Outgrowing Neurological Diseases: Microcircuits, Conduction Delay and Childhood Absence Epilepsy" with an introduction into the dynamical nature of the brain that is at the core of several neurological disorders. Milton et al. make a case study on **Childhood Absence Epilepsy** (CAE) as a dynamical disease caused by underlying channelopathy, i.e. disturbances in ion channels, that are often inherited. Most cases of CAE subside naturally with developmental changes accompanying the onset of adolescence, but the underlying neuronal correlates remain unknown. The model-based investigation implicates axonal conduction delay in local thalamo-cortical circuits (microcircuits) as a possible culprit. Furthermore, the model validates that developmental changes naturally rectify the aberrant biological conditions causing the delay. The paroxysmal (sudden onset, as opposed to periodic) oscillations that are hallmarks of EEG are presented from a dynamical systems perspective with supporting code for bifurcation and dynamical analysis (see the chapter Appendix).

The theme of channelopathy in epilepsy continues in chapter "Extracellular Potassium and Focal Seizures—Insight From in Silico Study" with a discussion on the role of extracellular potassium concentration in **Epileptic focal seizures**. Proposed in 1970, the potassium accumulation hypothesis suggests that an abnormal increase in neuronal firing leads to an excessive extracellular potassium accumulation while there is a lack of compensatory increases in potassium uptake. This leads to a further increase in neuronal firing, thus initiating an intrinsic synchronous oscillatory cycle constituting high-amplitude, low-frequency rhythmic spikes, marker of epileptic seizure. Suffczyński et al. make a model-based study to validate experimental results (obtained from whole guinea-pig brain in vitro model of epilepsy) that challenge the "traditional view of alteration in excitation/inhibition balance being the cause of epilepsy"; rather, abnormal increase in spiking activity of the inhibitory cortical interneurons are implicated as the causal factor in producing extracellular potassium accumulation. In an interesting series of model-based results, the authors raise the increasingly important issue of translational research in the context of using model-based predictions for novel therapeutic treatments; a plausible mechanism of need-based extra-cellular potassium absorption using nanoparticle therapy is proposed.

While still on epilepsy, chapter "Time Series and Interactions: Data Processing in Epilepsy Research" addresses an issue of prime importance: data processing in both time and frequency domain. Benkő et al. discuss the pros and cons of traditional information theoretic, statistical and signal processing approaches in context to **processing of Epileptic time-series data**. Essentially, the procedures discussed in the chapter can be safely claimed as generic to EEG, MEG and LFP time-series data obtained from patients with other neurological disorders. The authors make separate discussions on real epileptic data and model-based simulated data.

In chapter "Connecting Epilepsy and Alzheimer's Disease: Modelling of Normal and Pathological Rhythmicity and Synaptic Plasticity Related to Amyloidβ (Aβ) Effects", we are still touching on epilepsy, however, a novel concept of *correlating* the underlying ("hidden") neuronal pathologies of **Temporal Lobe Epileptic Activity** (TLEA) and **Alzheimer's disease** (AD) are discussed. "Where is the connection?" one may ask. Excerpts at the very onset in section "General Remarks" satiate such queries: "In AD, the incidence of convulsive seizures is ten times higher than in age-matched general population…Epilepsy is 87 times more frequent in patients with 'early onset' (of AD) …..occurs particularly early in familial AD…. " Furthermore, the authors reveal that the hippocampus is the common seat for both AD and TLEA. The destructive effects of Amyloid Beta (Aβ) accumulation in AD is now well-known; the hippocampus and entorhinal cortex are the earliest and worst affected areas leading to fast progressive cognitive losses in the disease. The scientific treatment in this chapter deals with kinetic mechanisms of synapses in the hippocampal cells, and looks into the effects of Aβ on synaptic plasticity, which in turn are functions of the calcium ion channel dynamics; the authors make an assumption, based on similar prior research, that NMDA (N-methyl D-aspartate)-receptors "are the primary sources of calcium".

This brings us to chapter "Dynamic Causal Modelling of Neurological Pathology: Using Neural Mass Models to Understand Dynamic Dysfunction in NMDA-Receptor Antibody Encephalitis", where Rosch et al. make a model based study on **NMDA-receptor Antibody Encephalitis**, an inflammation of the brain causing abnormal EEG activity similar to epileptic spike-wave discharges and sharp wave paroxysms. The authors start the chapter with a brief history of EEG in context to hallmarks of Epilepsy, and gently build a case towards the urgency in present day and time for a multi-hierarchical approach for computational model based studies in neurological disorders. Model parameterisation and optimisation are also discussed in detail for the underlying cortical framework.

In chapter "Oscillatory Neural Models of the Basal Ganglia for Action Selection in Healthy and Parkinsonian Cases", continuing with the computational modelling approach, Borisyuk et al. introduces the reader to research in **Parkinson's disease** and its correlation with abnormal dopaminergic synaptic levels in the underlying basal-ganglia circuit. Involuntary tremors in Parkinsonian disease have highlighted the lack of a holistic understanding of the underlying dynamics of the basal ganglia circuit; current research specifically implicate possible anomalies in beta band (12–30 Hz) oscillations. It is indeed surprising that although Deep Brain Stimulation (DBS) is a fairly successful *therapy* used for treating Parkinson's tremor, why and how exactly it works is still not clear. Borisyuk et al's presentation addresses this aspect using and validating a computational model. The model is biologically informed, and demonstrate the role of partial synchronisation in both "healthy and parkinsonian basal ganglia". In addition, the thoughts on translational aspects of computational model based research are also stressed here, along the lines expressed in chapter "Time Series and Interactions: Data Processing in Epilepsy Research".

Still on Parkinson's disease in chapter "Mathematical Models of Neuromodulation and Implications for Neurology and Psychiatry", Best et al. takes the reader to *pharmacological* solutions while addressing an important area that computational models seldom explore: **neuro-modulation**, an indirect means of synaptic information transmission for example "dopaminergic projection to the striatum in the basal ganglia from cells of the substantia nigra pars compacta"; "serotonergic projection to the striatum from the dorsal raphe nucleus". The authors present computational models of neuromodulation by both serotonin and dopamine, and demonstrates the importance of serotonin on the widely used levodopa treatment in **Parkinsonian tremor**. Once again, and echoing the thoughts in chapters "Time Series and Interactions: Data Processing in Epilepsy Research" and "Dynamic Causal Modelling of Neurological Pathology: Using Neural Mass Models to Understand Dynamic Dysfunction in NMDA-Receptor Antibody Encephalitis", the authors stress on the potential of computational models in informing translational research in both neurology and psychiatry.

In chapter "Attachment Modelling: From Observations to Scenarios to Design", Petters & Beaudoin further develop the theme of computational modelling by introducing the concept of **Attachment modelling** in computational neuropsychiatry, which in turn is based on the concept of Attachment Theory proposed and developed by John Bowlby and co-workers (see chapter for appropriate references). The authors translate Bolwby's theory into "contemporary cognitive architecture", where agent-based models can simulate attachments and interactions in humans, for example attachment of infants to their primary carers.

On this note, we move forward to chapter "Self Attachment: A Holistic Approach to Computational Psychiatry" where the author, A. Edalat, delves further into the realms of the mind, and links adult vulnerability to mental disorders with the insecurities that they faced as a child, mainly in the hands of the primary carer. The concept of 'attachment objects' is introduced, where a child or adult shows affinity and bonding with 'objects', for example a child with its mother, an adult with his/her religious beliefs. Imaging studies show that similar neural pathways are activated in three very different types of bond-making viz. romantic love, religious praying and maternal love. Based on these, the author introduces the concept of **'self-attachment'** where an adult forms an internal bond with the child in himself. The author discusses computational models that are used to study self-attachment viz. game-theoretic model and neural models.

Still on the computational psychiatric trail in chapter "A Comparison of Mathematical Models of Mood in Bipolar Disorder", Cochran et al. presents a mini-review of "mathematical models of **mood in bipolar disorder**". The mood swings that define bipolar disorder are irregular, unpredictable, and lack an objective biomarker. In this chapter, the authors explore basic modelling tenets for explaining how the subjective symptoms of mood change over time. Signatures of existing models are identified that could be compared with longitudinal data of mood. The authors contend that successful modelling of mood symptoms could help to describe precisely the symptoms of bipolar disorder, thereby facilitating better personalised treatment.

Thus far, we have discussed modelling in bespoke neurological and psychiatric disorders. One crucial aspect that provides a common thread to all of these models is the underlying mathematical definition of neuronal networks and their functional behaviour. Often, similar cortical and sub-cortical structures and their oscillatory dynamics are common to various disorders. On the other hand, modelling para-digms are often selected based on the problem at hand. Thus, networks of spiking neuron models are used to understand the spiking behaviour of neurons both as a single entity and as a part of the larger network. On the other hand, and as pre-viously mentioned, population and statistical neuronal models are often used to understand higher-level brain dynamics. Along these lines, chapters "Computational Neuroscience of Timing, Plasticity and Function in Cerebellum Microcircuits" to "A neural mass computational framework to study synaptic mechanisms underlying alpha and theta rhythms" showcase the modelling of specific brain structures that are vital in several brain disorders: In chapter "Computational Neuroscience of Timing, Plasticity and Function in Cerebellum Microcircuits", Diwakar et al. study a model of the Cerebellum, a structure that is known to be at the essence of timing in motor planning skills. In chapter "A Computational Model of Neural Synchronization in Striatum", Elibol and Sengor study a model of the Basal Ganglia, another subcortical structure that is at the core of decision making and action (discussed in previous chapters with context to Parkinson's disease). In chapter "A neural mass computational framework to study synaptic mechanisms underlying alpha and theta rhythms", Sen Bhattacharya et al.

present a population model of the thalamocortical structure to understand the underlying dynamics of EEG and LFP that bear signature of brain states in both health and disease.

Continuing with the theme of modelling, in chapter The role of simulations in neuropharmacology, Boutillier & Berger review the increasing importance of computational methodologies towards understanding the underlying pathology in both health and disease. From kinetic models of neurotransmitter receptors to spiking neurons to non-mechanistic models, the authors emphasise the potential and importance of computational models in the fast growing field of computational neuroscience. On an ending note, the authors summarise the current challenges in the field and provide some potential steps that should help in overcoming the obstacles, for example "using open standards and good modelling practices" to ensure "reproducible, iterative and collaborative modelling, consequently allowing iterative incorporation of additional findings while tuning and optimizing the rest of the model." This in turn will facilitate personalised treatments and drugs in neurological and psychiatric conditions.

Closing Remarks

> *If the human brain were so simple*
> *That we could understand it,*
> *We would be so simple*
> *That we couldn't.*
> *(Emerson M. Pugh, circa 1938 [29])*

Designing and building brain-inspired computational models to understand the brain has made steady advances over the years, and continue to gain increasing credibility among the highly interdisciplinary scientific community aspiring to understand brain dysfunction [19, 30]. While critics may cast doubt on the path forward, one may cite the greatest testimony of computational modelling in current times—the weather forecast system, which can be deemed as a hugely successful endeavour [31]. Nevertheless, unravelling the mysteries of the brain will be a daunting task, as predicted by Pugh [29], and call for a concerted effort worldwide. This book is a humble effort along these lines and towards breaking Lyall Watson's (in-) famous tag of 'Catch 22' [32] in brain science.

Acknowledgments The authors acknowledge the contribution of Fahmida N Chowdhury (Programme Director, National Science Foundation, USA) for reading the manuscript carefully and providing useful feedback that are subsequently implemented in the final version.

References

1. VS Ramachandran, "Phantoms in the Brain: Human nature and the architecture of the mind", Harper Perennial London, 2005.
2. (a) http://energeticsinstitute.com.au/spiritual-concept-of-third-eye/ (b) http://consciousreporter.com/spirituality/corruption-sacred-symbols-all-seeing-eye/.
3. EK Miller, DJ Freedman, JD Wallis, "The prefrontal cortex: categories, concepts and cognition", Philosophical Transactions of the Royal Society of London B, Vol 357 (1424), pp. 1123 – 1136, 2002.
4. Jeff Hawkins, "On Intelligence", Holt Paperbacks (NY), 2005.
5. Karl J Friston, "The free-energy principle: a unified brain theory?", Nature Review Neuroscience, Vol, 2010.
6. Greg Miller, "What is the Biological Basis of Consciousness", *In:* "What don't we know", Science (125th anniversary special issue), Vol 309 (5731), July 2005.
7. R. Restak, "The Brain - The Last Frontier", Warner Books Inc., (1st ed.) 1980.
8. Joint news release: WHO|World Bank news release, "Investing in treatment for depression and anxiety leads to fourfold return", http://www.who.int/mediacentre/news/releases/2016/depression-anxiety-treatment/en/, 13th April 2016.
9. D. Chisholm et al, "Scaling-up treatment of depression and anxiety: a global return on investment analysis", *The Lancet Psychiatry*, Vol 3 (5), pp. 415–424, May 2016. (published online: 12th April 2016).
10. P. Summergrad, "Investing in global mental health: the time for action is now", *The Lancet Psychiatry*, Vol 3 (5), pp. 390–391, May 2016. (published online: 12th April 2016).
11. Erol Basar, "Brain-body-mind in the nebulous Cartesian system: A holistic approach by oscillations", Springer 2011.
12. Aaron Sloman, "Computer revolution in Philosophy: Philosophy science and models of the mind", Harvester Press 1978.
13. Alex Boese, "Elephants on acid and other bizarre experiments", Pan Books, 2009.
14. HB van der Worp, DW Howells, ES Sena, MJ Porritt, S Rewell, et al, V O'Collins, MR Macleod, "Can animal models of disease reliably inform human studies?", PLoS Medicine Vol 7(3), e1000245, 2010.
15. (a) CJ Schwiening, "A brief historical perspective: Hodgkin and Huxley", Journal of Physiology, Vol 590(11), pp. 2571–2575, 2012. (b) https://www.nobelprize.org/nobel_prizes/medicine/laureates/1963/.
16. D. Sterrat, B. Graham, A. Gillies, D. Wilshaw, "Principles of Computational modelling in neuroscience", Cambridge University Press, 2011.
17. RD Traub, RK Wong, "Cellular mechanism of neuronal synchronization in epilepsy", Science, Vol 216 (4547), pp. 745–747, 1982.
18. Ivan Soltesz, Kevin Staley (Eds.) "Computational Neuroscience in Epilepsy", Academic Press 2008.
19. Basabdatta Sen Bhattacharya, Fahmida N Chowdhury (Eds) "Validating neuro-computational models of neurological and psychiatric disorders", Springer series in Computational Neuroscience (Ed) A. Destexhe, R. Brette, 2015.
20. Walter J. Freeman, "Mass action in the nervous system", 1975.
21. JD Cowan, J. Neuman, Wim van Drongelen, "Wilson-Cowan equations for neocortical dynamics", Journal of Mathematical Neuroscience, Vol 6 (1), 2016.
22. FH Lopes da Silva, A Hoeks, H Smits, LH Zetterberg, "Model of brain rhythmic activity, Kybernetik, Vol 15 (1), pp. 27–37, 1974.
23. AC Marreiros, KE Stephan, KJ Friston, "Dynamic Causal Modelling", Scholarpedia, Vol 5 (7), p. 9568.
24. Gyorgy Buzsaki, "Rhythms of the brain", Oxford University Press, 2006.
25. PR Montague, RJ Dolan, KJ Friston, P Dayan, "Computational psychiatry", Trends in Cognitive Science, Vol 16 (1), pp. 72–80, 2012.

26. KJ Friston, KE Stephan, R Montague, RJ Dolan, "Computational psychiatry: the brain as a phantastic organ", Lancet Psychiatry, Vol 1, pp. 148–158, 2014.
27. RA Adams, QJM Huys, JP Rosier, "Computational psychiatry: Towards a mathematically informed understanding of mental illness", Journal of Neurology, Neurosurgery & Psychiatry, Vol 87(1), pp. 53–63, 2016.
28. Thomas V. Wiecki, Jeffrey Poland, Michael J. Frank, "Model-based cognitive neuroscience approaches to computational psychiatry: clustering and classification", Clinical Psychological Science, Vol 3 (3), pp. 378–399, 2015.
29. George Edgin Pugh, "The Biological Origin of Human Values" (Chapter 7: Mysteries of the Mind, epigraph and footnote), Basic Books, New York, 1977.
30. Peter Lynch, "The origins of computer weather prediction and climate modelling", *Journal of Computational Physics* Vol 227 (7), pp. 3431–3444, 2008.
31. Nik Kasabov (Ed) "Handbook of Bio-/Neuro-informatics", Springer 2014.
32. Lyall Watson, 1979, while quoting [29]: Quoted in Marilyn Ferguson, *The Aquarian Conspiracy, Personal and Social Transformation in the 1980s,* Los Angeles: J.P. Tarcher, 1980.

Outgrowing Neurological Diseases: Microcircuits, Conduction Delay and Childhood Absence Epilepsy

John Milton, Jianhong Wu, Sue Ann Campbell and Jacques Bélair

Introduction

Physicians use the evolution of an illness to formulate a diagnosis and guide a treatment plan. Observations as to whether the disease onset is acute or sub-acute and the course is self-limited, relapsing-remitting or chronic progressive can be sufficient by themselves to significantly reduce the list of possibilities. On the other hand, the experiences of mathematicians and physicists emphasize that careful attention to how variables, namely something that can be measured, change as a function of time, referred to herein as dynamics, can uncover the identity of the underlying mechanisms. From this point of view many diseases may be *dynamical diseases* and arise in physiologial control mechanisms in which critical control parameters have

J. Milton (✉)
W. M. Keck Science Centre, The Claremont Colleges, Claremont, CA 91711, USA
e-mail: jmilton@kecksci.claremont.edu

J. Wu
Laboratory for Industrial and Applied Mathematics (LIAM), Department of Mathematics
and Statistics, York University, Toronto, ON M3J 1P3, Canada
e-mail: wujh@yorku.ca

S.A. Campbell
Department of Applied Mathematics and Centre for Theoretical Neuroscience,
University of Waterloo, Waterloo, ON N2L 3G1, Canada
e-mail: sacampbell@uwaterloo.ca

J. Bélair
Département de mathématiques et de statistique and Centre de recherches mathématiques,
Université de Montréal, Montréal, QC H3C 3J7, Canada
e-mail: belair@dms.umontreal.ca

J. Bélair
Centre for Applied Mathematics in Biosciences and Medicine, McGill University,
Montreal, QC H3G 1Y6, Canada

© Springer International Publishing AG 2017
P. Érdi et al. (eds.), *Computational Neurology and Psychiatry*,
Springer Series in Bio-/Neuroinformatics 6,
DOI 10.1007/978-3-319-49959-8_2

been altered [1–3]. Consequently important clues related to diagnosis and treatment may be embedded in the time-dependent changes in the relevant clinical variables, such as changes in temperature, weight, blood cells number, the electrical properties of the brain and heart, and so on. As wearable devices for the continuous, non-invasive monitoring of physiological variables become prevalent, it is likely that disease dynamics will increasingly become a focus of attention.

A potentially important application of continuous monitoring of physiological variables arises in the management of a patient with epilepsy. For these patients issues related to patient morbidity, and even mortality, are more often than not due to the unpredictability of seizure occurrence rather than to the seizure itself. For example, seizure occurrence while operating a moving vehicle could potentially be fatal. However, if seizure occurrence could be predicted, then maybe the seizures can be aborted [4–9]. At the very least it might be possible to give the patient enough time to make arrangements that minimize the effects of the impending seizure.

The study of inherited diseases of the nervous system which are characterized by recurring, paroxysmal changes in neurodynamics would be expected to shed light onto the answers to these questions [10–12]. In 1995, Milton and Black identified a number of familial disorders characterized by recurring episodes of abnormal neurodynamics (see Table 1 in [13]). Examples of the episodic changes in neurodynamics included epileptic seizures, headaches, paralysis and abnormal movements. They referred to these diseases as *dynamic diseases*. Herein we use the abbreviation DD to refer to both dynamical and dynamic diseases. Many of the DD's identified by Milton and Black were subsequently identified as *channelopathies* (Section "Dynamic diseases in neurology and psychiatry"). Channelopathies arise because of mutations in the genes that encode for the protein subunit components of ion channels of neurons and other exitable cells.

Paroxysmal changes in neurodynamics reflect transient losses in control by neurophysiological control mechanisms. The goal of this chapter is to identify possible mechanisms for paroxysmal seizure recurrence in childhood absence epilepsy (CAE). CAE is the most common and extensively studied epilepsy associated with a channelopathy (Table 1). There are two important issues: (1) How can a constantly present molecular defect give rise to the recurring seizures exhibited by the patient? (Sections "Paroxysmal seizure occurrence"–"Multistability in time delayed microcircuits") and (2) Why do seizures in CAE appear during childhood and then typically abate by late adolescence? (Section "Developmental aspects"). Our literature review links the paroxysmal nature of seizure occurrence to dynamical systems which contain time delays and which exhibit multistability. The tendency of children with CAE to outgrow their seizures is linked with changes in τ related to developmental changes in brain myelination. Finally we discuss our findings in Section "Discussion".

Table 1 Gene mutations in neurological DD's characterized by paroxysmal events

Dynamic disease[a]	Mutated gene[b]	Triggering events[c]
Channelopathies		
Andersen-Tawil syndrome	Kir2.1	None
Benign familial neonatal epilepsy	SCN2A	?
Childhood absence epilepsy	GABRA1, GABRA6, GABRB3, GABRG2, CACNA1H	Hyperventilation
Familial hemiplegic migraine	CACNA1A	Minor head trauma, cerebral angiography
Familial hyperplexia	GRAR1, GLRB	Unexpected auditory or tactile stimuli
Familial paroxysmal ataxia	CACNA1A, KCNA1, CACNB4	Stress, excitement
Hyperkalemic periodic paralysis	SCN4A	Fasting, exercise, K^+ foods
Hypokalemic periodic paralysis	CACNA1S, SCN4A	Insulin, glucose
Juvenile myoclonic epilepsy	DRD2, CACNB4, CLCN2, GABRA1, GABRD, EFHC1	Awakening
Nocturnal frontal lobe epilepsy	CHRNA4, CHRNB2, CHRNA2	Sleep I-II transition
Paroxysmal choreoathetosis/spasticity	SLC2A1	Alcohol, exercise, sleep deprivation, stress
Paroxysmal non-kinesigenic dyskinesia	MR-1	Alcohol, coffee, stress, fatigue
Paroxysmal kinesigenic dyskinesia	PRRT2	Sudden voluntary movement

[a]Clinical descriptions of these disorders and the identification of the gene mutations associated with these disorders can be found on the OMIM website (see text)
[b]Site of mutation: voltage-gated calcium channel (CACNA1A, CACNA1H, CACNA1S, CACNB4), cloride channel (CLCN2), dopamine receptor (DRD2), inward-rectifying potassium channel (Kir2.1), voltage-gated potassium channel (KCNA1), voltage-gated sodium channel (SCN4A), acetylcholine nictonic receptor (CHRNA4, CHRNA2, CHRNB2), glycine receptor (GLRA1, GLRB), $GABA_A$ receptor (GABRA1, GABRA6, GABRB2, GABRB3), acetylcholine nictonic receptor (CHRNA4, CHRNA2, CHRNB2), proline rich transmembrane protein (PRRT2), major histocompatibility complex related gene protein (MR-1), solute carrier gene (SLC2A1)
[c]The triggering events refer to stimuli and behaviors most often reported by patients as precipitants of "their attacks"

Dynamic Diseases in Neurology and Psychiatry

A practical problem for identifying the critical parameters and underlying control mechanisms for DD is that it is not often possible to monitor the patient at the time the dynamics change. For this reason patients in which paroxysmal events recur with

Table 2 Paroxysmal neurological dynamic diseases of childhood and adolescence which may be outgrown

Dynamic disease[a]	Mutated gene[b]	Triggering events	Outgrown by[c]
Epilepsy			
Benign familial infantile epilepsy	PRRT2		18 months
Benign familial neonatal epilepsy	SCN2A		7 years
Benign rolandic epilepsy	11p13	Awakening	Adolescence
Absence epilepsy	GABRA1, GABRA6, GABRB3, GABRG2, CACNA1H	Hyperventilation	Adolescence
Juvenile absence epilepsy	EFHC1	Awakening	3rd–4th decade
Juvenile myoclonic epilepsy	DRD2, CACNB4, CLCN2, GABRA1, GABRD, EFHC1	Awakening	3rd–4th decade
Occipital epilepsy	?	?	adolescence
Familial hyperekplexia	GLRA1, GLRB	Unexpected auditory or tactile stimuli	Childhood
Motor tics			
Tourette's syndrome	?	Anxiety, stress	Adolescence
Parasomnias			
Bed wetting	?		Adolescence
Night terrors	?		adolescence
Sleep walking	20q12-q13.12	Stress, alcohol, sleep deprivation	Adolescence
Sleep talking	?		Adolescence
Speech disorders			
Stuttering	?		Adolescence

[a]Clinical descriptions of these disorders and the identification of the gene mutations associated with these disorders can be found on the OMIM website
[b]Site of mutation: EF-hand domain containing protein 1 (EFHC1), gene mutation located on short arm of chromosome 11 (11p13), gene mutation long arm of chromosome 20 (20q12-q13.12). See also legend for Table 1
[c]These estimates are the most commonly observed age at which the troubling clinical signs disappear

a certain predictability are ideal candidates to characterize the nature of the DD transition. Thus it becomes possible, at least in principle, to use techniques such as multimodal imaging to document the structure of the brain and the physiological changes that occur at the time the changes in dynamics occur. The hope is that as more and more events are recorded, it may be possible to identify the common features and hence the critical control parameter(s).

Tables 1 and 2 summarize two groups of DD's of the nervous system that are potentially well suited for determination of how paroxysmal changes in signs occur [12]. The first group includes those diseases in which paroxysmal events can be repeatedly triggered with a certain predictability. The second group includes those neurological and psychiatric disorders that appear in infancy-childhood and then spontaneously disappear as the child gets older, typically by mid to late adolescence (Table 2). It should be noted that the clinical use of the words "periodic" and "paroxysmal" differs from their mathematical meaning. Typically physicians use the term "periodic" to mean that the signs recur "every so often" or "every once in a while". The term "paroxysmal" means that the onset of the signs occurs suddenly and without warning.

An exciting development has been the realization that many of the familial paroxysmal and periodic neurological diseases are channelopathies. Ion channels are the transmembrane pores which allow ions to flow across membranes in response to their electrochemical gradients. Although ion channels can be formed by a single protein (e.g., the transmembrane chloride conductance regulator in cystic fibrosis [14]), most often ion channels are formed from an assembly of protein subunits each encoded by a different gene. Over 400 ion channel genes have been identified: a useful resource is the Online Mendelian Inheritance in Man (OMIM) website: http://www.ncbi.nlm.nih.gov/omim.

Many of the DD's in Tables 1 and 2 are associated with gene mutations related to various ion channels including the voltage-gated Ca^{++}, Cl^-, K^+, and Na^+ channels and the ligand-gated acetylcholine nictonic and γ-aminobutyric acid A ($GABA_A$) receptors. These ion channels are the "excitable" in the term excitable cell. The work of Hodgkin and Huxley links the dynamics of excitable cells to the properties of the ion channels located in their membranes. These models take the general form

$$C\dot{V}(t) = -I_{ion}(V, W_1, W_2, \ldots, W_n) + I_0 \tag{1}$$

$$\dot{W}_i(t) = \beta \frac{[\hat{W}_i(V) - W_i]}{\Gamma(V)}$$

where $V(t)$ is the membrane potential, Γ is a time constant, C is the membrane capacitance, I_{ion} is the sum of V-dependent currents through the various ionic channel types, I_0 is the applied current, W_i describe the fractions of channels of a given type that are in various conducting states (e.g., open versus closed), $\hat{W}_i(V)$ describe the equilibrium functions and β is a time scale factor.

Over 50 years of work by mathematicians and neuroscientists have established a quantitative agreement between experimental observations on neurons and the predictions of (1) (for review see [15–20]). The predicted neuronal spiking dynamics include a variety of regular spiking and bursting patterns. It can be anticipated that it should be possible to draw analogies between abnormal neurodynamics and the clinical presentations. However, neurons are not the only type of excitable cell.

Cardiac and skeletal muscles and certain endocrine cells such as pancreatic β-cells are also excitable. Thus it is not difficult to appreciate the complexity of the clinical presentations and inheritance patterns of this group of diseases [21, 22].

Childhood Absence Epilepsy (CAE)

CAE is a channelopathy that exhibits both paroxysmal dynamics and a developmental pattern (Tables 1 and 2). Many families with absence epilepsy have a defect in one of the subunits of the γ-aminobutyric acid A (GABA$_A$) receptor. The GABA$_A$ receptor is an anion selective, ligand-gated ion channel [23]. The concept that CAE reflects a disturbance of inhibition is supported by both animal and human observations. For example, in cats, systemic injection of penicillin, a weak GABA$_A$ receptor antagonist, causes a dose-dependent transformation of sleep spindles to spike-wave discharges (SWD), the electro-encephalographic (EEG) signature of CAE [24, 25]. In human CAE, GABA-mimetic anti-epileptic drugs such as vigabatrin and tiagabine exacerbate absence seizures [26, 27].

Seizure onset is between the ages of 4–6 years and most commonly the seizures disappear by mid to late adolescence. There are no long term cognitive or behavioral sequelae. The seizures occur abruptly without warning and consist of brief spells of staring and unresponsiveness typically lasting 10–20 s. Minimal myoclonic jerks of the eyes and perioral automatisms can often be observed during the seizure. The frequency of the spells can be very high (100s per day). The EEG recorded using electrodes placed on the scalp changes during the seizure demonstrate the presence of generalized 3–4 Hz SWDs (Fig. 1a). Typically the seizure can be triggered at the bedside by having the child hyperventilate. Seizures in CAE can often be aborted using brief sensory stimuli, for example, the mother shaking or speaking to the child, a loud noise. The ability of sensory stimuli to abort seizures can also be observed in patients with atypical absence seizures characterized by 1.5–2.5 Hz SWDs (Fig. 1) [28].

Current debates concern the mode of onset of absence seizures in CAE [29, 30]. At the bedside, the classification of epileptic seizures was based on how a seizure begins in the first split second as determined by (1) direct observations of the clinical aspects of the seizure, and (2) correlation between the clinical features of the seizures and the changes detected in the EEG. If the seizure began in a focal area of the brain it was called a *partial epileptic seizure*. If the seizure appeared to begin everywhere at the same time it was called a *primary generalized seizure*. Thus historically seizures in CAE were considered to be primary generalized.

However, the use of scalp EEG recordings is not sufficient to rule out the possibility that absence seizures in CAE have a focal onset. For example, it would be very difficult to distinguish a generalized seizure from a focal onset seizure which rapidly generalizes. Indeed simple calculations based on estimates of seizure propagation velocities suggest that the fastest way to generalize a seizure is via reciprocal cortico-thalamic connections [31]. Depth electrode recordings in patients with

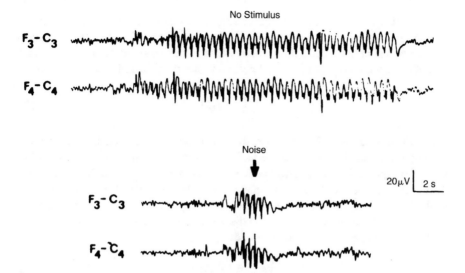

Fig. 1 *Top* Scalp EEG changes recorded during a generalized seizure in 16 year old with atypical absence epilepsy. *Bottom* The application of a brief sensory stimulus can shorten the length of the seizure. Figure reproduced from [28] with permission

generalized seizures were the first to demonstrate that seizure foci located in the frontal lobes could so rapidly generalize that a focal onset would be missed from scalp EEG recording [32, 33].

A cortical site for absence seizure onset has been identified in a rodent model for absence seizures [34]. It is located in peri-oral somatosensory cortex. Recently high resolution EEG-MEG studies together with advanced signal analysis techniques of absence seizures in human CAE have shown the presence of localized areas of pre-seizure activation in frontal cortex, orbito-frontal cortex, the mesial temporal lobe and the parietal lobe [35–37]. These observations are reminiscent of the concept developed for partial complex seizures that emphasizes the role of a spatially extended *epileptic system* that regulates the onset, maintenance and cessation of partial epileptic seizures [38, 39].

The above observations indicate that seizures in CAE are *secondarily generalized*. Here we focus our attention on the dynamics of seizure onset and do not consider how the epileptic activity spreads from the epileptic focus once the seizure is initiated.

Dynamical Systems Approaches to Seizure Onset

At the most basic level, a seizure represents a change in the activity of neurons. The cortical interictal state is primarily characterized by low frequency neuronal spiking [40, 41]. The hallmark of the onset of a seizure is a change in neural spiking rates.

Dynamics is concerned with the description of how variables, such as those related to neural spiking rates, change as a function of time. The fact that the magnitude of a variable in future time will be known once we know its initial value and rate of change per unit time is a fundamental property of the differential equation

$$\dot{x} \equiv \frac{dx}{dt} = f(x), \tag{2}$$

where x is, for example, the firing rate. The left-hand side of this equation re-iterates the importance of the change in the variable per unit time and the right-hand side states the hypothesis proposed to explain the time-dependent changes in the variable. In this chapter we are particularly interested in the role played by factors related to the physical separation of neurons on seizure onset. These factors include the axonal conduction velocity, v, and the distance, r, between neurons. Consequently (2) becomes the delay differential equation (DDE)

$$\dot{x} = f(x(t - \tau)), \tag{3}$$

where $\tau = r/v$ is the time delay. An introduction to the numerical methods available for the analysis of (3) is given in Appendix A. In order to obtain a solution to (3) it is necessary to specify an initial function, ϕ, on the interval $[-\tau, 0]$. These initial values can be changed using brief external stimuli. This observation is relevant to the clinical observation that brief sensory and electrical stimuli can abort an absence seizure. From a dynamical systems point of view this observation suggests *multistability*. To understand what is meant by the term multistability, we need to consider the nature of the solutions of (3).

Solutions of (3) can be classified by the qualitative nature of the changes in the variable as a function of time. Fixed point solutions are solutions where the values of the variables are fixed in time. Such solutions would correspond to a constant neural spiking rate. Periodic solutions are solutions which oscillate in time with some fixed period. Such solutions would correspond to oscillatory changes in spiking rate such as seen for a bursting neuron or population.

Another way that solutions can be classified is with respect to their response to perturbations. From this perspective a solution is called *stable* if solutions which have initial conditions close to the solution approach it in the longterm, otherwise it is called *unstable*. Stable solutions will be observed in numerical simulations and experiments. Unstable solutions will not persist in the longterm, but may be observed transiently. It is possible that an unstable solution may correspond to a seizure [12]. *Multistability* refers to the situation when there is more than one stable solution in the system: the long term behavior of the system then depends on the starting, or initial conditions and whether the system is subjected to any perturbations. If a system has multiple stable solutions, then it must also have unstable solutions.

A *parameter* is a variable which changes so slowly in comparison to the time scale of the variables of interest that it can be regarded as constant. Examples of parameters relevant for the occurrence of an absence seizure include τ, the number of GABA receptors, the receptor binding constant for GABA, the parameters that govern the gating of the Cl$^-$ channel, and so on. We will be particularly interested in how the number, type and stability of solutions change as one or more parameters are changed. This is called a *bifurcation*. Parameter values where this occurs are *bifurcation points*. When a bifurcation occurs in a system, the qualitative behavior of the system changes. For example, the system may transition from having a stable fixed point as the long-term behavior to having a periodic solution as the long-term behavior.

In summary, we use DDEs to describe the neurophysiological rules that govern the change rates of the system variables. These equations are often nonlinear and often involve neural physiological properties such as decay rates of action potentials and gains, and interconnectivity of the population such as synaptic connection weights, conduction velocities and time delays. These neurophysical properties, synaptic weights and conduction velocities remain constant in the time scale of the considered neurodynamics, and are called parameters. Naturally, solutions of the DDEs depend on both their initial values and the parameters. Understanding dynamic diseases in systems described by a DDE requires the examination of behaviors of solutions—evolutions with respect to time of the variables—for a wide range of plausible initial conditions and parameter values of the system. An important property of a dynamical system is the emerging long-term behaviors, in which solutions from a set of different initial conditions may converge to a particular solution which is called an attractor. The set of initial conditions for which the corresponding solutions converge to the attractor is called the basin of attraction. A dynamical system may have multiple attractors for a particular parameter value. This is called multistability. The same system with two different parameters may have different numbers and types of attractors, the critical values of parameters through which the system undergoes changes (bifurcations) in the numbers and types of attractors are called bifurcation points. Often these attractors take the form of fixed point solutions or limit cycles or periodic solutions.

In Section "Multistability: Hopfield model" we outline how to determine the number and stability of fixed points of a model, the nature of the possible bifurcations as a function of τ and the conditions for the occurrence of multistability. Our particular focus is on the situation when the time delay acts a bifurcation parameter. In order to improve the flow of the presentation of the mathematical results, we will not give references for all the standard results we use: these can be found in [42–44], and a more complete (and abstract) approach to the theory of delay differential equations can be found in [45] or [46].

Paroxysmal Seizure Occurrence

The hallmark of epilepsy is the paroxysmal nature of seizure occurrence [28, 47] which can even be observed in human neocortical slices [48]. Many computational models of absence seizures have examined topics related to the identification of the mechanism of action of anticonvulsant medications, the generation of the EEG and the nature of the mechanisms that recruit large populations of neurons into the evolving seizure (for reviews see [49–53]). Much less attention has been given to understanding how a seizure begins when it does and why a seizure, once started, eventually stops (for notable exceptions see [28, 47, 54]).

We take a dynamical systems approach. Our concern is on the dependence of the solutions of the governing differential equations as regulating parameters (e.g., nerve fiber length, conduction velocity) are changed and where these solutions start from. *Bifurcations* are qualitative changes in these behaviors. To be more specific, we briefly recall here some basic concepts relevant to dynamical systems in the context of DDEs. Under this paradigm, there are four general types of mechanisms in delay differential equations that can produce a paroxysmal change in dynamics (Fig. 2). Two of these mechanisms involve changes in parameters and two involve changes in variables.

The first mechanism proposes that sudden changes in dynamics arise because of a change in an important parameter such as a feedback gain or τ [1–3, 55].

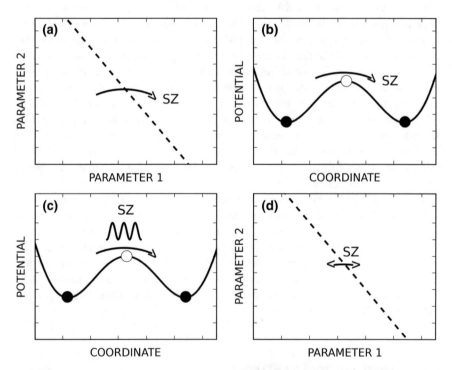

Fig. 2 Four mechanisms for producing a seizure (SZ). **a** A bifurcation caused by moving a parameter across a stability boundary (*dashed line*), **b** a change between two attractors caused by changes in the initial conditions, **c** a transient unstable oscillation which arises as the system moves from one attractor to another, **d** critical phenomena that arise when the dynamical system is tuned very close to a stability boundary (*dashed line*)

The changes in dynamics correspond mathematically to bifurcations in the relevant nonlinear equations which describe the affected physiological system. Examples of DD's which can be attributed to this mechanism include Cheyne-Stokes respiration [2] and blood cell diseases characterized by oscillations in blood cell number [56–58]. In terms of an explanation for an absence seizure this approach requires two parameter changes: one parameter change to explain the seizure onset, another to explain why the seizure stops.

The idea that seizure onset in CAE is related to changes in variables is based on clinical observations (Fig. 1). A brief sensory or electrical stimuli corresponds to a change in the initial function ϕ. This observation is suggestive of a multistable dynamical system [28, 47]. Figure 2b, c illustrates this concept of a dynamical system using the potential energy surface, $U(x)$. The minima ("valley") corresponds to a stable solution and the maxima ("hill") represent an unstable solution. In this interpretation, the onset of a seizure corresponds to a transition from one valley to the next, or in other words, a transition into a basin of attraction associated with a seizure. Brief stimuli abort the seizure by causing a transition from the seizure-related basin of attraction to one associated with healthy brain dynamics. This mechanism for an absence seizure also requires two changes in the variables: one change to explain seizure onset, another to explain why the seizure stops.

There are two mechanisms which incorporate both the onset and cessation of the seizure. The first involves modification of the multistability concept (Fig. 2c). It relies on the observation that in time-delayed dynamical systems, an unstable limit cycle can be associated with the separatrix that separates two stable attractors [12, 59–61], i.e. the "hump" between the "two valleys". It must be emphasized that the stable attractors do not correspond to the seizure, the seizure arises as a transient oscillation associated with the transition between the attractors [10, 12]. In the two-neuron microcircuits discussed in Section "Multistability: Hopfield model" delay-induced transient oscillations (DITOs) are associated with the presence of an unstable limit cycle [59, 60].

The final mechanism is based on the concept that the parameters of neural control mechanisms are tuned very close to the "edge of stability" (Fig. 2d). Indeed excitatory synaptic inputs to pyramidal neurons outnumber the inhibitory ones by 6.5 to 1 [62]. The fundamental concept is that a seizure might correspond to a phase transition. Thus near the bifurcation point, a dynamical system is expected to be characterized by collective behaviors for which it is not possible to define a specific correlation length. There are two clinical observations consistent with this hypothesis. First, the distribution of seizure sizes and times to occurrence exhibit power law behaviors [6, 63–65]. Second, even for individuals who do not have clinically-evident seizures, micro-seizures can be observed while the subject sleeps [66]. Thus from this point of view clinical epilepsy is a disease which is characterized by larger events [67]. However, dynamical systems tuned toward the edge of stability are also expected to generate a number of critical phenomena, such as critical slowing down and amplitude amplification [68]. These phenomena have not been observed for the majority of seizure occurrences [69].

Epileptic Micro-Circuits

Simultaneous recording of thalamic and cortical local field potentials during an absence seizure in CAE demonstrated that the oscillations are detected in the thalamus 1–2 s before SWDs are observed in the cortex [70]. Thus it is currently believed that the SWDs recorded by the scalp EEG during absence seizures are generated by the thalamo-cortical-thalamo circuit shown in Fig. 3 [71]. This network involves reticular thalamic neurons (nRT), thalamic relay neurons (TC) and cortical pyramidal neurons (CT). Inhibitory connections occur in both the thalamus and the cortex. This same circuit is involved in sleep and early investigators quickly recognized that the very mechanisms that generate sleep spindles are "hijacked" in CAE to generate the SWD [24, 25, 72].

It is useful to keep in mind that the thalamocortical circuit shown in Fig. 3 is for a rodent model of absence epilepsy [73]. However, much of the early work on absence seizure was done on feline brains [24, 25]. There are important physiological and anatomical differences between the thalamus of rodents and felines. For example, in rodents inhibitory interneurons are absent in almost all thalamic relay nuclei [74]. Thus intrinsic inhibition is absent in most of the thalamus and inhibition relies almost entirely on input from nRT. In contrast, in the feline thalamus, intrinsic GABAergic inhibitory interneurons are present throughout the thalamus, including its relay nuclei. Here the nRT provides an additional external inhibitory input. The thalamus in humans is much more developed that in rodents and felines [74, 75] and hence we can expect even more differences.

For unmyelinated axons, $v \sim \sqrt{d}$, and for myelinated axons, $v \sim dg\sqrt{\ln g}$, where d is the axon diameter and g is the ratio of d to overall fiber diameter. From Fig. 3 it can be seen that there is a distribution in the length of interneuronal axons: the length

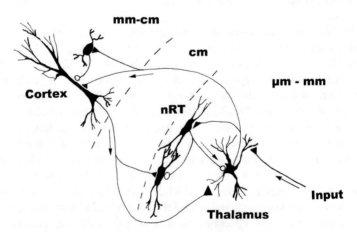

Fig. 3 Corticothalamic circuit involved in the genesis of absence seizures and sleep spindles. The cortical-thalamic distances between neurons are of the order of ~5 cm and those within the thalamus are <1 mm. The cortico-cortical distances range from <1 mm to 5–10 cm

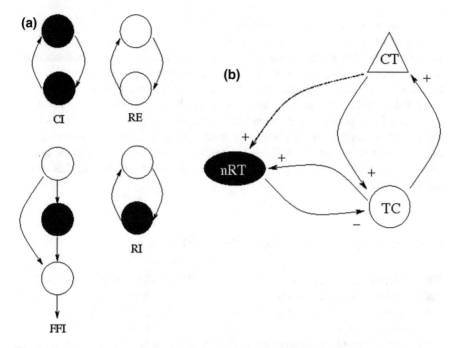

Fig. 4 Four neural micro-circuits important for the generation of epileptic seizures. **a** Recurrent inhibition (RI), counter inhibition (CI), recurrent excitation (RE) and feedforward inhibition (FFI). **b** Thalamocortical circuit important for CAE. The dark neurons are inhibitory and the white faced ones are excitatory

of axons associated with the intrathalamic connections (nRT ↔ TC) are shorter (< mm's) than those associated with thalamo-cortical connections (CT ↔ nRT and CT ↔ TC) (~ cm's). Since $\tau = r/v$, we can anticipate that there will be a bimodal distribution of τ [76]. This segregation in terms of short and long τ is further increased by the facts that (1) many of the cortico-thalamic axons are unmyelinated [77] (hence τ is increased) and (2) gap junctions exist between nRT neurons [78, 79] (hence τ is decreased).

It is increasingly being recognized that the behavior of large ensembles of neurons can be understood from smaller motifs involving 2–3 neurons [80–82]. Figure 4a shows four microcircuit motifs which have been emphasized for the generation of epileptic seizures [81]: (1) recurrent inhibition (RI), (2) counter inhibition (CI), (3) recurrent excitation (RE) and (4) feedforward inhibition (FFI). For CAE, the GABA$_A$ defect draws attention to those microcircuits which include an inhibitory component. Thus the thalamo-cortical ciruit shown in Fig. 3 can be interpreted as a FFI microcircuit (Fig. 4a, b). Indeed the FFI microcircuit has been considered to be critically important for the generation of SWD recorded by the EEG.

However, recent observations cast doubt on the importance of the FFI microcircuit for seizure onset [73]. First, seizures occur in the *Gria4$^{-/-}$* mice in which the connection between nRT and CT neurons has been deconstructed with optogenetic techniques [83]. Second, studies of olfactory cortex suggest that whereas FFI and excitation are balanced, RI dominates the intracortical excitations which are high-lighted in our analysis [84]. Finally, as already pointed out, absence seizures in CAE have a focal cortical onset. From a mathematical perspective, FFI requires a model that involves three neurons. Investigations of similar models [85] have shown that the delay effects we outline below likely do not play a role in such a circuit when it is isolated. More complex nonlinear effects involving interactions with other microcir-cuits [86] may be required to generate the multistable dynamics we emphasize here. Therefore in the discussion which follows we emphasize the role of RI, RE and CI for seizure onset.

Multistability in Time Delayed Microcircuits

The dominant theme of our discussion is that DDE models for the CI, RI and RE microcircuits readily generate multistable dynamics. This multistability is defined as the simultaneous co-existence of two or more stable states which may be fixed points (steady states) or periodic solutions. We demonstrate this observation with three types of models: (1) integrate-and-fire models, (2) Hopfield network models, and (3) Hodgkin-Huxley networks.

Multistability: Integrate-and-Fire Model

The simplest model for RI that illustrates the interplay between τ and multistability is the integrate-and-fire model whose dynamics are shown in Fig. 5a. This recurrent inhibitory loop involves a single excitatory neuron, E, and an inhibitory neuron, I [87]. The membrane potential, V, of E increases linearly at a rate, R, until it reaches the firing threshold, Π. When $V = \Pi$, E spikes and V is reset to the resting membrane potential, V_0. The period is $T = \Pi/R$. The spike generated by E excites I, which in turn after a time delay, τ, delivers an inhibitory post-synaptic potential (IPSP) to E. In general the effect of this IPSP will be to change the timing of the next spike generated by E by an amount δ, where δ is a function of the phase at which the IPSP arrives after E has fired. However, here we assume that δ is independent of the phase and hence the effect of the IPSP when $R > 0$ is to decrease V by an amount δ. This is equiva-lent to increasing the time that the next spike generated by E occurs by an amount δ. For simplicity we take $V_0 = 0$ and define the following dimensionless variables: $\tau^* = \tau/T, t^* = t/T, v^* = V/\Pi, \Delta = \delta/\Pi$, so that the dimensionless firing threshold, period and voltage growth rate are, respectively, $\Pi^* = 1, T^* = 1, R^* = \Pi^*/T^* = 1$. Dropping the asterisks we see that the dynamics of the recurrent loop depend only on two parameters, namely $\tau > 0$ and $\Delta \geq 0$. When $\tau < 1$, E spikes periodically with

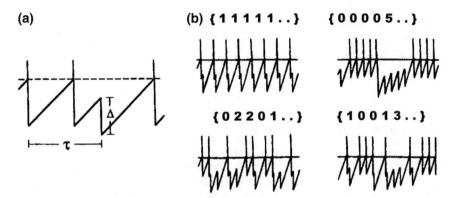

Fig. 5 a The time course of the membrane potential *v* for the integrate-and-fire neuron *E* in a time-delayed RI. The *dashed line* indicates the threshold. **b** Four co-existence periodic attractors that occur when $\tau = 4.1$ and $\Delta = 0.8$. The different patterns are described by the number of inhibitory pulses between two successive neuron spikings. Thus for the *lower left* pattern in **b** we have going from *left* to *right*, 0 inhibitory pulses between the first two spikes, 2 inhibitory pulses between the next two spikes, then 2 inhibitory spikes, then 0, then 1. After this the pattern repeats. Thus the shorthand label for this spike pattern is {02201}

period $1 + \Delta$. This is because decreasing the membrane voltage by an amount Δ is equivalent to increasing the interspike interval by $1 + \Delta$.

The essential condition for multistability in this dimensionless model is that $\tau > 1$ (Fig. 5b). Complex behaviors become possible since the inhibitory pulses are not necessarily the result of the immediately preceding excitatory pulse (Fig. 5a). It can be shown that the solutions which arise can be constructed from segments of length τ, where each segment satisfies an equation of the form

$$\tau = x + m + x\Delta,$$

where m, n are positive integers and $0 < x < 1$. For τ, Δ fixed, the total numbers of pairs that satisfy this relationship is $\lceil \tau/\Delta \rceil$, where the notation $\lceil . \rceil$ denotes the smallest integer greater than τ/Δ. Since the number of (m, n) segments is finite for a given τ and Δ, it follows that all solutions are periodic with period $S(1 + \Delta)$ where S is the number of excitatory spikes per period.

Despite the simplicity of this mechanism for generating multistability, it makes a number of predictions that may be relevant for CAE. First, this model draws attention to the importance of the long recurrent loops associated with long τ's in generating paroxysmal events. For a given recurrent inhibitory loop, multistability can arise either because T is decreased or because τ is increased. Increasing the excitatory drive to cortex by, for example, up regulation of excitatory synapses, decreases T and hence would be expected to produce multistability (seizures) as is observed experimentally [88, 89]. On the other hand, brain maturation is associated with increased myelination of neuronal axons which increases their conduction velocities (see Section "Developmental aspects"), thereby decreasing τ, and reducing the number of coexistent attractors. This observation could explain why this epilepsy is particularly common in children and why seizures tend to decrease in frequency, and even disappear altogether, as the child gets older.

Multistability: Hopfield Model

The next step is to examine models which describe the dynamics of two interacting neurons. In particular we explore the dynamics exhibited by the motifs in Fig. 4a using the equations for a Hopfield network. For 2-neuron circuit we have

$$\dot{x}_1 = -k_1 x_1(t) + \omega_{11} g_{11}(x_1(t - \tau_{11})) + \omega_{21} g_{21}(x_2(t - \tau_{21})) + I_1,$$
$$\dot{x}_2 = -k_2 x_2(t) + \omega_{22} g_{22}(x_2(t - \tau_{22})) + \omega_{12} g_{12}(x_1(t - \tau_{12})) + I_2. \quad (4)$$

In this model, the variables $x_j(t)$ $(j = 1, 2)$ are the spiking rates of the neurons at time t and the k_j represent a natural decay of activity in the absence of input. The parameters ω_{ij} represent the strength of the connections: ω_{11} and ω_{22} are the strengths of the self-connections; ω_{12} is the synaptic weight from x_2 to x_1, ω_{21} is the weight from x_1 to x_2. The sign of ω_{ij} determines whether the synapse is excitatory ($\omega_{ij} > 0$) or inhibitory ($\omega_{ij} < 0$). The parameters I_j $(j = 1, 2)$ are the external inputs to the neurons. The function $g(x)$ is sigmoidal and can be written in many ways, most commonly taken as $\tanh(cx)$, $x^n/(c + x^n)$, or $1/(1 + e^{-cx})$. Appendices B and C illustrate applications of the use of readily available computer software packages for the analysis of (4).

The CI, RE and RI micro-circuits depicted in Fig. 4a correspond to the following choices of signs in (4)

CI [12, 59–61, 90]: $\omega_{ij} < 0$
RE [90–94]: $\omega_{ij} > 0$
RI [87, 90, 95–100]: $\omega_{11}, \omega_{12} < 0, \omega_{21}, \omega_{22} > 0$

The fixed point solutions of a system with time delays, $\bar{x}$, are the same as those of the corresponding system with zero delay. Thus for (4) we obtain $(\bar{x}_1, \bar{x}_2)$ by setting $\dot{x}_1 = \dot{x}_2 = 0$ and solving

$$0 = -k_1 \bar{x}_1 + \omega_{11} g_{11}(\bar{x}_1) + \omega_{21} g_{21}(\bar{x}_2) + I_1, \quad (5a)$$
$$0 = -k_2 \bar{x}_2 + \omega_{22} g_{22}(\bar{x}_2) + \omega_{12} g_{12}(\bar{x}_1) + I_2. \quad (5b)$$

Since our model is two dimensional, we can also visualize the determination of the fixed points geometrically (Fig. 6). The fixed points are the intersection points of the curves defined by these equations. It is possible to analyze the equations in some detail to determine the number of possible fixed points see e.g. [101, 102].

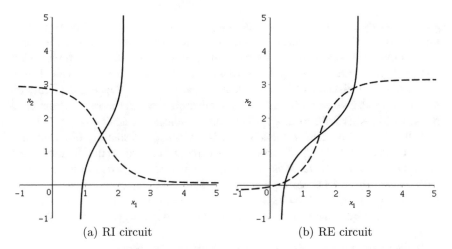

(a) RI circuit   (b) RE circuit

Fig. 6 Nullclines showing single fixed point in RI case and multiple fixed points in RE case. The *solid line* gives the x_1-nullcline determined from (5a) and the *dashed line* gives the x_2-nullcline determined from (5b). The nonlinearity is $g_{ij}(u) = \tanh(u - \theta)$. Parameter values are **a** $\theta = 1.5, k_1 = 1, \omega_{11} = -0.6, \omega_{21} = 1, I_1 = 1.5, k_2 = 1, \omega_{12} = -1, \omega_{22} = 0.5, I_2 = 1.5$ and **b** $\theta = 1.5, k_1 = 1, \omega_{11} = 0.2, \omega_{21} = 1, I_1 = 1.5, k_2 = 1, \omega_{12} = 1, \omega_{22} = 0.7, I_2 = 1.5$

To determine the stability of a fixed point we use linear stability analysis. First we linearize (4) about a fixed point, $\bar{x}$:

$$\dot{u}_1 = -k_1 u_1(t) + a_{11}u_1(t - \tau_{11}) + a_{21}u_2(t - \tau_{21}), \tag{6}$$
$$\dot{u}_2 = -k_2 u_2(t) + a_{22}u_2(t - \tau_{22}) + a_{12}u_1(t - \tau_{12}),$$

where $u(t) = x(t) - \bar{x}$ and $a_{ij} = \omega_{ij}g'_{ij}(\bar{x}_i)$. This model will tell us about the evolution of solutions which start *close enough* to a fixed point. To determine whether these solutions grow or decay in time, we consider trial solutions of the form $u \sim e^{\lambda t}$. Substituting this form into (6) and simplifying we arrive at the *characteristic equation*:

$$(\lambda + k_1 - a_{11}e^{-\lambda \tau_{11}})(\lambda + k_2 - a_{22}e^{-\lambda \tau_{22}}) - a_{12}a_{21}e^{-\lambda(\tau_{12} + \tau_{21})} = 0. \tag{7}$$

Any root λ of this equation leads to a solution of (6). The roots may be real or complex. If all the roots have negative real parts then all solutions of (6) decay to zero in the longterm. In this case the fixed point of (4) is stable. If at least one root has positive real part the some solutions of (6) will grow in time. In this case the fixed point of (4) is unstable. If any root has zero real part then the stability is not determined by the linearization.

If there are no delays in the model, then (7) is a quadratic polynomial, and has two roots which can be explicitly determined. The presence of the delays means that there are *an infinite number* of roots. Nevertheless, mathematical analysis can be used to determine if the equilibrium point is stable or not. In particular, one can show that all the roots except a finite number (possibly zero) have negative real parts. Of particular interest is the fact that the delays associated with the connections between the neurons only appear in the combination $\tau_{12} + \tau_{21}$. Thus, for the motifs we are considering, it is the *total* delay of the loop that is important not the individual components, as is well-known [85]. Note that the parameter a_{11} depends explicitly on ω_{11} but may also depend implicitly on the other ω_{ij} through the value of the fixed point, $\bar{x}_1$. Similarly for the other a_{ij}. This can complicate the analysis.

Bifurcations can occur in the system when a change of stability of an equilibrium point occurs. From the discussion above this corresponds to the situation when at least one root of the characteristic equation has zero real part, and the rest have negative real parts. To begin we focus on the simplest case, when the characteristic equation has a zero root ($\lambda = 0$). This situation is associated with a bifurcation that *creates or destroys fixed points*, thus can be important in the generation of multistability of fixed points.

In the micro-circuit model this type of bifurcation can occur if

$$(k_1 - a_{11})(k_2 - a_{22}) - a_{12}a_{21} = 0. \tag{8}$$

Recalling the definitions of the a_{ij} and the signs of the coupling, it is clear that this type of bifurcation is possible in all the micro-circuits, but only under some constraints, for example:

CI $a_{12} a_{21}$ sufficiently large, i.e., strong enough coupling between neurons
RE $a_{11} > k_1$ and $a_{22} > k_2$ or $a_{11} < k_1$ and $a_{22} < k_2$, i.e., similar self-coupling on both neurons either strong or weak
RI $a_{22} > k_2$ i.e., strong enough self-coupling on the inhibitory neuron

The next and perhaps most important case is when the characteristic equation has a pair of pure imaginary roots Ωi. Setting $\lambda = \Omega i$ in (7) and separating pure and imaginary parts yield the pair of equations

$$k_1 k_2 - \Omega^2 + a_{11}a_{22}\cos(\Omega[\tau_{11} + \tau_{22}]) - a_{12}a_{21}\cos(\Omega[\tau_{12} + \tau_{21}]) - \tag{9}$$
$$\Omega[a_{22}\sin(\Omega\tau_{22}) + a_{11}\sin(\Omega\tau_{11})] - k_1 a_{22}\cos(\Omega\tau_{22}) - k_2 a_{11}\cos(\Omega\tau_{11}) = 0,$$
$$(k_1 + k_2)\Omega - a_{11}a_{22}\sin(\Omega[\tau_{11} + \tau_{22}]) + a_{12}a_{21}\sin(\Omega[\tau_{12} + \tau_{21}]) - \tag{10}$$
$$\Omega[a_{22}\cos(\Omega\tau_{22}) + a_{11}\cos(\Omega\tau_{11})] + k_1 a_{22}\sin(\Omega\tau_{22}) + k_2 a_{11}\sin(\Omega\tau_{11}) = 0.$$

Fixing all the parameters except one, these equations can be solved for the value of the control parameter at which the pure imaginary roots occur and the corresponding value of Ω. Note that the equations are periodic with respect to each of the delays. Thus, fixing all of the parameters except one delay, say τ_{ij}, if (Ω^*, τ_{ij}^*) is a solution of these equations, then $(\Omega^*, \tau_{ij}^* + 2m\pi/\Omega)$, are also solutions for any integer value of m.

Alternatively, Eqs. (9)–(10) can be thought of as defining curves for two control parameters in terms of Ω and the other parameters. Figure 7 shows two examples where the control parameters are the combinations $\tau_{12} + \tau_{21}$ and $a_{12}a_{21}$.

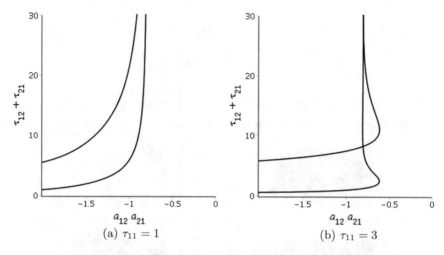

Fig. 7 Bifurcation curves for an RI circuit in connection strength-delay parameter space. The *curves* are determined by Eqs. (9)–(10). Due to the periodicity with respect to $\tau_{12} + \tau_{21}$ in these equations, there is an infinite set of curves, of which two are shown. The parameter values for k_j, ω_{jj}, I_j, $j = 1, 2$ are given in Fig. 6a, $\tau_{22} = 0$ and τ_{11} as shown. When the delay in the local loop of the inhibitory neuron, τ_{11}, is large enough intersection points can occur. The fixed point is stable to the *right* of the *curves*

Under appropriate conditions on the nonlinearities in the model (the functions g_{ij}) the system will undergo a *Hopf bifurcation* at these points, leading to the creation of a periodic solution. The stability of this solution also depends on the nonlinearities. See [43, 46, 103–105] for more details.

It is well known that the presence of delay in a model can facilitate the occurrence of Hopf bifurcations, this is known as a *delay-induced Hopf bifurcation*. In the microcircuit model, it is straightforward to show that Hopf bifurcations are not possible if there are no delays (i.e. $\tau_{ij} = 0$) regardless of the signs of the connection weights.

More complex bifurcations can occur if the characteristic equation has multiple roots with zero real part. In models with delay it has been shown that such behavior is quite prevalent if the system has multiple delays [103, 106, 107]. Such points, which correspond to intersection points on the bifurcation curves shown in Fig. 7, can lead to multistability and more complex dynamics [104, 105]. Examples of this behavior are illustrated in Fig. 8.

We briefly outline some situations that can occur. If the characteristic equation has a double zero root (a Bogdanov-Takens bifurcation point), it is possible to have multistability between a slowly varying periodic solution and one or more fixed points. This has been shown to occur in the RI [108] and CI [85, 107] microcircuits. If the characteristic equation has a zero root and a pure imaginary pair it is possible to have multistability between a periodic solution and one or more fixed points. If the characteristic equation has two pairs of pure imaginary eigenvalues without resonance, then it is possible to have bistability between periodic orbits with an unstable two

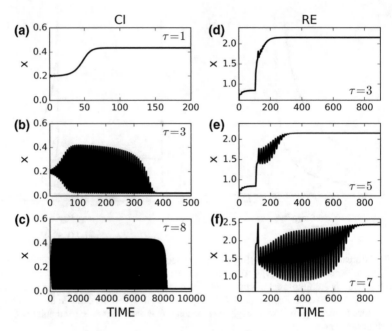

Fig. 8 Bistability between fixed points separated by an unstable limit cycle (DITO). **a–c** CI circuit. **d–f** RE circuit. Details of the CI circuit are given in [59] and the XPPAUT program `ditto.edo` can be downloaded from `faculty.jsd.claremont.edu/jmilton/Math_Lab_tool/Programs/XPPAUT`. The program and parameters for the RE circuit are given in Appendix B. The initial conditions are chosen to place the dynamics near the separatrix that separates the two co-existent stable fixed points. In each case we show only the activity of one of the neurons and $\tau = \tau_{12} = \tau_{21}$. Note that the duration of the DITO's is much longer than τ

torus (or the reverse). This has been shown to occur in the CI and RI microcircuits [85, 107].

Recalling our interpretation of this model in terms of firing rates, we can give biological meaning to these figures. For example, in Fig. 8e the neurons are firing at some steady rate when a brief stimulus switches this to a large amplitude oscillatory firing rate. This spontaneously disappears after some time and the system settles on a (different) steady firing rate. The large amplitude oscillation correponds to an unstable periodic orbit, which is present due to the delay in the system. With no delay, the system merely switches between two different firing rates (Fig. 8a). The oscillatory behavior is sometimes referred to as a delayed induced transient oscillation (DITO). Our main point is that the DITO behavior is very reminiscent of the seizure behavior observed Fig. 1.

Multistability: Hodgkin-Huxley Models with Delayed Recurrent Loops

The next step is to examine the effects of the ion channels on the dynamics of delayed recurrent loops. Foss et al. [87] described the membrane potential of the excitatory neuron E using the following Hodgkin-Huxley model (HH) by considering the effect of IPSP as self-feedback

$$\begin{cases} Cx'(t) = -g_{Na}m^3h(x(t) - E_{Na}) - g_Kn^4(x(t) - E_k) \\ \qquad\qquad -g_L(x(t) - E_L) - F(x(t - \tau)) + I_s(t), \\ m'(t) = \alpha_m(x)(1 - m) - \beta_m(x)m, \\ n'(t) = \alpha_n(x)(1 - n) - \beta_n(x)n, \\ h'(t) = \alpha_h(x)(1 - h) - \beta_h(x)h, \end{cases} \tag{11}$$

where $F(x)$ is the signal function which describes the effect of the inhibitory neuron I on the membrane potential of the excitatory neuron E, and τ is the time lag. Other variables and parameters include the membrane potential $(x(t))$, the membrane capacitance (C), the stimulus (I_s). Constants g_{Na} and g_K are the maximum conductance of sodium and potassium ion channels, the constant g_L is the conductance of leakage channel, constants E_{Na}, E_K and E_L are empirical parameters called the reversal potential. There are three (gating) variables (m, n, h) that describe the probability that a certain channel is open, and these variables evolve according to the aforementioned system of ordinary differential equations with functions α and β indexed by (m, n, h) appropriately. The initial function ϕ in the interval $[-\tau, 0]$ was assumed to have the form of neural spike trains. Namely, it is given by a sum of square pulse functions.

With sufficiently large I_s that makes the neuron fire successively, several coexisting periodic attractors were found [87, 102, 109] (Fig. 9). Solutions starting from basins of attraction of these periodic solutions exhibit exotic transient behaviors but eventually become periodic.

The corresponding linear integrate-and-fire model (LIF) and quadratic integrate-and-fire model (QIF) are given by

$$x'(t) = -\beta x(t) - F(x(t - \tau)) + I_s(t), \tag{12}$$

$$x'(t) = \beta(x - \mu)(x - \gamma) - F(x(t - \tau)) + I_s(t), \tag{13}$$

with the firing time t_f:

$$t_f : x(t) = \vartheta_1 \quad \text{and} \quad x'(t)|_{t=t_f} > 0,$$

and the firing threshold ϑ_1. These models can also exhibit multistability in terms of coexisting attractive periodic solutions, when the absolute refractoriness is incorporated. Each time the excitatory neuron fires a spike, a feedback is delivered at time τ later. The type of multistability not only depends on the time delay τ but also on the

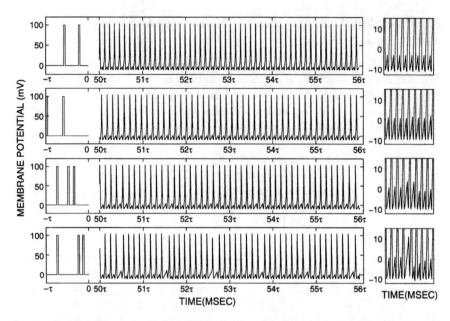

Fig. 9 Four coexisting attracting periodic solutions generated by the excitatory neuron E for the Hodgkin-Huxley model (HH) given by Eq. (11). The *right-hand side* is the blow up of the solutions in a given period (not delay τ) to clearly illustrate the patterns of solutions

effective timing of the feedback impacting on the excitatory neuron. The total timing of the feedback is the portion of the duration when the spike is above the firing threshold. An important factor determining the effective timing of the feedback is the absolute refractory period, a short period after the firing of a spike during which the neuron is not affected by inputs at all. Systematic analysis can be conducted for both the linear integrate-and-fire model and quadratic integrate-and-fire model to determine when multistability occurs, how many coexisting attractive periodic solutions appear and their patterns (inter-spike intervals and oscillatory patterns within these intervals). See [102].

We note that multistability is also observed in RI models which take into account the phase resetting properties of each neuron in the loop [97, 110]. The advantage of this approach is that the phase resetting curve can be measured experimentally and thus all parameters in the model are known.

Developmental Aspects

The developmental pattern of seizure recurrences in CAE suggests that there must be processes that are evolving on time scales of the order of years which modify the impact of the defect in the $GABA_A$ on neurodynamics at a given age. There are two main changes that occur in the human brain between 3–4 years of age and adolescence.

First there are changes in synaptic density, namely the number of synapses per unit volume of cortical tissue [111–113]. At birth the infant brain has a synaptic density nearly equal to that of the adult brain [112]. Beginning within the first year of life there is a 30–40 % increase in synaptic density which peaks in the frontal cortex between ages 3–4 [113]. This is followed by a process of synaptic elimination so that by adolescence the synaptic density is again approximately equal to that of the adult.

Second, there are changes in axonal myelination and hence axonal conduction velocities. The active period for axonal myelination in the brain begins during week 28 of gestation and then slows by mid to late adolescence. In some of the anterior association areas of the cortex, myelination continues until the 5th decade [114, 115]. The changes in myelination occur in a predictable spatial and temporal sequence [115]. Myelination proceeds in a posterior to anterior direction so that sensory pathways myelinate first, followed by motor pathways and then the association areas. Within a given functional circuit, sub-cortical structures myelinate before cortical regions. The corpus callosum, the major white matter bundle connecting the two hemispheres, begins myelinating at 4 months post-natally and is not complete until mid to late adolescence: the anterior part is the last to myelinate. The time course for the disappearance of absence seizures in CAE coincides with the myelination of the long association and commissural fibers in the anterior quadrants of the brain. Thus the ages during which seizure activity is highest corresponds to the time when synapses are being eliminated from the brain and the myelination of axons is increasing. In particular the disappearance of absence seizures coincides with the myelination of the long association and commissural fibers in the anterior quadrants of the brain which connect different regions of cortex within the same hemisphere (association fibers) and between the two hemisphere (commissural fibers).

It is not known whether changes in synapses and/or changes in axonal conduction velocities are most important for expression of absence seizures in CAE. However, the observation that there are no long term cognitive impairments in CAE patients and the intelligence of children with CAE is within normal limits provided that their seizures are well controlled suggests that it is unlikely that seizure generation is related to abnormalities in synaptic density. On the other hand, with the advent of diffusion tensor imaging (DTI) techniques, abnormalities in myelination have been identified in children with CAE, particularly in the anterior part of the corpus callosum [116]. Similar abnormalities have been reported in a rat model for absence epilepsy [117]. Although these associations do not prove causality, they do suggest the possibility that the dependence of axonal conduction velocities (and hence time delays) on myelination might be an important parameter for this dynamic disease.

In a similar manner, the developmentally dependent changes related to τ may also explain the bimodal incidence of all types of epilepsy shown in Fig. 10. Epileptic seizures are most common in the young and the elderly. Studies on aging monkeys suggest that increases in axonal conduction velocity are related to the death of oligodendrocyctes, namely the cell type responsible for myelinating axons in the

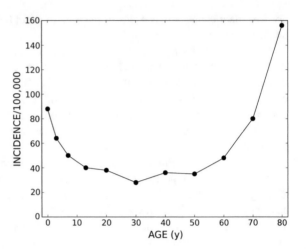

Fig. 10 Age-related incidence of epilepsy in industrialized societies. The data is from [118] and we have pooled data for males and females

brain. When an oligodendrocyte dies, other oligodendrocytes remyelinate the axon. However, the new myelin sheaths are thinner and the internode distances are shorter. Consequently, v is decreased and τ is increased.

Although the concept that the brain is most susceptible to generating seizures when τ is long is appealing, it may be an oversimplification. Indeed mean field estimates of cortical instability boundaries suggest that as v increases (τ decreases), the cortical model is able to reproduce SWDs [119].

Discussion

The brain is a complex dynamical system with a large number of state variables. This is because the number of neurons is large and signals between them are time delayed. The dynamics of the microcircuit building blocks of the brain, such as CI, RE and RI, all exhibit multistability. Thus it is not surprising that many authors have emphasized metaphors for brain dynamics that take the form of potential landscapes with many "hills" and "valleys" [28, 67, 120, 121]. Moreover this landscape itself continually changes as a result of changes in states of arousal, aging, and as the brain learns and adapts to its environment [12, 122]. Frequent transitions between the attractors ("hills") lead to mesoscopic states in which dynamics are bounded and time-dependent. A lower bound for neurodynamics is ensured because excitatory connections exceed inhibitory ones and the upper bound is the result of neural refractoriness, accommodation and limited energy resources.

Our approach has emphasized that clinically significant changes in neurodynamics, such as the occurrence of an epileptic seizure, may be more related to the unstable states that separate attractors ("hills") than to the stable attractors [28, 47]. This concept can be most readily understood in the context of the DDEs that describe the microcircuits which generate DITOs [12, 59–61, 93]; however, DITO-like phenomena can also arise in other contexts as well [123]. A DDE corresponds to an infinite dimensional dynamical system since the number of initial conditions that must be specified to obtain a solution is infinite. Each initial condition is associated with an eigenvalue. For the unstable solution that separates two attractors, there must be at least one positive eigenvalue. This means that in the long time the solutions must converge to one of the attractors. However, the short-time behaviors are influenced by the eigenvalues which have negative real parts. Consequently, the dynamical behaviors can be momentarily "trapped" in the vicinity of the unstable fixed-point. Our suggestion is that these "momentarily trapped" behaviors can sometimes be manifested as an epileptic seizure.

The study of the dynamics of neural microcircuits at the benchtop has a long history. Most often "hybrid" analogues of microcircuits are created in which a neuron interacts with an electronic device to collectively form a microcircuit (for a review see [124]). Modern day approaches use optogenetic techniques to manipulate neural microcircuits in situ [83]. However, it is not yet known how changes at the level of ion channels, e.g. the $GABA_A$ receptor in CAE, result in episodic seizure recurrences. It is possible that this will become clear as we understand the rules that relate the dynamics of single microcircuit to those of large ensembles of microcircuits. In this way the study of CAE may not only bring relief to its sufferers and their families, but also provide insights into how the brain functions.

Acknowledgements We thank B. Fortini for useful discussions. JM was supported by the William R. Kenan Jr. Charitable Trust. JB, SAC and JW acknowledge support from the National Sciences and Engineering Research Council of Canada. JW received additional support from the Canada Research Chairs Program. JB received additional support from Fonds de recherche du Québec - Nature et technologies (FRQNT).

Appendix A: Numerical Methods

Two numerical techniques which can aid in the study of delay differential equations such as (4), (11), (12) and (13) are numerical simulation and numerical bifurcation analysis.

Numerical simulation determines an approximate solution of a differential equation for a given choice of the initial state. Recall that for a delay differential equation, the initial state is a function which determines the value of the solution for t in $[-\tau, 0]$. For example, for the two-neuron Hopfield networks described by (4), the initial state is specified as follows

$$x_1(t) = \phi_1(t), \ x_2(t) = \phi_2(t), \quad -\tau \leq t \leq 0.$$

Typically ϕ_1, ϕ_2 are taken to be a constant, i.e.,

$$x_1(t) = x_{10}, \ x_2(t) = x_{20}, \quad -\tau \le t \le 0,$$

which is reasonable for most experimental systems. It should be noted that only solutions which are stable can be accurately approximated using numerical simulation. The presence of unstable solutions may be deduced from transient behavior, but details of the structure cannot be found through numerical simulation.

There are two commonly used programs for the numerical integration of delay differential equations. The free, stand-alone package XPPAUT [125] can perform numerical integration using a variety of fixed-step numerical methods. The program is run through a graphical user interface which is used not just to visualize results, but also to modify parameters, initial conditions and even the numerical integration method. The main benefit of this program is its flexibility and the ease with which different simulations can be compared. Information on how to download the package as well as documentation and tutorials are available at http://www.math.pitt.edu/~bard/xpp/xpp.html. The book [125] gives a overview of the package including many examples. XPPAUT code for the micro-circuit example (4) considered in this chapter is included in Appendix B. The MATLAB function, DDE23 [126], is a variable step size numerical integration routine for delay differential equations. A tutorial on this routine is available at http://www.mathworks.com/dde_tutorial and the DDE23 code for the micro-circuit example in Appendix B is given in Appendix C. A benefit of using DDE23 is that results may be visualized using the extensive graphing tools of MATLAB.

Numerical bifurcation analysis has two aspects: the approximation of a solution and the calculation of the stability of this solution. The approximation of a solution is done using numerical continuation, which uses a given solution for a particular parameter value to find a solution for a different (but close) parameter value. Numerical continuation can find both stable and unstable fixed point and periodic solutions. More complex solutions (such as tori) are not implemented in all packages. Once the continuation has found a fixed point solution to a desired accuracy, a numerical bifurcation program determines approximations for a finite set of the eigenvalues with the largest real part. The stability of periodic orbits can be numerically determined in a similar way. Numerical bifurcation packages generally track the stability of fixed points and periodic orbits, indicating where bifurcations occur.

One commonly used package that carries out numerical bifurcation analysis for delay differential equations is DDE-BIFTOOL [127], which runs in MATLAB. An overview of the numerical methods used in this package and some examples of applications can be found in [128]. The user manual and information on how to download the package are available at http://www.twr.cs.kuleuven.be/research/software/delay/ddebiftool.shtml.

A list of other software available for working with delay differential equations can be found at http://www.twr.cs.kuleuven.be/research/software/delay/software.shtml.

Appendix B: Delay Differential Equations Using XPPAUT

Here we illustrate the use of XPPAUT for simulating the neural micro-circuits shown in Fig. 4 using the Hopfield equations described by (4). Our focus is on how delay differential equations (DDE) are handled in XPPAUT and how stimuli in the form of pulses can be used to cause switches between attractors.

In our experience installing XPPAUT on PC computers is quite straight-forward. However, problems can arise when installing XPPAUT on Mac computers. The IT department at The Ohio State University has prepared a very useful installation guide for Mac users which can be accessed at https://docs.math.osu.edu/mac/how-tos/install-xpp-xppaut-mac/.

```
# Note: in ODE files, comments are preceded by #
#
# This program numerically integrates the
# Hopfield neural net equations with delay.
# We show the parameter choices for a RE circuit.
# However, the parameter choices we recommend for CI and  RI
# are given in Comments 3-4.  Note that the # command can be used
# to comment out lines of code which are not required.
#
# EQUATIONS

# See Comment 1

x1'= -k1*x1+w11*f1(delay(x1,tau11))+w21*f2(delay(x2,tau21))+I1+Istim1
x2'= -k2*x2+w12*f1(delay(x1,tau12))+w22*f2(delay(x2,tau22))+I2+Istim2
f1(x)=tanh(n1*(x-theta1))
f2(x)=tanh(n2*(x-theta2))

# See Comment 2

Istim1=I11*(heav(t-tstart1)-heav(t-tend1))
Istim2=I22*(heav(t-tstart2)-heav(t-tend2))

# PARAMETERS
# These parameters will reproduce Fig. 8 b)

# See Comments 3, 4

p k1=1,k2=1
p w11=0.5,w21=1,w22=0.5,w12=0.5
p tau11=1,tau12=7,tau21=7,tau22=15
p n1=1,n2=1,theta1=1.5,theta2=1.5
p I1=1.5,I2=1.5
p I11=0.805,tstart1=100,tend1=120
p I22=0,tstart2=100,tend2=120

# INITIAL CONDITIONS

See Comment 5
```

```
init x1=0.5,x2=0.6
x1(0)=0.5
x2(0)=0.6

# CHANGES FROM XPP'S DEFAULT VALUES

# See Comment 6

@ total=400,dt=.01,xhi=400,maxstor=2000000,delay=10

done
```

Comments:

1. Terms of the form $x(t - \tau)$ become `delay(x,tau)` in XPPAUT. See also Comment 5.
2. Switches between co-existing attractors are made by using square pulses. The magnitude of the pulse is given by `I11, I22`. The onset of the pulse occurs at `tstart` and the end of the pulse occurs at `tend`.
3. Counter inhibition (CI). Make the following parameter changes to the RE program (above):

   ```
   w12=w21=-1.2
   w11=-0.1,w22=-0.2
   tau11=tau22=6, tau12=tau21=4.5
   I11=I22=1
   tstart1=tstart2=300
   tend1=tend2=320
   ```

   with the initial conditions:

   ```
   init x1=1,x2=2.1
   x1(0)=1
   x2(0)=2.1
   ```

   For these choices of the parameters, there are three coexisting attractors: two stable fixed points and a stable limit cycle. As the time delays `tau12, tau21` are decreased the limit cycle disappears.
4. Recurrent inhibition (RI). Make the following parameter changes to the RE program (above):

   ```
   w11=-0.6, w12=-1, w21=1, w22=0.5
   tau11=3,tau12=4.5,tau21=4.5,tau22=0
   I11=0.45, tstart1=450 tend1=470
   I22=0
   ```

with the initial conditions:

```
init x1=2,x2=2.1
x1(0)=2
x2(0)=2.1
```

For these choices of the parameters, there is bistability between two limit cycles with an unstable torus in between. Changing the delays just a little eliminates the bistability.

5. In mathematics, the initial function, ϕ, for a DDE is defined on the interval $[-\tau, 0]$. However, in XPPAUT ϕ is divided into two parts: the initial condition at t=0, $\phi(0)$, and a function $\phi(s)$ where $s \in [-\tau, 0)$. The default choice is $\phi(s) = 0$. The commands x1(0)=0.1 and x2(0)=0.2 set $\phi(s)$ to a constant value. A look up table can be used to introduce an arbitrary $\phi(s)$ as shown in [59]. In running a XPPAUT program for a DDE it is necessary to open three windows Initial Data, Delay ICs and Parameters. The Initial Data panel will show the initial data, the Delay ICs will show $\phi(s)$ and the Parameters panel will show the parameter choices to be used for the simulation. An important point is that to run the simulation one must click on 'OK' for each panel and then click on 'Go' on one of these panels. Failure to do this will result in numerical errors since the initial function will not be handled correctly. Finally when determining the nullclines it is important to set all of the delays to 0.

6. The parameter delay should be greater than tau. The parameter delay reserves the amount of memory needed to store ϕ for each variable. Since one of the goals of the simulation is to see the effect of changing τ on the dynamics, it is convenient to set delay slightly higher that the largest delay anticipated. If the program runs out of memory, the amount of memory reserved for this purpose can be increased by using the command maxstor.

Appendix C: Delay Differential Equations Using Matlab's dde23

For users that have access to the latest version of Matlab, it is possible to integrate the Hopfield equations described by (4) using dde23. Here we give the code that integrates the RE model for the same parameters as described above. Note that two m files are required: delay_circuit.m supplies the parameters and performs the integration; DRHS.m gives the equations to be integrated. The symbol % comments out the parts of the code that are not needed for the RE circuit. In order to run this program, put both files in the same directory and then type delay_circuit()
DRHS.m

```
function yp = DRHS(t,y,Z);

global k W n theta I Istim tstart tend;

% Z(i,j) = xi(t-tauj)
% tau=[tau11,tau12,tau21,tau22]
ylag11 = Z(1,1); % x1(t-tau11)
ylag12 = Z(1,2); % x1(t-tau12)
ylag21 = Z(2,3); % x2(t-tau21)
ylag22 = Z(2,4); % x2(t-tau22)
yp = [
-k(1)*y(1) + W(1,1)*tanh(n(1)*(ylag11-theta(1))) ...
+ W(2,1)*tanh(n(2)*(ylag21-theta(2))) ...
+ I(1) + Istim(1)*(heaviside(t-tstart(1))-heaviside(t-tend(1)));
-k(2)*y(2) + W(1,2)*tanh(n(1)*(ylag12-theta(1))) ...
+ W(2,2)*tanh(n(2)*(ylag22-theta(2))) ...
+ I(2) + Istim(2)*(heaviside(t-tstart(2))-heaviside(t-tend(2)));
];
```

Note: The entries for yp are very long. The . . . is the Matlab code for breaking up long equations into shorter ones.

delay_circuit.m

```
function delay_circuit()
clear all;
close all;
clc;
global tau;
global k W n theta I Istim tstart tend;

% Initialization
% delays
tau11=0.001;
tau22=15;

% other parameters
w11=0.5;
w21=1;
w12=0.5;
w22=0.5;
theta=[1.5,1.5]';
I=[1.5,1.5]';
n=[1,1]';
k=[1,1]';
```

```
% stimulation parameters
I11=0.805;
tstart1=100;
tend1=120;
I22=0;
tstart2=100;
tend2=120;

% initial conditions
y10=0.5;
y20=0.6;

% start/end values of t
t0=0;
t1=1000;

% min/max for plotting
umin = 0;
umax = 3;

% matrix form of parameters
W=[[w11,w21]',[w12,w22]']
Istim=[I11,I22]'
tstart=[tstart1,tstart2]'
tend=[tend1,tend2]'
% initial conditions
yi =[y10,y20]'
% integration time
interval=[t0, t1];

% First plot
tau=[tau11, 1., 1., tau22]
sol = dde23('DRHS',tau,yi,interval);
fig1 = figure(1);
subplot(3,1,1);
plot(sol.x,sol.y(2,:),'-b','LineWidth',2);
title('\tau_{12}=\tau_{21}=1');
xlabel('time t');
ylabel('x_2(t)');
axis([t0 t1 umin umax]);
grid on;

% Second plot
tau=[tau11, 6., 6., tau22]
```

```
sol = dde23('DRHS',tau,yi,interval);
subplot(3,1,2);
plot(sol.x,sol.y(2,:),'-b','LineWidth',2);
title('\tau_{12}=\tau_{21}=6');
xlabel('time t');
ylabel('x_2(t)');
axis([t0 t1 umin umax]);
grid on;

% Third plot
tau=[tau11, 7., 7., tau22]
sol = dde23('DRHS',tau,yi,interval);
subplot(3,1,3);
plot(sol.x,sol.y(2,:),'-b','LineWidth',2);
title('\tau_{12}=\tau_{21}=7');
xlabel('time t');
ylabel('x_2(t)');
axis([t0 t1 umin umax]);
grid on;

end
```

References

1. L. Glass and M. C. Mackey. Pathological conditions resulting from instabilities in physiological control systems. *Ann. New York Acad. Sci.*, 316:214–235, 1979.
2. M. C. Mackey and L. Glass. Oscillation and chaos in physiological control systems. *Science*, 197:287–289, 1977.
3. M. C. Mackey and J. G. Milton. Dynamical diseases. *Ann. New York Acad. Sci.*, 504:16–32, 1987.
4. H. O. Lüders. *Deep brain stimulation and epilepsy*. Martin Dunitz, New York, 2004.
5. J. Milton and P. Jung. *Epilepsy as a dynamic disease*. Springer, New York, 2003.
6. I. Osorio and M. G. Frei. Real-time detection, quantification, warning and control of epileptic seizures: The foundation of a scientific epileptology. *Epil. Behav.* 16:391–396, 2009.
7. I. Osorio, M. G. Frei, S. Sunderam, J. Giftakis, N. C. Bhavaraja, S. F. Schnaffer, and S. B. Wilkinson. Automated seizure abatement in humans using electrical stimulation. *Ann. Neurol.* 57:258–268, 2005.
8. I. Osorio, M. G. Frei, and S. B. Wilkinson. Real-time automated detection and quantitative analysis of seizures and short-term prediction of clinical onset. *Epilepsia*, 39:615–627, 1998.
9. T. S. Salam, J. L. Perez-Velazquez, and R. Genov. Seizure suppression efficacy of closed-loop versus open-loop deep brain stimulation in a rodent model of epilepsy. *IEEE Trans. Neural Sys. Rehab. Eng.* 24(6): 710-719, 2016.
10. G. Milton J, A. R. Quan, and I. Osorio. Nocturnal frontal lobe epilepsy: Metastability in a dynamic disease? In I. Osorio, H. P. Zaveri, M. G. Frei, and S. Arthurs, editors, *Epilepsy: The intersection of neurosciences, biology, mathematics, engineering and physics*, pages 501–510, New York, 2011. CRC Press.

11. I. Osorio, H. P. Zaveri, M. G. Frei, and S. Arthurs. *Epilepsy: The intersection of neurosciences, biology, mathematics, engineering and physics*. CRC Press, New York, 2011.

12. A. Quan, I. Osorio, T. Ohira, and J. Milton. Vulnerability to paroxysmal oscillations in delay neural networks: A basis for nocturnal frontal lobe epilepsy? *Chaos*, 21:047512, 2011.

13. J. Milton and D. Black. Dynamic diseases in neurology and psychiatry. *Chaos*, 5:8–13, 1995.

14. J. M. Rommens, M. L. Iannuzzi, B. Kerem, M. L. Drumm, G. Melmer, M. Dean, R. Rozmahel, J. L. Cole, D. Kennedy, and N. Hidaka. Identification of the cystic fibrosis gene: chromosome walking and jumping. *Science*, 245:1055–1059, 1989.

15. S. Coombes and P. C. Bressloff. *Bursting: The genesis of rhythm in the nervous system*. World Scientific, New Jersey, 2005.

16. G. B. Ermentrout and D. H. Terman. *Mathematical Foundations of Neuroscience*. Springer, New York, 2010.

17. W. Gerstner and W. Kistler. *Spiking neuron models: single neurons, populations, plasticity*. Cambridge University Press, New York, 2006.

18. E. M. Izhikevich. *Dynamical systems in neuroscience: The geometry of excitability and bursting*. MIT Press, Cambridge, MA, 2007.

19. C. Koch and I. Segev. *Methods in Neuronal Modeling: From synapses to networks*. MIT Press, Cambridge, Ma, 1989.

20. J. Rinzel and G. B. Ermentrout. Analysis of neural excitability and oscillations. In C. Koch and I. Segev, editors, *Methods in Neuronal Modeling: From synapses to networks*, pages 135–169, Cambridge, MA, 1989. MIT Press.

21. J-B. Kim. Channelopathies. *Korean J. Pediatr.* 57:1–18, 2013.

22. D. M. Kullman and S. G. Waxman. Neurological channelopathies: new insights into disease mechanisms and ion channel function. *J. Physiol.*, 588.11:1823–1827, 2010.

23. E. Sigel and M. E. Steinman. Structure, function and modulation of GABA$_a$ receptors. *J. Biol. Chem.* 287:40224–40231, 2012.

24. P. Gloor, M. Avoli, and G. Kostopoulos. Thalamo-cortical relationships in generalized epilepsy with bilaterally synchronous spike-and-wave discharge. In M. Avoli, P. Gloor, R. Naquet, and G. Kostopoulos, editors, *Generalized Epilepsy: Neurobiological Approaches*, pages 190–212, Boston, 1990. Birkhäuser.

25. G. K. Kostopoulos. Spike-and-wave discharges of absence seizures as a transformation of sleep spindles: the continuing development of a hypothesis. *Clin. Neurophysiol.* 111 (S2):S27–S38, 2000.

26. L. Cocito and A. Primavera. Vigabatrin aggravates absences and absence status. *Neurology*, 51:1519–1520, 1998.

27. S. Knack, H. M. Hamer, U. Schomberg, W. H. Oertel, and F. Rosenow. Tiagabine-induced absence status in idiopathic generalized nonconvulsive epilepsy. *Seizure*, 8:314–317, 1999.

28. J. G. Milton. Epilepsy: Multistability in a dynamic disease. In J. Walleczek, editor, *Self-organized biological dynamics and nonlinear control*, pages 374–386, New York, 2000. Cambridge University Press.

29. G. van Luijtelaar, C. Behr, and M. Avoli. Is there such a thing as "generalized" epilepsy? In H. E. Scharfman and P. S. Buckmaster, editors, *Issues in Clinical Epileptology: A View from the Bench*, pages 81–91, New York, 2014. Springer.

30. G. van Luijtelaar and E. Sitnikova. Global and focal aspects of absence epilepsy: The contribution of genetic models. *Neurosci. Biobehav. Rev.* 30:983–1003, 2006.

31. J. Milton. Insights into seizure propagation from axonal conduction times. In J. Milton and P. Jung, editors, *Epilepsy as a dynamic disease*, pages 15–23, New York, 2003. Springer.

32. J. Bancaud. Physiopathogenesis of generalized epilepsies of organic nature (Stereoencephalographic study). In H. Gastaut, H. H. Jasper, J. Bancaud, and A. Waltregny, editors, *The Physiopathogenesis of the Epilepsies*, pages 158–185, Springfield, Il, 1969. Charles C. Thomas.

33. J. Bancaud. Role of the cerebral cortex in (generalized) epilepsy of organic origin. Contribution of stereoelectroencephalographic investigations (S. E. E. G.) to discussion of the centroencephalographic concept. *Presse Med.* 79:669–673, 1971.

34. H. K. M. Meeren, J. P. M. Pijn, E. L. J. M. Can Luijtelaar, A. M. L. Coenen, and F. H. Lopes da Silva. Cortical focus drives widespread corticothalamic networks during spontaneous absence seizures in rats. *J. Neurosci.* 22:1480–1495, 2002.

35. D. Gupta, P. Ossenblok, and G. van Luijtelaar. Space-time network connectivity and cortical activations preceding spike wave discharges in human absence epilepsy: a MEG study. *Med. Biol. Eng. Comput.* 49:555–565, 2011.

36. J. R. Tenney, H. Fujiwara, P. S. Horn, S. E. Jacobsen, T. A. Glaser, and D. F. Rose. Focal corticothalamic sources during generalized absence seizures: a MEG study. *Epilepsy Res.* 106:113–122, 2013.

37. I. Westmije, P. Ossenblok, B. Gunning, and G. van Luijtelaar. Onset and propagation of spike and slow wave discharges in human absence epilepsy: a MEG study. *Epilepsia*, 50:2538–2548, 2009.

38. S. A. Chkhenkeli and J. Milton. Dynamic epileptic systems versus static epileptic foci. In J. Milton and P. Jung, editors, *Epilepsy as a dynamic disease*, pages 25–36, New York, 2003. Springer.

39. J. G. Milton, S. A. Chkhenkeli, and V. L. Towle. Brain connectivity and the spread of epileptic seizures. In V. K. Jirsa and A. R. McIntosh, editors, *Handbook of Brain Connectivity*, pages 478–503, New York, 2007. Springer.

40. M. Abeles, E. Vaadia, and H. Bergman. Firing patterns of single units in the prefrontal cortex and neural network models. *Network*, 1:13–25, 1990.

41. P. Lennie. The cost of cortical computation. *Current Biol.* 13:493–497, 2003.

42. V.B. Kolmanovskii and V.R. Nosov. *Stability of functional differential equations*, volume 180 of *Mathematics in Science and Engineering*. Academic Press, London, England, 1986.

43. H. Smith. *An introduction to delay differential equations with applications to the life sciences*, volume 57. Springer Science & Business Media, 2010.

44. G. Stépán. *Retarded Dynamical Systems: Stability and Characteristic Functions*, volume 210 of *Pitman Research Notes in Mathematics*. Longman Group, Essex, 1989.

45. O. Diekmann, S. A. van Gils, S.M. Verduyn Lunel, and H.-O. Walther. *Delay Equations.* Springer-Verlag, New York, 1995.

46. J.K. Hale and S.M. Verduyn-Lunel. *Introduction to Functional Differential Equations.* Springer Verlag, New York, 1993.

47. F. H. Lopes da Silva, W. Blanes, S. Kalitizin, J. Parra Gomez, P. Suffczynski, and F. J. Velis. Epilepsies as dynamical diseases of brain systems: basic models of the transition between normal and epileptic activity. *Epilepsia*, 44 (suppl. 12):72–83, 2002.

48. C. M. Florez, R. J. McGinn, V. Lukankin, I. Marwa, S. Sugumar, J. Dian, L. N. Hazrati, P. L. Carlen, and L. Zhang. In vitro recordings of human neocortical oscillations. *Cerebral Cortex*, 25:578–597, 2015.

49. A. Destexhe. Spike-and-wave oscillations based on the properties of $GABA_B$ receptors. *J. Neurosci.* 18:9099–9111, 1998.

50. A. Destexhe. Corticothalamic feedback: A key to explain absence seizures. In I. Soltesz and K. Staley, editors, *Computational Neuroscience in Epilepsy*, pages 184–214, San Diego, 2008. Academic Press.

51. A. B. Holt and T. I. Netoff. Computational modeling of epilepsy for an experimental neurologist. *Exp. Neurol.* 244:75–86, 2013.

52. W. W. Lytton. Computer modeling of epilepsy. *Nat. Rev. Neurosci.* 9:626–637, 2008.

53. I. Soltesz and K. Staley. *Computational Neuroscience in Epilepsy*. Academic Press, San Diego, 2008.

54. G. Baier and J. Milton. Dynamic diseases of the brain. In D. Jaeger and R. Jung, editors, *Encyclopedia of Computational Neuroscience*, pages 1051–1061, New York, 2015. Springer.

55. L. Glass. Dynamical disease: challenges for nonlinear dynamics and medicine. *Chaos*, 25:097603, 2015.

56. C. Foley and M. C. Mackey. Mathematical model for G-CSF administration after chemotherapy. *J. Theoret. Biol.* 19:25–52, 2009.

57. M. C. Mackey. Periodic auto-immune hemolytic anemia: An induced dynamical disease. *Bull. Math. Biol.* 41:829–834, 1979.

58. J. S. Orr, J. Kirk, K. G. Gary, and J. R. Anderson. A study of the interdependence of red cell and bone marrow stem cell populations. *Brit. J. Haematol.* 15:23–34, 1968.

59. J. Milton, P. Naik, C. Chan, and S. A. Campbell. Indecision in neural decision making models. *Math. Model. Nat. Phenom.* 5:125–145, 2010.

60. K. Pakdaman, C. Grotta-Ragazzo, and C. P. Malta. Transient regime duration in continuous time neural networks with delay. *Phys. Rev. E,* 58:3623–3627, 1998.

61. K. Pakdaman, C. Grotta-Ragazzo, C. P. Malta, O. Arino, and J. F. Vibert. Effect of delay on the boundary of the basin of attraction in a system of two neurons. *Neural Netw.* 11:509–519, 1998.

62. C. Beaulieu, Z. Kisvarday, P. Somoygi, M. Cynader, and A. Cowey. Quantitative distribution of GABA-immunoresponsive and -immunonegative neurons and synapses in the monkey striate cortex (area 17). *Cerebral Cortex,* 2:295–309, 1992.

63. J. G. Milton. Epilepsy as a dynamic disease: A tutorial of the past with an eye to the future. *Epil. Beh.* 18:33–44, 2010.

64. I. Osorio, M. G. Frei, D. Sornette, J. Milton, and Y-C. Lai. Epileptic seizures: Quakes of the brain? *Phys. Rev. E,* 82:021919, 2010.

65. G. A. Worrell, C. A. Stephen, S. D. Cranstoun, B. Litt, and J. Echauz. Evidence for self-organized criticality in human epileptic hippocampus. *NeuroReport,* 13:2017–2021, 2010.

66. M. Stead, M. Bower, B. H. Brinkmann, K. Lee, W. R. Marsh, F. B. Meyer, B. Litt, J. Van Gompel, and G. A. Worrell. Microseizures and the spatiotemporal scales of human partial epilepsy. *Brain,* 133:2789–2797, 2010.

67. J. Milton. Neuronal avalanches, epileptic quakes and other transient forms of neurodynamics. *Eur. J. Neurosci.* 36:2156–2163, 2012.

68. W. Horsthemke and R. Lefever. *Noise-induced transitions: Theory and applications in physics, chemistry and biology.* Springer, New York, 1984.

69. P. Milanowski and P. Suffcznski. Seizures start without a common signature of critical transitions. *Int. J. Neural Sys.* 26: 1650053, 2016.

70. D. A. Williams. A study of thalamic and cortical rhythms in Petit Mal. *Brain,* 76:50–59, 1953.

71. M. Steriade. The GABAergic reticular nucleus: a preferential target of corticothalamic projections. *Proc. Natl. Acad. Sci. USA,* 98:3625–3627, 2001.

72. M. Steriade and D. Contreras. Relations between cortical and cellular events during transition from sleep patterns to paroxysmal activity. *J. Neurosci.* 15:623–642, 1995.

73. M. J. Gallagher. How deactivating an inhibitor causes absence epilepsy: Validation of a noble lie. *Epilepsy Curr.* 13:38–41, 2013.

74. E. G. Jones. *The Thalamus.* Plenum Press, New York, 1985.

75. C. Ajmone-Marsan. The thalamus. Data on its functional anatomy and on some aspects of thalamo-cortical organization. *Arch. Italiennes Biol.* 103:847–882, 1965.

76. Y. Choe. The role of temporal parameters in a thalamocortical model of analogy. *IEEE Trans. Neural Netw.* 15:1071–1082, 2004.

77. T. Tsumoto, O. D. Creutzfield, and C. R. Legendy. Functional organization of the corticofugal system from visual cortex to lateral geniculate nucleus in the cat. *Exp. Brain Res.* 32:345–364, 1978.

78. C. E. Landisman, M. A. Long, M. Beierlein, M. R. Deans, D. L. Paul, and B. W. Connors. Electrical synapses in the thalamic reticular nucleus. *J. Neurosci.* 22:1002–1009, 2002.

79. S-C. Lee, S. L. Patrick, K. A. Richardson, and B. W. Connors. Two functionally distinct networks of gap junction-coupled inhibitory neurons in the thalamic reticular nucleus. *J. Neurosci.* 34:12182–13170, 2014.

80. J. G. Milton. *Dynamics of small neural populations.* American Mathematical Society, Providence, RI, 1996.

81. J. T. Paz and J. R. Huguenard. Microcircuits and their interactions in epilepsy: is the focus out of focus? *Nat. Neurosci.* 18:351–359, 2015.

82. O. Sporns and R. Kötter. Motifs in brain networks. *PLoS Biol.* 2:e369, 2004.

83. J. T. Paz, A. S. Bryant, K. Peng, L. Fenno, O. Yizhar, W. N. Frankel, K. Deisseroth, and J. R. Huguenard. A new mode of corticothalamic transmission revealed in the Gria4(-/-) model of absence epilepsy. *Nat. Neurosci.* 14:1167–1173, 2011.

84. A. M. Large, N. W. Vogles, S. Mielo, and A-M. M. Oswald. Balanced feedfoward inhibition and dominant recurrent inhibition in olfactory cortex. *Proc. Natl. Acad. Sci. USA*, 113:2276–2281, 2016.

85. S.A. Campbell. Stability and bifurcation of a simple neural network with multiple time delays. In S. Ruan, G.S.K. Wolkowicz, and J. Wu, editors, *Fields Inst. Commun*, volume 21, pages 65–79. AMS, 1999.

86. A. Payeur, L. Maler, and A. Longtin. Oscillatory-like behavior in feedforward neuronal networks. *Phys. Rev. E*, 92:012703, 2015.

87. J. Foss, A. Longtin, B. Mensour, and J. Milton. Multistability and delayed recurrent feedback. *Phys. Rev. Lett.* 76:708–711, 1996.

88. K. D. Graber and D. A. Prince. A critical period for prevention of post-traumatic neocortical hyperexcitability in rats. *Ann. Neurol.* 55:860–870, 2004.

89. A. R. Houweling, M. M. Bazhenov, I. Timofeev, M. Steraide, and T. J. Sejnowski. Homeostatic synaptic plasticity can explain post-traumatic epileptogenesis in chronically isolated neocortex. *Cerebral Cortex*, 15:834–845, 2005.

90. K. L. Babcock and R. M. Westervelt. Dynamics of simple electronic networks. *Physica D*, 28:305–316, 1987.

91. N. Azmy, E. Boussard, J. F. Vibert, and K. Pakdaman. Single neuron with recurrent excitation: Effect of the transmission delay. *Neural Netw.* 9:797–818, 1996.

92. O. Diez Martinez and J. P. Segundo. Behavior of a single neuron in a recurrent excitatory loop. *Biol. Cybern.* 47:33–41, 1983.

93. K. Pakdaman, J-F. Vibert, E. Boussard, and N. Azmy. Single neuron with recurrent excitation: Effect of the transmission delay. *Neural Netw.* 9:797–818, 1996.

94. K. Pakdaman, F. Alvarez, J. P. Segundo, O. Diez-Martinez, and J-F. Vibert. Adaptation prevents discharge saturation in models of single neurons with recurrent excitation. *Int. J. Mod. Simul.* 22:260–265, 2002.

95. Y. Chen and J. Wu. Slowly oscillating periodic solutions for a delayed frustrated network of two neurons. *J. Math. Anal. Appl.* 259:188–208, 2001.

96. J. Foss, F. Moss, and J. Milton. Noise, multistability and delayed recurrent loops. *Phys. Rev. E*, 55:4536–4543, 1997.

97. J. Foss and J. Milton. Multistability in recurrent neural loops arising from delay. *J. Neurophysiol.* 84:975–985, 2000.

98. M. C. Mackey and U. an der Heiden. The dynamics of recurrent inhibition. *J. Math. Biol.* 19:211–225, 1984.

99. L. H. A. Monteiro, A. Pellizari Filho, J. G. Chaui-Berlinck, and J. R. C. Piquiera. Oscillation death in a two-neuron network with delay in a self-connection. *J. Biol. Sci.* 15:49–61, 2007.

100. R. E. Plant. A Fitzhugh differential-difference equation modeling recurrent neural feedback. *SIAM J. Appl. Math.* 40:150–162, 1981.

101. H.R. Wilson and J.D. Cowan. Excitatory and inhibitory interactions in localized populations of model neurons. *Biophys. J.* 12(1):1, 1972.

102. J. Ma and J. Wu. Multistability in spiking neuron models of delayed recurrent inhibitory loops. *Neural Comp.* 19:2124–2148, 2007.

103. J. Bélair and S.A. Campbell. Stability and bifurcations of equilibria in a multiple-delayed differential equation. *SIAM J. Appl. Math.* 54(5):1402–1424, 1994.

104. J. Guckenheimer and P.J. Holmes. *Nonlinear Oscillations, Dynamical Systems and Bifurcations of Vector Fields*. Springer-Verlag, New York, 1983.

105. Y.A. Kuznetsov. *Elements of Applied Bifurcation Theory*, volume 112 of *Applied Mathematical Sciences*. Springer-Verlag, Berlin/New York, 1995.

106. S.A. Campbell, J. Bélair, T. Ohira, and J. Milton. Limit cycles, tori, and complex dynamics in a second-order differential equation with delayed negative feedback. *J. Dyn. Diff. Eqns.* 7(1):213–236, 1995.

107. L. P. Shayer and S.A. Campbell. Stability, bifurcation, and multistability in a system of two coupled neurons with multiple time delays. *SIAM J. Appl. Math.* 61(2):673–700, 2000.

108. G. Fan, S.A. Campbell, G.S.K. Wolkowicz, and H. Zhu. The bifurcation study of 1:2 resonance in a delayed system of two coupled neurons. *J. Dyn. Diff. Eqns.* 25(1):193–216, 2013.

109. J. Ma and J. Wu. Patterns, memory and periodicity in two-neuron recurrent inhibitory loops. *Math. Model. Nat. Phenom.* 5:67–99, 2010.

110. M. Timme and F. Wolf. The simplest problem in the collective dynamics of neural networks: Is synchrony stable? *Nonlinearity*, 21:1579–1599, 2008.

111. H. T. Chugani, M. E. Phelps, and J. C. Mazziotta. Positron emission tomography study of human brain function development. *Ann. Neurol.*, 22:487–497, 1987.

112. P. R. Huttenlocher. Developmental changes of aging. *Brain Res.* 163:195–205, 1979.

113. P. R. Huttenlocher and A. S. Dabholkar. Regional differences in synaptogenesis in human cerebral cortex. *J. Comp. Neurol.*, 387:167–178, 1997.

114. N. Barnea-Goraly, V. Menon, M. Eckart, L. Tamm, R. Bammer, A. Karchemskiy, C. C. Dant, and A. L. Reiss. White matter development during childhood and adolescence: A cross-sectional diffusion tensor imaging study. *Cerebral Cortex*, 15:1848–1854, 2005.

115. G. Z. Tau and B. S. Peterson. Normal development of brain circuits. *Neuropschopharmacology*, 35:147–168, 2010.

116. J-S Liang, S-P Lee, B. Pulli, J. W. Chen, S-C Kao, Y-K Tsang, and K. L-C Hsieh. Microstructural changes in absence seizure children: A diffusion tensor magnetic resonance imaging study. *Ped. Neonatology* 57(4): 318-325, 2016.

117. H. Chahboune, A. M. Mishra, M. N. DeSalvo, L. H. Stocib, M. Purcano, D. Scheinost, X. Papademetris, S. J. Fyson, M. L. Lorincz, V. Crunelli, F. Hyder, and H. Blumenfeld. DTI abnormalities in anterior corpus callosum of rats with spike-wave epilepsy. *NeuroImage*, 47:459–466, 2009.

118. W. A. Hauser, J. F. Annegers, and L. T. Kurland. The incidence of epilepsy in Rochester, Minnesota, 1935–1984. *Epilepsia*, 34:453–468, 1993.

119. P. A. Robinson, C. J. Rennie, and D. L. Rowe. Dynamics of large-scale brain activity in normal arousal states and epileptic seizures. *Phys. Rev. E*, 65:041924, 2002.

120. W. J. Freeman. *Neurodynamics: An exploration of mesoscopic brain dynamics*. Springer Verlag, London, 2000.

121. W. J. Freeman and M. D. Holmes. Metastability, instability, and state transitions in cortex. *Neural Netw.* 18:497–504, 2005.

122. E. Sitnokova, A. E. Hramov, V. V. Grubov, A. A. Ovchinnkov, and A. A. Koronovsky. On-off intermittency of thalamo-cortical oscillations in the electroencephalogram of rats with genetic predisposition to absence epilepsy. *Brain Res.*, 1436:147–156, 2012.

123. M. Goodfellow, K. Schindler, and G. Baier. Intermittent spike-wave dynamics in a heterogeneous, spatially extended neural mass model. *NeuroImage*, 55:920–932, 2011.

124. J. Milton and J. Foss. Oscillations and multistability in delayed feedback control. In H. G. Othmer, F. R. Adler, M. A. Lewis, and J. C. Dallon, editors, *Case Studies in Mathematical Modeling: Ecology, Physiology and Cell Biology*, pages 179–198, Upper Saddle River, NJ, 1997. Prentice Hall.

125. G.B. Ermentrout. *XPPAUT 5.91 – the differential equations tool*. Department of Mathematics, University of Pittsburgh, Pittsburgh, PA, 2005.

126. L.F. Shampine and S. Thompson. Solving DDEs in MATLAB. *Appl. Num. Math.*, 37:441–458, 2001.

127. K. Engelborghs, T. Luzyanina, and G. Samaey. DDE-BIFTOOL v. 2.00: a MATLAB package for bifurcation analysis of delay differential equations. Technical Report TW-330, Department of Computer Science, K.U. Leuven, Leuven, Belgium, 2001.

128. K. Engelborghs, T. Luzyanina, and D. Roose. Numerical bifurcation analysis of delay differential equations using DDE-BIFTOOL. *ACM Trans. Math. Software*, 28(1):1–21, 2002.

Extracellular Potassium and Focal Seizures—Insight from In Silico Study

Piotr Suffczynski, Damiano Gentiletti, Vadym Gnatkovsky
and Marco de Curtis

Introduction

Epilepsy and seizures have been recognized as brain diseases in antiquity. Since then enormous progress in understanding pathophysiology of epilepsy has taken place. It led to the development of effective treatments, such as antiepileptic drugs [1], resective surgery [2], neuromodulation [3] and neurostimulation [4]. Despite these advances many patients still experience seizures or suffer from side effects of antiepileptic medications. Further progress in understanding mechanisms of epileptic brain activity is needed but the brain complexity at various scales of neuronal organization is the main challenge that needs to be overcome to achieve this goal. There is no existing technology that could possibly measure complex behaviours of thousands of individual neurons synchronizing their electrical activities during seizure. That's why, many properties of neurons, neuronal populations and activities of their networks can be best studied in computational models, where each neuronal element and overall activity may be observed simultaneously. Furthermore, in the case of the brain, a model as complex as the real system would probably be just as hard to investigate. Instead, a wise distinction between essential and irrelevant components of the observed phenomenon may lead to a simplified in silico representation of the complex neuronal system, that may be studied more efficiently. Computational epilepsy research is relatively new but rapidly growing field. Although seizure-like neuronal behaviour has been recreated in a number of

P. Suffczynski (✉) · D. Gentiletti
Department of Biomedical Physics, Institute of Experimental Physics,
University of Warsaw, Pasteura 5, Warsaw, Poland
e-mail: Piotr.Suffczynski@fuw.edu.pl

V. Gnatkovsky · M. de Curtis
Unit of Epileptology and Experimental Neurophysiology,
Fondazione IRCCS Istituto Neurologico Carlo Besta,
Via Giovanni Celoria, 11, Milan, Italy

© Springer International Publishing AG 2017
P. Érdi et al. (eds.), *Computational Neurology and Psychiatry*,
Springer Series in Bio-/Neuroinformatics 6,
DOI 10.1007/978-3-319-49959-8_3

49

computational models [5], their applications towards improved clinical practice has been very limited so far. One of the possible reasons is that most of the models of epileptic neuronal networks consider neurons and their connections isolated from the external neuronal environment. Although membrane and synaptic currents are generated by flow of ions across membranes, modelled intra- and extracellular ion concentrations are typically assumed, e.g. in Hodgkin and Huxley formalism [6], to be constant in time and in space. This assumption might hold during normal activity when ion fluxes are moderate and ion homeostasis can be maintained by a number of mechanisms including ion pumps and transporters, buffering and diffusion. However, during periods of increased neuronal discharges, such as seizures, activity-induced changes in ion concentrations may largely deviate from their baseline values [7]. These alterations in intra- and extracellular ion concentrations can impact a number of neuronal processes, such as maintenance of resting membrane potential, operation of voltage-gated channels, synaptic transmission, volume regulation and ion transport mediated by ionic pumps and cotransporters. In particular, extracellular K^+ concentration changes have been directly implicated in epilepsy and seizures [8]. Potassium accumulation hypothesis proposed by Fertzi-ger and Ranck [9] suggests that increased neuronal discharges lead to an increase of extracellular K^+ concentration, which depolarizes neurons leading to their increased firing, what in turn contributes to further increase in extracellular K^+. Such positive feedback mechanism was suggested to be at play during seizures development but also was deemed responsible for seizure termination through depolarization-induced inactivation of Na^+ currents. To date, a lot of experimental data on ion dynamics associated with seizure activity has been accumulated. Nonetheless, these findings are usually not incorporated into computational models, with few notable exceptions (e.g., Durand and Park [10], Kager et al. [11], Krishnan and Bazhenov [12], Wei et al. [13]). These highly realistic models are usually dedicated to particular epilepsy or seizure models. Accordingly, their electro-physiological and ionic patterns might not be similar and it is not yet clear, to what extent these individual results may be generalized.

In this chapter we explore the influence of ionic dynamics on specific seizure generation pattern using computational model with activity-dependent ion con-centration changes. The model is based on the experimental data obtained in in vitro isolated guinea pig brain preparation, which is considered an animal model of human focal epilepsy [14]. We show that combined experimental and modelling approach allowed to obtain new insight into functional role played by various ionic components, especially potassium, in focal seizure generation. We also suggest that such better understanding of basic mechanisms underlying epileptic seizures may advance translation of research findings into novel therapeutic applications.

The Scientific Problem

As mentioned above, among four main ions shaping electrical activities of neurons, i.e., Na^+, K^+, Cl^- and Ca^{2+}, dynamics of potassium has been suggested to play an important role in seizure generation. Therefore, we explain the scientific problem by first considering the mechanisms by which extracellular K^+ concentration ($[K^+]_o$) may modulate neuronal excitably. Next, we describe experimentally observed seizure pattern and associated $[K^+]_o$ changes and finally, we formulate the aim of this modelling study.

Effects of Extracellular K^+ Concentration on Neural Activity

Resting Potential

Resting membrane potential is given by the Goldman-Hodgkin-Katz equation (GHK):

$$V_m = \frac{RT}{F} \ln \frac{P_K[K^+]_o + P_{Na}[Na^+]_o + P_{Cl}[Cl^-]_i}{P_K[K^+]_i + P_{Na}[Na^+]_i + P_{Cl}[Cl^-]_o}.$$

For ion concentrations and permeability in typical mammalian cell [15], as shown in Table 1, the resting membrane potential is −66 mV.

Figure 1 shows the influence of concentrations of K^+, Na^+ and Cl^- on resting membrane potential, calculated according to GHK equation. On each graph, variation of the resting membrane potential is shown (solid line) when given concentration is being varied and all others are kept constant, at their baseline values. The reference value of a concentration being varied is shown on each graph by broken vertical line. Intersection of solid and broken lines on each graph marks the reference resting membrane potential, that is $V_m = -66$ mV. The slope of the solid line on each graph corresponds to sensitivity of the resting membrane potential to changes of the given concentration. One can see that resting membrane potential depends on all concentrations but it is most sensitive to change of extracellular potassium, as suggested by largest slope of the solid line in Fig. 1 left bottom panel, as compared to other panels. It suggests that even moderate increase of $[K^+]_o$ leads to significant change of the resting potential of the cell into depolarized direction.

Table 1 Ion concentrations in a typical mammalian cell	Ion	Inside (mM)	Outside (mM)
	K^+	140	5
	Na^+	10	145
	Cl^-	4	110
	Ca^{2+}	$1 * 10^{-4}$	2.5

Effects of ionic concentrations on resting potential

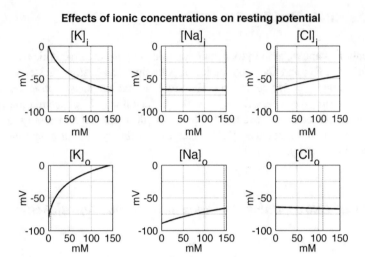

Fig. 1 Influence of K$^+$, Na$^+$ and Cl$^-$ concentration changes on the resting membrane potential. On each graph, *black solid line* shows the resting membrane potential calculated with GHK equation while *red vertical broken line* shows the reference value of the given concentration. Crossing of these lines marks the resting membrane potential for the reference conditions, that is about −66 mV. The resting membrane potential is sensitive to changes of all concentrations but it most sensitive to change of potassium extracellular concentration. Even minor change of extracellular K$^+$ concentration around its reference value leads to significant change of the resting membrane potential

Driving Force

The main role of all potassium currents is to control excitability and stabilize membrane potential around potassium reversal potential E_K, typically around −90 mV. Ionic current is proportional to the channel conductance and the ionic driving force, which is the difference between the membrane potential and the ion reversal potential. The ionic reversal potential in turn depends on the intra- and extracellular ion concentrations. When extracellular potassium increases, both potassium reversal potential and the resting membrane potential move to more positive level. However, E_K grows more considerably than membrane potential V_m, because change of V_m is limited by the presence of Na$^+$ and Cl$^-$ leak currents. As a result, E_K moves towards V_m leading to reduction of the driving force of all potassium currents. Accordingly, the cell's capability to resist membrane depolarization is diminished and may lead to increased action potential firing.

Synaptic Transmission

Effect of extracellular potassium on synaptic transmission is not trivial as both pre- and postsynaptic part of the synapse depend on membrane depolarization and therefore on $[K^+]_o/[K^+]_i$. It is established experimentally [16] that moderate increase of $[K^+]_o$ enhances excitatory synaptic transmission while above certain level of $[K^+]_o$ both excitatory and inhibitory synaptic transmission is abolished due to presynaptic depolarization-related inactivation of Na^+ and Ca^{2+} channels. Yet another effect is related to interaction between potassium and chloride. $GABA_A$ inhibitory potentials are associated with chloride influx that hyperpolarizes the cell. The Cl^- and K^+ ions are coupled by potassium-chloride co-transporter KCC2, which usually extrudes K^+ and Cl^- ions from neurons [17]. The KCC2 co-transporter operates at thermodynamic equilibrium defined by $[K^+]_o[Cl^-]_o = [K^+]_i[Cl^-]_i$ [18]. Under this condition, large increase of extracellular K^+ concentration must be compensated by increase of intracellular Cl^- and the direction of transport of K^+ and Cl^- ions reverses. It leads to intracellular Cl^- accumulation and elevation of the $GABA_A$ reversal potential. Accordingly, $GABA_A$ inhibitory potentials will be reduced or may even become depolarizing if E_{GABAA} surpasses membrane potential of the cell [19].

Focal Seizure Dynamics

Many types of epileptic seizures are not stationary processes but exhibit intrinsic dynamics. From the clinical point of view, seizure onset patterns are most important, as they are relevant for seizure early detection and localisation. However, seizure evolution may provide important clues regarding the mechanism involved in seizure generation. The typical pattern observed with intracranial electrodes in mesial temporal lobe epilepsy patients consists of low-voltage fast activity (>25 Hz) at the seizure onset, irregular activity phase with some increase of amplitude followed by synchronous quasi-rhythmic pattern that increases in spatial synchrony towards the end of the seizure. These focal seizure patterns can be reproduced and studied in animal in vitro models, which offer an additional advantage of performing intracellular and ion-selective recordings together with extracellular LFP measurements in the intact, whole brain networks.

In in vitro isolated guinea pig brain preparations [20] seizures may be induced acutely by arterial application of proconvulsant drugs [21]. Seizure patterns in this model resemble very much those observed in human focal epilepsy [14]. Intracellular recordings from entorhinal cortex (EC) neurons showed that seizures were initiated with increased firing of inhibitory interneurons and neuronal silence of principal cells, which correlated with low-voltage fast LFP oscillations (20–30 Hz). Neuronal firing of principal cells was subsequently restored with an acceleration-deceleration firing pattern followed by rhythmic burst activity. Increased firing of principal neurons correlated with ictal discharges in the LFP signal. An increase of extracellular potassium concentration was observed throughout the seizure. Typical intracellular seizure pattern and associated extracellular potassium concentration time course observed in isolated guinea pig brain is shown in Fig. 3a.

Aim of This Work

Despite the fact that the synaptic and non-synaptic mechanisms mediating seizure pattern observed in isolated guinea pig brain have been suggested, the specific roles played by various neural elements during seizures are not fully understood. During seizure states many mechanisms are interacting in complex ways, making it difficult to study experimentally. It is becoming recognized that studying these interactions may contribute to a better understanding of seizure generation. The aim of the present chapter is to investigate the link between ionic dynamics and experimentally observed seizure pattern, using a computational model, in order to provide such a synthetic view.

Computational Methods

Model Description

The model consists of two multicompartmental EC cells, a pyramidal neuron and an inhibitory interneuron, embedded in the common extracellular space (ECS) surrounded by a bath. The size of ECS was estimated by the extracellular volume fraction α defined by the ratio volume of extracellular space/volume of intracellular space. Ionic dynamics of K^+, Na^+, Ca^{2+} and Cl^- was incorporated and activity-dependent changes in their concentrations were computed. Concentration changes in a given extra- or intracellular compartment were dependent on a number of mechanisms such as: active and passive membrane currents, inhibitory synaptic $GABA_A$ currents, Na^+/K^+ pump, KCC2 cotransporter, glial K^+ buffering, radial diffusion between ECS and bath, and longitudinal diffusion between dendritic and somatic compartments. At each simulation step all ionic concentrations were updated and their reversal potentials were computed. The model setup is shown in Fig. 2. Simulations were performed using NEURON simulator with a fixed integration step of 0.05 ms. The parameters' values and units together with their sources are given in Gentiletti et al. [22]. Modifications in the present model, with respect to original publication include: (i) simplified cells' morphology, (ii) leak currents for K^+, Na^+, and Cl^- (instead of K^+, Na^+ and fixed leak), (iii) modification of radial diffusion coefficients to better reproduce experimental results, (iv) taking into account bicarbonate and Cl^- concentrations for calculation of E_{GABAA} (instead of considering only Cl^-). These changes are described in the subsequent sections, where modified parameter values are provided.

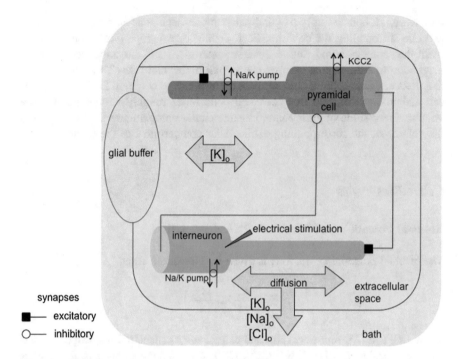

Fig. 2 Schematic diagram of the model. Two entorhinal cortex cells, a pyramidal cell and an interneuron are synaptically coupled and are embedded in the common extracellular space, which is surrounded by a bath. The model includes ionic regulation mechanisms: Na^+/K^+ pump in both cells, KCC2 cotransporter in pyramidal cell, glial buffering of K^+ ions and diffusion of K^+, Na^+ and Cl^- ions between compartments and between extracellular space and surrounding bath. The activity in the model can be triggered by excitatory synaptic input to pyramidal cell and current input to the interneuron

Cells' Morphology

The cells in our model were adapted from the compartmental model of the EC cells by Fransén et al. [23]. In the original model the pyramidal cell was composed of six compartments, one representing the soma, three representing the apical dendrite and two representing basal dendrites. The interneuron was composed of six compartments, one representing the soma, three representing a principal dendrite, and two representing remaining dendrites. In order to simplify the given models we lumped together all dendrites in both cells using Rall's rule. Therefore, the simplified model of pyramidal cell is composed of two compartments: one representing soma with length 20 μm and diameter 15 μm, and one representing lumped dendrites with length 450 μm and diameter 6.88 μm. The interneuron is also composed of two compartments, one representing soma with length 20 μm and diameter 15 μm and one representing lumped dendrites with length 700 μm and diameter 3 μm. The extracellular space is common for both cells, but is physically modelled around the interneuron. It consists of two cylindrical compartments, one surrounding the soma and one surrounding the dendrite. Each extracellular compartment communicates with the corresponding intracellular compartment of the modelled neurons, neighbouring extracellular compartment, glial buffer and bath. To account for the fact that each extracellular compartment corresponds to common ECS of pyramidal cell and interneuron and that each cell contributes roughly similar extracellular volume, the volume of each common extracellular compartment is given by twice the volume of the corresponding extracellular compartment of the interneuron.

Cells' Biophysics

Reversal Potentials and Passive Electrical Properties

The reversal potential of the ions is given by Nernst equation:

$$E_X = \frac{RT}{zF} \ln\left(\frac{[X]_o}{[X]_i}\right)$$

where $[X]_i$ and $[X]_o$ represent intra and extracellular concentrations, respectively of the ionic species $X = \{Na^+, K^+, Ca^{2+}, Cl^-\}$, F is Faraday's constant, R is the gas constant, z is the valence of the species X and $T = 32\ ^\circ C$ is temperature (as specified in Gnatkovsky et al. [24]). Leakage currents were present in all the compartments of both cells and had the following, standard expressions:

$$I_{Na, leak} = g_{Na, leak}(V_m - E_{Na})$$
$$I_{K, leak} = g_{K, leak}(V_m - E_K)$$
$$I_{Cl, leak} = g_{Cl, leak}(V_m - E_{Cl})$$

The resting membrane potential was set to -65 mV in the pyramidal cell and -75 mV in the interneuron. In pyramidal cell we set initial potassium and chloride concentrations at the thermodynamic equilibrium of KCC2, i.e., $[K^+]_o[Cl^-]_o = [K^+]_i[Cl^-]_i$. In this way, potassium and chloride cotransporter currents were zero initially and membrane potential was more stable. Specific axial resistance in both cells was set to $R_a = 100$ Ohm $\times$ cm and specific membrane capacitance was set to $C_m = 1 \mu F/cm^2$, as in Fransén et al. [23]. It should be noted that under background excitatory synaptic input, ionic concentrations and membrane potentials deviated slightly from the equilibrium conditions.

Active Currents

The pyramidal cell model included the following active membrane currents: the Na^+ and K^+ currents present in both compartments and responsible for fast action potentials (I_{NaT} and I_{Kdr}, respectively); a persistent Na^+ current I_{NaP} in the soma; a high-threshold Ca^{2+} current I_{CaL} in both compartments; a calcium-dependent K^+ current I_{KAHP} in both compartments; a fast calcium- and voltage-dependent K^+ current I_{KC} in both compartments; and a noninactivating muscarinic K^+ current I_{KM} in the soma. The interneuron model had the Na^+ and K^+ currents, I_{NaT} and I_{Kdr}, in both compartments, responsible for fast action potentials. All the models of active currents were based on Fransén et al. [23]. Some changes of the parameters were required in order to compensate the additional ionic regulation mechanisms that were not present in the original model. Implemented modifications together with all the currents' formulas and parameters are given in Gentiletti et al. [22].

Synaptic Connections and Network Inputs

The pyramidal cell and inhibitory interneuron are synaptically coupled, as shown in Fig. 2. An inhibitory synapse is placed in the middle of somatic compartment of the pyramidal cell, and an excitatory synapse is placed in the middle of dendritic compartment of the interneuron. The synaptic conductances are modelled according to NEURON's built-in Exp2Syn mechanism with a double-exponential function of the form:

$$g = g_{max}[\exp(-t/\tau_2) - \exp(-t/\tau_1)]$$

The τ_1 and τ_2 are the rise and decay time constants respectively, taking values 2 ms and 6 ms for all synapses. The reversal potential of excitatory postsynaptic currents (EPSC) was set to 0 mV. In order to investigate the impact of chloride concentration changes the inhibitory $GABA_A$ postsynaptic currents were explicitly mediated by chloride influx. Because $GABA_A$ receptor pore conducts both Cl^- and HCO_3^- in a 4:1 ratio [25], E_{GABAA} was calculated using the Goldman-Hodgkin-Katz equation [26]:

$$E_{GABAA} = \frac{RT}{F} \ln \left(\frac{4[Cl^-]_i + [HCO_3^-]_i}{4[Cl^-]_o + [HCO_3^-]_o} \right)$$

For simplicity, HCO_3^- concentrations were assumed to be constant and equal to $[HCO_3^-]_i = 15$ mM and $[HCO_3^-]_o = 25$ mM. Excitatory and inhibitory synaptic weights w_e, w_i were equal to 0.001 and 0.002 μS, respectively. With these settings, the unitary excitatory postsynaptic potential had amplitude of ~ 2.3 mV at the soma, while the unitary inhibitory postsynaptic potential had amplitude of ~ 1 mV at the soma. A synaptic response was generated in a postsynaptic cell when presynaptic membrane potential in the soma crossed the threshold of -10 mV.

Pyramidal cell received excitatory background synaptic input via dendritic excitatory synapse activated by a Poisson spike train with 66 Hz rate. In order to reproduce enhanced interneuronal firing at seizure onset, the inhibitory interneuron was stimulated by the somatic injection of a depolarizing current with initial amplitude of 1.3 nA at second 13 that was linearly decreasing toward 0.5 nA at second 60.

Ion Accumulation

Changes in ion concentrations due to transmembrane currents are described by the following equation:

$$\frac{d[X]}{dt} = \frac{\sum I_X}{zFV},$$

where $[X]$ is the concentration of the ionic species X, $\sum I_X$ is the net ion transmembrane current, F is Faraday's constant, V the compartment volume, and z the valence of the species X. The right-hand side is positive if the net flow is outward (for positive z), negative, if inward. The longitudinal diffusion contribution is calculated using the built-in mechanism in the NEURON simulation environment:

$$\frac{d[X_i]}{dt} = D_X \sum_n \frac{([X_j] - [X_i])S_{ij}}{L_{ij}V_i},$$

where $[X_i]$ is the ion concentration in the compartment i, D_X is the longitudinal diffusion constant of the ionic species X, S_{ij} is the flux area between the adjacent compartments i and j, L_{ij} is the distance between the centers of the compartments i and j, and V_i is the volume of the compartment i. The right-hand side is positive if the net flow is increasing the ion concentration of the given compartment, negative otherwise. The sum is made over the total number of contributions to ionic concentration in the i-th compartment. For our two compartmental cells and ECS n is equal to 1. The ions accumulated in the extracellular space can diffuse radially to the bath representing the extracellular space and vasculature not included in the model. Radial diffusion represents the net effect of various processes such as extracellular diffusion to more distant areas of the brain, potassium transport into capillaries and potassium regulation by the network of glial cells coupled by gap junctions. The time constant of these joint processes is likely to be much slower than that of lateral diffusion. Therefore radial diffusion constant is assumed to be equal to the longitudinal diffusion constant D_X, divided by a scale constant $s = 5000$. Bath concentrations are assumed to be constant and equal to the initial extracellular concentrations. For the sake of simplicity, radial ion diffusion was modelled only for K^+, Na^+ and Cl^-, as Ca^{2+} fluxes were much smaller than these of other ion types. The equation implemented to account for radial diffusion is as follows:

$$\frac{d[X_i]_o}{dt} = \frac{1}{s} D_X \frac{([X]_{bath} - [X_i]_o) S_i}{(dr/2) V_i},$$

where $[X]_{bath}$ is ion X concentration in the bath, S_i is radial flux area corresponding to the i-th compartment, V_i is the volume of the i-th extracellular compartment, and dr is the thickness of the extracellular space. The values of radial diffusion constants (in [μm^2/ms]) for K^+, Na^+ and Cl^- were 1.96, 0.226 and 2.03, respectively. The electrostatic drift of ions was neglected, as the ion movement due electrical potential gradient in extracellular space is small compared to diffusion.

Na⁺/K⁺ Pump

The Na^+/K^+ pump is moving Na^+ and K^+ ions across the membrane in order to maintain their concentration gradients and exchanges 2 K^+ ions for 3 Na^+ ions. The Na^+ and K^+ pump currents are as is [27]:

$$I_{Na} = 3 I_{max} flux([Na^+]_i, [K^+]_o)$$
$$I_K = 2 I_{max} flux([Na^+]_i, [K^+]_o)$$
$$flux([Na^+]_i, [K^+]_o) = \left(1 + \frac{Km_K}{[K^+]_o}\right)^{-2} \left(1 + \frac{Km_{Na}}{[Na^+]_i}\right)^{-3}$$

using maximal flux $I_{max} = 0.0013$ mA/cm², $Km_K = 2$ mM and $Km_{Na} = 10$ mM.

Glial Buffer

Potassium buffering is modelled with a first-order reaction scheme simulating glial potassium uptake system. It involves three variables: $[K^+]_o$—extracellular potassium concentration, $[B]$—free buffer, $[KB]$—bound buffer, and backward and forward rate constants k_1 and k_2, respectively. The set of differential equations solved is as follows [27]:

$$\frac{d[K^+]_o}{dt} = -k_2 [K^+]_o [B] + k_1 [KB]$$
$$\frac{d[B]}{dt} = -k_2 [K^+]_o [B] + k_1 [KB]$$
$$\frac{d[KB]}{dt} = -k_2 [K^+]_o [B] - k_1 [KB]$$

KCC2 Cotransporter

In adult hippocampal cells low intracellular Cl^- concentration is maintained by means of potassium-chloride cotransporter KCC2 that mediates $K - Cl$ cotransport across the membrane. The transport process involves one for one extrusion of Cl^- ion together with K^+ ion. The KCC2 cotransporter currents are modelled according to Wei et al. [13]:

$$I_K = \gamma U_{KCC2} \ln \left(\frac{[K^+]_i [Cl^-]_i}{[K^+]_o [Cl^-]_o} \right)$$

$$I_{Cl} = -I_K$$

where $U_{KCC2} = 0.3$ mM/s is the cotransporter strength and $\gamma = S_i/(FV_i)$ is a conversion factor from the concentration per second units (mM/s) to the current density units (mA/cm^2). For pyramidal compartments, soma and dendrite, γ takes the value γ_s and γ_d, respectively. The parameters S_i, V_i, are the total surface area and total intracellular volume of the respective cell compartment and F is Faraday's constant.

Results

In Silico Test of the Whole Brain In Vitro Hypothesis

Experimental studies of seizures in the entorhinal cortex of the in vitro isolated guinea pig brain [14, 24] showed that onset of ictal episodes is associated with strong discharge of inhibitory interneurons and initial silence of principal cells, which are subsequently recruited into progressively synchronized discharges. It was proposed that interneuronal firing-related increase of extracellular potassium together with intracellular chloride accumulation and reduction of GABA$_A$ inhibition results in hyperexcitability of principal cells and progression of a seizure. We tested this hypothesis using our computational model. A comparison of experimental and modelling results is shown in Fig. 3. In the model the seizure-like event was initiated by depolarizing current applied to soma of the inhibitory interneuron as shown by red trace in Fig. 3b middle panel. Current stimulation triggered strong interneuronal discharge at the initial rate of about 300 Hz. The membrane potential of the pyramidal cell transiently decreased and the cell ceased firing following onset of strong inhibitory drive (considered as the beginning of the seizure). After gap in firing lasting about 7 s, pyramidal cell resumed its activity reaching maximal firing rate of about 8 Hz in about 11 s from seizure onset. Throughout the seizure, inhibitory interneuron exhibited tonic discharges with gradually decreasing spike amplitude. These changes were accompanied by increase of extracellular potassium concentration, which increased sharply, reached maximal value of 9 mM within first seconds of the seizure and slightly decreased afterwards. These simulation

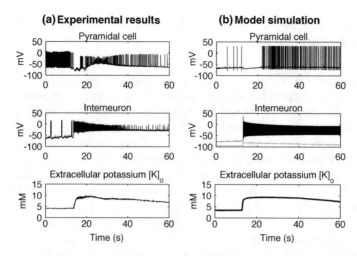

Fig. 3 Comparison between experimental data (**a**) and model simulation (**b**), Experimental recordings obtained in entorhinal cortex of in vitro isolated guinea pig brain show intracellular traces of pyramidal cell (*top panel*), interneuron (*middle panel*) and extracellular potassium (*bottom panel*). Seizure onset is associated with increased firing of the inhibitory interneuron and cessation of firing in pyramidal cell for few seconds. Afterwards, principal cell resumes its tonic activity that progressively transforms into bursts. Extracellular potassium is elevated up to 9 mM within first few seconds of the seizure and remains elevated throughout the seizure. Activity patterns of both cells (except bursting) and extracellular potassium time course are reproduced by the model. In the model discharge of the interneuron is triggered by somatic injection of the depolarizing current, shown in *red* in part (**b**), *middle panel*

results agree qualitatively and quantitatively in many respects with the experimentally observed seizures in the isolated guinea pig brain (Fig. 3a). To make the role of inhibitory interneurons more evident the pyramidal cell shown in Fig. 3a was stimulated by a steady positive current delivered via the intracellular pipette. The resulting depolarization was responsible for fast discharge present before the seizure. Initial discharge of interneurons at a rate about 400 Hz transiently inhibited pyramidal cells, which stopped firing but resumed their activity within few seconds, reached maximal firing rate of 9 Hz about 13 s from seizure onset and gradually decreased afterwards. Strong firing of the interneuron gradually decreased and exhibited reduction of spike amplitude. Extracellular potassium concentration attained its maximal value of 9.5 mM few seconds after seizure onset and slowly decreased subsequently.

End of the seizure, characterized by bursting firing pattern of pyramidal cells, is not yet captured by the model. If the simulation is prolonged beyond second 60, the interneuron enters depolarization block (around second 80) and pyramidal cells exhibits slow and steady decrease of the firing rate, reaching initial background firing rate around second 120.

Relationship Between Intracellular Seizure Patterns and Ion Concentration Changes

The distinctive firing patterns observed in the model during seizure-like events result from the interplay between synaptic mechanisms, membrane currents and ion concentration changes. Summary of changes of concentration gradients and corresponding reversal potentials of K^+, Na^+ and Cl^- is shown in Fig. 4. Seizure onset is associated with fast rise of extracellular potassium concentration and increase of potassium reversal potentials in both interneuron and pyramidal cell (Fig. 4a, b, first column). Initially pyramidal cell is inhibited by strong $GABA_A$ receptor-mediated currents and ceases firing. Slowing down of interneuron discharge and intracellular chloride accumulation in pyramidal cell (Fig. 4a, third column) both contribute to reduction of IPSPs and increase of membrane potential of pyramidal cell. Additionally, amplitude of action potentials generated by the interneuron gradually declines due to decrease of sodium reversal potential (Fig. 4b, second column). When the spike amplitude in the interneuron falls below -10 mV the inhibitory synaptic transmission is prevented. An increase of extracellular potassium

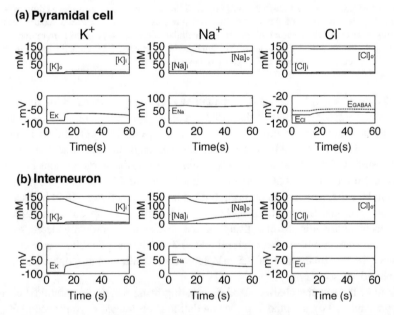

Fig. 4 Summary of changes of ionic activities in pyramidal cell (**a**) and interneuron (**b**) during modelled seizure episodes. For each cell and ion type, intra- and extracellular concentrations are shown in upper panel and reversal potentials are shown in bottom panel. Seizure-like activity is associated with significant ion concentration changes in both cells. Accordingly, in pyramidal cell potassium, chloride and GABAA reversal potentials are elevated, while that of sodium is reduced. In interneuron potassium reversal potential increases while sodium potassium reversal exhibits markedly declines

concentration together with loss of inhibition causes pyramidal cell to resume its activity within few seconds from seizure onset. Initially strong pyramidal cell discharge slows down as a result of slow decrease of extracellular potassium leading to decrease of potassium reversal potential and associated increase of outward potassium leak current.

Neuronal $[K^+]_o$ Sources and Regulation

The dynamic increase of extracellular potassium concentration seems to be a primary factor leading to development of seizure-like events in the model. Accordingly, we investigated the sources of $[K^+]_o$ accumulation in pyramidal cell and interneuron. To this end, we computed time integral of various potassium membrane currents and divided the result by the volume of the corresponding extracellular space. The contributions of $[K^+]_o$ from the interneuron soma are shown in Fig. 5, while those from the pyramidal cell soma are shown in Fig. 6. One can notice that the dominant $[K^+]_o$ contribution comes from action potential firing in the interneuron (Fig. 5, upper panel, red diamonds), while contribution from potassium leak current in the interneuron is negligible (Fig. 5, upper panel, blue circles). Potassium current due to Na^+/K^+ pump contributes to $[K^+]_o$ removal (Fig. 5, upper panel, green stars), but cannot balance potassium outflow and the net result is positive (Fig. 5, lower panel). In the pyramidal cell, contributions from leak and all voltage-gated potassium membrane currents are negligible. The dominant influence comes from KCC2 cotransporter, which due to high $[K^+]_o$ conditions, lowers $[K^+]_o$

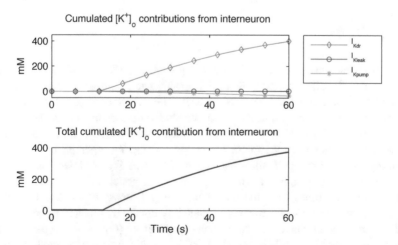

Fig. 5 *Upper panel* cumulated potassium fluxes in the interneuron. The largest outward flux corresponds to I_{Kdr} current (*red diamonds*), while the inward flux is mediated by potassium current generated by the Na^+/K^+ pump (*green stars*). *Lower panel* total accumulated potassium in extracellular space due to potassium fluxes in the interneuron

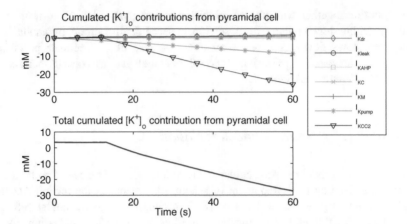

Fig. 6 *Upper panel* cumulated potassium fluxes in the pyramidal cell. The outward fluxes mediated by potassium membrane currents are negligible. The inward fluxes are mediated by potassium current generated by the Na$^+$/K$^+$ pump (*green stars*) and KCC2 (*black triangles*). Lower panel: total cumulated potassium is negative, what means that potassium is up-taken from extracellular space by pyramidal cell

by cotransport of K$^+$ and Cl$^-$ ions into the cell (Fig. 6, upper panel, black triangles). The transport of K$^+$ by Na$^+$/K$^+$ pump in pyramidal cell is inward and reduces [K$^+$]$_o$ as in interneuron (Fig. 6, upper panel, green stars). Because in pyramidal cell potassium inward flux is of larger magnitude than the outward flux the net amount of [K$^+$]$_o$ regulated by that cell is negative (Fig. 6, lower panel).

The [K$^+$]$_o$ Balance

Apart from membrane contributions and regulation of [K$^+$]$_o$, potassium accumulation is additionally regulated by a glial uptake, lateral diffusion to neighbouring compartment and radial diffusion to the bath. The net effect of membrane currents and these additional mechanisms is responsible for time course of [K$^+$]$_o$ observed during model activity. The individual components shaping [K$^+$]$_o$ time course in the somatic extracellular compartment are shown in Fig. 7 (upper panel), while the resulting extracellular potassium balance is shown in Fig. 7 (lower panel). This trace is equivalent to that shown in Fig. 3b (lower panel). The dominant contribution to [K$^+$]$_o$ comes from the interneuron (Fig. 7, upper panel, red diamonds). Extracellular potassium clearance is mediated by uptake by glia and pyramidal cell and by lateral and radial diffusion. Initially glial uptake (Fig. 7, upper panel, green stars) plays a dominant role but as [K$^+$]$_o$ builds up and remains elevated, diffusion processes (Fig. 7, upper panel, pink circles and black crosses) become main mechanisms of potassium regulation. Despite both the inward and outward flows of potassium produce cumulated extracellular concentrations of the order of hundreds

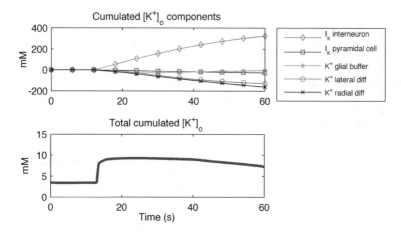

Fig. 7 *Upper panel* cumulated potassium fluxes mediated by the pyramidal cell, interneuron, glial buffer and diffusion. The outward flux is mediated by potassium action potential currents in the interneuron (*red diamonds*). Potassium clearance is mediated initially by glial buffer (*green stars*) and subsequently by lateral- and radial diffusion (magenta circles and *black crosses*). *Lower panel* total cumulated potassium due to all potassium fluxes in the model

of mM, the total cumulated extracellular potassium is of the order of 10 mM, showing that the processes of potassium release and clearance operate at a fine balance.

Selective Impairment of Potassium Homeostasis Mechanisms

To investigate the effects of various components of potassium homeostasis implemented in the model, the potassium clearance mechanisms were selectively removed and simulations of such altered models were performed. These hypothetical manipulations show how impairment of a certain neuronal process may affect behaviour of cells and extracellular potassium.

When the glial buffer was removed, $[K^+]_o$ increased rapidly up to around 25 mM causing depolarization block in pyramidal cell during the whole simulation (Fig. 8, top trace). Interneuron continued to fire but within few seconds spike amplitude was severely reduced (Fig. 8, middle trace), limiting further release of potassium. Consequently $[K^+]_o$ started to decrease due to clearance by Na^+/K^+ pump and diffusion (both lateral and radial), reaching a final value of $\sim$10 mM.

The effect of removal of potassium diffusion in the model is shown in Fig. 9. In this case, $[K^+]_o$ increased rapidly up to around 10 mM causing depolarization block in pyramidal cell during the whole simulation (Fig. 9, top trace). Interneuron fired with gradually decreasing spike amplitude until second 40 and entered depolarization block afterwards (Fig. 9, middle trace). $[K^+]_o$ was increasing throughout the simulation showing limited ability of glial buffer and Na^+/K^+ pump to clear

Fig. 8 Model simulations with removed glial uptake mechanisms. Soon after seizure onset $[K^+]_o$ attains high value of about 25 mM and pyramidal cell enters permanent depolarization block. Interneuronal activity is maintained but with drastically reduced spike amplitude

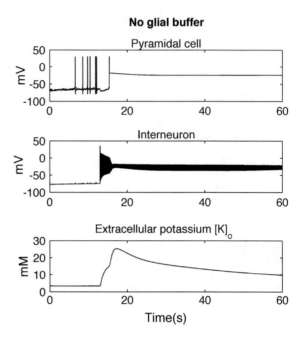

potassium effectively (Fig. 9, bottom trace). When glial buffer was filled-up around second 40, $[K^+]_o$ rose to the level of around 25 mM causing depolarization block of the interneuron.

Finally, Fig. 10 shows the model behaviour in the absence of the Na^+/K^+ pump. Absence of the pump caused abnormal firing of the pyramidal cell (Fig. 10, top trace), associated with small build-up of $[K^+]_o$ even before the application of the depolarizing current to the interneuron (Fig. 10, bottom trace). When interneuron started to fire, $[K^+]_o$ rose sharply to a value above 10 mM causing depolarization block in pyramidal cell. Interneuron fired with gradually decreasing spike amplitude until second 45 and entered depolarization block afterwards (Fig. 10, middle trace). The overall time course of $[K^+]_o$ was similar to that obtained for the reference conditions but $[K^+]_o$ levels were slightly higher. Here $[K^+]_o$ reached maximal and final value of ~10.5 mM and ~7.5 mM, respectively (vs. ~9.3 mM and ~7.2 mM for the intact model).

In summary, removal of any of the potassium regulatory mechanisms had significant effect on the model dynamics. Blocking of the diffusion process or glial buffering system led to depolarization block of pyramidal cell and high extracellular potassium reaching concentration above 25 mM, what may correspond to spreading depression episode. Impairment of the Na^+/K^+ pump led to depolarization block of both cells but extracellular potassium concentration was similar to that in the intact model. It shows that in the model, diffusion process and glial buffering is more effective in potassium clearance that the Na^+/K^+ pump. This result is in agreement with earlier observations of relative role of potassium regulatory mechanisms, illustrated in Fig. 7.

Fig. 9 Model simulations with removed diffusion mechanisms. Soon after seizure onset pyramidal cell enters permanent depolarization block. $[K^+]_o$ gradually increases and reaches maximal value about 25 mM around second 40. Interneuron exhibits firing with decreasing spike amplitude for about 30 s and enters depolarization block afterwards

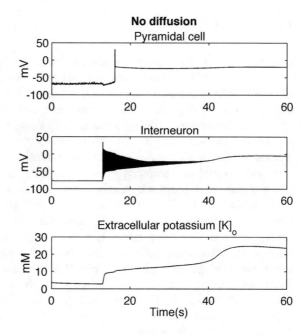

Fig. 10 Model simulations with removed Na^+/K^+ pump. Absence of the pump causes slight intial increase of $[K^+]_o$ and associated strong firing of pyramidal cell. Soon after seizure onset pyramidal cell enters permanent depolarization block, similar to one observed in Figs. 8 and 9. Interneuron enters depolarization block around second 45, while extracellular potassium time course is similar and only slightly elevated as compared to simulation of the intact model

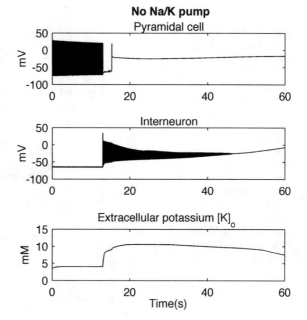

Future Therapies

Currently 38% of newly diagnosed adult patients with localisation related epilepsy are drug resistant [28]. This number remains almost unchanged since the discovery of the first modern antiepileptic drug over hundred years ago [29]. It shows that new concepts and investigation of other therapeutic strategies is necessary. Using the model we investigated whether introduction of artificial extracellular potassium regulation mechanism might successfully control neuronal excitability. Potassium regulation was implemented as additional potassium buffer with the activation threshold 3.5 mM and buffering speed 0.8 mM/ms. Such a mechanism could be possibly realized in practice by a nanoparticle system as innovative applications of nanomaterials in medicine continue to emerge [30]. Model simulations with nanoparticle buffering agent present in the extracellular space are shown in Fig. 11. Strong interneuronal firing (Fig. 11, middle trace) that would normally cause increase of extracellular potassium and abnormal pyramidal cell discharges was ineffective in triggering seizure-like episode in the presence of the artificial buffer. Potassium remained at its baseline level (Fig. 11, bottom trace) and when inhibition decreased following $[Cl^-]_i$ accumulation, pyramidal cell resumed normal activity (Fig. 11, top trace). It shows that such artificial $[K^+]_o$ regulation mechanism might lead to successful seizure control.

Fig. 11 Model simulations with additional nanoparticle potassium clearance mechanism included. Despite strong discharge of the interneuron, extracellular potassium remains close to its reference level and pyramidal cell resumes normal firing when $GABA_A$ mediated inhibition decreases due to chloride accumulation. These results illustrate that hypothetical artificial potassium clearing mechanisms could lead to effective antiepileptic treatment

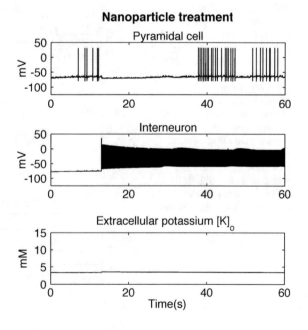

Fig. 12 Simulation results of the model with constant ionic concentrations. Strong discharge of the interneuron (*middle panel*) inhibits the pyramidal cell, which stops firing (*top panel*) while extracellular potassium concentration is stable at the initial value 3.5 mM. These results illustrate that in order to account for experimental results shown in Fig. 3a, ionic dynamics is essential

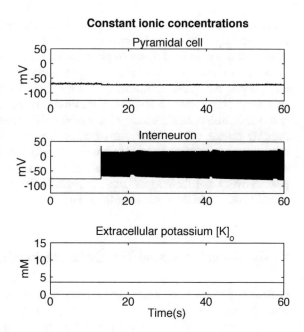

The Model with Constant Ion Concentration Gradients

In the absence of extracellular space, i.e., with constant ion concentrations, the cells communicate only through the synaptic connections and $[K^+]_o$ remains constant throughout the whole simulation and is equal to the initial value of 3.5 mM (Fig. 12, bottom trace). The inhibitory interneuron exerts strong inhibition on the pyramidal cell, which remains hyperpolarized and silent (Fig. 12, top trace). It shows that firing patterns observed experimentally and in reference simulation (Fig. 3a, b) indeed depend on activity-induced changes in intra- and extracellular ion concentrations.

"Take Home" Message for Neurologists, Psychiatrists

This computational study confirms experimentally based hypothesis that focal seizures may be triggered by strong discharges of the inhibitory interneurons. When firing-associated potassium release exceeds potassium clearance, potassium accumulates in the extracellular space, leading to generation of pathological discharges in pyramidal cells and seizure progression. This scenario challenges the traditional view of alteration in excitation/inhibition balance being the cause of epilepsy. Our study points to importance of non-synaptic mechanisms in epilepsy, especially changes in extracellular potassium. Additionally, simulations reveal the very fine balance that exists between potassium release and uptake and identify dominant processes responsible for observed overall potassium dynamics. Insight into these mechanisms would be difficult to grasp by purely experimental observations.

We also show proof-of-concept of the feasibility of seizure control by a novel antiepileptic nanoparticle treatment. The proposed hypothetical mechanism would be able to recognize the level of extracellular potassium concentration in a surrounding medium, buffer excess of potassium if it exceeded a certain threshold and possibly release accumulated potassium back to the neuronal environment if the normal extracellular potassium level was restored by natural potassium homeostasis mechanisms. Nanoparticle therapies already exist as alternative means for drug and gene delivery and their applications towards ion absorption, like the one hypothesized here, might be possible in the future.

"Take Home" Message for Computationalists

Our modelling study, as well as some other models mentioned earlier, suggest that accurate modelling of seizure activity should include activity dependent changes in intra- and extracellular concentrations and ion homeostasis mechanisms. Significant shifts in ion concentration gradients are observed during epileptic seizures [7] and it is reasonable to assume that seizure initiation, maintenance and termination are causally related to the ionic dynamics. Our model with constant ion concentrations cannot reproduce experimental results (Fig. 12) and couldn't contribute to understanding the key mechanisms that determine seizure initiation and evolution. It shows that in order to explain pathophysiology of seizures in terms of realistic mechanisms, models should incorporate ion concentration dynamics. Such models may help to elucidate the complex interactions between neurons and their environment during seizures and may ultimately lead to development of new therapeutic strategies targeting regulation of ion concentration gradients.

References

1. Duncan JS, Sander JW, Sisodiya SM, Walker MC. 2006. Adult epilepsy. Lancet 367 (9516):1087–100.
2. Téllez-Zenteno JF, Hernández Ronquillo L, Moien-Afshari F, Wiebe S. Surgical outcomes in lesional and non-lesional epilepsy: A systematic review and meta-analysis. Epilepsy Research 89.2 (2010): 310–318.
3. Boon P, Vonck K, De Reuck J, Caemaert J. Vagus nerve stimulation for refractory epilepsy. Seizure 2001;10:448–455.
4. Fisher, Robert, et al. "Electrical stimulation of the anterior nucleus of thalamus for treatment of refractory epilepsy." Epilepsia 51.5 (2010): 899–908.
5. Lytton B. Computer modelling of epilepsy. Nat Rev Neurosci. 2008;9(8):626–37.
6. Hodgkin AL, Huxley AF. A quantitative description of membrane current and its application to conduction and excitation in nerve. J Physiol. 1952;117(4):500–44.

7. Raimondo JV, Burman RJ, Katz AA, Akerman CJ. Ion dynamics during seizures. Front Cell Neurosci. 2015, 21;9:419.
8. Fröhlich, F., Bazhenov, M., Iragui-Madoz, V., and Sejnowski, T. J. (2008). Potassium dynamics in the epileptic cortex: new insights on an old topic. Neuroscientist 14, 422–433. doi:10.1177/1073858408317955.
9. Fertziger, A. P., and Ranck, J. B. Potassium accumulation in interstitial space during epileptiform seizures. Exp. Neurol. 1970, 26, 571–585. doi:10.1016/0014-4886(70)90150-0.
10. D.M. Durand and E.H. Park, Role of potassium lateral diffusion in non-synaptic epilepsy: a com- putational study. J. of Theoretical Biology 238 (2005) 666–682.
11. H. Kager, W.J. Wadman and G.G. Somjen, Computer simulations of neuron-glia interactions mediated by ion flux. J. Comput. Neurosci. 25 (2008) 349–365.
12. Krishnan, G. P., and Bazhenov, M. Ionic dynamics mediate spontaneous termination of seizures and postictal depression state. J. Neurosci., 2011, 31;8870–8882.
13. Y. Wei, G. Ullah, and S.J. Steven, Unification of Neuronal Spikes, Seizures, and Spreading Depression. J. Neurosci. 34(35) (2014) 11733–11743.
14. M. de Curtis and V. Gnatkovsky, Reevaluating the mechanisms of focal ictogenesis: the role of low-voltage fast activity. Epilepsia 50(12) (2009) 2514–2525.
15. Johnston D and Wu SM, Foundations of Cellular Neurophysiology, MIT Press, 1995.
16. Somjen G.G. Ions in the brain: normal function, seizures, and stroke. Oxford UP, New York (2004)
17. J.A. Payne, C. Rivera, J. Voipio and K. Kaila, Cation-chloride co-transporters in neuronal com- munication, development and trauma. *TRENDS in Neurosciences* Vol. 26 No.4 (2003).
18. Peter Blaesse, Matti S. Airaksinen, Claudio Rivera, and Kai Kaila. Cation-Chloride Cotransporters and Neuronal Function. Neuron 61, 820–838, 2009.
19. Cohen, I., Navarro, V., Clemenceau, S., Baulac, M. & Miles, R. On the origin of interictal activity in human temporal lobe epilepsy in vitro. Science 298, 1418–1421 (2002).
20. M. de Curtis, A. Manfridi and G. Biella. Activity-dependent pH shifts and periodic recurrence of spontaneous interictal spikes in a model of focal epileptogenesis. J Neurosci. 18(18) (1998) 7543–51.
21. Laura Uva, Laura Librizzi, Fabrice Wendling, and Marco de Curtis. Propagation Dynamics of Epileptiform Activity Acutely Induced by Bicuculline in the Hippocampal–Parahippocampal Region of the Isolated Guinea Pig Brain. *Epilepsia*, 46(12):1914–1925, 2005.
22. Gentiletti D, Gnatkovsky V, de Curtis M, Suffczynski P. Changes of Ionic Concentrations During Seizure Transitions – a Modelling Study. IJNS (2017), in press.
23. E. Fransén, A.A. Alonso, and E.H. Michael, Simulations of the role of the muscarinic-activated calcium-sensitive nonspecific cation current INCM in entorhinal neuronal activity during delayed matching tasks. J. Neurosci. 22(3) (2002) 1081–1097.
24. Gnatkovsky V, Librizzi L, Trombin F, de Curtis M. Fast activity at seizure onset is mediated by inhibitory circuits in the entorhinal cortex in vitro. Ann Neurol 2008, 64:674–686.
25. J. Voipio, M. Pasternack, B. Rydqvist B and K. Kaila, Effect of Gamma-Aminobutyric- Acid on Intracellular pH in the Crayfish Stretch-Receptor Neuron. J. Exp. Biol. 156 (1991) 349–361.
26. Doyon N, Prescott SA, Castonguay A, Godin AG, Kröger H, De Koninck Y (2011) Efficacy of Synaptic Inhibition Depends on Multiple, Dynamically Interacting Mechanisms Implicated in Chloride Homeostasis. PLoS Comput Biol 7(9): e1002149. doi:10.1371/journal.pcbi.1002149.
27. Kager H., Wadman W.J. and Somjen G.G., Simulated seizures and spreading depression in a neuron model incorporating interstitial space and ion concen- trations. J. Neurophysiol. 84(1) (2000) 495–512

28. Mohanraj R, Brodie MJ. (2005) Outcomes in newly diagnosed localization-related epilepsies. Seizure 14:318–323.
29. Löscher, W., Schmidt, D. Modern antiepileptic drug development has failed to deliver: ways out of the current dilemma. Epilepsia 2011;52(4):657–78.
30. Nuria Sanvicens and M. Pilar Marco. Multifunctional nanoparticles – properties and prospects for their use in human medicine. Trends in Biotechnology Vol.26 No. 8, 425–433, 2008.

Time Series and Interactions: Data Processing in Epilepsy Research

Zsigmond Benkő, Dániel Fabó and Zoltán Somogyvári

Introduction

Computational methods can have significant contribution to epilepsy research not only through modeling, but through data analysis as well. Data analysis in epileptology has a varied set of aims that relate to possible treatment methods. For surgical treatment, data analysis aims to find any marker to localize the epileptic tissue, the seizure onset zone (SOZ). For deep or intracranial stimulations, the aim is to predict seizures and the effect of stimulation. For pharmacological treatment, the aim is to understand the roles of different channels to the cellular and network dynamics, and their possible modulation. In this review, we will focus on the first aim: data analysis for SOZ determination, which supports surgical planing, thereby resulting in more restricted resections, better outcomes, and less side effects.

A new era of brain research was opened with the vast amount of neural data that is now available. New data analysis methods are needed to take full advantage of the available resources. The traditional way to determine SOZ relies on human's natural skills in pattern matching or mismatch recognition to try to identify the first pathological patterns at the initiation of the epileptic seizures. Pathological patterns typically differ in their frequency and wave-shape, but they could appear in different forms. This plurality of wave-shapes, as well as the possibility of possible multiple or

Z. Benkő · Z. Somogyvári (✉)
Wigner Research Centre for Physics of the Hungarian Academy of Sciences,
Department of Theory, Konkoly-Thege M. St. 29-33, Budapest 1121, Hungary
e-mail: somogyvari.zoltan@wigner.mta.hu

Z. Benkő
e-mail: benko.zsigmond@wigner.mta.hu

D. Fabó · Z. Somogyvári
Department of Neurology and Epileptology, National Institute
of Clinical Neurosciences, Amerikai St. 57., Budapest 1145, Hungary
e-mail: fabo.daniel@gmail.com

© Springer International Publishing AG 2017   73
P. Érdi et al. (eds.), *Computational Neurology and Psychiatry*,
Springer Series in Bio-/Neuroinformatics 6,
DOI 10.1007/978-3-319-49959-8_4

deep origins and a quick spreading of signals through intracortical pathways, makes it difficult to identify the SOZ, even for seizures that are precisely recorded. The challenges of SOZ identification arise in three settings, outlining a roadmap for data analysis in epilepsy:

- Identification of the SOZ based on invasive methods (i.e. high density, subdural electrode recordings) during the initiation of the epileptic seizure.
- Identification of the SOZ based on invasive recordings without epileptic seizures.
- Identification of the SOZ with noninvasive methods.

While identification of the SOZ with noninvasive methods seems to be achievable only in long run, the first two aims are reachable in many cases. The precision of the identification, however, strongly depends on mathematical methods. In this chapter, detection and analysis methods for possible markers of the epileptic tissue, either ictal and interictal, are discussed, starting with detection algorithms for interictal spikes and high frequency oscillations and open questions for these algorithms. We continue along with methods for analyzing continuous signals, such as time-frequency analysis and entropy calculations, and finish with methods for determining causal interactions among signals and their usage in locating epileptic foci.

Short Pathological Events as Biomarkers

Interictal Epileptic Spikes

Interictal epileptic spikes are the first candidate we discuss for a biomarker of epilepsy, promising to identify the SOZ in the absence of ictal recordings. From the point of view of data analysis, interictal epileptic spikes are relatively easy to detect, due to their large amplitude and stereotyped waveforms. The only major difficulty is caused by movement and other artifacts. However, these spikes occur in regions well beyond the limits of the actual epileptic tissue, and so, intensity maps of these spikes are poor indicators of seizure onset zone. Presumably, interictal spikes are more widely observed, because they can be evoked by synaptic activity arriving from (perhaps a distant) seizure onset zone. This observation raises the question. Are there reliable morphological differences between a spike that originated locally from one that originated in remote origin? If there are reliable differences between them, then this information could delimit the seizure onset zone without recorded seizures.

In the SOZ (proven post hoc) and nearby tissue, chronic laminar microelectrode recordings revealed differences in the laminar profile between locally-generated and remotely-evoked neocortical epileptic spikes [1]. Although the ECoG signals on the surface of the brain are only projections of the whole laminar profile, the different laminar profile could be reflected in differences on the surface potentials as well, forming a possible basis for the possible distinction of the central and peripheral spikes.

High Frequency Oscillations—Ripples

Ripples are short duration, small amplitude sinusoidal high frequency oscillations (HFO) in the nervous system. Although HFOs are physiological in some brain areas, such as CA regions of the hippocampus, they are considered pathological and linked to the seizure onset zone in other cortical areas. They are often accompanied by epileptic interictal spikes. HFO events are now divided into ripples whose characteristic frequency peak is between 80–250 Hz and fast ripples in the 250–500 Hz frequency regime. These two types may differ not only in their peak frequency but the underlying physiological mechanisms as well.

Detecting HFOs by a human observer is very time consuming, thus it is of high importance to develop automatic detection algorithms. These algorithms will not only disburden the human researchers from this task, but also provide more objective and comparable results. Comparing HFOs within channels is important since this forms the basis HFO intensity maps and supports the surgical planing process. Comparing HFOs between patients is also important for developing conclusions that generalize across patients.

HFO identification has a number of challenges. First, identifying any transient patterns in the electrocorticography (ECoG) signal is affected by movement and muscle artifacts and different noise levels of different channels. Movement and muscle artifacts lead to signals with variable frequencies of high amplitude, so they can deceive even sophisticated detectors. Second, in contrast to interictal spikes, the amplitude of HFOs are low, comparing to the ongoing EEG activity. Thus, their identification mainly based on the emergence of specific frequencies only, but these specific frequencies can vary between brain areas and patients. Third, the specific frequencies of interest emerge only for short durations, and hence, the signal should be decomposed using time-frequency methods, such as wavelets, instead of more traditional frequency spectra methods. Last, the signal is accompanied by sharp spikes that appear independently and have wide range frequency content, which after high-pass filtering, can result in ripple-like patterns that can deceive not only an automatic detector algorithm, but even a well trained neurologist as well. This observation emphasizes that, although the filtering can help the neurologist to find the ripple candidate events, it is highly important to check the original signal as well and accept the event as ripple only if the sinusoid oscillation is observable superimposed on the accompanying spike [2].

The first automatic detector algorithms were based on few simple criteria, extracted from the cumulative experience of neurologists. These simple criteria try to formulate the known properties of HFOs, which discriminate HFOs from the normal ECoG. Such criteria included increases in power and the number of cycles corresponding to specific frequency regimes [3]. These approaches were validated using a dataset of HFO events detected by human experts. Later works compared different criteria [4] or tried to adapt criteria to the background activity or noise level of the actual ECoG channels by applying relative power criteria [5] or iteratively adapting thresholds for channels of high HFO background [6]. A promising new direction of finding proper features for HFO detection was applied by [7]. They calculated the approximate entropy of the signal, expressing the unpredictability or disorder of the signal, and characterized HFOs by elevated entropy.

Application of these simple rules has the advantage that it is clearly interpretable, but it does not guarantee to get close to the optimal results from the detection point of view. Considering the difficulty of the HFO detection, more principled approaches are required.

An alternative approach was to use supervised learning methods from the artificial neural network literature to optimize the performance of detection algorithms. The set of events detected by experts were used as a training set for the algorithms. A set of features is calculated over all events, and then the features of events detected by experts are used to train the detection algorithm. During the supervised learning procedure, a learning algorithm is applied to features in the training set to optimize decision boundaries for approximating the performance of the human experts. For detection, application of low dimensional feature spaces, such as the predefined criteria in the above mentioned algorithms, is disadvantageous since it creates an informational bottleneck. Low dimensional problems can be embedded into higher dimensional feature spaces, even if at random. This embedding tends to transform problems which may be linearly non-separable in the lower dimensional space into problems that are linearly separable in higher dimension. In simpler terms, lower dimensions require more complex decision boundaries, and thus more complex learning methods, whereas higher dimensions could use simpler methods. As an example for low dimensional feature space but complex learning process, [7] used a feature set determined by tree consecutive value of approximate entropy and then trained an artificial neural network with recurrent hidden layer using Cubature Kalman Filter to determine decision boundaries.

In these learning approaches and in other optimization procedures, it is important to divide the available hand-sorted data set to a distinct training and test set to avoid overfitting. Overfitting can arise when the fitted decision boundary or the neural network is highly complex with a large number of free parameters to fit. In this case, one could get near-perfect or perfect recognition performance on the training set, but because the decision boundary does not generalize to other sets, poor results on the test set. Reliable algorithms can only be developed by using independent datasets to train and to test performance [8].

In contrast to the above algorithms, which use supervised learning, data driven approaches can also use unsupervised learning or clustering methods and do not require ground truth in the form of a training set of events detected by experts [9, 10]. In these approaches, pre-selection is first applied to remove short, high-frequency candidate epochs from the continuous signal. Next, the candidate epochs are clustered and the optimal number of clusters is determined. [9] found a description with 4 clusters to be optimal (Fig. 1). They then characterized the clusters using the previously applied criteria, such as power ratio, peak and centroid frequency and line length, and found that the 4 clusters corresponded to ripples (36.5 %), fast ripples (8.5 %), spike-like artifacts (24 %) and mixed frequency signals (31 %).

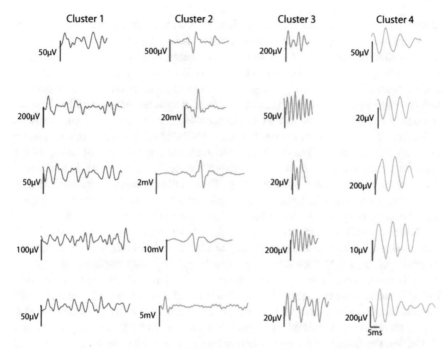

Fig. 1 Example ripple waveforms from the 4 HFO clusters identified by Blanco et al. [9]. They found that the 4 clusters corresponded to ripples (Cluster 4, 36.5 %), *fast* ripples (Cluster 3, 8.5 %), spike-like artifacts (Cluster 2, 24 %) and mixed frequency signals (Cluster 1, 31 %). Reproduced from [9] with permission

Independent of the method of detection, the final goal is to relate the occurrence of different types of HFOs to the epileptic cortical mechanisms and make them useful in identification of seizure onset zone [11].

Analysis of the Continuous Signal

Frequency Analysis

Since many epilepsy-related phenomena are defined by its specific frequency content, it is of natural interest to determine the dominant frequencies in the signal. Although it sounds simple, it is not always so simple. There are two main problems related to the frequency analysis. The first is that the frequency content of the signal changes over time. The second is related to the Heisenberg uncertainty principle: from a data processing point of view, it says that the amplitude of low frequencies can be determined only from appropriately long time series; thus, the temporal resolution for low frequencies is inherently low. On the other hand, shorter time series are enough to determine high frequencies, but the frequency resolution is proportional to the length of the analyzed segments; thus, shorter segments results in worse frequency resolution. As a result, there is a trade off between the temporal and frequency resolution: longer segments results in better frequency resolution but worse temporal resolution and vice versa. Thus, in the case of temporally-changing frequency content, one has to find a compromise between frequency and temporal resolution. The traditional Fourier-transform has maximal frequency resolution, but no temporal resolution. The Wavelet-transformation applies optimal tilling of the time-frequency space. It starts from a mother wavelet function, which is a localized wave packet either in time and frequency. If the mother wavelet is upscaled in temporal length according to a geometrical series and then convolved with the signal, then the result is to decrease the temporal resolution but inversely increase frequency resolution towards the lower frequencies. As we can see, the short duration of HFO events means inherently wrong frequency resolution independently from the applied method: 100 ms length results no more than 10 Hz frequency resolution at best.

Epileptic seizures are typical examples for a spectrum changing over time. Figure 2 shows wavelet analysis for ECoG of evoked seizures in rats: A and D shows normal ECoG activity and its time-frequency spectrum by Morlet-wavelet and B and E shows seizure evoked by 4-Aminopyridin. Wavelet-transformation clearly shows that the dominant frequency decreasing rapidly during the first period of the seizure, while the amplitude of the oscillation increases. During the second and the third phase of the seizure the dominant frequency remained relatively constant for a while. Together with the highest amplitude peak, this may imply a resonance phenomenon in the cortex. Subfigures C and F shows an epileptic seizure evoked by 4-Amionopyridin after Trans-Amino-1,3-dicarboxycyclopentane (ACPD, a metabotropic glutamate receptor agonist) treatment, which caused higher seizure susceptibility and amplitude. Note that wavelet-transform reveals not only the frequency decay during the seizure, but a short period dominated by an increasing frequency peak at the onset of the seizure.

An additional problem is determining significant peaks of the calculated time-frequency maps. The existing mathematical results on significant peak detection, either based on dominant peak selection or analytical (typically Gaussian function) model fitting are rarely applied in epilepsy research. An interesting direction of peak detection is based on building up a signal from an overcomplete set of time-frequency basis functions called Gabor-atoms, along with the matching pursuit algorithm. Gabor-atoms are actually wavelets, but in this case the algorithm uses more basis functions than is minimally necessary to fit any continuous functions. The use of overcomplete set of the basis functions results in more than one (actually infinite) possible decompositions, and thus, freedom to choose among them. The matching pursuit fitting algorithm choses the Gabor-atom which has the largest projection to the signal, thereby implicitly building up a more sparse representation of the signal than the wavelet-transformation. The sparse decomposition typically provides more clear peaks either in frequency and in time.

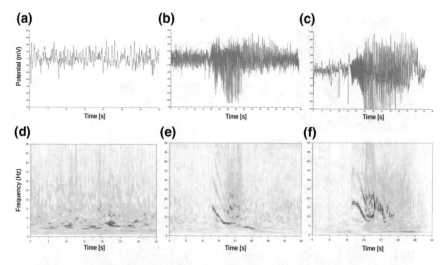

Fig. 2 Wavelet analysis of ECoG of evoked seizures in rats. **a** and **d** normal ECoG activity and its time-frequency spectrum by Morlet-wavelet. **b** and **e** seizure evoked by 4-Aminopyridin. Wavelet-transformation clearly shows that the dominant frequency decreasing rapidly during the first period of the seizure, while the amplitude of the oscillation increases. During the second and the third phase of the seizure the dominant frequency remained relatively constant for a while. **c** and **f** Epileptic seizure evoked by 4-Amionopyridin after Trans-ACPD treatment, which caused higher seizure susceptibility and amplitude. Note that the wavelet-transform reveals not only the frequency decay during the seizure, but a short period dominated by an increasing frequency peak at the onset of the seizure. Modified from [12]

Entropy of the Signal

Entropy of a signal captures its complexity or unpredictability. Although entropy (H) is well-defined theoretically, it is not straightforward to approximate, depending on the definition of the state space in which the probability is measured. Entropy H is given by

$$H(X) = - \sum_i p\left(x_i\right) \log p\left(x_i\right),$$

where X is some discrete random variable, x_i are possible values of X, and $p(x_i)$ is the probability that $X = x_i$?

There are two widespread methods to calculate the entropy of a signal: spectral entropy (SE) and approximate entropy (AE). The spectral entropy measures the flatness of the spectrum,and hence, its closeness to the spectrum of the white noise. Approximate entropy quantifies unpredictability by measuring how likely nearby states in the state space remain close after an iteration [13]. As show in Fig. 3, AE increases significantly during the initial low amplitude phase of the seizure, but then decreases below the baseline during the high amplitude phase of the seizure. The

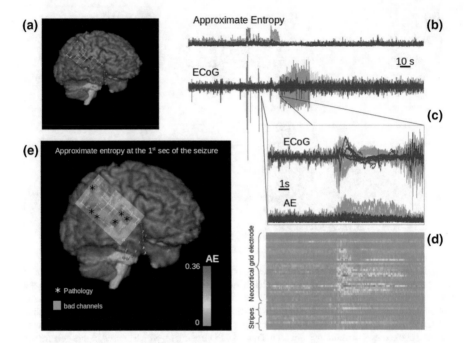

Fig. 3 Approximate entropy (AE) of the ECoG during seizure initialization: **a** The position of the subdural grid electrodes on the brain surface. **b** *Colored lines* represents the 48 channel during an epileptic seizure, parallel to the corresponding AE values. AE is significantly increased solely during the initial, low amplitude phase of the seizure, then AE is decreased below the baseline during the high amplitude phase of the seizure. **c** Temporal zoom of the initiation of the seizure. **d** *Color coded* representation of the AE values on each channels (*horizontal lines*) during the initial period of the seizure. Time scale is the same as in graph **c** and *color code* is as on graph (**e**). AE have been increased only on specific channels. **e** *Color coded* AE calculated from the first second of the seizure is projected to the surface of the brain. The positions of the increased AE values during the first sec of the seizure corresponds very well to the seizure onset zone determined by experts and marked by *stars*

positions of the increased AE values during the first second of the seizure corresponds very well to the seizure onset zone. AE also served as a basis for seizure prediction systems [14].

Analysis of Interactions Among Signals

Correlation

Linear correlation coefficient (ρ_{xy}) is the simplest and most natural measure to quantify association between two signals:

$$\rho_{xy} = \frac{\sum_t (x_t - \mu_x) (y_t - \mu_y)}{N\sigma_x\sigma_y}$$

where x_t and y_t are the data points of two time series, μ_x and μ_y are their respective mean values, σ_x and σ_y are the respective standard deviations, and N is the number of data points.

This correlation method assumes that signals depend on each other linearly and that there is no time delay between the analyzed signals; linear dependency effectively means that the signals have the same waveforms in the case of time series.

In the case of macroscopic electric signals from the brain, high correlation, and thus high instantaneous linear coupling, is often the consequence of electric crosstalk between electrodes and wires or the effect of far fields. If high correlation is observed, care should be taken to eliminate these effects.

Besides the assumption of instant and linear interaction, the correlation method has a third important constraint: it is a symmetric measure, and hence, an undirected measure of connection strength. There are methods to exceed these limitations and overcome these three constraints, but there is no one perfect method. The simplicity of the linear correlation is always an advantage because it not only requires less computational resource, but more importantly less data and is also less sensitive to noise.

Mutual Information

The most important symmetric connection measure, which can reveal any types of nonlinear dependencies, is mutual information. It is based on the entropy measure of a signal and quantifies the information known about a random variable with knowledge of an other random variable. Specifically, it calculates the difference between the entropy of the joint distribution of the two variable ($H(X, Y)$) and the entropies of the individual variables ($H(X), H(Y)$):

$$I(X; Y) = H(X) + H(Y) - H(X, Y)$$

Although it can reveal nonlinear connections, its calculation requires constructing probability density functions for each marginal variable and for the joint variable, typically by means of histograms. Much more data is typically required to determine the full probability distribution function with the necessary precision, especially when compared to the simpler linear correlation method. Generally, calculation of mutual information requires more data and more computational efforts, but rarely gives proportionally more information about the existence of the interaction. It is because the majority of the interactions, even the nonlinear forms can be approximated linearly to some extent, thus can be revealed by linear correlation, and those forms of interactions are rare which can not be approximated linearly at all, thus can not be revealed by correlation.

Cross Correlation Function and Coherence

The cross correlation function measures the linear correlation coefficient between
one signal and a time-delayed version of an second signal as a function of the time
delay between the two signals, thereby overcoming the instataneous assumption of
the simple linear correlation:

$$\rho_{xy}(\tau) = \frac{\sum_t (x_t - \mu_x)(y_{t+\tau} - \mu_y)}{N\sigma_x\sigma_y}.$$

Here, $\rho_{xy}(\tau)$ is the correlation coefficient for τ time lag and $y_{t+\tau}$ is the τ-lagged
version of the timeseries y_t.

The cross correlation function is also used to determine two functions: cross-
spectrum $(S_{xy}(\omega))$ and its normalized version called the coherence $(C_{xy}(\omega))$ given
by

$$C_{xy}(\omega) = \frac{|S_{xy}(\omega)|^2}{S_{xx}(\omega) S_{yy}(\omega)}$$

Specifically, the cross spectrum $S_{xy}(\omega)$, also called cross-spectral density, is the
Fourier transform of the cross-correlation function.

Coherence expresses the stability of the phase lag on different frequency regimes
between the two time series. As a Fourier-spectrum, coherence represents phase relations with complex numbers. High amplitude in a given frequency regime means that
the two oscillations follow each other by a constant phase lag through the majority of
the time, whereas small amplitude means that the phase leg changes randomly in that
frequency regime. The specific phase lag between the two signals is contained by the
argument of the complex number. By using complex wavelet transform, instead of
Fourier transform, even temporal changes in the phase coherence and phase lag can
be described.

An interesting application of coherence can be found in the micro electric imaging concept introduced in [15]. Coherence has been calculated between local field
potential (LFP) recordings of a high channel count microelectrode array system,
chronically implanted into the hippocampus of a rat. It was assumed that the electrodes, which monitor the same synaptic layer of the hippocampus, will show high
coherence, since they receive the same input, albeit at different delays. Thus, coherence was calculated between signals from each possible pair of electrodes to measure their similarity, and a clustering method was applied to the set of coherences
measured between electrodes. Clustering can then lead to an anatomical dissection
of the hippocampus, where electrodes belonging to the same cluster are inervated
by the same synaptic pathway. As Fig. 4 shows, high coherence in the gamma frequency regime (20–90 Hz) marked almost all synaptic layers of the hippocampus.
This synaptic layer structure can be complemented by a high frequency (300 Hz)

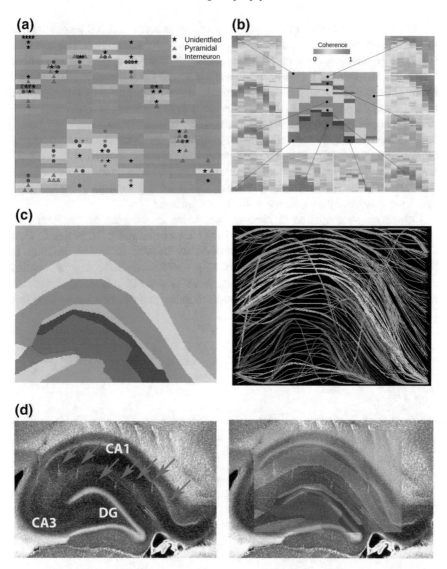

Fig. 4 Electroanatomy of the hippocampus. **a** Distribution of high-frequency power (300 ± 10 Hz) on each of the 256 sites of the silicon probe. The 32 × 8 color matrix is a representation of the 256-site probe. Each rectangle represents a 300 μm (intershank distance) by 50 μm (*vertical* intersite distance) area to mimic the 2-dimensional geometry of the probe coverage. Clustered neurons, assigned to the largest amplitude recording sites, are superimposed on the power map. **b** Coherence maps of gamma activity (30–90 Hz). The 10 example sites (*black dots*) served as reference sites, and coherence was calculated between the reference site and the remaining 255 locations for a 1 s long recording segment.**c** Composite figure of the combined coherence maps. *Left* 2-dimensional combined map of gamma coherence and high-frequency power distribution. *Right* coastline map of layer-specific coherence contours. **d** Histological reconstruction of the recording tracks (*arrows*). The shifting of the tracks in the neocortex is due to a slight displacement of the neocortex/corpus callosum relative to the hippocampus during the tissue sectioning process. DG, dentate gyrus. *Right* physiology-based map superimposed on the recording tracks. From [15] with permission

power map to denote the somatic layers, resulting in an almost full reconstruction of the hippocampal structure, based simply on the LFP recordings. The structure was not fully marked, because the outer layer of the startum moleculare of Dentate Gyrus and the stratum lacumosum-moleculare of the CA1 were not distinguished by the coherence method, presumably because they are innervated by same synaptic input pathway innervates, i.e. the perforant path from the entorhinal cortex.

Applying this coherence clustering method to neocortical micro-electrode recordings resulted in a less clear structure, but supragranular, granular and infragranular layers can be identified clearly on the micro-electroanatomical structure (Fig. 5). Even the different (granular versus agranular) structure of S1 and M1 cortical area and the borderline between them is recognizable in the left hemisphere, as the granular layer (marked by white) ends at the transition zone.

Coherence were applied to identify differences in subtypes of temporal lobe epilepsies [16]. Walker et al. [17] found excessive delta and theta power in slow foci in all intractable patients, and hypocoherence in theta in 75 % of patients. In this study, the aim for neurofeedback training was to restore the normal, healthy coherence pattern between EEG channels. When power and coherence abnormalities could be restored by the neurofeedback training, patients become seizure free [17].

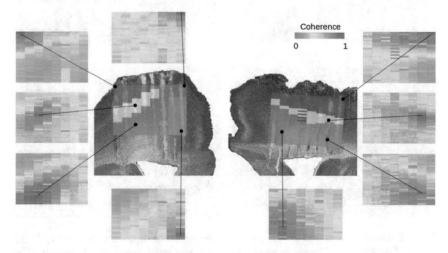

Fig. 5 Electroanatomy of the neocortex. **a** Combined coherence map of gamma activity (30–90 Hz) as in Fig. 4b. Each site served as a reference, and coherence was calculated between the reference site and the remaining 255 locations. The resulting combined map is superimposed on the histologically reconstructed tracks in the sensorimotor cortex. Note reliable separation of layer IV, superficial, and deep layers and the lack of a layer IV coherence band in the adjacent motor cortex (shanks 6–8). Modified from [15]

Causality Analysis

The cross correlation function, as well as coherence spectrum, could be used to determine directional effects between the analyzed signals based on the phase of the peak, corresponding to the time delay between the signals in a given frequency (eq. phase leg index). However, this approach builds on two main assumptions. First, it assumes that there are no difference in the observational delays for the two signals; second, it assumes similar wave-shapes on the two channels, as it is based on the linear correlation coefficient implicitly. Although both assumptions can be valid in many situations, it is hard to verify them in general.

The majority of causality analysis methods are based on Norbert Wiener's principle on predictability: a time series is said to be causal to an other, if it's inclusion makes the prediction of the caused time series more precise [18]. The first practical and applicable implementation of this principle is the Granger-causality introduced by Clive Granger in 1969 [19]. The Granger formalization is based on autoregressive (AR) models, where the next element of a time series is approximated by the linear mixture on the recent elements. Specifically, the formalization can written as follows:

$$x_t = \sum_{i=1}^{p} a_i x_{t-i} + \varepsilon_t$$

$$x_t = \sum_{i=1}^{p} b_i x_{t-i} + \sum_{j=1}^{q} c_j y_{t-j} + \eta_t$$

$$F_{y \to x} = \log \left(\frac{\sigma_\varepsilon^2}{\sigma_\eta^2} \right)$$

where a_i, b_i and c_j are the AR coefficients; p and q control the order of the models; ε_t and η_t are error residuals; σ-s are the variances of residual errors and $F_{y \to x}$ is the directed Granger-causality index. Inclusion of the recent past of the other time series into the autoregressive model does not necessarily results same improvement of prediction error in both directions, thus $F_{y \to x}$ and $F_{x \to y}$ are generally non-equal. The original Granger method quantifies the effect of including the other time series using the log-ratio of the variance of the residual error signal between a prediction with the other time series and a prediction without the other time series.

Since then, numerous versions and improvements of the original Wiener-Granger idea exists. Directed transfer function and Partial Directed Coherence solves the Granger problem in the frequency domain instead of time domain [20–22]. Conditional or multivariate Granger causality includes non-pairwise comparisons and there are nonlinear extensions as well. A version of Granger-causality called short-time directed transfer function has been adapted to analyze the event related activity, developed by [23] and applied to reveal information transmission during visual stimulation of the brain.

A non-parametric translation of Norbert Wiener's original idea to information theory's language is Transfer Entropy introduced by Thomas Schreiber in 2008 [24]. Transfer Entropy quantifies the predictive information transfer, the Mutual Information between present X values and past Y states (Y^-) conditioned on past X states (X^-):

$$TE\,(Y \to X) = I\,(X; Y^-\,|X^-)$$

Transfer Entropy and Granger causality are equivalent in the case of jointly gaussian variables. TE was used to reconstruct interaction delays in turtle visual system [25]. There are several toolboxes for the computation of TE, for example JIDT with python bindings or TRENTOOL for Matlab.

Sugihara's Causality

A dynamical system perspective on causality detection was invented by George Sugihara in 2012 [26]. A dynamical system is a system whose state (Z) changes with time. From the current state of a system, one can predict all the coming future states, if time evolution rules are known:

$$Z_t = \Phi_t\left(Z_0\right)$$

The actual state is a point in state space, the space with state variables on each axe. As time evolves, the system's state traces out a trajectory in state space. In many cases, the system's state is attracted to a lower dimensional subspace of state space and the points form a manifold.

Sugihara's idea is based on Takens theorem [27] which claims that the state of a chaotic dynamical system can be restored (reconstructed) with the aid of one time series measurement from that system by a process called time delay embedding. The method has two parameters the embedding dimension (m) and the embedding delay (τ) with

$$x_t \to X_t = \left(x_t, x_{t-\tau}, x_{t-2\tau}, \ldots, x_{t-(m-1)\tau}\right)$$

According to Takens theorem, the time delay procedure for $m \geq 2d + 1$ is an embedding where d is the dimensionality of the original state space. That is to say, for every point, there is an invertible smooth mapping (whose derivative is also injective) between the reconstructed and the original state space [27]. It follows that the manifold formed by the points in the reconstructed state space is topologically equivalent to the manifold in the original state space, meaning that every point has the same set of neighboring points in both spaces.

Deyle and Sugihara [28] generalized Takens theorem into a multivariate form, when not only the different time lags of a time series provides the state-space embedding, but different observation functions of the same dynamical systems as well. Based on this theorem, they found a new principle for causality analysis: The new idea is that if two time series measurements (X, Y) were from the same dynamical system, then the reconstructed state spaces can be mapped back to the same original manifold, and so, there should also be a smooth mapping between the two reconstructed state spaces. In this case, one can identify causality between the two variable as well.

An asymmetrical relationship between the variables can also be revealed when their original state-spaces are not the same, but one of them is a lower dimensional sub-manifold, a (not necessarily linear) projected version of the other. The mapping works in one direction, but is non-invertible. In this case, one can speak about unidirectional causal relationship, wherein one variable causes the other variable but not the reverse.

If there are no such mappings between the two reconstructed manifolds, they do not belong to the same dynamical system. In this case, one can say that there is no causality between the two variables. Convergent Cross-Mapping is a procedure which tests the existence of this mapping. It is considered a cross-mapping procedure, because it estimates the mapping between the two reconstructed manifolds; it is considered convergent because this estimate converges to the true mapping as one lengthens the time series.

Sugihara's method is able to detect the direction of causality, without needing to assume the two signals are observed at the same time. However, if this assumption holds, the method could be extended to detect delays in the interaction between variables [29]. Parallel work of Schumacher et al. is based on similar principles and also contains time delay detection and in addition they applied their method to neural data [30].

There are various methods to test the existence of the smooth mapping between the reconstructed trajectories. Ma et al. used a feed forward neural network to explicitly estimate the smooth mapping between the embedded times series. When the mapping error was sufficiently small, they detected a causal relationship, otherwise they said that the two time series were independent in the time segment [31].

Sugihara's method works well on deterministic data and when there is direct causality between variables, but it cannot detect hidden common causes. So far, we found only one attempt in the data analysis literature to distinguish direct causality from common cause, but this method has yet to be applied neural data [32]. Moreover the crosstalk or linear mixing, which always appears between multiple electric signals recorded in a volume conductor restricts the applicability of Sugihara's method on raw extracellular signals. Due to linear mixing, each of the recorded signals contains the effect of all neural sources with different weights. Thus, all the sources can be reconstructed from all recorded mixtures to some extent, resulting false detection of circular causal connections between all recordings. To avoid these false positive detections, determination of the individual sources from the measured potentials is necessary and the causality analysis should be applied between the reconstructed time series of the sources, in stead of potentials. This could be achieved by current source density calculation [33–37] or linear demixing due to independent component analysis [38].

Causality of Epilepsy

Causality analysis is a natural tool for revealing functional connections between brain areas during epileptic events. It is also natural to assume that functional connections change as one develops an epileptic seizure. However, the nature of this change is less clear.

The general assumption, based on the spreading nature of the seizure, is that the SOZ should play the role of a causal source during the initiation of a seizure. However, it is possible that the brain areas connected to the active focus (the cortical umbra) which are 'attacked' by a massive synaptic bombardment during the initial phase of the seizure, produce a strong, possibly inhibitory, 'defensive' synaptic answers towards the SOZ. Such an answer would also be a strong effect and thus, measurable by the causality analysis methods. Even if we can accept the more likely assumption, that the SOZ should work as a causal source during the initiation and spread of the seizure, it is less clear what kind of causal changes should precede the seizure. Do the connections between areas strengthen as the system gets closer to the seizure onset, or just the opposite, do the physiologically-well connections stop their activity if one of its connected nodes operates outside its normal regime? Perhaps both types changes occur and perhaps they occur through spatial or temporal separation. As some observations suggest, spatial separation could result from the epileptic network becoming more modular as local connections strengthen and long range connections weaken. Regardless, causal analysis methods are well suited to answer these numerous questions, which require collecting and systematizing the results of the causal analysis.

Sabesan et al. [39] identifies the SOZs by calculating the spatially averaged net transfer entropy, where net transfer entropy is the difference between the outgoing and the incoming causality for a node. Thus, they assumed, that the SOZ corresponds to a net entropy source. Similar idea was implemented by Epstein et al. [40], but by using Granger-causality in the high frequency regime between 80–250 Hz. They found that the net causality outflow from the SOZ area increased only for short bursts before the seizure with 2–42 s, instead of keeping high value continuously or building up monotonously.

Conclusions

Despite the significant arsenal of analytical methods applied to localize the SOZ based on ECoG signals, a general and robust solution is still missing. Although, we have good reasons to believe that increasing information content of the high channel-count subdural surface and deep electrodes will make precise SOZ determination increasingly possible, provided this additional information can be exploited by the proper mathematical analysis. It is also clear that the increasing data flow could not be handled by the traditional way. Analytical and advanced visualization tools are also necessary to evaluate the results and to ensure the results can be incorporated into surgical practice. Maybe the most promising research directions apply causality analysis, which not only reveals pathological alternations from the normal behavior, but could also provide deeper understanding of the underlying mechanisms due to the quantification of the neural interactions and network structure of brain areas under normal and epileptic conditions.

Acknowledgements ZS and BZS were supported by National Research, Development and Innovation Fund under grants NKFIH-K113147 and NKFIH-NN118902. DF and ZS were supported by the Hungarian Brain Research Program - Grant No. KTIA 13 NAP-A-IV/1,2,3,4,6.

References

1. Ulbert, I., Heit, G., Madsen, J., Karmos, G., Halgren, E.: Laminar analysis of human neocortical interictal spike generation and propagation: current source density and multiunit analysis in vivo. Epilepsia **45 Suppl 4**, 48–56 (2004)
2. Benar, C.G., Chauviere, L., Bartolomei, F., Wendling, F.: Pitfalls of high-pass filtering for detecting epileptic oscillations: a technical note on "false" ripples. Clin Neurophysiol **121**(3), 301–310 (2010)
3. Staba, R.J., Wilson, C.L., Bragin, A., Fried, I., Engel, J.: Quantitative analysis of high-frequency oscillations (80-500 Hz) recorded in human epileptic hippocampus and entorhinal cortex. J. Neurophysiol. **88**(4), 1743–1752 (2002)
4. Birot, G., Kachenoura, A., Albera, L., Benar, C., Wendling, F.: Automatic detection of fast ripples. J. Neurosci. Methods **213**(2), 236–249 (2013)
5. Salami, P., Levesque, M., Gotman, J., Avoli, M.: A comparison between automated detection methods of high-frequency oscillations (80–500 Hz) during seizures. J. Neurosci. Methods **211**(2), 265–271 (2012)
6. Zelmann, R., Mari, F., Jacobs, J., Zijlmans, M., Dubeau, F., Gotman, J.: A comparison between detectors of high frequency oscillations. Clin Neurophysiol **123**(1), 106–116 (2012)
7. Lopez-Cuevas, A., Castillo-Toledo, B., Medina-Ceja, L., Ventura-Mejia, C., Pardo-Pena, K.: An algorithm for on-line detection of high frequency oscillations related to epilepsy. Comput Methods Programs Biomed **110**(3), 354–360 (2013)
8. Bishop, C.M.: Pattern Recognition and Machine Learning. Springer (2006)
9. Blanco, J.A., Stead, M., Krieger, A., Stacey, W., Maus, D., Marsh, E., Viventi, J., Lee, K.H., Marsh, R., Litt, B., Worrell, G.A.: Data mining neocortical high-frequency oscillations in epilepsy and controls. Brain **134**(Pt 10), 2948–2959 (2011)
10. Blanco, J.A., Stead, M., Krieger, A., Viventi, J., Marsh, W.R., Lee, K.H., Worrell, G.A., Litt, B.: Unsupervised classification of high-frequency oscillations in human neocortical epilepsy and control patients. J. Neurophysiol. **104**(5), 2900–2912 (2010)
11. van 't Klooster, M.A., Zijlmans, M., Leijten, F.S., Ferrier, C.H., van Putten, M.J., Huiskamp, G.J.: Time-frequency analysis of single pulse electrical stimulation to assist delineation of epileptogenic cortex. Brain **134**(Pt 10), 2855–2866 (2011)
12. Somogyvari, Z., Barna, B., Szasz, A., Szente, M.B., Erdi, P.: Slow dynamics of epileptic seizure: Analysis and model. Neurocomputing **38-40**, 921–926 (2001)
13. Pincus, S.M.: Approximate entropy as a measure of system complexity. Proc. Natl. Acad. Sci. U.S.A. **88**(6), 2297–2301 (1991)
14. Zhang, Z., Chen, Z., Zhou, Y., Du, S., Zhang, Y., Mei, T., Tian, X.: Construction of rules for seizure prediction based on approximate entropy. Clin Neurophysiol **125**(10), 1959–1966 (2014)
15. Berenyi, A., Somogyvari, Z., Nagy, A.J., Roux, L., Long, J.D., Fujisawa, S., Stark, E., Leonardo, A., Harris, T.D., Buzsaki, G.: Large-scale, high-density (up to 512 channels) recording of local circuits in behaving animals. J. Neurophysiol. **111**(5), 1132–1149 (2014)

16. Bartolomei, F., Wendling, F., Vignal, J.P., Kochen, S., Bellanger, J.J., Badier, J.M., Le Bouquin-Jeannes, R., Chauvel, P.: Seizures of temporal lobe epilepsy: identification of subtypes by coherence analysis using stereo-electro-encephalography. Clin Neurophysiol 110(10), 1741–1754 (1999)

17. Walker, J.E.: Power spectral frequency and coherence abnormalities in patients with intractable epilepsy and their usefulness in long-term remediation of seizures using neurofeedback. Clin EEG Neurosci 39(4), 203–205 (2008)

18. Wiener, N.: The theory of prediction. In: E. Beckenbach (ed.) Modern Mathematics for Engineers, vol. 1. McGraw-Hill, New York (1956)

19. Granger, C.W.J.: Investigating causal relations by econometric models and cross-spectral methods. Econometrica 37, 424–438. (1969)

20. Baccala, L.A., Sameshima, K.: Partial directed coherence: a new concept in neural structure determination. Biol Cybern 84(6), 463–474 (2001)

21. Kaminski, M., Ding, M., Truccolo, W.A., Bressler, S.L.: Evaluating causal relations in neural systems: granger causality, directed transfer function and statistical assessment of significance. Biol Cybern 85(2), 145–157 (2001)

22. Kaminski, M.J., Blinowska, K.J.: A new method of the description of the information flow in the brain structures. Biol Cybern 65(3), 203–210 (1991)

23. Liang, H., Ding, M., Nakamura, R., Bressler, S.L.: Causal influences in primate cerebral cortex during visual pattern discrimination. Neuroreport 11(13), 2875–2880 (2000)

24. Schreiber, T.: Measuring information transfer. Phys. Rev. Lett. 85(2), 461–464 (2000)

25. Wibral, M., Vicente, R., Lizier, J.T. (eds.): Directed Information Measures in Neuroscience. Understanding Complex Systems. Springer-Verlag Berlin Heidelberg (2014)

26. Sugihara, G., May, R., Ye, H., Hsieh, C.H., Deyle, E., Fogarty, M., Munch, S.: Detecting causality in complex ecosystems. Science 338(6106), 496–500 (2012)

27. Takens, F.: Detecting strange attractors in turbulence. In: D.A. Rand, L.S. Young (eds.) Dynamical Systems and Turbulence, Lecture Notes in Mathematics, vol. 898, p. 366–381. Springer-Verlag (1978)

28. Deyle, E.R., Sugihara, G.: Generalized theorems for nonlinear state space reconstruction. PLoS ONE 6(3), e18,295 (2011)

29. Ye, H., Deyle, E.R., Gilarranz, L.J., Sugihara, G.: Distinguishing time-delayed causal interactions using convergent cross mapping. Sci Rep 5, 14,750 (2015)

30. Schumacher, J., Wunderle, T., Fries, P., Jakel, F., Pipa, G.: A Statistical Framework to Infer Delay and Direction of Information Flow from Measurements of Complex Systems. Neural Comput 27(8), 1555–1608 (2015)

31. Ma, H., Aihara, K., Chen, L.: Detecting causality from nonlinear dynamics with short-term time series. Sci Rep 4, 7464 (2014)

32. Hirata, Y., Aihara, K.: Identifying hidden common causes from bivariate time series: a method using recurrence plots. Phys Rev E Stat Nonlin Soft Matter Phys 81(1 Pt 2), 016,203 (2010)

33. Mitzdorf, U.: Current source-density method and application in cat cerebral cortex: investigation of ecoked potentials and EEG phenomena. Physiol Rev 65, 37–100 (1985)

34. Nicholson, C., Freeman, J.A.: Theory of current source-density analysis and determination of conductivity tensor for anuran cerebellum. J Neurophysiol 38, 356–368 (1975)

35. Pettersen, K.H., Devor, A., Ulbert, I., Dale, A.M., Einevoll, G.T.: Current-source density estimation based on inversion of electrostatic forward solution: Effect of finite extent of neuronal activity and conductivity discontinuites. Journal of Neuroscience Methods 154(1-2), 116–133 (2006)

36. Potworowski, J., Jakuczun, W., Leski, S., Wojcik, D.: Kernel current source density method. Neural Comput 24(2), 541–575 (2012)

37. Somogyvári, Z., Cserpán, D., István, U., Péter, E.: Localization of single-cell current sources based on extracellular potential patterns: the spike csd method. Eur J Neurosci 36(10), 3299–313 (2012)

38. Vigario, R., Oja, E.: BSS and ICA in neuroinformatics: from current practices to open challenges. IEEE Rev Biomed Eng 1, 50–61 (2008)

39. Sabesan, S., Good, L.B., Tsakalis, K.S., Spanias, A., Treiman, D.M., Iasemidis, L.D.: Information flow and application to epileptogenic focus localization from intracranial EEG. IEEE Trans Neural Syst Rehabil Eng **17**(3), 244–253 (2009)
40. Epstein, C.M., Adhikari, B.M., Gross, R., Willie, J., Dhamala, M.: Application of high-frequency Granger causality to analysis of epileptic seizures and surgical decision making. Epilepsia **55**(12), 2038–2047 (2014)

Connecting Epilepsy and Alzheimer's Disease: Modeling of Normal and Pathological Rhythmicity and Synaptic Plasticity Related to Amyloidβ (Aβ) Effects

Péter Érdi, Takumi Matsuzawa, Tibin John, Tamás Kiss and László Zalányi

Alzheimer's Disease and Epilepsy: The Big Picture

General Remarks

Hidden links have been demonstrated between neurodegeneration due to Alzheimer's disease (AD) and temporal lobe epileptic activity (TLEA).

Alzheimer's disease is a devastating neurodegenerative disorder likely affecting millions of people. Neurodegeneration in early AD primarily affects the hippocampal formation in the medial temporal lobe, leading to severe memory loss [1]. This region is also the main focus of TLEA. In AD the incidence of convulsive seizures is ten times higher than in age-matched general population [2, 3]. Epilepsy is 87 times more frequent in patients with 'early onset' disease and occurs particularly early in familial AD, as an integral part of the pathophysiology [4]. There is accumulating evidence [2, 5, 6] that seizures in the cortico-hippocampal system might contribute to cognitive decline. Cognitive decline starts 5.5 years earlier in AD patients with epilepsy than in those without [7]. In a mouse model of AD, combined video and electroecephalograpic (EEG) recordings revealed abundant non-convulsive seizures characterized by cortical and hippocampal spikes and sharp waves [8]. Together these lines of evidence indicate a strong association between the mechanisms of AD and epilepsy, but one that may often be masked by the covert, non-convulsive nature of seizures. Common mechanisms of epilepsy and dementia extend to other neurodegenerative disorders as well. For example, Lewy body dementia (LBD) is the second most frequent cause of dementia in the elderly and often co-exists with both

P. Érdi (✉) · T. Matsuzawa · T. John
Center for Complex Systems Studies, Kalamazoo College, Kalamazoo, MI, USA
e-mail: perdi@kzoo.edu

P. Érdi · T. John
Wigner Research Centre for Physics, Institute for Particle and Nuclear Physics, Hungarian Academy of Sciences, Budapest, Hungary

P. Érdi · T. Kiss · L. Zalányi
Wigner Research Centre for Physics, Hungarian Academy of Sciences, Budapest, Hungary

© Springer International Publishing AG 2017
P. Érdi et al. (eds.), *Computational Neurology and Psychiatry*,
Springer Series in Bio-/Neuroinformatics 6,
DOI 10.1007/978-3-319-49959-8_5

sporadic and familial AD. LBD is characterized by EEG abnormalities, including epileptiform activity, and can sometimes show an overlap of clinical phenotype with epileptic seizures [9]. A better understanding of possible common neural mechanisms underlying AD, LBD and TLEA is crucial for an efficient clinical management of these conditions.

Many facts have been accumulated to support hypotheses that link the elevated level of **human amyloid precursor protein (hAPP) related** β-**amyloid** (Aβ) to pre-clinical and clinical observations related to AD [5, 10]. The most significant elements of a working hypothesis assume:

- Aβ alters hippocampal rhythmicity (sections "Modeling Hippocampal Rhythm Generation and Control: Aβ Pathology" and "Aβ Overproduction and Hippocampal Network Dysfunction: Modeling the Age-Dependent Effects").
- Aβ alters long term synaptic plasticity by several mechanisms, enhance long-term depression (LTD) and impair long-term potentiation (LTP)) (section "Two-Way Relationship Between Altered Synaptic Activity and Neuronal Dysfunction")
- Elevated Aβ implies neuronal dysfunction resulting from an impaired balance between positive and negative feedback loops in modulation of synaptic transmission (section "Two-Way Relationship Between Altered Synaptic Activity and Neuronal Dysfunction").
- Non-convulsive, subclinical partial seizures worsen the memory and behavioral symptoms in AD (section "Non-convulsive, Subclinical Partial Seizures Worsen the Memory and Behavioral Symptoms in AD").
- Antiepileptic drugs can reduce the deteriorating effects of epileptiform activity in AD (section "Antiepileptic Drugs Can Reduce the Deteriorating Effects of Epileptiform Activity in AD").

Computational modeling is an appropriate tool to test the hypothesis. Figure 1 summarizes the big picture to explain the multiple and multilevel effects of Aβ: from altered synaptic plasticity via network dysfunction to cognitive deficit. The figure contains three connected columns. The left column is about the relationship between hippocampal structure and normal and pathological rhythms. Based on the classical knowledge going back to Cajal on the morphology and trisynaptic circuit of hippocampal formation a computational model has previously been constructed to simulate the generation of gamma-related theta-frequency resonance and pharmacological control of the septo-hippocampal theta rhythm [11–13].

Several neuron populations are modeled using individual conductance-based cell models. These biophysically realistic conductance-based single cell models correspond to resistor-capacitor circuits with time and voltage-dependent parallel resistors representing different types of experimentally verified channels in each cell type. A previous study [12] demonstrated a correlation between the effects of experimental and computational manipulations on theta power. With the view that this model is thus validated with respect to physiologically-relevant septo-hippocampal theta rhythm generation, the single compartment models consisting of uniform channel densities for hippocampal basket interneurons, horizontal oriens interneurons, and GABAergic neurons of the medial septum are further used without modification.

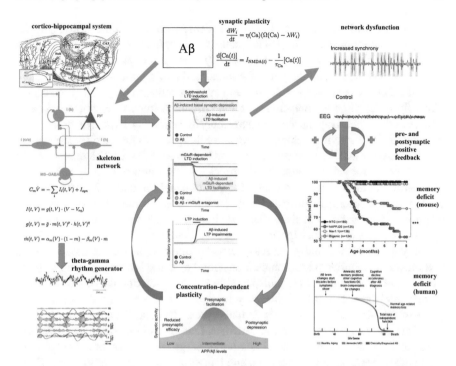

Fig. 1 Hypothetical causal chain to explain the multiple and multilevel effects of Aβ: from altered synaptic plasticity via network dysfunction to cognitive deficit. A skeleton network of the hippcampal system generates gamma and theta rhythms. Aβ concentration-dependent altered synaptic plasticity implies network dysfunction including epileptiform activity. This activity contributes to cognitive deficit by positive feedback cellular mechanisms.

The multi-compartmental model of a CA1 pyramidal cell, capable of demonstrating the many firing patterns of this neuron, is also used without change unless otherwise specified.

The second column is about altered synaptic plasticity due to the effect of β-amyloid. The starting point of the analysis is the three middle plots, demonstrating the dual qualitative effects of elevated Aβ, i.e. LTD facilitation and LTP impairment. The bottom diagram shows a hypothetical relationship between Aβ concentration and synaptic activity, the details will be discussed in section "Two-Way Relationship Between Altered Synaptic Activity and Neuronal Dysfunction". The top panel shows a computational model of bidirectional plasticity [14]. In this paper we offer an extension of the model to take into account the multiple effects of Aβ to synaptic plasticity.

The third column displays two interconnected sub-processes. Pathologically elevated Aβ and altered synaptic activity implies abnormal synchronized activity exhibiting (maybe subclinical) epileptic seizures. We believe that no computational models have addressed the problem of falsifying/supporting hypotheses on the causal relationship between Aβ-related synaptic depression and aberrant network activity. There might be a two-way relationship between synaptic activity and network dysfunction. One main hypothesis to be tested assumes that Aβ-induced increases in excitatory network activity lead to synaptic depression by a **homeostatic plasticity** compensatory mechanism. (Homeostatic plasticity is interpreted as "staying the same through change"). Homeostatic plasticity is a neural implementation of a feedback control strategy with the goal of stabilizing the firing rate by changing synaptic parameters such as receptor density and synaptic strength [15, 16]. Homeostatic plasticity is supposed to compensate for the unstable features of Hebbian synapses. Failure of this stabilizing mechanism may imply hyperactivity, hypersynchronization and epileptiform activities. However, we leave this problem for another publication. Altered network activity at least is correlated to cognitive deficit. The main working hypothesis is that seizures amplify the process of AD progression by some positive feedback mechanisms involving Aβ deposition and cell death [2]: both pre- and postsynaptic mechanisms provide the molecular bases for modeling of such kinds of positive feedback mechanisms. The two bottom panels show results of memory tests for mouse and human experiments. In a later phase of this project we will build memory tests of neural networks into the computational platform to model normal and impaired cognitive performance. Behavioral tests used to assess memory functions in AD mouse models are indispensable for characterizing the degree and type of memory deterioration [17]. Memory functions associated with different subregions of the hippocampus, namely dentate gyrus, CA3 and CA1, were tested both experimentally and by computational models. Different subregions may implement different functions, such as spatial and temporal pattern separation, short-term or working memory, pattern association, and temporal pattern completion [18]. These domain-specific memory performance tests will be implemented and used.

Our big goal is to provide insight using the tools of computational neuroscience on how cellular and synaptic level effects of Aβ accumulation translate across spatial scales into network level changes in theta and gamma rhythms [19], and aberrant network synchronization leading to cognitive deficits. Our multi-level model considers the brain as a hierarchical dynamical system. To specify a dynamical system, characteristic **state variables** and **evolution equations** governing the change of state must be defined. The dynamic laws at the molecular level can be identified with chemical kinetics, at the channel level with biophysical detailed equations for the membrane potential, and at the synaptic and network levels with learning rules to describe the dynamics of synaptic modifiability, see Table 1. Overall, our perspective on multi-level hippocampal modeling is summarized here [20].

Modeling Hippocampal Rhythm Generation and Control: Aβ Pathology

Skeleton Network Model

The skeleton network model (Fig. 2) of the hippocampal CA1 region and the septum introduced in [12] consisted of five cell populations: pyramidal cell, basket cells, two types of horizontal neurons and the septal γ-Aminobutyric acidergic (GABAergic) cells.

Table 1 The brain as a hierarchical dynamical system. The possible state variables and dynamical equations are shown for different levels in the hierarchy

Level	Variables	Equations
Molecule	Chemical composition	Reaction kinetics
Membrane	Membrane potential	Hodgkin–Huxley-type equations
Cellular	Cellular activity	Integrate-and-fire neurons
Synapse	Synaptic efficacy	Elementary learning rules (short term and long term plasticity)
Network	Synaptic weight matrix	Hebbian and homeostatic learning rules

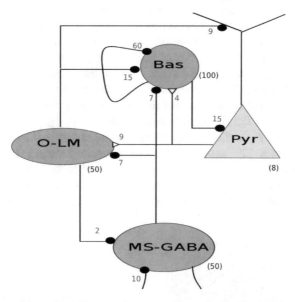

Fig. 2 Structure of septo-hippocampal network model. *Red symbols* and *black circles* indicate inhibitory populations and synapses, respectively; *yellow symbols* and open *triangles* indicate excitatory populations and synapses. Representative connectivity specified quantitatively with divergence numbers (*blue*) defining the number of cells in the target population that each cell in the presynaptic population innervates. Each target cell was chosen from a uniform distribution over the target population cells for each network instantiation. Similarly, convergence numbers (*red*) define the number of presynaptic neurons innervating each cell in target population. Total simulated cell numbers in each population are given in parentheses

Connections within and among cell populations were created by faithfully following the hippocampal structure. The main excitatory input to horizontal neurons is provided by the pyramidal cells via alpha-amino-3-hydroxy-5-methyl-4-isoxazolepropionic acid (AMPA) mediated synapses [21]. Synapses of the septally projecting horizontal cells [22] and synapses of the other horizontal cell population, the oriens-lacunosum moleculare (O-LM) cell population innervating distal apical dendrites of pyramidal cells [23] are of the $GABA_A$ type synapses are taken into account. O-LM neurons also innervate parvalbumin containing basket neurons [24]. Basket neurons innervate pyramidal cells at their somatic region and other basket neurons [25] as well. Septal GABAergic cells innervate other septal GABAergic cells and hippocampal interneurons [26, 27] (Fig. 2).

The above described model captures several elements of the complex structure of the hippocampal CA1 and can be used to account for very precise interactions within this region. However, when the focus of interest is instead on general phenomena taking place during rhythm generation, modelers might settle for a simpler architecture. In [11] we described gamma [28] related theta oscillation generation in the CA3 region of the hippocampus. The architecture of the model is exceedingly simplified: only an interneuron network is simulated in detail. This simplification, however, allowed the authors to introduce an extrahippocampal input and study its effect on rhythm generation. As a result, the model is able to account for basic phenomena necessary for the generation of gamma related theta oscillation.

We plan to adapt, extend and combine our models of the hippocampal rhythm generation for describing mechanisms not yet studied by computational models. The extended version will be able to take into account **aberrant changes** in cellular and synaptic morphology, intrinsic membrane and synaptic parameters to study the causal chains between $A\beta$ induced structural changes and **pathological rhythms**. Recently, Scott et al. [29] observed an age-dependent reduction in the amplitude of a slow oscillation in the extracellular electric potential of the hippocampus of mice overexpressing $A\beta$, as Fig. 3 shows. The goal of our computational model [30] was to demonstrate how $A\beta$ affects the ability of neurons in hippocampal networks to fire

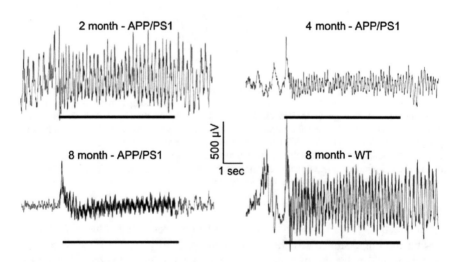

Fig. 3 Age-dependent reduction in the amplitude of theta rhythm elicited by electrical stimulation of the brainstem nucleus pontis oralis for transgenic mice (APP/PS1) exhibiting increasing $A\beta$ plaque loads from 2 months to 8 months. No theta power reduction was observed in wildtype (WT) mice, which do not exhibit $A\beta$ plaques. Data from [29]

in unison at theta frequencies to reduce the amplitude of theta rhythm. For the results see section "Aβ Overproduction and Hippocampal Network Dysfunction: Modeling the Age-Dependent Effects", also in [31].

Two-Way Relationship Between Altered Synaptic Activity and Neuronal Dysfunction

In a seminal paper entitled *Synaptic Depression and Aberrant Excitatory Network Activity in Alzheimer's Disease: Two Faces of the Same Coin?* [32] discusses the intricate (presumably two-way) relationship between altered synaptic activity and network dysfunction. Elevated Aβ implies neuronal dysfunction due to the consequence of the impaired balance between positive and negative feedback loops in modulation of synaptic transmission. Two main possibilities have been suggested: (i) "...depression of excitatory synaptic activity could lead to network disinhibition, if it affected inhibitory interneurons more than principal excitatory cells.."; (ii) "...Aβ–induced increases in excitatory network activity lead to synaptic depression through homeostatic or compensatory mechanisms."

As suggested by Palop and Mucke [5], Aβ might have a concentration-dependent dual control effect on excitatory synapses: reduced efficacy of the presynaptic component, presynaptic facilitation and postsynaptic depression appear for low, intermediate and high Aβ concentrations, respectively. Since pathologically elevated Aβ impairs LTP and enhances LTD related to the partial block of N-Methyl-D-aspartate (NMDA) receptors, it is assumed that small increases in postsynaptic calcium trigger LTD, whereas large increases induce LTP [14, 33].

Non-convulsive, Subclinical Partial Seizures Worsen the Memory and Behavioral Symptoms in AD

There is accumulating evidence [2, 5, 6, 34] that seizures in the cortico-hippocampal systems might contribute to cognitive decline. Seizures in individuals with AD have been described previously and by today there is a robust data-based foundation to support clinical comorbidity. The question now is as it was formulated by [2]: "Two Disorders or One?" Both mouse models and human data support the structural connection between the two disorders. Impairment of inhibitory mechanisms (the phenomenon called disinhibition) may destabilize network oscillatory activity at early stages of the disease. Hyperexcitability and hypersynchrony in cellular and circuit activities may imply subclinical seizures in the temporal lobe and aggravate memory loss, as it was identified in mouse models of AD. There is a remarkable regional overlap in human AD and TLEA, and it looks to be a testable working hypothesis that subconvulsive seizures due to plasticity within hippocampal circuitry contribute to the memory impairment of AD. Therefore, one might cautiously assume that these animal models of AD may have temporal lobe epilepsy-like syndromes. Similarities and differences between epilepsy and AD from studying mechanisms of hyperexcitability and seizures have been analyzed by [6]. Seizures facilitate production of Aβ and can cause impairments in cognition and behavior in both animals and humans. There seems to be a correlative relationship between duration of epilepsy and degree of impairments in episodic memory.

A recent review [35] on co-morbidity of AD and seizures hypothesizing common pathological mechanism highlights that (i) clinical data from familial and sporadic AD patients reveal increased seizure risk; (ii) many APP-linked AD mouse models develop seizures and other EEG abnormalities; (iii) APP and/or APP-derived peptides may link AD pathology to epileptiform activity; and (iv) epileptiform activity in AD mouse models can be rescued independent of Aβ reduction.

Our longer term plan is to use a computational model to test possible mechanisms, which suggest that high levels of Aβ imply aberrant (epileptiform) activity and the failure of compensatory inhibitory responses contributes to the emergence of cognitive deficits associated with AD.

Antiepileptic Drugs Can Reduce the Deteriorating Effects of Epileptiform Activity in AD

There is also growing evidence suggesting that some antiepileptic drugs (such as levetiracetam (LEV)) can reduce abnormally enhanced electrical activity, and slow down or even reverse hippocampal synaptic dysfunction, and cognitive deficits in hAPP mice, or even in human patients [36–38].

It seems to be convincing that perturbations of brain network activity are observed in AD patients and aberrant network activity might contribute to AD-related cognitive decline. Human APP transgenic mice simulate key aspects of AD, including pathologically elevated levels of Aβ peptides in brain, aberrant neural network activity, remodeling of hippocampal circuits, synaptic deficits, and behavioral abnormalities. Computational modeling seems to be an indispensable tool to study whether there is any causal mechanism to connect these elements.

To explore whether Aβ-induced aberrant network activity contributes to synaptic and cognitive deficits, [38] treated patients with amnestic mild cognitive impairment (aMCI; a condition in which memory impairment is greater than expected for a person's age and which greatly increases risk for Alzheimer's dementia) with different antiepileptic drugs. It was shown that very low doses of the drugs calm hyperactivity in patients' brain. Among the drugs tested, LEV effectively reduced abnormal spike activity detected by electroencephalography. Chronic treatment with LEV also reversed hippocampal remodeling, behavioral abnormalities, synaptic dysfunction, and deficits in learning and memory in hAPP mice. These findings support the hypothesis that aberrant network activity contributes causally to synaptic and cognitive deficits in hAPP mice. Consequently, LEV might also help ameliorate related abnormalities in people who have or are at risk for AD.

There are more and more recent studies [34, 35], which support the view that some antiepileptic drugs also reverse memory deficits at least in aMCI patients, adding further support to the hypothesis that neuronal network hyperactivity may causally contribute to cognitive impairment in both aMCI and AD mouse models.

Dynamical system theory helps us to understand the neural mechanisms of temporal and spatio-temporal neural activities. The discipline of **computational neuropharmacology** [39] emerges as a new tool of drug discovery by constructing biophysically realistic mathematical models of neurological and psychiatric disorders. Nowadays the term "Quantitative Systems Pharmacology" is used and defined as an approach to translational medicine that combines experimental and computational methods to apply new pharmacological concepts to the development of new drugs.

We have adapted methods of computational neuroscience to the emerging field of computational neuropharmacology [12, 30, 39–46]. The computational framework was developed for testing hypotheses related to pharmacotherapy of anxiety and schizophrenia. Subsequently, based on a similar but extended model of septohippocampal theta rhythm generation and control further preliminary results were obtained related to AD [47]. Specifically, it was shown that reduced density of fast sodium current channels in basket cells and oriens-lacunosum-moleculare interneurons, a possible physiological effect of Aβ, resulted in an increase in theta rhythm amplitude in the 8–10 Hz range. Furthermore, distinct effects on single cell firing patterns included a reduction of action potential amplitude in basket cells as reported experimentally [48].

Aβ Effects on Synaptic Plasticity: Brief Summary of the Experimental Background

It is well known that glutamatergic synaptic transmission is controlled by the number of active NMDA receptors (NMDARs) and AMPA receptors (AMPARs) at the synapse. NMDAR activation has a central role, as it can induce either LTP or LTD, depending on the intracellular calcium rise in dendritic spines. Activation of synaptic NMDARs and large increases in calcium concentration are required for LTP, whereas internalization of synaptic NMDARs, activation of perisynaptic NMDARs and lower increases in intracellular calcium concentration are necessary for LTD. LTP induction implies recruitment of AMPARs and growth of dendritic spines, while LTD induces spine shrinkage and synaptic loss.

The multiple effects of pathologically elevated Aβ on synaptic plasticity are being studied [5, 49]. Generally speaking, it now seems to be accepted that Aβ impairs LTP and enhances LTD. Most likely soluble oligomers rather than plaques are the major cause of synaptic dysfunction and ultimately neurodegeneration [50], for a review see [51].

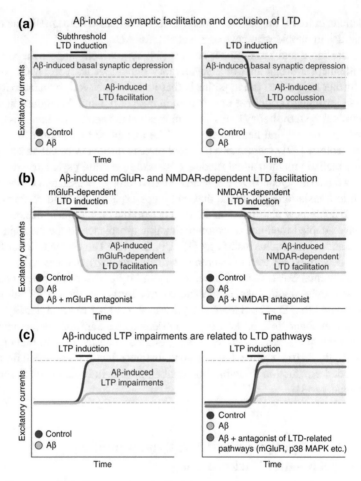

Fig. 4 Summary of $A\beta$ effects on synaptic plasticity

The detailed mechanism underlying $A\beta$-induced LTP and LTD modifications are not fully understood. Computational models of normal bidirectional synaptic plasticity could be supplemented with $A\beta$-induced effects to elaborate on this question. Both phenomenological and more detailed receptor-kinetic models (taking into account receptor internalization, desensitization, etc.) should be studied. This paper makes a step into this direction.

Figure 4 is the reproduction of Fig. 2 of [5]

Below is the list of stylized facts based on [5] to be explained by model studies in a consistent way:

- $A\beta$ suppresses basal excitatory synaptic transmission
- $A\beta$ facilitates LTD after subthreshold LTD inductions
- $A\beta$ occludes LTD

- Aβ facilitates LTD by inducing activation of metabotropic glutamate receptors (mGluRs) and NMDARs
- Aβ-induced facilitation of mGluR-dependent LTD is suppressed by mGluR antagonists
- Aβ-induced facilitation of NMDAR-dependent LTD is suppressed by NMDAR antagonists
- Aβ-induced LTP deficits depend on activation of LTD pathways. Aβ potently inhibits LTP
- Blocking LTD-related signaling cascades with mGluR5 antagonists or an inhibitor of p38 MAP prevents Aβ induced LTP impairment

Model Construction

We suggest here that the findings and hypotheses should be explained within the framework of calcium dependent models of bidirectional synaptic plasticity [14].

Modeling Modulation of Synaptic Transmission by Aβ

As Shankar and Walls [52] wrote: "How Aβ mediates its effects on synaptic plasticity may take many years to fully understand...". It seems likely that Aβ influences the feedback loop that controls neuronal excitability. Palop and Mucke [5], suggests that reduced presynaptic efficacy, presynaptic facilitation and postsynaptic depression may occur at small, intermediate, and large Aβ concentrations, respectively. An often used simple implementation of the **calcium control hypothesis** [14, 53] is given by Eq. 1:

$$\frac{dW_i(t)}{dt} = \eta([Ca^{2+}(t)]) \left(\Omega([Ca^{2+}(t)]) - \lambda W_i(t) \right). \tag{1}$$

Here W_i is the synaptic weight of synapse i, the value of the function Ω depends on calcium concentration, and determines both sign and magnitude of the change of synaptic strength. η is the learning rate, and also depends (typically monotonously increasingly) on calcium concentration, λ is a decay constant. To complete the model we need an equation which prescribes calcium dynamics. A simple assumption is that the source of calcium depends on the NMDA current, as Eq. 3 defines:

$$\frac{d\left[Ca^{2+}(t)\right]}{dt} = I_{\text{NMDA}}(t) - \frac{1}{\tau_{\text{Ca}}}\left[Ca^{2+}(t)\right], \tag{2}$$

where $[Ca^{2+}(t)]$ is the calcium concentration at the spine, $I_{\text{NMDA}}(t)$ is the NMDA current, and τ_{Ca} is the decay time constant of calcium in the spine. The details of the

calculation of the NMDA currents are given in [53] based on the assumption that
NMDARs are the primary sources of calcium.

$$I_{\text{NMDA}} = P_0 G_{\text{NMDA}} \left[I_f \theta(t) e^{\frac{-t}{\tau_f}} + I_s \theta(t) e^{\frac{-t}{\tau_s}} \right] H(V) \tag{3}$$

where I_f and I_s are the relative magnitude of the slow and fast component of the
NMDAR current. $I_f + I_s = 1$ is assumed. $H(V)$ is the general form of the voltage
dependence. $\theta = 0$ if $t < 0$ and $\theta = 1$ if $t \geq 0$. P_0 is the fraction of NMDARs in the
closed state, and set to be 0.5.

The functional form selected for Ω function is based on experimental data of [54]
(for the underlying mathematical details see also Supporting Information of [14]) and
given as

$$\Omega\left(\left[Ca^{2+}(t)\right]\right) = \frac{e^{\beta_2(\left[Ca^{2+}(t)\right]-\alpha_2)}}{1+e^{\beta_2(\left[Ca^{2+}(t)\right]-\alpha_2)}} - \gamma \frac{e^{\beta_1(\left[Ca^{2+}(t)\right]-\alpha_1)}}{1+e^{\beta_1(\left[Ca^{2+}(t)\right]-\alpha_1)}} + \gamma \tag{4}$$

The function is visualized with the blue line of Fig. 5.

Construction of Ω Function to Implement $A\beta$ Effects

The Ω function (Fig. 5) plays a crucial role in the calcium control hypothesis, and
determines the behavior of the synaptic weight at different calcium concentration.
Essentially, the shape of the Ω function determines when LTP or LTD occurs. This
implies that abnormal synaptic plasticity can be modeled by modifying the Ω function in Eq. 1.

A new Ω function was constructed to incorporate the effect of $A\beta$. As reviewed
in [5] $A\beta$ affects only the LTD-related pathways and yet besides strenghtening LTD
it also impairs LTP. In the original paper by Shouval et al. [14] the Ω function contains two terms in which LTP and LTD processes are phenomenologically described
in a mixed form. This way the selective effect of $A\beta$ on LTD-related pathways can
not be incorporated. Moreover, as reviewed in [55] LTP and LTD processes uti-
lize partially separate molecular pathways. According to these findings the new Ω
function assumes competition of LTP and LTD processes. In Eq. 5 Ω_{LTP} describes
the onset of LTP as a sigmoid function with threshold set at α_3. In contrast to Ω_{LTP},
Ω_{LTD} consists of two sigmoid functions capturing the onset and the offset of the LTD
process. Here the threshold parameters are functions of $A\beta$ concentration. These two
processes supposed to be equal in strength providing the possibility to cancel each
other, which is one possible way to eliminate LTP in the model according to the
findings reviewed in [5]. The two processes are balanced but not weighted equally,
a synapse can be potentiated three times stronger than its basal level but can only
be weakened to zero. To achieve this weighting in the normalized synaptic process
model a sigmoid function is composed with the two competing processes with the
ability to set the basal synaptic strength level arbitrarily.

$$\Omega_{\text{new}}([Ca^{2+}(t)], A\beta) = \frac{e^{\beta(k_1\Omega_{\text{LTP}}-k_2\Omega_{\text{LTD}})-\epsilon}}{1 + e^{\beta(k_1\Omega_{\text{LTP}}-k_2\Omega_{\text{LTD}})-\epsilon}}$$

$$\Omega_{\text{LTP}}([Ca^{2+}(t)]) = \frac{e^{\beta_3([Ca^{2+}(t)]-\alpha_3)}}{1 + e^{\beta_3([Ca^{2+}(t)]-\alpha_3)}}$$

$$\Omega_{\text{LTD}}([Ca^{2+}(t)], A\beta) = \frac{e^{\beta_1([Ca^{2+}(t)]-\alpha_1(A\beta))}}{1 + e^{\beta_1([Ca^{2+}(t)]-\alpha_1(A\beta))}} - \frac{e^{\beta_2([Ca^{2+}(t)]-\alpha_2(A\beta))}}{1 + e^{\beta_2([Ca^{2+}(t)]-\alpha_2(A\beta))}} \quad (5)$$

In section "Altered Synaptic Plasticity" it will be shown that by using the constructed new Ω function the simulation results are in accordance with the experimental data on altered synaptic plasticity. In section "Biophysical Backgrounds of the Effects of Aβ on Ω Function" some explanation is given for the kinetic basis of altered synaptic plasticity.

Simulation Results

Aβ Overproduction and Hippocampal Network Dysfunction: Modeling the Age-Dependent Effects

The simulations described here investigate possible mechanisms behind the experimentally observed age-dependent reduction in theta rhythm power in Aβ-overproducing transgenic mice (Fig. 3).

Aβ Pathology Initially Decreases Hippocampal Theta Power with Subsequent frequency Reduction and Power Normalization

Age-dependent neurophysiological effects on both cellular and network levels observed in mice models of amyloid pathology approximately between the ages of 2 and 8 months were incorporated to describe what may determine age-dependent reduction of elicited theta power. These age-dependent changes appear in sodium conductance, and in the number and connectivity of O-LM cells (Table 2). Reduction in theta power and eventual reduction in theta frequency (each determined from a power spectrum) were observed when these progressive effects were applied to the computational model (Fig. 6).

Pyramidal Cells Exhibit Reduced Population Synchrony Dependent on Cellular Events

The initially declining local field potential (LFP) theta power suggests further investigation of whether reduced temporal coordination amongst the cells or fewer active cells participating in the rhythm occurs. Since the source of LFP patterns is thought

Table 2 Simulation of progressive neurophysiological changes observed in Aβ overproducing mice. These represent approximate parameter changes corresponding to age [56–58] in the neural network model

Simulated age (mo.)	K_{DR}, g_{max} (% increase)	Na, g_{max} (% increase)	No. of OLM cells (% loss)	Divergence, OLM cells (% increase)
2	100	0	0	0
4	100	0	40	33.33
6	100	50	50	50
8	100	50	60	75

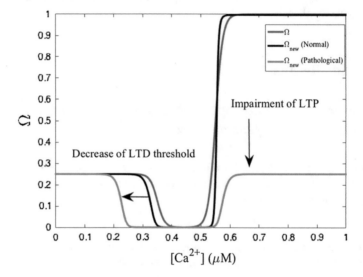

Fig. 5 The Ω (*blue*) and Ω_{new} (*black and red*) functions are plotted against the intracellular calcium concentration. Ω_{new} with a pathological parameter set decreases the LTD threshold and impairs LTP by weakening the LTP strength

to be synaptic activity acting along pyramidal cells, the population-level spike timing of pyramidal cells can provide clues as to the source of theta rhythm attenuation observed in Fig. 6, and can be observed in this computational model. The correlation amongst spike times in pyramidal cells was reduced by the incorporated cellular effects of Aβ accumulation in addition to theta power changes in the local field potential, pointing toward alterations in mutual spike timing regulation mechanisms in the network model (Fig. 7).

Average zero time-lag cross-correlations using a synchrony time window of 10 ms for each pair of spike trains in pyramidal cells revealed a significant reduction in this measure of coherence across the simulated mice groups (Fig. 8).

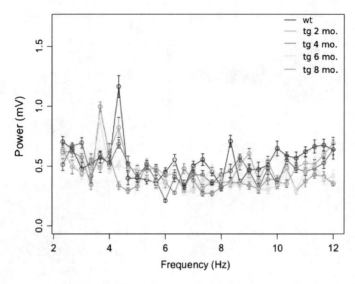

Fig. 6 Synthetic local field potential power spectra time course due to integrated age-dependent neurophysiological effects. Progressive reduction in power of theta rhythm at about 4 Hz for 2- and 4-month old simulated transgenic mice networks is followed by reduction in peak frequency to below 4 Hz accompanied by subsequent increase in power for 6- and 8-month old mice simulated transgenic networks. Error bars denote standard error of the mean

Pyramidal Cells Exhibit Increased Spiking Period Variability

To investigate the loss of correlation amongst action potential timings at theta frequency caused by amyloid pathology effects, the time intervals between spikes of individual neurons were analyzed for variability using a Poincaré plot. This plot relates each spike period to its preceding period for all pyramidal cells in representative network instantiations, revealing greater variation around the line-of-identity when amyloid pathology induced effects were implemented, suggesting spiking variability occurs primarily on the time-scale of consecutive pyramidal cell spikes (Fig. 9).

Pyramidal Cells Exhibit Altered Resonance Properties Dependent on Cellular Events

Since the intrinsic cell property of resonance or frequency preference can contribute to spike timing of individual cells in a rhythmic network [59], we tested resonance properties of the pyramidal cell model after cellular alterations corresponding to Table 2. This was performed at a range of transmembrane potentials seen in wildtype-network simulations, since resonance in pyramidal cells has been shown to be voltage-dependent [60]. Injecting a ZAP current into these model cells, we

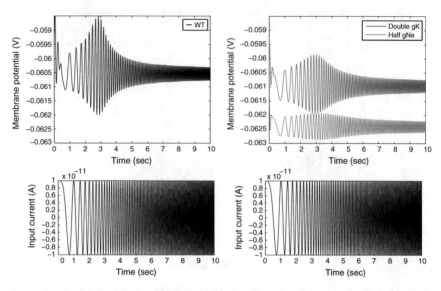

Fig. 7 Subthreshold membrane potential response of WT pyramidal cell model (*upper left*) during ZAP current stimulus with a DC component of 0 A and an AC amplitude of 10 pA, varying linearly in frequency from 0 to 20 Hz over 10 s (both *left and right lower panel*, reproduced twice for clarity). This stimulus was chosen such that the response matched the range of membrane potentials exhibited by the model cell in the full theta-generating network (centered around –60 mV). The same stimulus was provided to model simulating ion channel expression modifications (*top right*) corresponding to changes observed in mouse models overproducing Aβ, exhibiting a weaker response at theta frequency stimuli around 6 Hz (at 3 s) relative to other frequencies, in addition to exhibiting an overall hyperpolarized response, both of which potentially contribute to theta rhythm attenuation in the full network

found a lowering of response to theta frequencies of the stimulus in cell models with potassium channel gain and sodium channel loss (Fig. 7). This reduction in frequency preference could contribute to loss of rhythmicity in spikes in pyramidal cells in the context of the oscillating network, and the lower baseline response at all frequencies after sodium channel loss could contribute to the reduction in frequency of the network (Fig. 7). These results suggest cellular events in pyramidal cells that affect sodium and potassium channel expression in early Alzheimer's disease mouse models may contribute to the early reduction in theta rhythm observed experimentally as in (Figs. 3, 5 and 10).

Altered Synaptic Plasticity

The constructed function Ω_{new} alters synaptic plasticity. This is done by simply replacing Ω in Eq. 4 with Ω_{new} in Eq. 5. *LTP* and *LTD* in Eq. 5 are interpreted as activity of each process. In Eq. 5, $\alpha_{1,2,3}$ characterize calcium concentration when

LTP and LTD processes are active. LTD process is active when the intracellular calcium concentration is between α_1 and α_2. When the calcium concentration is higher than α_3, LTP process is active. $k_{1,2}$ is activity coefficient for LTP and LTD respectively. If LTD process were blocked entirely, k_2 would become zero. β and $\beta_{1,2,3}$ determine the steepness of the sigmoid functions. ϵ is related to the initial synaptic strength. Note that there are only 5 parameters in Eq. 4: $\alpha'_{1,2}$, $\beta'_{1,2}$ and γ'. We refer to them using primed letters to distinguish them from parameters of Ω_{new}. For Ω_{new}, however, there are 10 parameters: $\alpha_{1,2,3}$, $\beta_{1,2,3}$, $k_{1,2}$, β, and ϵ. We can construct the parameter set which exhibits normal synaptic behavior by choosing $\alpha'_1 = \alpha_1$, $\alpha'_{2,3} = \alpha_2$, $\beta'_{1,2,3} = \beta_{1,2}$, and $\epsilon = \log(\frac{1}{\gamma'} - 1)$. The next step is to find a parameter set that exhibits the pathological synaptic plasticity. The pathological parameter set should produce results which correspond to the stylized facts summarized in the section "Aβ Effects on Synaptic Plasticity: Brief Summary of the Experimental Background". The Aβ-induced basal synaptic depression can be modeled by decreasing ϵ. To enhance the LTD, α_1 is decreased and α_2 is increased. This change widens the region of the calcium level in which LTD occurs. The simulation results for both normal and pathological cases are shown in Figs. 11, 12, 13 and 14. Aβ facilitates LTD after a subthreshold LTD induction (Fig. 11). Figure 12 illustrates LTD for both normal and pathological cases. The pathological parameter set exhibits stronger LTD than the normal one. When the activity coefficient for the LTD process is small, LTD disappears even if LTD is induced. Figure 14 illustrates LTP for both normal and pathological cases. The induced LTP for the pathological parameter set is clearly impaired compared to the normal parameter set. If LTD-related pathways were blocked, LTP would be no longer impaired. Again, the key idea of Ω_{new} is a

Fig. 8 *Top panel* represents firing raster plots of the pyramidal cell population in wildtype simulations, and bottom panel shows firing for 8 month effects exhibiting reduced synchrony. Time windows in which two spikes are considered synchronous was 10 ms. Significantly lower coherence was observed across all simulated transgenic mice group in comparison to wildtype simulations (ANOVA, F4,35 = 3.72, p = 0.0127, not showed here)

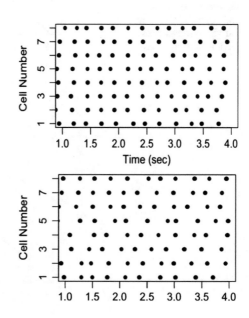

competition between the LTD and LTP processes. $A\beta$ enhances the LTD but not the
LTP process, leading to an impairment of the LTP. Blocking the LTD pathways may
remove the LTP impairment.

Biophysical Backgrounds of the Effects of $A\beta$ on Ω Function

Kinetic Models: Some Remarks

Many kinetic models have been constructed to explain the molecular mechanisms
behind bidirectional synaptic plasticity. Since there are many (the order of magni-

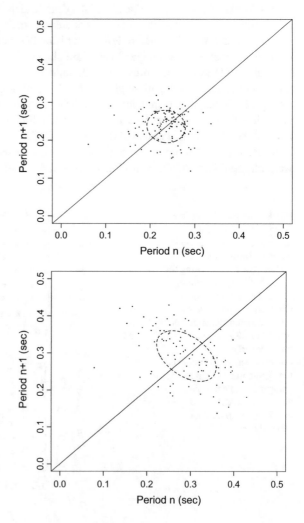

Fig. 9 Greater variation perpendicular to the line of identity representing greater consecutive period variation was observed in addition longer interspike intervals when effects of $A\beta$ at the 8 month stage are implemented (*lower*) than in the baseline network (*upper*). Axes of ellipses represent standard deviation in each direction. Points represent spike intervals of eight pyramidal cells in individual representative random network trials

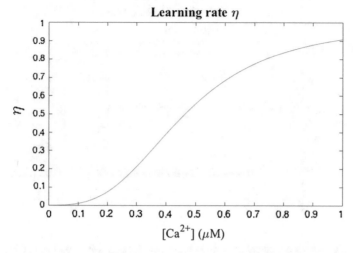

Fig. 10 The learning rate function η in Eq. 1 is plotted against the intracellular calcium concentration

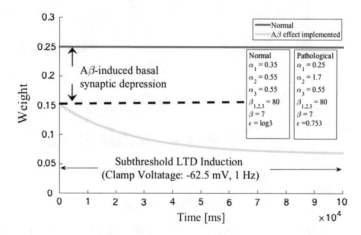

Fig. 11 Subthreshold LTD induction produces no change in the weight when Ω_{new} with a normal parameter set replaced Ω in Eq. 1; however, it induced LTD for Ω_{new} with a pathological parameter set

tude is about hundred) molecules involved in the biochemical pathways of synaptic plasticity, a huge number of mathematical models have been suggested (for reviews, see [61–63]). A subset of these models explain bidirectionality.

In [64] both a phenomenological kinetic model and a biophysical (but simplified) model of the phosphorylation cycle were given to derive the functional form of the Ω function. The phenomenological model is based on the insertion/removal of postsynaptic AMPARs. Both models paved the road towards the understanding of the

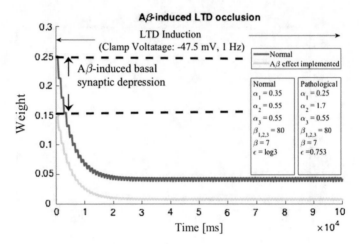

Fig. 12 LTD induction protocol properly induced LTD for Ω_{new} with a normal and pathological parameter set

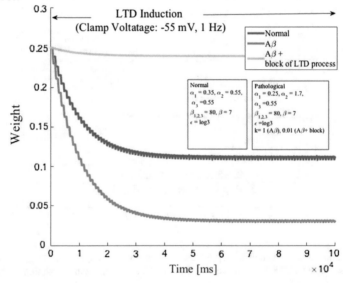

Fig. 13 LTD induction protocol properly induced LTD for Ω_{new} with a normal and pathological parameter set. Block of the LTD process (k = 0.01) properly prevented the LTD induction

molecular basis of the calcium control hypothesis, but did not give a full description.

In our own studies we specifically chose to use three kinetic models with increasing complexity studied by [65]. These models have been constructed to grasp the requirements of calcium-induced bidirectional synaptic plasticity. The models use a signal molecule S^* (the Ca^{2+}/calmodulin complex), which activates two different pathways to control the production of either the active conformation R^* or the inac-

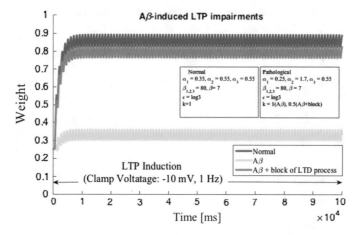

Fig. 14 LTP was induced for Ω_{new} with a normal and pathological parameter set. LTP was observed for the normal, and the LTP was impaired for the pathological case. The LTP was once again observed when the LTD process was blocked

tive conformation R of a response molecule, respectively. The two pathways can be identified as the phosphorylation and dephosphorylation pathways associated with potentiation and depression of the AMPAR activity, respectively.

Kinetic Modeling of Normal and Pathological Ω Function

Figure 15 shows the three kinetic models with increasing complexity (from [65]). The Model I and II consist of an activator A and an inhibitor I which are modulated by a signal molecule S*. If the activation and inactivation of a response molecule

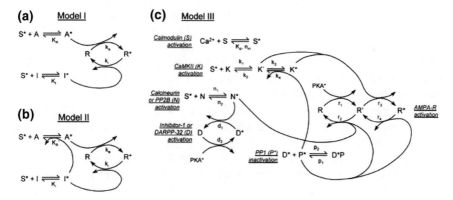

Fig. 15 d'Alcantara's kinetic model of phosphorylation

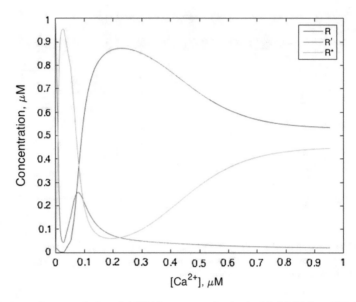

Fig. 16 A steady state solution of AMPAR concentration in the Model III from [65] is *plotted* against calcium concentration. The Ω function can be obtained by scaling the R^* *curve*

R were regulated through independent pathways (Model I), the system would be biphasic [65]. The concentration of the response molecule in an active conformation would be modeled by a sigmoid function. If the stimulation were regulated through interdependent pathways (Model II), the system would exhibit a U-shaped biphasic response [65].

The Model III in Fig. 15 is an extension of Model II, and would also exhibit a U-shaped response (Fig. 16). Assuming that the binding of Ca^{2+} to calmodulin is always at equilibrium, the stationary solution of the system can be calculated. See [65] for the details. We may construct the Ω function by scaling the R^* curve in Fig. 16. The Ω function constructed by this process is comparable to the Ω function in Eq. 5. Remember that the Ω function in Eq. 5 assumes the competition of LTP and LTD activity. In the current model, the phosphorylation and dephosphorylation pathways regulate the potentiation of AMPAR activity. It is noteworthy that the interpretation of the Ω function in Eq. 5 is consistent, and the current kinetic model produces a curve similar to the Ω function.

$A\beta$ impairs the LTP and enhances LTD while it affects only LTD-related pathways. There are essentially 9 kinetic parameters to solve for a steady-state solution of the Model III: K_d, n_H, k_1/k_2, k_3/k_4, r_1/r_2, r_3/r_4, n_2/n_1, d_1/d_2, and p_1/p_2. Out of these 9 parameters, r_1/r_2, r_3/r_4, n_2/n_1, d_1/d_2, and p_1/p_2 are relevant parameters for dephosphorylation. To ease computation, we have only focused on r_3/r_4, n_2/n_1, and p_1/p_2 to implement the effect of $A\beta$. We may strengthen the depression of the AMPAR activity by decreasing r_3/r_4 and n_2/n_1, and increasing p_1/p_2. Figure 17 shows the R^* against calcium concentration for the normal and pathological cases.

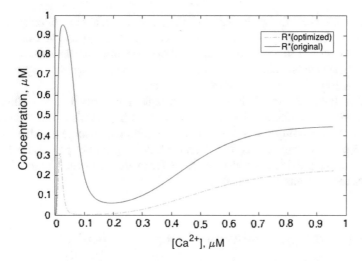

Fig. 17 Decreasing $r3/r4$ widens the region of LTD, and impairs LTP strength. (This does not explain the subthreshold LTD induction). Increasing $p1/p2$ achieves the subthreshold LTD induction

The parameters for the pathological R^* are optimized so that the Ω function should achieve certain features (1) decrease in the LTD threshold (2) impairment of LTP. The optimized parameter set does not only achieve these features, but also increase the LTD strength. The depression of the AMPAR activity with the pathological parameter set is greater than that with the normal set. This is not originally considered in the Ω function in Eq. 5; however, it agrees with the stylized facts because it enhances LTD.

Further Plans

This chapter constitutes the first step toward the implementation of the proposal sketched in section "General Remarks". We have a plan to build a two-stage model.

First, the model should connect synaptic events to altered network dynamics. Computational simulations will support the existence of a **causal relationship** between two Aβ induced phenomena, namely (1) reduced excitatory transmission and plasticity at the synaptic level; (2) epileptiform activity at the network level. According to our best knowledge, no computational models have addressed the problem of falsify/support hypotheses on the causal relationship between synaptic depression and aberrant network activity. One main hypothesis to be tested assumes that Aβ-induced increases in excitatory network activity lead to synaptic depression by a **homeostatic plasticity** compensatory mechanism. (Homeostatic plasticity is interpreted as "staying the same through change"). Homeostatic plasticity is a neural implementation of a feedback control strategy with the goal of stabilizing the firing rate by changing such synaptic parameters as receptor density and synaptic strength [15, 16]. Homeostatic plasticity is supposed to compensate for the unstable features of Hebbian synapses. Failure of this stabilizing mechanism may imply hyperactivity, hypersynchronization and epileptiform activities.

Second, the model should be extended to explain the transition from altered network activity to cognitive deficit. The main working hypothesis of this second stage is that seizures amplify the process of AD progression by some positive feedback mechanisms involving Aβ deposition and cell death [2]: both pre- and postsynaptic mechanisms provide the molecular bases for modeling of such kinds of positive feedback mechanisms.

Messages for Neurologists and Computer Scientists

A significant amount of data and hypotheses about the neural mechanisms of the interaction of these diseases have been accumulated. Computational models have proved to be efficient tools to test working hypotheses about normal and pathological neural mechanisms. Such kinds of models offer an integrative perspective to organize scattered data obtained by methods of anatomy, electrophysiology, brain imaging, neurochemistry, behavioral studies, etc. into a coherent picture.

Our specific message for neurologists is that computational platform under development constitutes an appropriate tool to test the hypotheses related to the potential mechanisms and **multiple effects** of **elevated levels of human amyloid precursor protein related β-amyloid** (Aβ) [5, 10].

The main specific message for computer (better saying computational) scientists is that combining different neural models, such as compartmental techniques, phenomenological and biophysically detailed descriptions of synaptic plasticity including biochemical kinetic models, network models of synchronized activity, memory models will undoubtedly help uncovering the hidden links between epilepsy and Alzheimers Disease.

Acknowledgements PE thanks to the Henry Luce Foundation to let him serve as a Henry R. Luce Professor.

References

1. Weintraub S, Wicklund AH, Salmon DP. The Neuropsychological Profile of Alzheimer Disease. Cold Spring Harbor Perspectives in Medicine. 2012; 2(4):a006171. doi:10.1101/cshperspect.a006171.
2. Noebels JL. A Perfect Storm: Converging Paths of Epilepsy and Alzheimer's Dementia Intersect in the Hippocampal Formation. Epilepsia 2011; 52(Suppl 1):39–46.

3. Romanelli MF, Morris JC, Ashkin K, Coben LA. Advanced Alzheimer's disease is a risk factor for late-onset seizures. Arch Neurol. 1990; 47(8):847–850.
4. Scarmeas N, Honig LS, Choi H, et al. Seizures in Alzheimer Disease: Who, When, and How Common? Archives of neurology. 2009; 66(8):992–997. doi:10.1001/archneurol.2009.130.
5. Palop JJ, Mucke L.: Amyloid-β Induced Neuronal Dysfunction in Alzheimer's Disease: From Synapses toward Neural Networks. Nature neuroscience 13(812–818), 2010.
6. Chin J and Scharfman HE.: Shared cognitive and behavioral impairments in epilepsy and Alzheimer's disease and potential underlying mechanisms. Epilepsy Behav. 2013 Mar; 26(3):343–51.
7. Vossel KA., Beagle AJ, Rabinovici GD, Shu H, Lee SE, Naasan G, Hegde M, Cornes SB, Henry ML, Nelson AB, Seeley WW, Geschwind MD, Gorno-Tempini ML, Shih T, Kirsch HE, Garcia PA, Miller BL, Mucke L. Seizures and Epileptiform Activity in the Early Stages of Alzheimer Disease. JAMA Neurol. 2013 September 1; 70(9): 1158–1166. doi:10.1001/jamaneurol.2013.136
8. Palop JJ, Chin J, Roberson ED, Wang J, Thwin MT, Bien-Ly N, Yoo J, Ho KO, Yu GQ, Kreitzer A, Finkbeiner S, Noebels JL, Mucke L. Aberrant excitatory neuronal activity and compensatory remodeling of inhibitory hippocampal circuits in mouse models of Alzheimer's disease. Neuron, September 2007 doi:10.1016/j.neuron.2007.07.025
9. Magaki S, Yong WH, Khanlou N, Tung S, Vinters HV. Comorbidity in dementia: update of an ongoing autopsy study. J Am Geriatr Soc. 2014 Sep; 62(9):1722–8. doi:10.1111/jgs.12977. Epub 2014 Jul 15.
10. Palop JJ, Mucke L.:Epilepsy and cognitive impairments in Alzheimer disease. Arch Neurol. 2009 Apr; 66(4):435–40.
11. Orbán G, Kiss T, Lengyel M, Érdi P. Hippocampal rhythm generation: gamma-related theta-frequency resonance in CA3 interneurons. Biol Cybern. 2001 Feb; 84(2):123–32.
12. Hajos M, Hoffmann WE, Orbán G, Kiss T and Érdi P: Modulation of septo-hippocampal θ activity by GABA$_A$ receptors: An experimental and computational approach. *Neuroscience*, **126** (2004) 599–610
13. G. Orbán, T. Kiss and P. Érdi: Intrinsic and synaptic mechanisms determining the timing of neuron population activity during hippocampal theta oscillation. Journal of Neurophysiology 96(6) (2006) 2889–2904
14. Shouval, H. Z., Bear, M. F., and Cooper, L. N. (2002). A unified theory of NMDA receptor-dependent bidirectional synaptic plasticity. Proc. Natl. Acad. Sci. USA, 99:10831–6.
15. Turrigiano GG, Leslie KR, Desai NS, Rutherford LC, Nelson SB (1998). "Activity-dependent scaling of quantal amplitude in neocortical neurons." Nature. 391:892–6.
16. Turrigiano G and Neslon S:Homeostatic plasticity in the developing nervous system Nature Reviews Neuroscience 5, 97–107, 2004.
17. Götz J and Ittner LM: Animal models of Alzheimer's disease and frontotemporal dementia. Nature Reviews Neuroscience 9, 532–544, July 2008
18. Kesner RP, Gilbert PE and Wallensteing, GV: Testing neural network models of memory with behavioral experiments. Current Opinion in Neurobiology 10(260–265), 2000.
19. Pena-Ortega F, Amyloid beta protein and neural network dysfunction, Journal of Neurodegenerative Diseases, July 2013; (2013), 657470
20. Érdi P, Kiss T and Ujfalussy B : Multi-level models. In Hippocampal Microcircuits, eds. Vassilis Cutsuridis, Bruce Graham, Stuart Cobb, and Imre Vida. Springer 2010. pp. 527–554.
21. Lacaille JC, Mueller AL, Kunkel DD, and Schwartzkroin PA: Local circuit interactions between oriens/alveus interneurons and CA1 pyramidal cells in hippocampal slices: electrophysiology and morphology. *J Neurosci*, **7** (1987) 1979–1993
22. Jinno S and Kosaka T: Immunocytochemical characterization of hippocamposeptal projecting gabaergic nonprincipal neurons in the mouse brain: a retrograde labeling. *Brain Res*, **945** (2002) 219–231
23. Lacaille JC, Williams S: Membrane properties of interneurons in stratum oriens-alveus of the CA1 region of rat hippocampus in vitro. *Neuroscience* **36** (1990) 349–359

24. Katona I, Acsády L, Freund TF: Postsynaptic targets of somatostatin-immunoreactive interneurons in the rat hippocampus. *Neuroscience* **88** (1999) 37–55
25. Freund TF and Buzsáki G: Interneurons of the hippocampus. *Hippocampus*, **6** (1996) 347–470
26. Freund TF and Antal M: Gaba-containing neurons in the septum control inhibitory interneurons in the hippocampus. *Nature*, **336** (1998) 170–173
27. Varga V, Borhegyi Zs, Fabo F, Henter TB, and Freund TF: In vivo recording and reconstruction of gabaergic medial septal neurons with theta related firing. *Program No. 870.17. Washington, DC: Society for Neuroscience.*, 2002.
28. Wang XJ and Buzsáki G: Gamma oscillation by synaptic inhibition in a hippocampal interneuron network model. *J Neurosci*, **16** (1996) 6402–6413
29. Scott L, Feng J, Kiss T, Needle E, Atchison K, Kawabe TT, Milici AJ, Hajos-Korcsok E, Riddell D, Hajos M (2012) Age-dependent disruption in hippocampal theta oscillation in amyloid-beta overproducing transgenic mice. Neurobiology of aging 33(7):1481 e1413–1423.
30. John T, Kiss T, Lever C and Érdi P: Anxiolytic Drugs and Altered Hippocampal Theta Rhythms: The Quantitative Systems Pharmacological Approach. Network: Computation in Neural Systems March-June 2014, Vol. 25, No. 1–2 , Pages 20–37.
31. John, T: Early Cellular Changes Related to Overproduction of Alzheimer's Amyloid-β Alter Synchronized Network Activity in Computational Model of Hippocampal Theta Rhythm Generation. (BA thesis, Kalamazoo College 2015)
32. Palop JJ, Mucke L.: Synaptic depression and aberrant excitatory network activity in Alzheimer's disease: two faces of the same coin? Neuromolecular Med. 2010 Mar; 12(1):48–55.
33. Lisman, J. A. (1989). A mechanism for the Hebb and the anti-Hebb processes underlying learning and memory. Proc Natl Acad Sci USA, 86:9574–9578.
34. Stargardt A, Swaab DF, Bossers K. Storm before the quiet: neuronal hyperactivity and A*beta* in the presymptomatic stages of Alzheimer's disease. Neurobiol Aging. 2015 Jan; 36(1):1-11. doi:10.1016/j.neurobiolaging.2014.08.014.
35. Born HA: Seizures in Alzheimer's disease. Neuroscience. 2015 Feb 12; 286:251–63. doi:10. 1016/j.neuroscience.2014.11.051. Epub 2014 Dec 4.
36. Sanchez PE, Zhu L, Verret L, Vossel KA, Orr AG, Cirrito JR, Devidze N, Ho K, Yu GQ, Palop JJ, Mucke L.: Levetiracetam suppresses neuronal network dysfunction and reverses synaptic and cognitive deficits in an Alzheimer's disease model. Proc Natl Acad Sci U S A. 2012 Oct 16; 109(42):E2895–903.
37. Vossel KA, Beagle AJ, Rabinovici GD, Shu H, Lee SE, Naasan G, Hegde M, Cornes SB, Henry ML, Nelson AB, et al. Seizures and epileptiform activity in the early stages of Alzheimer disease JAMA Neurol. 2013 Sep 1; 70(9):1158–66.
38. Bakker A, Krauss G.L., Albert M.S., Speck C.L. , Jones L.R., Stark C.E., Yassa M.A., Bassett S.S.,Shelton A.L., Gallagher M. i Reduction of hippocampal hyperactivity improves cognition in amnestic mild cognitive impairment Neuron, 74 (2012), pp. 467–474
39. I. Aradi and P. Érdi: Computational neuropharmacology: dynamical approaches in drug discovery. Trends in Pharmacological Sciences 27(5) (2006) 240–243
40. P. Érdi, Zs. Huhn, T. Kiss: Hippocampal theta rhythms from a computational perspective: code generation, mood regulation and navigation Neural Networks 18(9) (2005) 1202–1211
41. Kiss T, Orbán G and Érdi P: Modelling hippocampal theta oscillation: applications in neuropharmacology and robot navigation. International Journal of Intelligent Systems 21(9) (2006) 903–917 (2006)
42. P. Érdi and J. Tóth: Towards a dynamic neuropharmacology: Integrating network and receptor levels. In: Brain, Vision and Artifical Intelligence. M. De Gregorio, V. Di Maio, M. Frucci and C. Musio (eds). Lecture Notes in Computer Science 3704, Springer, Berlin Helidelberg 2005, pp. 1–14.
43. T. Kiss and P.Érdi: From electric patterns to drugs: perspectives of computational neuroscience in drug design. BioSystems 86(1–3) 46–52 (2006)
44. Érdi P, Kiss T, Tóth J, Ujfalussy B and Zalányi L: From systems biology to dynamical neuropharmacology: Proposal for a new methodology. IEE Proceedings in Systems Biology 153(4) 299–308 (2006)

45. B. Ujfalussy, T. Kiss, G. Orbán, WE. Hoffmann, P. Érdi and M. Hajós: Pharmacological and Computational Analysis of alpha-subunit Preferential GABAA Positive Allosteric Modulators on the Rat Septo-Hippocampal Activity. Neuropharmacology 52(3) (2007) 733–743
46. P. Érdi, B. Ujfalussy, L. Zalányi, VA. Diwadkar: Computational approach to schizophrenia: Disconnection syndrome and dynamical pharmacology. In: A selection of papers of The BIO-COMP 2007 International Conference L. M. Ricciardi (ed.) Proceedings of the American Institute of Physics 1028, 65–87
47. Érdi P, John T, Kiss T. and Lever CP: Discovery and validation of biomarkers based on computational models of normal and pathological hippocampal rhythms. In: Validating Neuro-Computational Models of Neurological and Psychiatric Disorders. Eds: Bhattacharya B and Chowdhury F. Springer, Computational Neuroscience series (in press)
48. Verret L, Mann EO, Hang GB, Barth AM, Cobos I, Ho K, Devidze N, Masliah E, Kreitzer AC, Mody I, Mucke L, Palop JJ.: Inhibitory interneuron deficit links altered network activity and cognitive dysfunction in Alzheimer model. Cell. 2012 Apr 27; 149(3):708–21.
49. Esposito Z, Belli L, Toniolo S, Sancesario G, Bianconi C, Martorana A Amyloidβ, Glutamate, Excitotoxicity in Alzheimer's Disease: Are We on the Right Track? CNS: Neuroscience & Therapeutics 19(549–555), August 2013
50. Li S, Hong S, Shepardson NE, Walsh DM, Shankar GM, Selkoe D. Soluble oligomers of amyloid Beta protein facilitate hippocampal long-term depression by disrupting neuronal glutamate uptake. Neuron. 2009 Jun 25;62(6):788–801. doi:10.1016/j.neuron.2009.05.012.
51. Danysz W, Parsons GC. Alzheimer's disease, beta amyloid, glutamate, NMDA receptors and memantine. Br J Pharm 2012; 167:324–352
52. Shankar GM and Walls GM: Alzheimer's disease: synaptic dysfunction and $A\beta$. Molecular Neurodegeneration 2009, 4:48
53. Castellani GC, Quinlan EM, Cooper LN and Shouva HZ: A biophysical model of bidirectional synaptic plasticity: Dependence on AMPA and NMDA receptors. Proc. Natl. Acad. Sci. USa, 98:12772–7.
54. Cormier R, Greenwood AC, Connor JA (2001) Bidirectional synaptic plasticity correlated with the magnitude of dendritic calcium transients above a threshold. J Neurophysiol 85:399–406
55. Lisman J. (1994): The CaM kinase II hypothesis for the storage of synaptic memory. Trends Neurosci. 1994 Oct;17(10):406–12.
56. Wykes R, Kalmbach A, Eliava M, et al. Changes in physiology of CA1 hippocampal pyramidal neurons in preplaque CRND8 mice. Neurobiology of aging. 2012; 33: 1609–1623. doi:10.1016/j.neurobiolaging.2011.05.001
57. Ramos B, Baglietto-Vargas D, Carlos del Rio J, et al. Early neuropathology of somatostatin/NPY GABAergic cells in hippocampus of a PS1xAPP transgenic model of Alzheimers disease. Neurobiology of aging. 2006; 27:1658–1672. doi:10.1016/j.neurobiolaging.2005.09.022
58. Kerrigan TL, Brown JT, Randall AD. Characterization of altered intrinsic excitability in hippocampal CA1 pyramidal cells of the ABeta-overproducing PDAPP mouse. Neuropharmacology. 2014; 79: 515–524. doi:10.1016/j.neuropharm.2013.09.004
59. Hutcheon B, Yarom Y. Resonance, oscillation and the intrinsic frequency preferences of neurons. Trends Neurosci.2000; 23:216–222.
60. Hu H, Vervaeke K, Storm J. Two forms of electrical resonance at theta frequencies generated by M-current, h-current and persistent Na+ current in rat hippocampal pyramidal cells. Journal of physiology. 2002; 545(3):783–805. doi:10.1113/jphysiol.2002.029249
61. Manninen T, Hituri K, Kotaleski JH, Blackwell KT, Linne M-L. (2010): Postsynaptic Signal Transduction Models for Long-Term Potentiation and Depression Comput Neurosci. 4: 152.
62. Manninen T, Hituri K, Toivari E, Linne M-L. Modeling Signal Transduction Leading to Synaptic Plasticity: Evaluation and Comparison of Five Models. EURASIP Journal on Bioinformatics and Systems Biology. 2011; 2011(1):797250. doi:10.1155/2011/797250.
63. He Y, Kulasiri D, Samarasinghe S. (2014): Systems biology of synaptic plasticity: A review on N-methyl-d-aspartate receptor mediated biochemical pathways and related mathematical models. BioSystems 122(7–17)

64. Shouval, H. Z., Castellani, G. C, Blais, B. S., Yeung , L. C.,Cooper, L. N. (2002). Converging evidence for a simplified biophysical model of synaptic plasticity. Biol. Cyb. 87: 383–91.
65. D'Alcantara P, Schiffmann S and Swillens S (2003): Bidirectional synaptic plasticity as a consequence of interdependent Ca2+-controlled phosphorylation and dephosphorylation pathways European Journal of Neuroscience 17 2521–2528

Dynamic Causal Modelling of Dynamic Dysfunction in NMDA-Receptor Antibody Encephalitis

Richard E. Rosch, Gerald Cooray and Karl J. Friston

Introduction

The invention of electroencephalography (EEG) in the early 20th century facilitated a revolutionary insight into the dynamic behaviour of the human brain. For the first time, clinicians and researchers were able to examine direct evidence of brain function, in both normal human participants and patients with neurological conditions [36]. Clinically this new way of assessing brain function has had significant impact on our understanding of a number of neurological and psychiatric conditions, but none more so than epileptic seizure disorders.

Epilepsy is the label given to a number of heterogeneous conditions characterised by an enduring risk of epileptic seizures—because of their heterogeneity, these conditions are now often referred to as 'the epilepsies' [51]. The term epileptic seizure describes the transient occurrence of signs and symptoms caused by abnormally excessive or synchronous activity in the brain [17]. Whilst the concept that abnormal electrical activity causes epileptic seizures predates the invention of

R.E. Rosch (✉) · G. Cooray · K.J. Friston
Wellcome Trust Centre for Neuroimaging, Institute of Neurology,
University College London, London, UK
e-mail: r.rosch@ucl.ac.uk

G. Cooray
e-mail: Gerald.Cooray@ki.se

K.J. Friston
e-mail: k.friston@ucl.ac.uk

R.E. Rosch
Centre for Developmental Cognitive Neuroscience,
Institute of Child Health, University College London, London, UK

G. Cooray
Department of Clinical Neurophysiology, Karolinska University Hospital,
Stockholm, Sweden

© Springer International Publishing AG 2017
P. Érdi et al. (eds.), *Computational Neurology and Psychiatry*,
Springer Series in Bio-/Neuroinformatics 6,
DOI 10.1007/978-3-319-49959-8_6

EEG, much of our current physiological understanding and clinical decision making is based on EEG recordings from patients suffering from epilepsy [16].

Descriptions and analyses of EEG recordings have remained virtually unchanged since its conception. Its clinical use largely rests on the description of visually recognisable features and their phenomenological categorisation, with the exception of some recently adopted advanced source localisation algorithms [63]. But relying just on these visually apparent pathological patterns does not capture the entire breadth of information that is available in an EEG recording.

One of the main advantages of EEG (which it shares with magnetoencephalography, MEG) over other methods assessing brain function is its temporal resolution, which still remains unparalleled when it comes to investigating the human brain in vivo. This results in rich datasets, which capture interacting fluctuations of electric activity across frequencies that may be two or three orders of magnitude apart. Whilst surface EEG recorded non-invasively from the scalp has a limited spatial resolution, it does allow for the simultaneous recording of neuronal activity across almost the entirety of the cortical surface. Furthermore, the spatial resolution limitations have been addressed by developing invasive EEG recording devices that can be implanted neurosurgically where a better understanding of the spatial origins of an EEG signal are deemed clinically necessary.

Much of the information contained within these datasets is not accessible through visual inspection alone, but rather needs to be elicited utilising more quantitative analysis methods [59]. Applying such quantitative analysis methods has led to the description of a wide variety of novel electrophysiological findings. For example, analysing the correlation between the time series of individual EEG channels will yield a matrix of channel-to-channel correlation measures. These can be read as indicators of *functional connectivity*, with the results interpreted in a graph theory framework as *functional network*; analysis [5]. Another example can be found in the recent emergence of cross-frequency coupling as a potentially important mechanism for neuronal computation: Quantitative analysis of power at different EEG frequencies in humans has shown that amplitude fluctuations are measurably modulated and time locked to the phase of concurrent slower frequency oscillation, known as phase-amplitude cross-frequency coupling [6].

At the same time as advances in the detailed quantitative analysis of macroscopic EEG signals in health and disease, there has been an exponential increase in our understanding of the molecular, and to some extend cellular basis of many of the epilepsies [30, 58]. Increasingly, knowledge of associated molecular abnormalities, such as the presence of relevant gene mutations or specific autoantibodies against synaptic targets, influences prognosis, clinical management and specific treatment decisions for patients with epilepsy.

Features of these disease-associated molecular abnormalities have also lead to a putative understanding of the pathophysiological mechanisms underlying different epileptic seizure disorders. For example, the frequency of mutations in ion channel genes has led to the concept of epilepsies and other paroxysmal neurological disorders being *channelopathies*, i.e. disorders in neuronal ion channel function [56], which can be further investigated in appropriate animal model systems.

However, with the increase in knowledge, new challenges arise. The availability and increased clinical use of testing for specific mutation and autoantibodies has quickly led to the realisation that even apparently specific molecular abnormalities are associated with a wide variety of disease pictures in human patients. The same mutation in the voltage gated sodium channel gene *SCN1A* for example, can cause a diverse selection of phenotypes within the same family: ranging from comparatively mild phenotypes consisting of childhood febrile seizures, to a severe epileptic disorder characterised by difficult to control frequent daily seizures associated with global developmental delay [46]. Similarly, mutations in the *GRIN2A* gene, coding for a subunit of the N-methyl-D-aspartate receptor (NMDAR), can cause a range of electroclinical syndromes, even within the same family [40].

This opens an explanatory gap: On one side there is an increased understanding of the putative molecular and cellular causes of dynamic disorders of the brain; on the other there are macroscale measures of abnormal brain function that, whilst loosely associated with some microscale abnormalities, do not allow a direct one-to-one mapping. Bridging this gap is likely to require an intermediate step—a conceptual bridge that can link information about molecular dysfunction with its expression in neuronal function at an intermediate level, the *mesoscale*, in order to understand the emergence of phenotypic variability observed in human patients.

In other fields within neuroscience, this approach is emerging as a necessary step for linking observations at the microscale (e.g. cellular neuronal circuits) with the observations at the macroscale (e.g. organism behaviour). Whilst descriptions of the cellular circuitry may include too many particular details to understand their functional contributions to the overall behaviour, evaluating only a behaviour in the whole organism may not sufficiently represent the complexity of the underlying neuronal processes. Bridging this gap requires the identification recurrent themes at an intermediate functional level, such as neuronal computations, which may be implemented through differing neuronal circuits, but produce similar effects [7].

In the context of clinical neurology, a similar approach would suggest that in order to link putative molecular causes (microscale) with diverse disease phenotypes observed in patients (macroscale), we need to consider the intermediate step: Dysfunction at the level of neuronal computations (mesoscale, Fig. 1). Linking molecular abnormalities to in vivo neuronal dysfunction is now a standard component of identifying disease mechanisms in emerging genetics and autoimmune conditions [50]. Attempts in relating macroscale findings to models of mesoscale neuronal dynamics have been less forthcoming, and will be the focus here.

This chapter will discuss the use of computational models of neuronal circuit function to link abnormalities in clinical neurophysiological measurements to pathophysiological mechanisms. An introduction of population models of neuronal function will be followed by an in-depth discussion of how such models can be used to make inference on the mechanism underlying specific electrophysiological changes. This approach will then be illustrated using an exemplar clinical case with a known underlying molecular diagnosis of NMDAR antibody encephalitis.

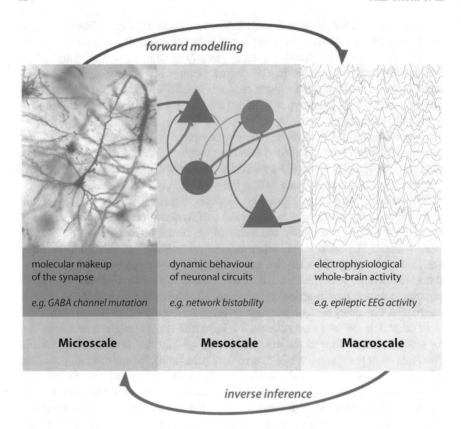

Fig. 1 Understanding epileptic dynamics at different scales: Different lines of evidence lead to descriptions of pathology on different scales. Clinical syndromes often rely on the description of recognisable phenotype at the macroscale. Recent advances in understanding associated molecular abnormalities have improved our pathophysiological understanding of many diverse epilepsies, but robustly linking clinical phenotypes with microscale abnormalities has proven difficult. Including an intermediate consideration of network dynamics may aid both prediction, and allow for addressing the inverse problem of inferring pathophysiology from whole-brain electrophysiological measurements

The Scientific Problem

The functional validation of possible molecular causes of neurological diseases is an essential step in any description of new pathophysiological mechanisms. In order to increase confidence that a genetic mutation that is epidemiologically linked with a specific phenotype could actually play a causative role in the disease, some evidence that this can have an effect on neuronal function is considered a standard requirement [53].

This approach usually relies on replicating the molecular abnormality in a model organism or system and evaluating the model for any resultant deficits, particularly in regards to neuronal function. For example, to provide evidence for the direct pathogenicity of NMDAR antibodies in the recently described NMDAR-antibody associated encephalitis [10], the antibody-rich patient cerebrospinal fluid was applied to murine hippocampal slices prepared for voltage-clamp recordings in order to measure the effects of antibody exposure on glutamate transmission. This provided evidence both for acute antagonism of the NMDAR by the antibody, as well as chronic reduction of NMDAR associated with antibody exposure [33].

Whilst this approach is powerful, and necessary in order to evaluate candidate molecular causes of neurological disorders in the context of neuronal function, several problems remain unsolved when relying on this approach alone:

- **Animal models for human disease**: Model systems used to assess the pathological effects of molecular abnormalities are usually non-human organisms or tissues, leaving uncertainty as to whether similar effects would be evident in the human brain.
- **Emergent properties at different scales**: There may be a gap between individual cell and small circuit abnormalities assessable in an experimental model system, and the inference drawn on larger networks and systems. These models are particularly prone to neglecting emergent properties at different scales (e.g. bistability of a network), and the effects of unknown modulators in the whole system that may enhance or suppress the observed microscale abnormality
- **Human phenotypic variability**: An unexpected result of the recent increase in molecular diagnoses in neurology is the discovery of large phenotypic variability even where a molecular cause has been identified and well characterized [31]. Functional investigations in homogeneous model systems do not address the mechanisms underlying phenotypic diversity.

Issues of disease pathology in humans, understanding whole-organism, and delineating relevant categories within phenotypically diverse groups are essential for translating basic neuroscientific findings into clinically relevant advances. In order to start addressing these issues, the inverse problem has to be addressed: How do macroscale abnormalities relate to underlying pathophysiology?

The EEG signal, despite containing a lot of rich information, is a poor measure of neuronal function at the cellular, or synaptic level: because of the spatial inaccuracies and the summation of many million individual neurons' activity into a composite signal, there are an infinite number of possible neuronal constellations that could cause the same measureable EEG signatures. Attempting to relate this composite, diffuse signal to underlying neuronal dysfunction is thus an ill-posed problem, where no unique solution exists.

Ill-posed problems are common in neuroscience, both in terms of problems researchers encounter when investigating nervous systems (e.g. the source localization problem for EEG signals, [21]), and problems that nervous systems themselves have to address (e.g. the feature binding problem, [42]). These problems are not impossible to solve, as is evident in the successful application of source-reconstruction algorithms in identifying epileptogenic brain areas for surgery [39], and the brain's successful and reliable decoding of visual information [37].

With underdetermined ill-posed problems, providing constraints to the possible solutions is crucial [20]. Constraining the problem reduces the space of possible solutions and makes the problem more tractable. These constraints also help in keeping inverse solutions more interpretable and relevant to the scientific question at hand.

One way to constrain such inverse problems in neuroscience is the use of computational models of neuronal populations as a mesoscale representation of neuronal dynamics. This casts the inverse problem of attempting to infer microscale causes of macroscopic phenomena into a more restricted problem: Assuming basic mechanisms of neuronal function and organization are met (i.e. the neuronal model applies), which setup of the known circuitry could produce an observed effect (i.e. which specific set of model parameters can produce the observed macroscale response?)

In the following, we will use advanced computational methods to address this problem, inferring underlying neuronal circuitry abnormalities from clinical EEG recordings of a patient with a known molecular abnormality: Specifically we will use mesoscale neuronal modelling to assess what synaptic abnormalities underlie paroxysmal EEG abnormalities in a paediatric patient with NMDAR antibody encephalitis.

NMDAR antibody encephalitis is a recently described autoimmune encephalitis [10], i.e. an inflammatory condition of the brain associated with, and most likely caused by, autoantibodies against molecular targets relevant for synaptic function. EEG abnormalities are commonly associated with the condition and have some diagnostic value [25], but are very varied between patients [18] and evolve over time [49]. In paediatric patients the common abnormalities described include non-specific sharp wave paroxysms, longer runs of rhythmic activity and more clearly epileptic spike and wave discharges (see Fig. 2 for examples from our own clinical cohort).

In the following sections we will discuss mesoscale models of neuronal function and how they can be used to explain observed EEG phenomena by constraining the inverse problem. We will then highlight a specific computational approach— *dynamic causal modelling* (DCM)—and apply the method to EEG abnormalities observed in one NMDAR antibody encephalitis patient. We will then discuss our results in terms of their relation to other findings regarding the pathophysiology in NMDAR, as well as further implications for computational methods in the age of exponential discovery of candidate molecular mechanisms in neurology.

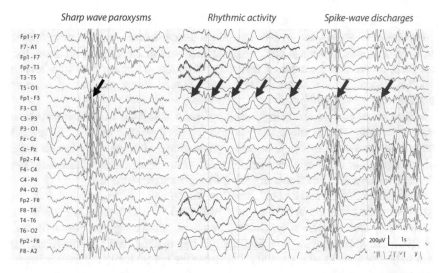

Fig. 2 EEG abnormalities observed in paediatric patients with NMDAR antibodies: The figure collates three different EEG findings from separate paediatric NMDAR antibody encephalitis patients. Abnormalities range from non-specific sharp wave paroxysms (*left panel*), to rhythmic activity with or without impairment of consciousness (*middle panel*) to clearly epileptic spike-wave activity (*right panel*)

Computational Methods

In this section we will discuss how computational models of neuronal function can be used to make inference on causative mechanisms underlying phenomena observed in human EEG recordings. This will be done in three parts—the first will give a short overview of generative models of neuronal function and different approaches to linking them to empirical data. In the second part we will introduce the approach applied to our empirical case—namely *dynamic causal modelling*—in a little more detail. And in the concluding part of this section we will illustrate how dynamic causal modelling can be used to fit a generative neuronal model to empirical EEG data and make inference on underlying mechanisms using an illustrative case of NMDAR encephalitis.

Generative Models of Neuronal Population Activity

Neuronal systems are highly nonlinear coupled systems [61]. This means that predicting input-output relationships is challenging and often counterintuitive. One of the great strengths of computational models is that they can be used to explore input-output relationships systematically and help in identifying some of the unexpected effects produced by nonlinear interactions.

The pioneering work by Hodgkin and Huxley [32] produced one of the first such computational models, and by some measures the most successful one developed so far. Using empirical voltage clamp measurements from the giant squid axon, they elegantly developed a model of neuronal membrane dynamics based entirely on voltage dependent ion channels, that could predict many patterns of neuronal behaviour observed empirically.

Since then, models are being developed on a multitude of different neuronal scales, ranging from subcellular compartment dynamic models, to models describing the output of whole neuronal populations. Because of the spatial scales of measurement, those models that represent whole neuronal populations are particularly informative when relating them to EEG measurements.

One of the earliest such models was the Wilson and Cowan neural mass model [62]—they describe the behaviour of a whole set of interconnected neurons not individually but as whole aggregates, based on similar approaches in particle physics. They also provide a justification for this method based, interestingly, not just in its computational tractability, but rather the conceptually different inference this approach enables:

> It is probably true that studies of primitive nervous systems should be focused on individual nerve cells and their precise, genetically determined interactions with other cells. [...] [S]ince pattern recognition is in some sense a global process, it is unlikely that approaches which emphasize only local properties will provide much insight. Finally it is at least a reasonable hypothesis that local interactions between nerve cells are largely random, but that this randomness gives rise to quite precise long-range interactions [62].

Using this approach they arrive at a system of two ordinary differential equations describing a neuronal oscillator consisting of two different populations, one excitatory, one inhibitory:

$$\tau_e \frac{dE}{dt} = -E + (k_e - r_e E)S_e(c_1 E - c_2 I + P) \tag{1}$$

$$\tau_i \frac{dI}{dt} = -I + (k_i - r_i I)S_i(c_3 E - c_4 I + Q) \tag{2}$$

This system describes two neuronal populations, whose current states (E, I: proportion of cells in the population firing) influence each other through weighted connections (c_{1-4}, weights of population connections, see Fig. 3a). This coupling is mediated through a sigmoid activation function ($S_{e,i}$), which acts like a switch integrating all incoming synaptic influences and translating them into a postsynaptic state change within a defined dynamic range (i.e. 0–1). The sigmoid functions are population specific (and can therefore be parameterised independently) and are the source of non-linearity in the model.

Even despite the extreme simplification of these models of neuronal function, a whole range of dynamic behaviours can be reproduced with WC-type models at the scale of neuronal populations or cortical patches [29, 45, 60]. Because of the

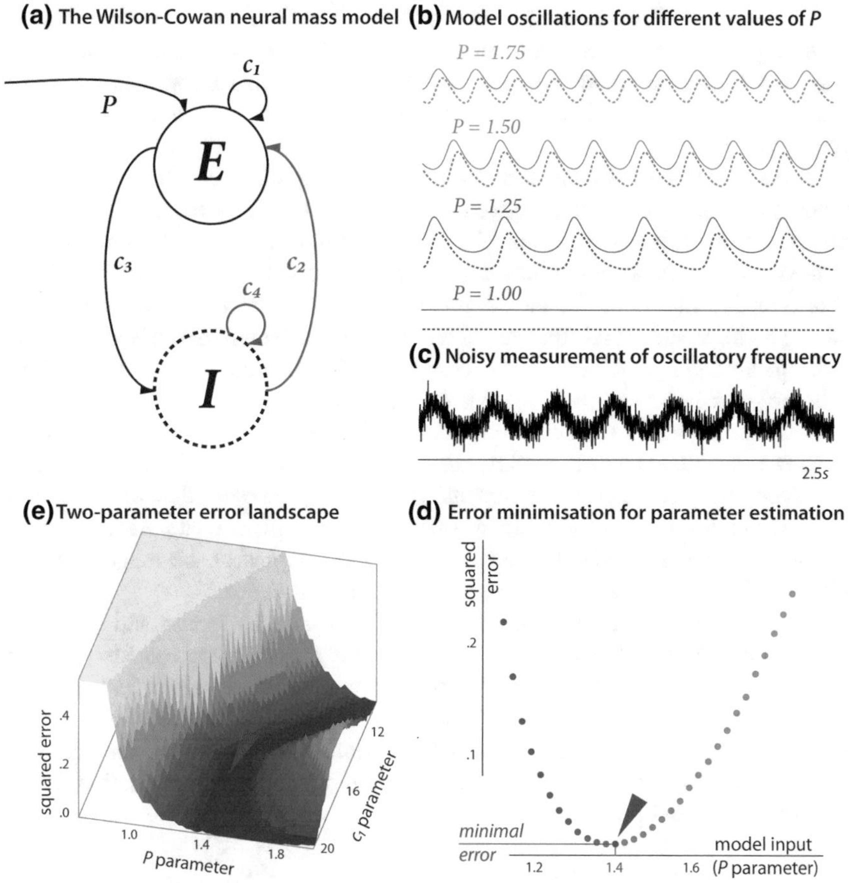

(a) The Wilson-Cowan neural mass model **(b)** Model oscillations for different values of P

$P = 1.75$

$P = 1.50$

$P = 1.25$

$P = 1.00$

(c) Noisy measurement of oscillatory frequency

2.5s

(e) Two-parameter error landscape   **(d)** Error minimisation for parameter estimation

Fig. 3 The Wilson-Cowan neural mass model **a** The model consists of one inhibitory and one excitatory neuronal population, coupled through synaptic connections of a specific connection strength (c_i parameters). These can be excitatory (*black*) or inhibitory (*red*). The system receives external stimulating current input (P parameter) and acts as a neuronal oscillator. **b** The model generates particular oscillation patterns for different parameter constellations—this figure illustrates steady state oscillatory responses with decreasing values for the input parameter P. Excitatory populations are represented by the *solid lines*, inhibitory populations by *dashed lines*. **c** Synthetic data illustrating a noisy measurement of neuronal population oscillation driven with $P = 1.4$. We show how *oscillatory frequency* alone can be used to derive the P-parameter from noisy measurements such as this: **d** Estimates of steady state oscillatory frequency can be derived from the model for a range of different values for P. Plotting the squared difference between estimated frequencies and that derived from the noisy synthetic signal, we can identify the P value that produces the minimal error. This approach identifies $P = 1.4$ as the value producing the minimal error (indicated by the *red arrow*). **e** If more than one parameter is allowed to vary (e.g. input P, and self-excitation strength c_1) the error landscape becomes more complex and error minimisation alone does not produce unambiguous results—the *red arrow* indicates the same parameter constellation identified in 3D). The model specifications were taken directly from the model's original description [62]. Parameters for the modelling were taken from one of the known oscillatory states and unless otherwise stated were: $c_1 = 16$, $c_2 = 12$, $c_3 = 15$, $c_4 = 3$, $P = 1.25$

coupled nonlinearities, however, 'forward' predictions of model behaviour given specific parameters is non-intuitive and usually requires simulation of the model response. Recurrent simulations for varying parameter values can then be used to establish a link between model parameterisation and overall dynamic response (Fig. 3b).

These parameter/response relationship can be exploited to make inference on model parameters underlying a given observations. Faced for example, with noisy measurements of a population oscillation (Fig. 3c), one can use systematic variations of a model parameter to identify the specific parameter value that best fits the data (Fig. 3d, illustrated for the stimulating current parameter P). However, even adding a single additional free parameter (e.g. the connection parameter c_1) creates a complex model prediction error landscape that is much more difficult to optimise (Fig. 3e)—an important problem that we will return to later.

WC models allow for generation of complex dynamics that remain computationally tractable enough to explore different parameter compositions and attempt inference on parameter combinations producing a certain dynamic response. However, in the original formulation consisting of a single excitatory and inhibitory population, they are limited in how well they can represent the range and complexity of cortical dynamics observed in the laminate cortex.

A major extension of the WC model was introduced by Jansen and Rit in [34] building on an extant literature of adaptations of the WC-type models [44]. The Jansen and Rit (JR) model explicitly models dynamics of a local cortical circuit by ascribing different neuronal populations to specific cortical lamina and describing their dynamics in terms of differential equations. In this model, an additional excitatory neuronal population is added allowing separate parameterisation for two excitatory neuronal populations.

$$\tau_e \frac{dx_1}{dt} = H_e(P + S_1(v_2)) - 2x_1 - \frac{v_1}{\tau_e} \tag{3}$$

$$\frac{dv_1}{dt} = x_1 \tag{4}$$

$$\tau_i \frac{dx_2}{dt} = H_i S_2(v_3) - 2x_2 - \frac{v_2}{\tau_i} \tag{5}$$

$$\frac{dv_2}{dt} = x_2 \tag{6}$$

$$\tau_e \frac{dx_3}{dt} = H_e S_3(v_1 - v_2) - 2x_3 - \frac{v_3}{\tau_e} \tag{7}$$

$$\frac{dv_3}{dt} = x_3 \tag{8}$$

This constellation of neuronal populations allows a diverse spectrum of frequency mixtures to be modelled, and is capable of producing a host of response dynamics also observed in empirical measurements of cortical potential fluctuations [1, 27, 34]. An additional benefit that emerges from the laminar specificity of the JR model, is that it relates naturally to commonly available brain recordings in humans—specifically MEG and EEG. The electromagnetic activity measurable at the scalp is thought to mainly reflect postsynaptic currents in the apical dendrites of populations of pyramidal cells [43]. These are explicitly modelled in the JR model, so that their selective contribution to EEG/MEG measurements can be distinguished from the activity of other cell populations.

The laminar specificity and the wide range of physiological frequencies that can be modelled mean that JR-type models are commonly employed in computational models of cortical function. They currently form one of model-based approaches for the analysis of large scale brain dynamics, such as the dynamic causal modelling (DCM) framework, which will be discussed in more detail later [11]. Because they aim to represent biophysical connectivity patterns found in actual cortical micro-circuits, their architecture is also congruent with computational motifs thought to be the basis of cortical processing (e.g. predictive coding, [3]).

Within the DCM framework, several extensions of existing neuronal models have been developed to address specific hypotheses regarding neuronal function [47]. One of the extensions to the classical JR model employed in DCM is the so-called 'canonical microcircuit' or CMC (Fig. 4). Here the single JR pyramidal cell population is separated into distinct 'superficial' and 'deep' pyramidal cells. This allows not only for afferent and efferent projections to be separated into distinct cortical laminae, but also accommodates differences in spectral output among different layers of the same cortical column - that are seen empirically in invasive measurements [4].

The model consists of a simple extension of the differential equations given in Eqs. 3–8. Using the mean-field approximation the model can also be reconceptualised in terms of average membrane potentials and firing rates interacting through specific kernels that summarise the activity-dependent integration of input at the postsynaptic membrane, and the nonlinear transformation of all input into an output firing rate (Fig. 4b) [12].

Because the architecture of the CMC model represents neuroanatomical features of the cortex, most of the modelling parameters are neurophysiologically meaningful and thus easily interpretable. The model parameters can be directly manipulated to reproduce many different dynamic behaviours—increasing the degree of self inhibition in superficial pyramidal cells for example will produce high frequency oscillations (Fig. 5, [52]).

Clearly the more intriguing question is whether inference on the model parameters can be made from empirical measurements, to identify which functional abnormality in the microcircuitry produced an abnormal measurement. This problem usually has more than one possible solution—meaning that many different possible constellations of parameters may cause identical appearing measurements, particularly where only some of the system's states are measureable, or observable

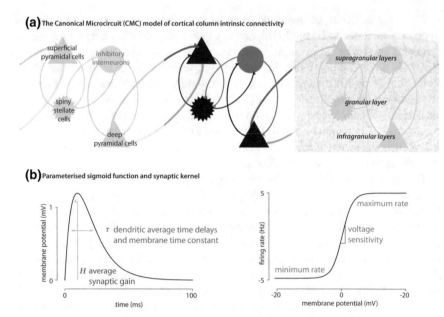

(a) The Canonical Microcircuit (CMC) model of cortical column intrinsic connectivity

superficial pyramidal cells
inhibitory interneurons
spiny stellate cells
deep pyramidal cells
supragranular layers
granular layer
infragranular layers

(b) Parameterised sigmoid function and synaptic kernel

membrane potential (mV)

τ dendritic average time delays and membrane time constant

H average synaptic gain

time (ms)

firing rate (Hz)

maximum rate

voltage sensitivity

minimum rate

membrane potential (mV)

Fig. 4 The Canonical Microcircuit (CMC) Model: **a** This extension of the Jansen-Rit model consists of four neuronal populations (*left panel*) mapping onto different cortical laminae (*right panel*). The *middle panel* shows the intrinsic excitatory (*black*) and inhibitory (*red*) connections contained in the model (for simplicity, recurrent self-inhibition present for each population is not shown). **b** Two operators define the evolution of population dynamics: First a synaptic kernel performs a linear transformation of presynaptic input into an average postsynaptic potential, dispersed over time (*left panel*). This is parameterised by synaptic gain parameters and averaged time constants. Second there is a nonlinear transformation of average membrane potential into population firing rates, described as a parameterised, population-specific sigmoid function (*right panel*)

(e.g. local field potentials), whilst many remain hidden (e.g. intracellular ion concentration fluctuations); the problem is ill-posed. This becomes particularly problematic where many different parameters can be used to explain a limited set of observed states. Even for the WC models with very few free parameters, simple optimisation routines as indicated in Fig. 3d, e quickly become intractable, with complex multidimensional error landscapes that cannot be comprehensively mapped. The flexibility afforded by the increased number of free parameters in the CMC model comes at the cost of increased complexity of the space of possible solutions, making it difficult to evaluate which best explains observed behaviours.

There are several possible approaches to addressing this ill-posed problem, many of which have been employed in the computational modelling of epileptic seizures and EEG abnormalities. In the following, we will introduce a few of these approaches, with a focus on dynamic causal modelling. This will then be applied to address the question as to what abnormalities in the functional architecture can explain the paroxysms observed in patients with NMDAR antibody encephalitis.

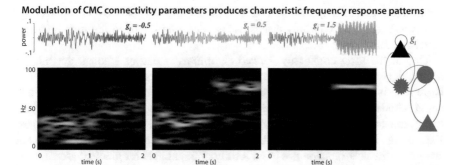

Fig. 5 Changes in intrinsic connectivity produce characteristic responses: gradual changes to the recurrent self-inhibition gain parameter g_i are introduced to a CMC model of the cortical column at around 1 s, changing the intrinsic modulation from 0 to −0.5, 0.5, and 1.5 respectively (*left, middle, right panels*). These changes produce characteristic signatures in the spectral output, apparent in both the time traces (*top panels*), and the spectral densities (*bottom panels*), with increases in self-inhibition leading to high power high frequency oscillations

Model Inversion for EEG Abnormalities

One of the most intuitive strategy to inversely link observed EEG features to changes in the parameters of an underlying generative model—i.e. to *invert* the model—is to systematically vary the parameters and evaluate how well the model simulations then fit the observed measurements. This can be done 'by hand', choosing individual parameter ranges and assessing the individual modelling outcomes (illustrated in Fig. 3d, e for the simple case of estimating input currents producing a specific frequency output in the WC model). This approach can be informative, even in complex models of laminar cortical connectivity [15], but is limited to a small numbers of varying parameters if comprehensive parameter mapping is attempted.

This limitation can be overcome by finessing (1) how the space of possible parameter values is explored to find a model that explains the data better (termed *optimization algorithms*), and (2) by utilising different measures to rank models against each other, i.e. to evaluate which model is the 'better' model (e.g. using a *cost function*). A large number of different approaches to both of these issues exist, of which a variety have been employed in inverting models of neuronal function to fit EEG data.

Optimisation Algorithms

Optimisation algorithms describe computational strategies to identify parameter constellations within a range of possible values that produce a model output that best matches the observed results. There is a large literature regarding competing optimisation methods in a whole host of different areas of science and engineering, so in this section we will only discuss a few algorithms applied in fitting neuronal models to EEG data.

One of the most commonly applied algorithms is that of gradient descent (or ascent, depending on whether one is attempting to find minima or maxima). The basic idea is that from a random starting point, in order to find a minimum, one could iteratively take steps following the direction of the steepest downward gradient until no more changes are made at each step, i.e. the algorithm converges. Because the local gradient is defined by the first derivative of the cost function at any point, the cost function has to be locally differentiable for gradient ascent to be applicable. This approach is intuitive and easy to apply to a range of optimisation problems, such as seizure classification [57], or to refine aspects of EEG source reconstruction [28].

There are a two major limitations to this approach, however, which apply to the problem at hand—namely inverting complex, multi-parameter neural mass model to fit EEG data:

(1) The gradient descent approach relies on the cost function to be smooth and continuous in order to be able to calculate the derivatives. Furthermore, in systems where there are unobserved variables in addition to unknown parameters that need to be inferred, calculating the derivatives directly is often not possible because of recursive dependencies between variables and parameters.

(2) It is designed to identify a local optimum, not the global optimum. Where the cost function is complex and has multiple local extrema, the local optimum identified in this approach may be far from the global optimum possible in the parameter ranges.

There are several alternative optimisation algorithms that address these problems. Genetic algorithms for example resemble the process of natural selection by producing random parameter variations and propagating the most 'successful' ones. After iteratively varying some of the parameters (introducing mutations) and then choosing the best variants (selection), the algorithm will converge to the best global solution, without requiring estimation of local gradients for its progression. This has been applied to fitting parameters of a detailed phenomenological model of individual EEG abnormalities in clinical EEG recordings, identifying patient-specific differences in the transition through parameter space [48]. Similarly, algorithms such as particle swarm optimization, or simulated annealing use direct search strategies that do not rely on knowledge of the gradients. Thee algorithms converge to a global maximum without getting stuck in local optima. A variety of these have been used in model based analysis of EEG signals [13, 26, 54].

Therefore we have two broad classes of algorithms: (1) global direct search strategies, that yield robust convergence to global optima but come at a high computational cost, and (2) gradient descent algorithms that are more computationally efficient but may get stuck in local optima and not yield a global resolution.

The balance of these competing limitations dictates which optimisation algorithm is most appropriate in a given situation.

When making inference on models with relatively few parameters, it is often possible to use one of the global algorithms for a model inversion, as the computational requirements for inverting a model of only a few parameters are usually manageable. However, in models, such as the CMC, where there are many free parameters that need to be fitted, the computational expense of these stochastic algorithms can be prohibitive and the more efficient gradient descent algorithms are called upon. In this setting, prior constraints are used to ensure model inversion is less susceptible to arresting in local optima. Paradoxically, the local minima problem can also be finessed by have many free parameters (as 'escape routes' are more likely to be present where there are many different dimensions of parameter space).

The gradient descent approach can be further finessed to address some of the remaining problems: local linearization can be used to estimate gradient where the underlying cost function is expensive to calculate; expectation-maximisation (EM) algorithms can be employed to invert probabilistic models where not all variables are observed [14], hierarchical model inversion can help to avoid local extrema [24]. Each of these strategies is employed within the DCM framework, but crucially hinge on the cost-function employed—i.e. what is being optimised.

Cost Functions

In order to apply optimisation routines and improve how well a model represents data, we need to define which measure should be optimised. Often the most intuitive approach is to calculate the difference between the numerical predictions of the model states and the empirical measurements, and try and reduce the sum of squared errors between model prediction and empirical measurement. This approach has been successfully applied to EEG in a variety of ways [2, 55].

If closeness of the model fit is the only criterion for the optimisation function, all free parameters within the models will be adjusted in order to produce the best model fit. Especially in models with many free parameters, this can lead to idiosyncratic results that resemble specific features of a given dataset, but show poor generalisability across different, similar datasets—a problem that has been termed *overfitting*. Several strategies can be employed to avoid overfitting and ensure generalisability of the modelling results.

One such approach has emerged naturally from reformulating the cost function not in terms of an absolute error that needs to be reduced, but rather in terms of the Bayesian model evidence (also known as the marginal likelihood) that needs to be maximised. The evidence is simply the probability of getting some data under a model of how those data were caused. This is generally evaluated by trying to estimate the underlying parameters of a model. In more detail: within the Bayesian framework, one estimates the probability of a given parameterisation ϑ, given a set of observations, or data y, by assuming that these were produced from a model m as follows:

$$p(\vartheta|y, m) = \frac{p(y|\vartheta, m)p(\vartheta, m)}{p(y|m)} \tag{10}$$

This posterior probability is not easy to estimate directly, but various approaches can be used to approximate it. Variational Bayes is a generic approach to the analysis of posterior probability densities. In this approach, the *free energy* represents a bound on the log of the model evidence and can therefore be used in optimisation routines to identify optima in the model evidence distribution [22]. The (log-) evidence or marginal likelihood is defined as follows (where D (||) denotes the Kulback-Leibler, or *KL* divergence—a measure of the difference between two probability distributions; y denotes data; m denotes the model; ϑ denotes a set of model parameters; $q(\vartheta)$ denotes the variational density, i.e. the approximating posterior density which is optimised; thus $-\langle ln\ q(\vartheta) \rangle_q$—denotes the entropy and $\langle L(\vartheta) \rangle_q$ denotes the expected energy; F denotes the free energy):

$$\ln p(y|m) = F + D(q(\vartheta)||p(\vartheta|y, m)) \tag{11}$$

$$F = \langle L(\vartheta) \rangle_q - \langle \ln q(\vartheta) \rangle_q \tag{12}$$

The log evidence itself can be split into an accuracy and a complexity term, and thus automatically contains a penalty for overly complex models that are prone to overfitting. In the context of DCM the complexity of the model is established on the basis of how far parameters deviate from their prior values. Therefore, maximizing this Bayesian estimate of model evidence provides a compromise between goodness of model fit, and the generalizability of the model.

Specifically in regards to epilepsy there are further specific problems that need to be addressed: Often the changes of a parameter that varies with time are of interest (for example whilst trying to track network changes during the transition into a seizure). If no account were taken of the temporal contiguity between individual time steps, the already computationally expensive model inversion needs to be fully repeated at each time step, treating each window as independent sample.

For dynamic systems, where there is a temporal dissociation between fast varying states and more slowly changing underlying model parameters, this problem can be overcome through optimization approaches that take into account the temporal dependencies between parameter values at neighboring time points. One of the most successful of these approaches it the Kalman filter. This was originally developed for linear systems, but soon extended to nonlinear systems [35]. The Kalman approach has been used very successfully to estimate parameters underlying transitions into seizure state, where it has proved to benefit from its ability to estimate unobserved (hidden) states [19].

A similar (and mathematically equivalent) approach can be implemented within the DCM framework, where each time step receives the preceding model inversion posteriors as prior expectations, resulting in evidence accumulation (also known as Bayesian belief updating) across the whole modelling time [9]. More recently, a generic approach to estimating parameters at two modelling levels has allowed to accommodate arbitrary relationships between individual model inversion steps in a computationally efficient way (parametric empirical Bayesian approach, [23].

In summary, a Bayesian framework for the cost function allows incorporation of the required constraints to solve the inverse problem as *prior beliefs* regarding the parameters (consisting of expected value, and uncertainty measures). The use of priors furthermore allows the model evidence to be cast directly in terms of *accuracy—complexity*, and therefore preventing overfitting of excessively complex models. Furthermore, several computationally efficient techniques are available to accommodate modelling of time series data. We will now illustrate these procedures using a worked example.

Workflow for Analysis of NMDAR Antibody-Related Paroxysms

Patients with NMDAR antibody encephalitis show a whole variety of apparently different EEG paroxysms. The aim of the subsequent analysis is to identify possible causative mechanisms of how the molecular pathology is translated into an observable abnormal dynamic state. In order to address this aim, we call on the computational mechanisms introduced above.

Specifically, we utilise recent advances in parametric empirical Bayes within the DCM framework [23, 24, 41], which allows for a two-stage modelling approach:

1. Fit parameters of canonical microcircuit neural mass model to both background and paroxysmal conditions separately, in order to find the parameter constellation that provides the best fit
2. Estimate the evidence for models of reduced complexity (i.e. fewer free parameters) to identify subset of parameters that explain most of the changes between background and paroxysms using Bayesian model comparison

The workflow for the analysis of an individual patient is illustrated in Fig. 6. Note that in line with recent advances in dynamic causal modelling, first a 'full' model is inverted—i.e. all typically changing parameters are freed up and are allowed to change in order to explain observed data. Bayesian model comparison is conducted between models with reduced complexity, where the differences between conditions are explained by only a pre-defined subset of parameters. This second step allows for direct comparison of competing hypotheses (about which specific synaptic parameters mediate seizure onset) within the Bayesian framework.

The specific hypotheses tested in this analysis are founded in the existing knowledge of the molecular mechanism associated with abnormal EEG features: NMDA receptor antibodies affect glutamatergic, excitatory connections. Here we

(a) – Categorisation and source localisation

Background EEG   *Paroxysmal EEG*

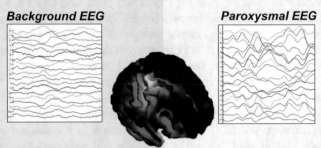

(b) – Dynamic causal modelling

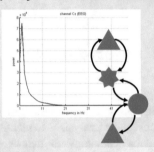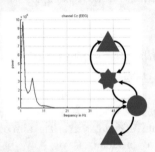

(c) – Parametric empirical Bayes

Differences between background and paroxysm

Changes in synaptic dynamics   *Changes in connection strengths*

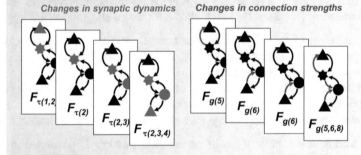

$F_{\tau(1,2)}$ $F_{\tau(2)}$ $F_{\tau(2,3)}$ $F_{\tau(2,3,4)}$ $F_{g(5)}$ $F_{g(6)}$ $F_{g(6)}$ $F_{g(5,6,8)}$

(d) – Bayesian model comparison

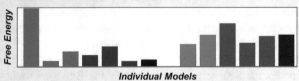

Fig. 6 Analysis workflow: **a** Abnormal paroxysmal activity is visually identified on the EEG and source localised to a single source [9] in order to extract a single 'virtual electrode' local field potential-like trace that contains the main spectral features of the activity. A matching number of time windows of background EEG activities is selected from artefact-free portions of the EEG. **b** Dynamic causal modelling is used to respectively fit a single source canonical microcircuit to the paroxysmal, and the background activity. **c** Parametric empirical Bayes is used estimate the free energy for models that explain both background, and paroxysmal activity with changes in only a subset of free parameters. **d** Bayesian model comparison estimates the evidence for each of the reduced models from the previous step and is used to decide which model is most likely to have caused the observed data features

want to explore whether the resultant effects in the microcircuit mainly have an effect on the time constant of specific neuronal populations (parameterised as time constant, τ), or on the connection strength of excitatory connections (parameterised as g). The specific parameters of interest are summarised in Table 1.

Thus, the Bayesian model comparison in the second stage of the analysis will be used to decide whether changes in the *connection strength* of excitatory connections, or the *temporal integration dynamics* of individual neuronal subpopulations best explain the observed transitions from background EEG to paroxysmal abnormalities.

Table 1 Model parameters and combinations for reduced model comparison

Model parameters evaluated in reduced models			
τ_1	Superficial pyramidal cells	g_1	Connection from ss to sp
τ_2	Spiny stellate cells	g_2	Connection from dp to ii
τ_3	Inhibitory interneurons	g_3	Connection from ss to ii
τ_4	Deep pyramidal cells		
Reduced models (i.e. combination of free parameters to explain both conditions)			
Model 1	τ_1	Model 16	g_1
Model 2	τ_2	Model 17	g_2
Model 3	τ_3	Model 18	g_3
Model 4	τ_4	Model 19	g_1, g_2
Model 5	τ_1, τ_2	Model 20	g_1, g_3
Model 6	τ_1, τ_3	Model 21	g_2, g_3
Model 7	τ_1, τ_4	Model 22	g_1, g_2, g_3
Model 8	τ_2, τ_3		
Model 9	τ_2, τ_4		
Model 10	τ_3, τ_4		
Model 11	τ_1, τ_2, τ_3		
Model 12	τ_1, τ_2, τ_4		
Model 13	τ_2, τ_3, τ_4		
Model 14	$\tau_1, \tau_2, \tau_3, \tau_4$		

Results

Here we report a single case analysis of fitting a neuronal mass model of cortical dynamics to paroxysmal abnormalities in a patient with NMDA receptor encephalitis. Utilising the laminar and cell-type specificity of the canonical microcircuit and the computational efficiency of the variational Bayes approach to fitting the parameters, allows estimating the most likely constellations of bio-physically relevant parameters that explain the observed EEG patterns.

Time windows of 2 s around visually defined episodic EEG activity (Fig. 7a shows an example) is source localised to a single point and their power spectral densities are averaged across time windows (total number = 12). The same number of time windows is randomly selected from artefact free background activity and source activity is estimated at the same cortical source for estimation of the power spectral densities.

In the first stage of the analysis, a single source canonical microcircuit is fitted to the empirical spectral densities for each of the two conditions separately.

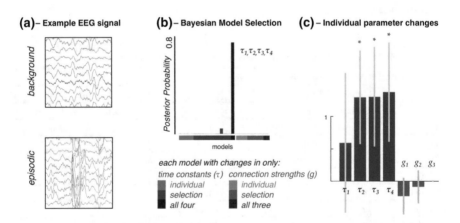

Fig. 7 Bayesian model selection and parameter averages: **a** Two-second windows around all paroxysms (total: 12) and an equal number of background EEG windows were selected visually from the whole-scalp EEG—an exemplar single time window is shown here for background (*top*) and episodic (*bottom*) activity. **b** Single source canonical microcircuits were fitted to average source localised background, and episodic power spectral densities respectively. Models that explained the difference between background and episodic activity with a limited set of free parameters (cf. Table 1) were then compared through Bayesian Model Selection—the winning model here was one where changes in all time constants were required to explain the difference between background EEG and episodic EEG. **c** Bayesian model averaging was performed to estimate the parameter differences that explain the transition from background to episodic activity. A significant increase of time constants was estimated in spiny stellate, inhibitory interneuron and deep pyramidal cell populations. **Variables.** *time constants*: τ_1—superficial pyramidal cells, τ_2—spiny stellate cells, τ_3—inhibitory interneurons, τ_4—deep pyramidal cells; *connection strengths*: g_1—spiny stellate to superficial pyramidal cells, g_2—deep pyramidal cells to inhibitory interneurons, g_3—spiny stellate to inhibitory interneurons

This results in two fully fitted DCMs, where all parameters are allowed to change between background and episodic activity. In order to assess which of these parameter changes is necessary for the observed differences conditions, Bayesian model selection was performed over a set of reduced model, where only a subset of parameters are allowed to change between episodic and background activity. This model space is laid out in detail in Table 1—and broadly divides models into those with different combinations of changes in the synaptic dynamics (i.e. time constants, models 1–14), and the synaptic connection strengths (models 16–22). Comparing these models, the model with changes in all time constants provides the best fit (posterior probability ~0.76), followed by the model with changes in τ_2, τ_3, and τ_4 only (posterior probability ~0.13) as shown in Fig. 7b.

Bayesian model averaging further provides estimates of the size and direction of changes of each parameter between background and episodic activity, shown in Fig. 7c. Because the Bayesian model averaging takes into account uncertainty over specific parameter estimates, this allows for the calculation of a Bayesian confidence interval, and inference whether any given parameter change has a probability exceeding a certain significance threshold (here 99 %). According to these estimates, we find that the episodic spectral densities are associated with a significant increase of time constants in the spiny stellate (τ_2), inhibitory interneuron (τ_3) and deep pyramidal cell (τ_4) populations.

The model fits can be seen in Fig. 8. Figure 8a, b show the fits of the independently inverted full DCMs, whilst Figure c and d show the model fits for the reduced winning model where parameters across both conditions are identical apart from the time constants τ_1, τ_2, τ_3, and τ_4. The paroxysms have a clear frequency peak in the low beta frequency range, which are present in both the full model fits (8B) and the reduced fits (8D) of that condition. Whilst the model fits for the full model are better for both he episodic and the background activity, most of the important differences between them are preserved well even in the reduced models where only a small subset of parameters contributes to explaining the differences seen. Most notably, the emergence of an additional frequency component in the beta range with an identical peak frequency is modelled well, whilst the relative power of high and low frequencies is not preserved as well in the reduced model prediction.

These findings are interesting in two ways. Firstly, identifying changes in parameters that carry biophysical relevance means that results from human EEG measurements can be used to evaluate hypotheses that emerge from molecular findings. Specifically for our case, a significant body of work has already established that antibodies against NMDAR have direct effects on glutamate transmission dynamics in a mouse model of NMDAR antibody encephalitis [33].

These experiments investigating the blockade of NMDAR transmission show that (1) it affects mostly the temporal dynamics of the glutamate response, and not its size (as the latter is largely determined by the preserved AMPA-receptor response); and (2) the most significant effect on NMDAR availability in the mice treated with NMDAR antibody positive CSF was not within the dendritic spines (the sites of classical synaptic transmission), but on extrasynaptic NMDARs.

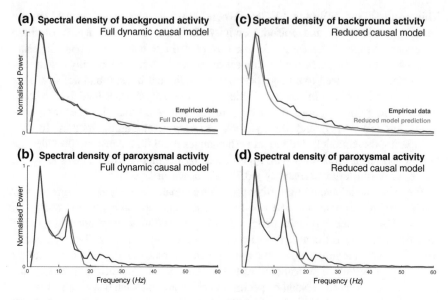

Fig. 8 Model fits for power spectral densities: model fits are shown for the full DCM inversions (**a, b**) and the winning reduced models (**c, d**), where the differences between episodic and background activity are explained by changes in the time constants only. The *top panels* show the background activity, whilst the *bottom panels* show the paroxysmal, episodic abnormality. Model predictions and power estimation range from 4–60 Hz

Our hypothesis space for the analysis presented was specifically designed to address the first aspect: In humans, can dynamic abnormalities on the EEG caused by NMDAR antibody exposure be best explained by changes in the time constants (as predicted from mouse models), or by a change in excitatory connection strengths. The findings from the Bayesian model selection in fact support changes in temporal dynamics underlying the observed EEG abnormalities, providing the first evidence from human studies that the mechanism observed in other model systems may explain the pathological features seen in patients.

However, our analysis pathway tried to explain transitions between background and paroxysmal activity within an individual patient (and during a single EEG recording) through changes in network parameters, even though the presumed underlying cause—the NMDAR antibody—are present and active throughout the recording. Our interpretations of the findings therefore do not suggest that NMDAR cause a permanent change in the time constants, but rather promote a volatility in time constants that facilitates the transient appearance of paroxysmal abnormalities.

Time constants themselves are known to be a composite measures that depend on particular physiological states and are therefore not actually constant, but are themselves dynamic in their expression [38]. One identified mechanism of activity-dependent changes in the temporal profile of postsynaptic integration is the recruitment of extrasynaptic NMDAR during excessive stimulation at

AMPAR-only synapses [8]. The observations made in animal models suggest that NMDAR antibodies change the balance between extrasynaptic and synaptic NMDAR, and are therefore likely to change the dynamics of time-constant changes governing physiological synaptic transmission, making the transient changes in temporal dynamics described in our model a plausible pathophysiological mechanism.

The main effect as estimated from the canonical microcircuit is an increase of the time constants in a variety of neurons. The increased time constants may facilitate temporal integration of neuronal signal and therefore result in an increase in the coupling between superficial and deep pyramidal cells, potentially explaining the high amplitude paroxysmal activity observed on the EEG.

Limitations

In this study we applied DCM to episodic abnormalities in patient EEGs. In the process of this study, many simplifying assumptions are required to render such an empirical inversion tractable.

We have chosen to investigate intrinsic changes in neuronal population coupling and dynamics, based on a single trace extracted from a 'virtual electrode' at a source location estimated from the paroxysmal abnormalities. Thus we have not specifically addressed any larger scale topographic heterogeneity or network level interactions in this study.

Furthermore, the current analysis describes only the state switching between short paroxysmal abnormalities and the patient-specific background, and not the transition into pathological EEG patterns at the onset of illness (as this data is rarely available). This means that the inference we draw is one regarding fluctuations in an already pathological state, that may explain the variations in the EEG phenotypes observed.

The study also is specifically designed to investigate different mechanisms intrinsic to a single cortical source—we therefore do not model differences in input or extrinsic connectivity. However, these are likely to contribute to state switching between the different dynamical states illustrated here, and will be the focus of further modelling research.

Take Home Message *for Clinicians*

This chapter offers an introduction to using empirical electrophysiological data to inform the parameterisation of advanced mesoscale neuronal models. This approach is particularly suited to link conditions where long-lasting, or even permanent pathologies (such as a lesion, or a molecular abnormality) find their pathophysio-logical expression only transiently in abnormal neuronal dynamics. The prime example for this is epilepsy, where abnormalities in the neuronal network produce intermittent and unpredictable abnormal states—epileptic seizures; but the same approach is also relevant to neuropsychiatric conditions, or the encephalitides, as discussed in this chapter.

Linking macroscopic, and often transient observations—such as clinical features, or EEG measurements—to underlying causes, even where they are understood in some detail is far from intuitive. Whilst the discovery of NMDAR binding antibody in the context of clinical autoimmune encephalitis clearly suggest a direct pathophysiological role for the antibody, understanding how it affects synaptic transmission in order to produce the abnormalities observed in neuronal states still remains difficult. This currently limits the prognostic and diagnostic value of EEG recordings, as well as hindering the development of targeted therapies.

The chapter introduces mesoscale computational modelling as a possible link between molecular or microstructural pathology, and macroscale phenotypes. Exploiting recent advances in both neuronal models of cortical function, and the fitting of parameters to empirical data within the well-established framework of Dynamic Causal Modelling allows for the testing of specific mechanistic hypotheses. The approach presented here allows researchers to directly address specific questions emerging from other disease models and evaluate whether evidence for similar mechanisms can be identified in human patients.

These computational models can facilitate a thorough understanding of the dynamic effects of apparently static abnormalities within an organism. Whilst they are not set up to reproduce the complexity of whole organisms, they allow the mapping of changes in the model parameters and dynamic outputs of the model. They are therefore an ideal tool to further explore hypotheses derived from newly identified genetic mutations, other molecular causes, or animal models of specific conditions.

Regarding the example of NMDAR encephalitis, the computational approach presented here provides empirical evidence for electrophysiological abnormalities being caused by changes in the temporal dynamics of synaptic transmission, rather than changes in connection strength. This replicates findings from animal models, providing converging lines of evidence that the observations made in the animal models is in fact related to the dynamic abnormalities we see in human patients.

Take Home Message *for Computationalists*

Advanced computational modelling in the analysis of electrophysiological signals is currently limited to a few source localisation algorithms in routine clinical practice. Yet recent advances both in machine learning algorithms, and the increased availability of computational resources provide an opportunity to integrate advanced computational analysis of neuronal signals into clinical practice.

The approach presented in this chapter is deliberately using empirical data to parameterise an existing, full generative model of neuronal population function (as opposed to more data driven machine learning approaches): The generative model both constrains the inverse problem we face by attempting to make inference on mesoscale mechanisms from macroscale recordings. But crucially it also forces the results of the computational analysis to be cast within biophysically plausible terms.

The work presented here should not be seen in isolation, but instead provides a novel, and necessary perspective on an existing scientific question. When attempting to identify causative mechanisms in NMDAR encephalitis, neither computational nor animal-based approaches will give us the full answer. Rather the strength of the evidence lies in the use of existing evidence from other model systems to constrain the computational analysis to only address specific, competing hypotheses. This 'evidence accumulation' is easiest where all lines of evidence refer to similar neurophysiological concepts (e.g. connection strengths, time constants, gain parameters). Indeed on can foresee the application of dynamic causal modelling to data from animal models to provide a formal integration of animal and human measurements.

The Dynamic Causal Modelling approach presented here furthermore has the benefit that it will provide estimates of model evidence (to decide between competing hypotheses) as well as individual parameter estimates (to evaluate specific effects), derived from fitting the model to empirical data. This combines the benefits of data driven analysis: the DCM can provide direct empirical measures for, or against specific hypotheses, as well as being utilised as a generative model of neuronal dynamics whose parameter space can be explored in detail.

The example presented shows that the approach is uniquely flexible and can be applied to a wide variety of contexts. All software used here, including model inversion techniques, canonical microcircuitry models and classical EEG analysis modules, is freely available as part of the Statistical Parametric Mapping (SPM) academic freeware (www.fil.ion.ucl.ac.uk/spm).

Acknowledgments RER is funded by a Wellcome Trust Clinical Research (106556/Z/14/Z). KJF is funded by a Wellcome Trust Principal Research Fellowship (088130/Z/09/Z).

References

1. Aburn, M.J. et al., 2012. Critical Fluctuations in Cortical Models Near Instability. *Frontiers in Physiology*, 3, p. 331.
2. Babajani-Feremi, A. & Soltanian-Zadeh, H., 2010. Multi-area neural mass modeling of EEG and MEG signals. *NeuroImage*, 52(3), pp. 793–811.
3. Bastos, A.M. et al., 2012. Canonical Microcircuits for Predictive Coding. *Neuron*, 76(4), pp. 695–711.
4. Buffalo, E.A. et al., 2011. Laminar differences in gamma and alpha coherence in the ventral stream. *Proceedings of the National Academy of Sciences*, 108(27), pp. 11262–11267.
5. Bullmore, E. & Sporns, O., 2009. Complex brain networks: graph theoretical analysis of structural and functional systems. *Nature Reviews Neuroscience*, 10(3), pp. 186–198.
6. Canolty, R.T. & Knight, R.T., 2010. The functional role of cross-frequency coupling. *Trends in Cognitive Sciences*, 14(11), pp. 506–515.

7. Carandini, M., 2012. From circuits to behavior: a bridge too far? *Nature Neuroscience*, 15(4), pp. 507–509.
8. Clark, B.A. & Cull-Candy, S.G., 2002. Activity-dependent recruitment of extrasynaptic NMDA receptor activation at an AMPA receptor-only synapse. *Journal of Neuroscience*, 22 (11), pp. 4428–4436.
9. Cooray, G.K. et al., 2016. Dynamic causal modelling of electrographic seizure activity using Bayesian belief updating. *NeuroImage*, 125, pp. 1142–1154.
10. Dalmau, J. et al., 2008. Anti-NMDA-receptor encephalitis: case series and analysis of the effects of antibodies. *The Lancet Neurology*, 7(12), pp. 1091–1098.
11. David, O. et al., 2006. Dynamic causal modeling of evoked responses in EEG and MEG. *NeuroImage*, 30(4), pp. 1255–1272.
12. David, O. & Friston, K.J., 2003. A neural mass model for MEG/EEG: *NeuroImage*, 20(3), pp. 1743–1755.
13. Van Dellen, E. et al., 2012. MEG Network Differences between Low- and High-Grade Glioma Related to Epilepsy and Cognition D.R. Chialvo, ed. *PLoS ONE*, 7(11), p.e50122.
14. Do, C.B. & Batzoglou, S., 2008. What is the expectation maximization algorithm? *Nature Biotechnology*, 26(8), pp. 897–899.
15. Du, J., Vegh, V. & Reutens, D.C., 2012. The Laminar Cortex Model: A New Continuum Cortex Model Incorporating Laminar Architecture L. J. Graham, ed. *PLoS Computational Biology*, 8(10), p.e1002733.
16. Eadie, M.J. & Bladin, P.F., 2001. *A Disease Once Sacred. A History of the Medical Understanding of Epilepsy* 1st ed., New Barnet: John Libbey & Co Ltd.
17. Fisher, R.S. et al., 2014. ILAE Official Report: A practical clinical definition of epilepsy. *Epilepsia*, 55(4), pp. 475–482.
18. Florance, N.R. et al., 2009. Anti-N-methyl-D-aspartate receptor (NMDAR) encephalitis in children and adolescents. *Annals of Neurology*, 66(1), pp. 11–18.
19. Freestone, D.R. et al., 2014. Estimation of effective connectivity via data-driven neural modeling. *Frontiers in Neuroscience*, 8, p. 383.
20. Friston, K., 2005. A theory of cortical responses. *Philosophical Transactions of the Royal Society B: Biological Sciences*, 360(1456), pp. 815–836.
21. Friston, K. et al., 2008. Multiple sparse priors for the M/EEG inverse problem. *NeuroImage*, 39(3), pp. 1104–1120.
22. Friston, K. et al., 2007. Variational free energy and the Laplace approximation. *NeuroImage*, 34(1), pp. 220–234.
23. Friston, K., Zeidman, P. & Litvak, V., 2015. Empirical Bayes for DCM: A Group Inversion Scheme. *Frontiers in Systems Neuroscience*, 9(November), pp. 1–10.
24. Friston, K.J. et al., 2016. Bayesian model reduction and empirical Bayes for group (DCM) studies. *NeuroImage*, 128, pp. 413–431.
25. Gitiaux, C. et al., 2013. Early electro-clinical features may contribute to diagnosis of the anti-NMDA receptor encephalitis in children. *Clinical Neurophysiology*, 124(12), pp. 2354–2361.
26. Gollas, F. & Tetzlaff, R., 2005. Modeling brain electrical activity in epilepsy by reaction-diffusion cellular neural networks. In R. A. Carmona & G. Linan-Cembrano, eds. p. 219.
27. Goodfellow, M., Schindler, K. & Baier, G., 2012. Self-organised transients in a neural mass model of epileptogenic tissue dynamics. *NeuroImage*, 59(3), pp. 2644–2660.
28. Hansen, S.T. & Hansen, L.K., 2015. EEG source reconstruction performance as a function of skull conductance contrast. In *2015 IEEE International Conference on Acoustics, Speech and Signal Processing (ICASSP)*. IEEE, pp. 827–831.
29. Heitmann, S., Gong, P. & Breakspear, M., 2012. A computational role for bistability and traveling waves in motor cortex. *Frontiers in Computational Neuroscience*, 6, p. 67.
30. Helbig, I. et al., 2008. Navigating the channels and beyond: unravelling the genetics of the epilepsies. *The Lancet Neurology*, 7(3), pp. 231–245.

31. Hildebrand, M.S. et al., 2013. Recent advances in the molecular genetics of epilepsy. *Journal of Medical Genetics*, 50(5), pp. 271–279.
32. Hodgkin, A.L. & Huxley, A.F., 1952. A quantitative description of membrane current and its application to conduction and excitation in nerve. *Journal of Physiology*, 117(4), pp. 500–44.
33. Hughes, E.G. et al., 2010. Cellular and Synaptic Mechanisms of Anti-NMDA Receptor Encephalitis. *Journal of Neuroscience*, 30(17), pp. 5866–5875.
34. Jansen, B.H. & Rit, V.G., 1995. Electroencephalogram and visual evoked potential generation in a mathematical model of coupled cortical columns. *Biological Cybernetics*, 73(4), pp. 357–66.
35. Julier, S.J. & Uhlmann, J.K., 2004. Unscented Filtering and Nonlinear Estimation. *Proceedings of the IEEE*, 92(3), pp. 401–422.
36. Jung, R. & Berger, W., 1979. Hans Bergers Entdeckung des Elektrenkephalogramms und seine ersten Befunde 1924–1931. *Archiv fuer Psychiatrie und Nervenkrankheiten*, 227(4), pp. 279–300.
37. Kawato, M., Hayakawa, H. & Inui, T., 1993. A forward-inverse optics model of reciprocal connections between visual cortical areas. *Network*, 4(4), p. 4150422.
38. Koch, C., Rapp, M. & Segev, I., 1996. A brief history of time (constants). *Cerebr Cortex*, 6, pp. 93–101.
39. Lantz, G., Grouiller, F. & Spinelli, L., 2011. Localisation of Focal Epileptic Activity in Children Using High Density EEG Source Imaging. *Epileptologie*, 28, pp. 84–90.
40. Lesca, G. et al., 2013. GRIN2A mutations in acquired epileptic aphasia and related childhood focal epilepsies and encephalopathies with speech and language dysfunction. *Nature Genetics*, 45(9), pp. 1061–1066.
41. Litvak, V. et al., 2015. Empirical Bayes for Group (DCM) Studies: A Reproducibility Study. *Frontiers in Human Neuroscience*, 9(Dcm), pp. 1–12.
42. Di Lollo, V., 2012. The feature-binding problem is an ill-posed problem. *Trends in Cognitive Sciences*, 16(6), pp. 317–321.
43. Lopes da Silva, F.H., 2010. *MEG: An Introduction to Methods*, Oxford University Press.
44. Lopes da Silva, F.H. et al., 1974. Model of brain rhythmic activity. The alpha-rhythm of the thalamus. *Kybernetik*, 15(1), pp. 27–37.
45. Meijer, H.G.E. et al., 2015. Modeling Focal Epileptic Activity in the Wilson–Cowan Model with Depolarization Block. *The Journal of Mathematical Neuroscience*, 5(1), p. 7.
46. Miller, I.O. & Sotero de Menezes, M.A., 2014. SCN1A-Related Seizure Disorders. *GeneReviews*.
47. Moran, R., Pinotsis, D. a & Friston, K., 2013. Neural masses and fields in dynamic causal modeling. *Frontiers in Computational Neuroscience*, 7(May), p. 57.
48. Nevado-Holgado, A.J. et al., 2012. Characterising the dynamics of EEG waveforms as the path through parameter space of a neural mass model: Application to epilepsy seizure evolution. *NeuroImage*, 59(3), pp. 2374–2392.
49. Nosadini, M. et al., 2015. Longitudinal Electroencephalographic (EEG) Findings in Pediatric Anti-N-Methyl-D-Aspartate (Anti-NMDA) Receptor Encephalitis: The Padua Experience. *Journal of Child Neurology*, 30(2), pp. 238–245.
50. Pal, D. & Helbig, I., 2015. Commentary: Pathogenic EFHC1 mutations are tolerated in healthy individuals dependent on reported ancestry. *Epilepsia*, 56(2), pp. 195–196.
51. Panayiotopoulos, C., 2005. *The Epilepsies* 1st ed., Oxford: Bladon Medical Publishing.
52. Papadopoulou, M. et al., 2015. Tracking slow modulations in synaptic gain using dynamic causal modelling: Validation in epilepsy. *NeuroImage*, 107, pp. 117–126.
53. Quintáns, B. et al., 2014. Medical genomics: The intricate path from genetic variant identification to clinical interpretation. *Applied & Translational Genomics*, 3(3), pp. 60–67.
54. Shirvany, Y. et al., 2012. Non-invasive EEG source localization using particle swarm optimization: A clinical experiment. In *2012 Annual International Conference of the IEEE Engineering in Medicine and Biology Society*. IEEE, pp. 6232–6235.
55. Sitnikova, E. et al., 2008. Granger causality: Cortico-thalamic interdependencies during absence seizures in WAG/Rij rats. *Journal of Neuroscience Methods*, 170(2), pp. 245–254.

56. Spillane, J., Kullmann, D.M. & Hanna, M.G., 2015. Genetic neurological channelopathies: molecular genetics and clinical phenotypes. *Journal of Neurology, Neurosurgery & Psychiatry*, p.jnnp–2015–311233.
57. Thomas, E.M. et al., 2008. Seizure detection in neonates: Improved classification through supervised adaptation. In *2008 30th Annual International Conference of the IEEE Engineering in Medicine and Biology Society*. IEEE, pp. 903–906.
58. Thomas, R.H. & Berkovic, S.F., 2014. The hidden genetics of epilepsy—a clinically important new paradigm. *Nature Reviews Neurology*, 10(5), pp. 283–292.
59. Tong, S. & Thakor, N.V., 2009. *Quantitative EEG Analysis Methods and Clinical Applications*, Artech House.
60. Wang, Y. et al., 2014. Dynamic Mechanisms of Neocortical Focal Seizure Onset B. Ermentrout, ed. *PLoS Computational Biology*, 10(8), p.e1003787.
61. Werner, G., 2007. Metastability, criticality and phase transitions in brain and its models. *Biosystems*, 90(2), pp. 496–508.
62. Wilson, H.R. & Cowan, J.D., 1972. Excitatory and inhibitory interactions in localized populations of model neurons. *Biophysical journal*, 12(1), pp. 1–24.
63. Zschocke, S. & Hansen, H.-C., 2012. *Klinische Elektroenzepalographie* 3rd ed., Berlin, Heidelberg: Springer-Verlag.

Oscillatory Neural Models of the Basal Ganglia for Action Selection in Healthy and Parkinsonian Cases

Robert Merrison-Hort, Nada Yousif, Andrea Ferrario and Roman Borisyuk

Introduction

The Basal Ganglia and Parkinson's Disease

The neuronal circuits of the Basal Ganglia (BG) play an important role in movement control and it is known that without correctly functioning BG, the cortex's ability to control movements is diminished. Parkinson's disease (PD), for example, is primarily a disease of the basal ganglia. One aspect of pathological activity in the parkinsonian basal ganglia is an increase in synchronous oscillatory firing patterns [1].

The basal ganglia are a group of nuclei found in the subcortical area of the brains of vertebrates. These nuclei include the putamen and caudate nucleus (together called the dorsal striatum, neostriatum, or here simply "the striatum"), the globus pallidus (GP), subthalamic nucleus and the substantia nigra pars compacta (SNc) and pars reticulata (SNr). In primates, the globus pallidus is divided into two segments: a medially-located "internal" segment (GPi) and more laterally-located "external" segment (GPe). In rodents, on the other hand, the globus pallidus is generally considered a single structure, with connections to other nuclei that make it similar to that of the GPe in primates. The rodent entopeduncular nucleus (EP), meanwhile, is

R. Merrison-Hort · A. Ferrario · R. Borisyuk (✉)
School of Computing and Mathematics, Plymouth University,
Drake Circus, Plymouth PL4 8AA, UK
e-mail: rborisyuk@plymouth.ac.uk

N. Yousif
Division of Brain Sciences, Neuromodulation Group,
Faculty of Medicine, Imperial College London, London, UK

R. Borisyuk
Institute of Mathematical Problems of Biology, Keldysh Institute
of Applied Mathematics of Russian Academy of Sciences,
Pushchino 142290, Russia

© Springer International Publishing AG 2017   149
P. Érdi et al. (eds.), *Computational Neurology and Psychiatry*,
Springer Series in Bio-/Neuroinformatics 6,
DOI 10.1007/978-3-319-49959-8_7

usually considered homologous to the primate GPi. In this paper we will typically treat the term "GP" as being equivalent to "GPe" and "EP" as being equivalent to "GPi", although when discussing experimental results from specific species we will use the correct terminology for that species. The primary input to the BG is the cortex, and their main output is the thalamus. Figure 1 shows the main connections between the nuclei of the basal ganglia, including the so-called "direct", "indirect" and "hyper-direct" pathways from cortex to thalamus through the BG.

Parkinson's disease is the second most common neurological disease after Alzheimer's disease, with an estimated 1 % of people in industrialized countries over the age of 60 suffering from it [2]. The disease encompasses a wide range of motor and non-motor symptoms which vary from patient to patient, however it is characterized by three main motor abnormalities that are seen to some degree in nearly all patients [3]:

- Resting tremor, in which a 4–6 Hz tremor is seen when the patient is at rest, usually unilaterally.
- Bradykinesia, slowness in initiating or executing movements. Voluntary (i.e. non-cued) movements are particularly affected. Around 47 % of patients also experience "freezing": a temporary inability to move certain limbs that can occur spontaneously during normal movements, such as walking.
- Rigidity, in which increased muscle tone causes high resistance of limb movement, postural deformities and instability.

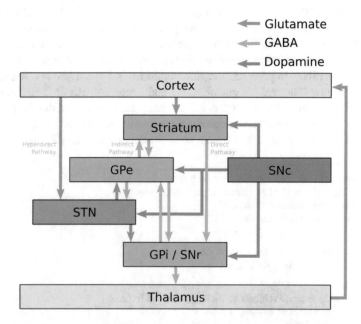

Fig. 1 Schematic overview of the main connections between nuclei of the basal ganglia. *Colours* correspond to the predominant neurotransmitter that released by neurons in each module

The main pathological feature of Parkinson's disease is the death of the dopaminergic neurons in the SNc (for review see [4]). Most pharmacological treatments aim to reverse the reduction in dopamine that results from the loss of these neurons. This is accomplished either directly, through the dopamine precursor l-DOPA (levodopa) or indirectly, through dopamine agonists or drugs that prevent the metabolism of dopamine. The UK National Institute for Health and Clinical Excellence (NICE) guidelines recognize levodopa as being the most effective drug for the reduction of the primary motor symptoms of PD [5] and it is therefore often the first choice for pharmacological treatment of these symptoms. Long-term levodopa use is, however, associated with a number of adverse side effects including uncontrollable abnormal movements (a form of dyskinesia) and unpredictable fluctuations in effectiveness (the so-called "on-off" phenomenon [6]. For patients with more severe symptoms, perhaps with serious side-effects from high levodopa doses, an increasingly common surgical treatment is deep-brain stimulation (DBS). In DBS, an implanted electrode provides constant high frequency (~120 Hz) electrical stimulation to either the STN or GPi [7]. The precise effects of this stimulation on neuronal activity in the basal ganglia are not fully understood and are likely to be many and varied (reviewed in [8]).

The "rate model", originally described by DeLong [9], offers a simple but persuasive explanation for the hypokinetic symptoms of Parkinson's disease (e.g. bradykinesia). This model arose from experimental evidence showing that when a monkey is rendered Parkinsonian there is a clear increase in the average firing rate in the GPe and STN, along with a decrease in the GPi [10, 11]. This theory holds that the basal ganglia are organised into two parallel pathways, a pro-kinetic "direct" pathway (striatum → GPi/SNr) and an anti-kinetic "indirect" pathway (striatum → GPe → STN → GPi/SNr). The fact that direct pathway MSNs express D1-type dopamine receptors whereas indirect pathway MSNs express D2-type receptors suggests that under Parkinsonian conditions the inhibitory output from the direct pathway becomes stronger, whereas that from the indirect pathway becomes weaker (since dopamine is excitatory at D1 receptors and inhibitory at D2 receptors). The direct/indirect pathway model of Parkinsonism explains the surprising result that lesioning the STN reverses the effects of MPTP poisoning [12].

The rate model of PD remains popular and is often used to explain the motor symptoms of the disease in medical textbooks. Despite its popularity, it is clear that the model has some important limitations. Since it predicts a reduction in motor activity under Parkinsonian conditions it cannot explain how tremor arises; this appears to be related to an increase in rhythmic bursting at tremor frequencies in the STN [13]. More recent discoveries about basal ganglia connectivity suggest that the simple division into two feed-forward pathways is overly simplistic. For example, it appears that the majority of GPe neurons project to the GPi and/or SNr in addition to the STN. If this is the case then it is not clear what the role of the STN is within the rate model. Similarly, the model also fails to explain the functional significance of the connection back from STN to GPe, and the "hyper-direct" cortical projection to the STN, while other recent findings suggest that the inter-striatal connections are important that the direct and indirect pathways are intertwined [14, 15]. Finally, the effectiveness of high frequency electrical deep brain stimulation of the subthalamic nucleus at ameliorating the motor symptoms of Parkinson's disease [16–18] is difficult to explain using the rate model. While DBS was originally envisioned to act as a "reversible lesion", more recent evidence has shown that high frequency stimulation actually increases the rate of spiking in efferent STN axons [19, 20]; according to the rate model this should cause an exacerbation of hypokinetic symptoms.

One aspect of pathological activity in the parkinsonian basal ganglia is an increase in synchronous oscillatory firing patterns [1], particularly in the so-called beta frequency band (12–30 Hz). There are a number of reasons why widespread pathological oscillations may cause motor deficits, for example they may impair the ability to relay information [21]. It has also been proposed that, in health, sporadic beta oscillations act as a global signal for maintenance of the current motor activity [22]. Also, there are several hypothesis related to synchronization of neural activity. For example, Hutchison et al. [23] put forward the hypothesis that beta oscillations are enhanced in PD and prevent generation of voluntary movements. Results reported in Bar-Gad et al. [24] showed a complex phase-locking between a stimulus and its response in most neurons in the globus pallidus. Some neurons increased their firing rate but the majority of neurons displayed partial inhibition during the stimulus train. The activities of simultaneously recorded neurons display rate correlation but no spike-to-spike correlation. The authors hypothesize that the effect of DBS on the GP is not complete inhibition but rather a complex reshaping of the temporal structure of the neuronal activity within that nucleus.

When considering the available experimental data regarding the causes and treatment of Parkinson's disease, it is important to remember that much of this research makes use of animals in which conditions similar to the disease have been induced (see Betarbet et al. [25] for a review). For example, many of the classical results in non-human primates attempt to replicate the disease by using the neurotoxin 1-methyl-4-phenyl-1, 2, 3, 6-tetrahydropyridine (MPTP) to selectively kill the dopaminergic neurons of the SNc. Similarly, rodent studies often use the chemical 6-hydroxydopamine (6-OHDA) to damage the SNc, either uni- or bi-laterally. One must keep in mind that these animal preparations only induce Parkinson's-like symptoms, and differ from the true disease in some important ways. Tremor, for example, is absent in 6-OHDA lesioned mice and most species of MPTP lesioned monkeys [25].

Parkinson's Disease and Deep Brain Stimulation

Deep brain stimulation (DBS) involves the surgical implantation of electrodes into disorder-specific target regions, via which the neural tissue is stimulated using trains of electrical pulses. For Parkinson's disease, the subthalamic nucleus is the most common target for DBS, and, interestingly, is the same target for ablative surgery which preceded DBS treatment. The decision to allow a patient to undergo

DBS implantation depends on a number of factors, and is usually made only after trying the non-surgical option of drug treatment to replace the lost dopamine. Since DBS was pioneered almost 30 years ago [26] there have been over 125,000 implantations worldwide (Medtronic, Minneapolis, US), with 69 % of patients showing total or significant suppression of tremor (Medtronic DBS Therapy for Parkinson's Disease and Essential Tremor Clinical Summary, [27].

However, although DBS is widely used and is successful at achieving therapeutic benefits, the precise way in which the injected electrical current affects the electrical activity of the brain is not fully understood. One difficulty is that although we can image where the implanted electrode is inside the brain using magnetic resonance imaging (MRI) or computed tomography (CT), we are limited in our ability to simultaneously stimulate the brain and record exactly how the current spreads in the tissue, and how this current interacts with the neural activity.

One approach to addressing this problem is to use mathematical modelling to better understand how the current influences the brain's activity and predict how to use DBS more effectively [28]. For example, such work with theoretical models can explain the difference in the electric fields created by two commonly used stimulation approaches, and therefore can help clinicians to better reach the targeted brain region. The final challenge will be to use such models within routine clinical practice in order to predict the best settings for the current applied to each individual patient, as and when they require the intervention.

As the use of this procedure spreads to new ailments such as epilepsy, depression, and bipolar disorder, the number of patients who may benefit from this surgical intervention will also increase. In order to understand more about how the electrical current is achieving the observed effects, theoretical research hand in hand with clinical research needs to be undertaken.

The Scientific Problem

In this chapter we consider modelling of the basal ganglia, concentrating on approaches from mathematical and computational neuroscience. These approaches are grounded in the available neurobiological data about the BG, as well as medical data regarding Parkinson's disease and its treatment by deep brain stimulation. Although significant progress has been made in neuroscience towards understanding how neuronal circuits control movement, many questions remain unanswered. Specifically, we are far from understanding the critical neuronal mechanisms which allow these circuits to function. We develop mathematical and computational models of oscillatory neuronal activity to explain how the neuronal circuits of the BG can select actions and how they can switch between them.

As discussed in the previous section, oscillatory activity in the basal ganglia is thought to underlie many of the symptoms of Parkinson's disease. It is known that the dynamics of a neuronal network comprised of excitatory and inhibitory neurons can be oscillatory. For example, a study of spiking integrate-and-fire type neurons with random connections demonstrated that an oscillatory regime is stable and exists for a broad range of parameter values [29]. How can an oscillatory regime appear in such a system of interactive neurons? Mathematics provides a description of how low amplitude oscillations can appear from a stable equilibrium under parameter variation, a scenario known as an Andronov-Hopf bifurcation. Borisyuk [29] found that this bifurcation can occur in a randomly connected system of integrate-and-fire neurons. Parameter variation provides a valuable correspondence between regions in parameter space and the rhythmic activity of some particular frequency. In fact, in some circumstances the parameter space can even be partitioned into zones of rhythmic activities in a particular band: a region of alpha-rhythm, a region of beta-rhythm, etc. These results provide the possibility of controlling both the dynamical mode and the frequency of oscillations, as well as finding the parameter values that correspond to particular dynamics. Our studies of neural network models of spiking neurons with random connectivity have shown that a repertoire of dynamical regimes is possible, including: (1) a regime of stationary activity (low or high); (2) a regime of regular periodic activity where all oscillators are highly synchronized and demonstrate coherent pattern of spiking; (3) a regular bursting regime; (4) a partial synchronization regime where some patterns of individual neurons are irregular and stochastic but the average activity of population is periodic; (5) an irregular spiking regime where both individual neuron firing patterns and average activity are stochastic. Our study shows that the regime of partial synchronization is of interest for modelling of different brain functions such as perception, attention, novelty detection and memory [30–32]. In this chapter we study partial synchronization in the neural circuits of the BG and demonstrate how these dynamics can be used for action control.

As mentioned above, another part of the scientific problem is to understand the mechanism by which DBS is able to effectively treat various different neurological conditions. For example, why do high frequency (>100 Hz) and high amplitude (several volts) electrical pulses prevent the low frequency oscillations that are related to Parkinsonism? Theoretical neuroscience plays an important role in solving this very difficult problem, and can be used, for example, to model and visualize the spread of current in three dimensional space near the tip of stimulating electrode [33], to study oscillatory activity in BG circuits in healthy conditions and PD pathology [34, 35], as well as to identify conditions for pathological synchronization and methods of desynchronization [36]. It was shown that using mathematical and computational methods of nonlinear dynamics is useful for study of neuronal mechanisms of DBS and suppression of pathological brain rhythms [37, 38]. In general, computational and mathematical modelling can be used to study the mechanism by which DBS might de-synchronize or "reset" pathological rhythms, as well as how it might repair connectivity within neuronal circuits that have been damaged by disease.

Computational Methods

Mathematical and computational modelling of neuronal activity in the BG is a hot topic and there are many models considering different aspects of BG activity, mostly in relation to PD and DBS. Existing neural network models of the BG can be divided into two classes:

- Spiking models, in which equations represent the activity of individual neurons, possibly within one or more synaptically coupled populations.
- Models of interactive nuclei, where each nucleus (or population of neurons) is described by averaged population-level equations [39].

In this section we will briefly review these two approaches to computationally modelling neuronal activity, and describe how each has been applied to the study of the BG.

Spiking Models

Hodgkin-Huxley and Multi-compartment Modelling

In the early 50s Alan Hodgkin and Andrew Huxley derived a set of non-linear equations that describe the dynamics of the squid giant axon—work for which they received the Nobel prize. The cell membrane is thought as a circuit with capacitance C and three ionic fluxes (sodium: Na, potassium: K and non-specific leak: LK) that act as parallel resistors (channels) and generate currents flowing across the membrane. Each ionic current has intrinsic dynamics and it is regulated by variables that describe the proportion of activation and inactivation of each channel. The model is able to reproduce highly non-linear phenomena in the voltage called action potentials, which are at the basis of neural transmission in the brain. According to the Hodgkin-Huxley model, the voltage across the membrane varies according to the following equation:

$$C\frac{dV}{dt} = g_{LK}(E_{LK} - V) + m^3 h g_{Na}(E_{Na} - V) + n^4 g_K(E_K - V) + I_{ext}$$

Here I_{ext} is an external source of current, g_{Na}, g_K, g_{LK} and E_{Na}, E_K, E_{LK} are the maximal conductance and equilibrium potentials of the sodium, potassium and leakage currents, respectively. The variables m, h and n are the gating variables of the sodium and potassium channels and their integer powers represent the number of molecules involved in the dynamics of each channel. Each of these gating variables evolves according to an equation of the following form (where X is m, h or n):

$$\frac{dX}{dt} = \alpha_X(V)(1 - X) - \beta_X(V)X$$

The function α_X is the forward rate function (specified in $msec^{-1}$) at which the corresponding gating molecule moves from its configuration where ions are blocked to its configuration where ions can flow. Conversely β_X are backwards rate functions, which determine how quickly gating molecules move from the unblocked to the blocked configuration. The functions α_X and β_X follow the form:

$$f(V) = \frac{A + BV}{C + \exp(\frac{V+D}{E})}$$

where A, B, C, D and E are parameters determined using experimental voltage clamp protocols.

The Hodgkin-Huxley equations, which are sometimes classed as a "conductance-based" model, are biologically realistic and can show a variety of dynamics seen in experimental recordings, including tonic spiking and bursting. Although being biologically realistic the four dimensional Hodgkin-Huxley system usually has computational limitations, especially if extended with multiple ionic currents. For this reason, several simplified models have been developed. The Morris-Lecar [40] and FitzHugh-Nagumo [41] models are examples of two dimensional systems of ordinary differential equations that demonstrate many of the same basic features of the Hodgkin-Huxley model, but they cannot reproduce complex behaviors like busting. Existing two-dimensional models are able to reproduce a different firing patterns (including bursting), but take into account a reset condition. Examples of the previously mentioned dynamics are Izhikevich neuron model [42] and the adaptive exponential leaky integrate and fire model [43].

The simplest form of Hodgkin-Huxley model (the so-called "single compartment" approach) considers the voltage across the membrane to be identical everywhere in the cell, but this is not the case in reality. The electrical activity spreading from the dendrites to the soma and then on to the axon depends on the spatial distribution of the neurites (axon and dendrites). Neurons receive inputs from thousands of synapses at different dendritic locations and the propagation of the activity is attenuated and delayed until it reaches the soma and axon. To incorporate this into a model, the neuron can be approximated as multiple connected cylinders in parallel. This more detailed "multi-compartmental" modelling gives extra an extra level of biological realism, but requires significantly more computational power and requires that much more experimental data (for example neuron's spatial geometry) is available to constrain the model.

Spiking Models of the Basal Ganglia

Probably the most well-known computational model of the basal ganglia is that of Terman et al. [44]. This is a conductance-based spiking model of the interconnected subthalamic nucleus (STN) and globus pallidus (GP), where each neuron is of single compartment Hodgkin-Huxley type, with a (simplified) set of ion channels chosen to represent the main electrophysiological features of the neurons in each nucleus. This model was used to investigate the emergence of different patterns of spiking behaviour, such as rhythmic bursting, under conditions of simulated Parkinson's disease. In a later paper the model was improved with parameter changes and the addition of thalamo-cortical relay neurons, and this revised model was used to consider the effects of deep-brain stimulation (DBS) on the activity in the network [45]. A more advanced single compartment model of 500 neurons within the basal ganglia network was built by Hahn and McIntyre [46], and was similarly used to investigate the effects of STN-DBS. These models have many advantages: since they are built of reasonably biologically-realistic conductance-based neurons, their outputs can be easily compared with electro-physiological data, and because they consist of networks of such neurons they can be used to examine how different patterns of synaptic connectivity produce different outputs. However, these models do not make use of any anatomical information about the spatial distribution of their constituent neurons. This limits their utility for studying the effects of spatially heterogeneous stimulation, such as DBS, and makes it difficult to compare the results of simulations with spatially averaged experimental recordings, such as local field potentials.

A well-known larger scale spiking model of the basal ganglia is that developed by Humphries et al. [47]. This model comprises 960 spiking neurons across five basal ganglia nuclei, where each nucleus is divided into three parallel "channels". Each neuron in the model is based on the leaky integrate-and-fire formalism, with additional currents added to neurons in each sub-population to better reflect their physiological behaviour. Although the model does not contain any synaptic plasticity mechanisms, and cannot therefore exhibit learning, it is able to demonstrate robust selection of outputs based on simulated cortical input, in line with the proposed action selection role of the basal ganglia. Furthermore, this model is able to reproduce a number of experimental characteristics of the basal ganglia under both healthy and parkinsonian conditions. Chersi et al. [48] developed a model that is similar to that of Humphries et al. [47], but with the addition of cortical and thalamic populations and a much larger number of neurons: 14,600 in total. These neurons are also of the leaky integrate-and-fire type, and are similarly organized into distinct channels. The main improvement in this model is the addition of spike timing dependent plasticity (STDP). Chersi et al. claim that their model may represent how the basal ganglia are able to facilitate learning of habitual motor responses to sensory input, and they demonstrate this by showing that a "virtual primate" is able to learn to perform a simple behavioral task when driven by the model. While models of the type described in Chersi et al. [48] and Humphries et al. [47] are valuable for studying the function of networks of spiking elements arranged in a fashion that is broadly similar to that of the basal ganglia function, they lack biological realism in several key ways, which limits their utility. Firstly, the use of integrate-and-fire neurons in both models limits the extent to which experimental data from real neurons can be used to tune the model, since the model parameters and measured variables cannot be directly related to results of biological experiments. Secondly, the models contain relatively few neurons in comparison to the real basal ganglia and have equal numbers of neurons in each nucleus, despite the actual size of these populations differing by orders of magnitude—for example the primate subthalamic nucleus (STN) and striatum are estimated to contain approximately 104 and 107 neurons, respectively [49]. Finally, because these models do not include any spatial information, they cannot be used to model the effects of spatially heterogeneous electrical stimulation (i.e. DBS), and their results cannot be compared with experimental recordings that are dependent on the spatial relationship between neurons and the recording apparatus (e.g. local field potential or MEG recordings).

Several successful attempts have been made to produce detailed multi-compartmental models of different types of neurons within the basal ganglia. Günay et al. [50] used a database-driven methodology to build models of neurons within the globus pallidus (GP), and several later studies have made use of their results to study in detail the effects that different distributions of ion channel proteins have on the activity of pallidal neurons (e.g. [51]. Such studies are typically performed in tandem with in vitro experiments using real neurons, with the experimental and modelling work closely informing each other—a key advantage that is made possible by such a detailed modelling approach. Beyond the globus pallidus, Gillies and Willshaw [52, 53] developed a multi-compartmental model of a subthalamic neuron, and this model has been particularly popular for examining the effects of deep-brain stimulation on STN activity. Since multi-compartmental models include information about the three dimensional morphology of neurons, several studies have used the Gillies and Willshaw [52, 53] model to demonstrate how the effects of electrical stimulation differ based on the spatial location and orientation of neurons [54, 55]—a feature that is not typically captured by simpler single-compartment models. Other groups have also produced multi-compartment models of other basal ganglia neuron types, such as the striatal "medium-spiny" neurons [56]. Unfortunately, due to the complexity of constructing and simulating models of this nature, they are typically restricted to simulating single neurons or very small networks within one nucleus, although Miocinovic et al. [54] built a model which contained 50 multi-compartmental STN neurons alongside 100 simulated axons from the GPi and cortico-spinal tract.

Mesoscopic Models

Population-Level Modelling

Although reduced spiking models such as the Izhikevich model require less computational power than more realistic models, even relatively small regions in relatively small brains contain a huge numbers of neurons and synapses. For this reason computational and mathematical analysis of large spiking neural networks is usually computationally infeasible. As an alternative to the relative complexity of the previously described models, it is possible to use another class of computational models do not model the detailed spiking activity of individual neurons, but instead consider only the average activity (e.g. mean spiking rate) in individual neurons or, more commonly, entire populations of neurons. Each model measures the average activity of ensembles of interacting cells in different ways: some models measure the proportions of cells firing in a homogeneous ensemble of excitatory and inhibitory neurons [57], Amari et al. [58], other models measure the average membrane potential [59], or the average dendritic field [60].

We will describe the model developed by Wilson and Cowan [57], which describes the proportion of active neurons in a pair of synaptically coupled excitatory and inhibitory subpopulations. Let $E(t)$ and $I(t)$ be the proportion of excitatory and inhibitory neurons firing at time t. The equations are proportional to the probability of each cell in each population being sensitive (not refractory) during an absolute refractory period r multiplied by the probability of each cell being active when receiving an average level of excitation. A key feature of the model is the idea of "subpopulation response curve", $Z(x)$, which determines the activation probability as a function of synaptic excitation. A frequently used example of $Z(x)$ is the logistic (or sigmoid) curve. It is a nonlinear, monotonically increasing function assuming values in $(0, 1)$ given by:

$$Z(x) = \frac{1}{1 + e^{-a(x-\theta)}} - \frac{1}{1 + e^{a\theta}}$$

Here a is the maximum slope of the sigmoid and θ is its position on the horizontal axis. Parameter θ represents the common threshold value for all the cells of a subpopulation type (excitatory or inhibitory). The average level of excitation arriving to each neural subpopulation is given by the weighted sum of the proportion of active excitatory and inhibitory neurons, where the weights represent the directed strength of the connections between the two populations. The complete system is given by:

$$\tau_E \frac{dE}{dt} = -E + (1 - r_E E) \cdot Z_1(w_{11}E + w_{21}I + J_E)$$

$$\tau_I \frac{dI}{dt} = -I + (1 - r_I I) \cdot Z_1(w_{12}E + w_{22}I + J_I)$$

where $J_{E,I}$ represents external inputs to each population, $\tau_{E,I}$ the populations' average membrane time constants, and $r_{E,I}$ the refractory periods of the subpopulations. The terms w_{ij} represent the synaptic strength from population i to population j, for $i,j = E, I$. Wilson and Cowan based their model on the idea that "all nervous processes of any complexity are dependent upon the interaction of excitatory and inhibitory cells" [57], and therefore considered the first population to be an excitatory ($w_{11,12} > 0$) and the second to be an inhibitory ($w_{22,21} < 0$); in fact it can be shown that oscillations cannot exist unless these conditions are true. Borisyuk and Kirillov [61] used bifurcation analysis in two dimensions in order to understand the dynamical regimes that are possible for different values of J_1 and w_{12}. This approach elucidated the way in which the different dynamics described by Wilson and Cowan (steady state, bi-stability and oscillations) arise as the parameters are varied and revealed an interesting new regime in which there is bi-stability between a stable fixed point and a stable limit cycle.

Mesoscopic Models of the Basal Ganglia

The circuit formed by the reciprocally connected subthalamic nucleus and globus pallidus has been particularly well studied using population-level models [62, 63] as this circuit has been hypothesized to act as a neuronal pacemaker that generates pathological rhythms in Parkinson's disease. The advantage of relatively simple mathematical models such as these is that they permit detailed mathematical analysis of the network, which can be used to determine, for example, the conditions under which oscillatory activity can occur. Another advantage of averaged models is that because they are computationally straightforward to simulate and typically have only a small number of parameters, they can be used alongside optimization techniques in order to fit experimental measurements, such as average spike rates. This allows macroscopic-scale characteristics of the network, for example the relative overall synaptic connection strengths between populations, to be determined [39]. However, since these models are averaged over time and often represent the activity of many neurons as a single equation, they are unable to address many questions that are likely to be very important for understanding basal ganglia (dys)function. For example, averaged models cannot be used to study the information carried by precise spiking patterns, or to unravel the role played by the circuits formed between neurons of the same nucleus. Similarly, population-level models cannot be used to investigate the effects of stimulation that has a varying effect at different points in space within a particular nucleus, such as DBS.

Modelling of DBS

Deep brain stimulation (DBS) is a successful therapy for movement disorders such as PD [64]. DBS has also been used to treat epilepsy, chronic pain, and, more recently, psychiatric disorders (e.g. depression) and Alzheimer's disease. Although it is widely accepted that DBS is an effective treatment, there remains a lack of understanding of the fundamental neuronal mechanisms which underlie it [55].

Why do high frequency (>100 Hz) and high amplitude (several volts) electrical pulses prevent the low frequency oscillations that are related to Parkinsonism? Theoretical neuroscience plays an important role in solving this very difficult problem, and can be used, for example, to model and visualize the spread of current in three dimensional space near the tip of stimulating electrode [55], to study oscillatory activity in BG circuits in healthy conditions and PD pathology [34, 35], as well as to identify conditions for pathological synchronization and methods of desynchronization [36]. It was shown that using mathematical and computational methods of nonlinear dynamics is useful for study of neuronal mechanisms of DBS and suppression of pathological brain rhythms [37, 38]. In general, computational and mathematical modelling can be used to study the mechanism by which DBS might de-synchronize or "reset" pathological rhythms, as well as how it might repair connectivity within neuronal circuits that has been damaged by disease.

We now briefly illustrate the potential for computational approaches to improve DBS treatment with one example that relates to our own work. The increase in use and applications of DBS brings the desire to optimize the treatment in order to maximize the benefit to the individual patient. For example by choosing the best possible site for stimulation within a target region, or predicting the stimulation parameters to maximize the beneficial effect and minimize unwanted side effects. However, such optimization remains difficult as the mechanisms underlying the clinical improvement induced by DBS remain unclear. It even remains unclear whether these stimulating pulses are excitatory or inhibitory, and whether they primarily affect cells or axons [65]. The technical challenges of directly visualizing the current spread and effect on neurons is one obstacle to this process, and has led to the alternative approach of using computational models to calculate the spread of current around the DBS electrode and estimate the impact on the surrounding neuronal structures [19, 33, 66, 67, 68, 69, 70, 71]. To date, the majority of such studies focus on quantifying the region of tissue around the electrode in which models of axons would be induced to fire action potentials with each DBS pulse applied [72–74]. Such an approach has now been incorporated into the planning software for DBS surgery offered by the medical device companies [75] (http://www.vercise.com/vercise-and-guide-dbs-systems/guide-dbs/).

The studies of DBS tissue activation mentioned so far are focused on modelling axons because they have a lower firing threshold than cell bodies [76]. We have previously shown however, that DBS can cause changes in spontaneous firing of model STN neurons which cannot be distinguished using axons [55, 77]. We used a finite element model (FEM) to calculate the electric field induced by DBS in the tissue surrounding an implanted electrode and applied this electrical potential distribution as a stimulus for multi-compartment neuronal models. The FEM approach consists of defining a geometry, which in our case consisted of the implanted quadripolar DBS electrode (model 3389, Medtronic, MN, USA) and a cylinder of surrounding homogenous brain tissue. The modelling package COMSOL Multi-physics 3.3 (COMSOL AB, Stockholm, Sweden) was used to create, mesh and solve the Laplace equation within this three-dimensional geometry.

The calculated potential was applied to a previously published compartmental model of an STN projection neuron [52, 53]. The morphology of this model neuron consists of a soma and three identical dendritic trees, totalling in 189 compartments. The cell has both passive properties and a number of Hodgkin-Huxley based ionic channels. We simulated the model using the software NEURON 7.2 [78] and we used the set of parameters defined by Gillies and Willshaw [52, 53]. This model cell is able to fire in two modes, a tonic or single spike mode and a bursting mode.

We found that in the tonic mode, when the cell spontaneously fires at a rate of 38 Hz, therapeutic DBS frequencies (100–150 Hz) drove the cell to fire action potentials at the DBS frequency, whereas sub-optimal frequency stimulation could not induce such a shift. In burst mode however, the relationship between DBS frequency and firing frequency changed. A stimulus which can drive a neuron in tonic mode to fire at stimulation frequency, can increase the firing frequency of the bursting cell, but does not cause the same linear relationship. Interestingly, on the offset of high frequency stimulation we found that the cell exhibited a silent period where no action potentials were fired, and subsequently resumed spontaneous bursting as before. We manipulated the ionic properties of the neuron and found that in the absence of the Ih current the silent period following a period of stimulation was absent.

Finally, we validated the model by estimate the region of tissue influenced by DBS, i.e. how DBS can influence neurons located at different positions in the surrounding region. We modelled 50 STN neurons as before in a 2-dimensional grid. The cells were stimulated extracellularly using the calculated potential convolved with a time dependent square wave. For every position in this grid we plotted the number of action potentials induced by the DBS. Simulating the clinical parameter settings used for a Parkinsonian patient and plotting the results in a patient specific anatomical model, we found that the region of tissue in which the spontaneous neuronal firing was altered by DBS coincides with clinical observations of the optimal DBS site.

Results: Modelling of Action Selection via Partial Synchronization in the Basal Ganglia

Population Level Modelling

Our paper [35] describes a population level model of two interactive populations: one of excitatory STN neurons and one of inhibitory GPe neurons. Population level modelling does not include detailed spiking activity; instead, only one variable is used to describe the average population activity. Thus, in our model we use a two dimensional dynamical system: the variable $x(t)$ represents the average activity in the STN population and $y(t)$ represents the average activity in the GPe population. In particular, to describe population dynamics we use Wilson-Cowan equations [79]. Connections between populations include:

- An excitatory connection from STN to GPe, with average connection strength w_{sg};
- Self-inhibition of GPe neurons inside the population, with average connection strength w_{gg};
- An inhibitory connection from GPe to STN, with average connection strength w_{gs};
- Self-excitation of neurons in the STN population, with average connection strength w_{ss}.

Thus, the following equations describe the STN-GPe dynamics:

$$
\begin{aligned}
\tau_s \dot{x} &= -x + Z_s(w_{ss}x - w_{gs}y + I) \\
\tau_g \dot{y} &= -y + Z_g(w_{sg}x - w_{gg}y),
\end{aligned}
\tag{1}
$$

where, Z_s, Z_g are sigmoid functions for the STN and GPe populations respectively, τ_s, τ_g are time constants for the STN and GPe, I is a constant external input to the STN population, which we assume represents drive from the cortex to BG via the hyper-direct pathway. Some parameter values of the model were chosen on the basis of our previous studies of the Wilson-Cowan model [61] and some from another population-level model of the BG [63]. Nevado-Holgado et al. [63] determined two sets of parameter values, corresponding to healthy and Parkinsonian conditions, and we also use these two sets of values.

As it was already reported by several modelling studies (see e.g. [62]) there is a puzzling question about the value of the parameter w_{ss}. Experimental data does not confirm any significant self-excitation in STN population, i.e. it is likely that there is no self-excitation in the population of STN neurons ($w_{ss} = 0$). If we assume this condition is true then it is possible to prove that no oscillatory regime exists. However, we have reported above that there are many experimental and clinical studies which demonstrate oscillations both in healthy and Parkinsonian cases, for example, oscillations in beta range have been considered in many papers as a correlate of Parkinsonian conditions. Thus, the puzzle is: how can oscillations appear in the STN-GPe model if there is no excitatory self-connectivity in the STN?

To solve the puzzle, Holgado et al. suggest including synaptic connection delays into the model. From mathematical point of view this suggestion means considering a system of delayed differential equations (DDEs) instead of ordinary differential equations (ODEs). It is known that systems of DDEs have a wider repertoire of dynamical modes than systems of ODEs. For example, in a one dimensional ODE system oscillations are not possible but in a one dimensional DDE system they are.

A disadvantage of this approach is that experimental data on connectivity delays is mostly unknown. Therefore we prefer to use a system of ODEs for building our oscillatory model of the BG.

Another approach was been used in Gillies et al. [62], where it was suggested that there STN self-connections exist but are rather sparse and therefore difficult to find in experiments. To clarify this possibility we use the bifurcation analysis of our ODE model, assuming that all parameters are fixed except the strength of the external cortical input to the STN population (I) and the strength of STN self-excitation (w_{ss}). We found that there is a minimal threshold value w^{THR} for the existence of oscillations: if $w_{ss} < w^{THR}$ then oscillations are not possible (Fig. 2 Bifurcation diagram for the two dimensional model consisting of an STN and GPe population, under Parkinsonian conditions. Oscillations are possible for parameter values inside the regions C and D). However, for $w_{ss} > w^{THR}$ there are regions in parameter space where stable oscillatory activity is possible (regions C and D in Fig. 2). While our modelling suggests that the oscillatory regime dominates for Parkinsonian parameters, oscillatory regions do not exist or are very small when the healthy parameter values were used.

Based on our study of dynamical regimes, we suggest a new solution for the puzzle of oscillatory activity in STN-GPe. We assume that populations of STN-GPe neurons are organised in such a way that they can be considered as a collection of interactive "micro-channels". This channel hypothesis is popular and it has been

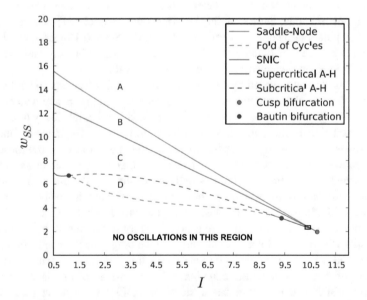

Fig. 2 Bifurcation diagram for the two dimensional model consisting of an STN and GPe population, under Parkinsonian conditions. Oscillations are possible for parameter values inside the regions C and D

used to describe movement control in the BG [47, 48, 80, 81]. Thus, we consider our model of two populations as a model of sub-populations of STN-GPe and call it as a micro-channel model. We assume that there is no self-connectivity in the STN ($w_{ss} = 0$) and, therefore, a single micro-channel model cannot generate oscillations.

To generalize the BG model, we consider N micro-channels and include local interaction between them. To introduce local coupling between micro-channels we assume that all channels are located either on a line (with mirroring conditions for two ending micro-channels) or a circle (therefore, there are no ends). Also, we assume that all micro-channels are identical and each channel is coupled with the right and left neighbors with the same connection strength α. Thus, interactions between micro-channels is characterized by one parameter α. The following dynamical system describes the dynamical behavior of the interactive micro-channels' activity:

$$
\begin{aligned}
\tau_s \dot{x}_k &= -x_k + Z_s(w_{ss}x_k - w_{gs}y_k + I_k) \\
\tau_g \dot{y}_k &= -y_k + Z_g(w_{sg}x_k - w_{gg}y_k + \alpha(y_{k-1} + y_{k+1})), \quad k = 1, \ldots N,
\end{aligned}
\tag{2}
$$

where α is the local connection strength for the neighboring GPe populations; all other parameters and functions have the same meaning as in the system (1). For the case of populations on a line, the two channels at the beginning and the end ($k = 1$, $k = N$) receive double coupling from existing neighbors, for the case of populations on a circle, the index k is considered modulo-N. For parameter values and sigmoid function specifications see [35].

If $\alpha = 0$ then the channels are independent and there are no oscillations in the network. If coupling between channels is weak then the oscillation are also absent. However, if the connection strength between micro-channels increases then oscillations can appear. For $\alpha > \alpha_{cr}$ oscillations appear in the system of coupled channels and the critical parameter value α_{cr} corresponds to a supercritical Andronov-Hopf bifurcation in the system with symmetry.

The homogeneous system of coupled micro-channels demonstrates an interesting behavior. Figure 3a–c show the activity in STN populations when 799 channels arranged in a line with connection strength $\alpha = 1$ are simulated. Due to coupling, the end points of the system begin to oscillate (Fig. 3a) and oscillations propagate along the line from both ends (Fig. 3b) resulting in a complex spatio-temporal pattern of population activities (Fig. 3c).

The parallel channel hypothesis of the BG [80] assumes that a set of particular channels can be considered as a "code" of some movement and the activation of a set of channels results in the movement execution. Therefore, a non-homogeneous system of interactive parallel channels is a good candidate for modelling of action selection. Of course, the separate channel should be complex enough and should include an oscillatory mode. Thus, the first step in developing a neural network model for action selection is to find a model of an "oscillatory channel" which can be used as an elementary unite for building a network. We suggest that the system

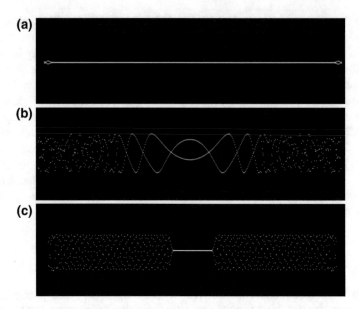

Fig. 3 Activities of 799 STN populations shown at different times. **a** Time near beginning of simulation. **b** Later in the simulation, when activity has almost propagated from the end points to the centre. **c** Even later in the simulation, showing complex activity of interactive channels

of two interactive micro-channels is the simplest possibility to compose the oscillatory channel: two pairs of STN-GPe populations coupled by inhibitory connection between the GPe subpopulations (see Fig. 4). This system is able to generate oscillations; therefore, we call it an oscillatory channel and the following equations describe the dynamics:

$$\tau_s \dot{x}_l = -x_l + Z_s(w_{ss}x_l - w_{gs}y_l + I)$$
$$\tau_g \dot{y}_l = -y_l + Z_g(w_{sg}x_l - w_{gg}y_l + \alpha y_r)$$
$$\tau_s \dot{x}_r = -x_r + Z_s(w_{ss}x_r - w_{gs}y_r + I)$$
$$\tau_g \dot{y}_r = -y_r + Z_g(w_{sg}x_r - w_{gg}y_r + \alpha y_l),$$
$$x = x_l + x_r; \quad y = y_l + y_r,$$

Here indexes l, r relate to the left and right populations, respectively; all parameters and functions are the same as above; $x(t)$, $y(t)$ are the output oscillatory channel activities for the STN and GPe respectively.

Figure 4 shows the oscillatory channel where the excitatory output is the summed activity of two STN populations (red curve) and the inhibitory output is the summed activity of two GPe populations. Figure 5 shows dynamics of output activities of the oscillatory channel: the total excitatory activity of STN populations versus time ($x(t)$ in red) and the total inhibitory activity of for and GPe populations versus time ($y(t)$ in blue).

Fig. 4 Graphical representation of oscillatory channel: *left* and *right* pairs of mini-channels are coupled by inhibitory connection between GPe populations; the output excitatory activity from the STN is shown by the *red wave*, and the output inhibitory activity from the GPe is shown by the *blue wave*

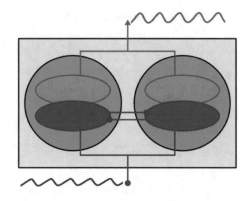

Fig. 5 Dynamics of output activities of the oscillatory channel: STN populations in *red* and GPe populations versus in *blue*

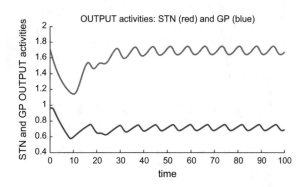

Figure 6 shows some additional details on micro-channel activities: STN populations of two microchannels oscillate in antiphase as do those in GPe populations.

Partial Synchronization as a Mechanism for Action Selection

Our oscillatory model of action selection employs a mechanism of partial synchronization. It was shown both experimentally and theoretically that partial synchronization between oscillating neural entities can be found in many brain structures and this mechanism can be considered as an important theoretical concept for understanding of many cognitive processes [31].

The model design is based on the following hypotheses. (1) We consider a set of oscillatory channels with different frequencies. Selection of some particular action corresponds to a partial synchronization of oscillatory channels, i.e. a small set of oscillatory channels demonstrate synchronous activity (the same frequency of oscillations) but all other oscillate with different frequencies. (2) The mechanism of partial synchronization is based on interaction of oscillatory channels with a special

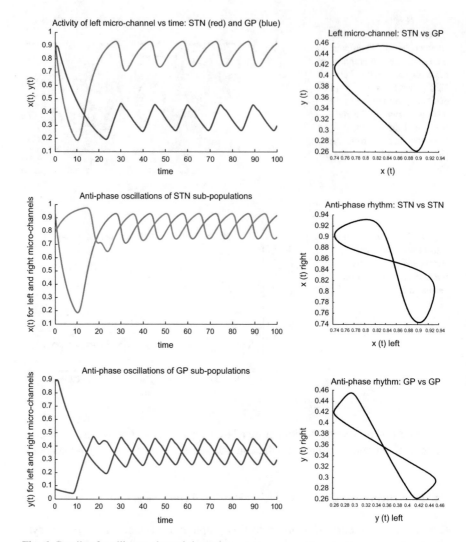

Fig. 6 Details of oscillatory channel dynamics

Central Oscillator (CO) which is itself an oscillatory channel. Figure 7 shows a diagram of the network with a CO which is shown at the top. This CO influence activities of all other oscillatory channels and they also influence the activity of the CO. Thus, the CO is coupled by mutual connections with each oscillatory channel. This type of connectivity is referred to as a star-like neural network (SLNN). (3) The activity dynamics of the CO depends on the total activity of all oscillatory channels. Also, the activity dynamics of each oscillatory channel depends on the activity of the CO. The following equations describe the dynamics of the system:

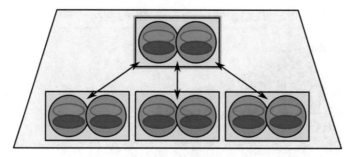

Fig. 7 Star-like architecture of interactive oscillatory channels

$$\tau_s\dot{x}_{COl} = -x_{COl} + Z_s\left(w_{ss}x_{COl} - w_{gs}y_{COl} + I - \beta_1 \sum_{k=1}^{N} y^k\right)$$

$$\tau_g\dot{y}_{COl} = -y_{COl} + Z_g\left(w_{sg}x_{COl} - w_{gg}y_{COl} + \beta_2 \sum_{k=1}^{N} x^k - \beta_3 \sum_{k=1}^{N} y^k\right)$$

$$\tau_s\dot{x}_{COr} = -x_{COr} + Z_s\left(w_{ss}x_{COr} - w_{gs}y_{COr} + I - \beta_1 \sum_{k=1}^{N} y^k\right)$$

$$\tau_g\dot{y}_{COr} = -y_{COr} + Z_g\left(w_{sg}x_{COr} - w_{gg}y_{COr} + \beta_2 \sum_{k=1}^{N} x^k - \beta_3 \sum_{k=1}^{N} y^k\right)$$

$$x_{CO} = x_{COl} + x_{COr}, \quad y_{CO} = y_{COl} + y_{COr},$$

$$\tau_s\dot{x}_l^k = -x_l^k + Z_s\left(w_{ss}x_l^k - w_{gs}y_l^k + I - \gamma_1 y_{CO}\right)$$

$$\tau_g\dot{y}_l^k = -y_l^k + Z_g\left(w_{sg}x_l^k - w_{gg}y_l^k + \gamma_2 x_{CO} - \gamma_3 y_{CO}\right)$$

$$\tau_s\dot{x}_r^k = -x_r^k + Z_s\left(w_{ss}x_r^k - w_{gs}y_r^k + I - \gamma_1 y_{CO}\right)$$

$$\tau_g\dot{y}_r^k = -y_r^k + Z_g\left(w_{sg}x_r^k - w_{gg}y_r^k + \gamma_2 x_{CO} - \gamma_3 y_{CO}\right)$$

$$x^k = x_l^k + x_r^k, \quad y^k = y_l^k + y_r^k, \quad k = 1, 2, \ldots, N,$$

where the CO variables x_{COl}, y_{COl}, x_{COr}, y_{COr} describe activities of STN and GPe populations for left and right micro-channels of the CO, respectively; variables x_l^k, y_l^k, x_r^k, y_r^k describe activity in the STN and GPe populations for left and right micro-channels of the kth oscillatory channel; parameters β_1, β_2, β_3 describe connection strengths from the oscillatory channels to the CO; parameters γ_1, γ_2, γ_3 describe connection strengths from the CO to each oscillatory channel; other parameter have been specified in Eq. (1).

The results of model simulations are shown in Figs. 8 and 9. First we consider a set of 50 uncoupled oscillatory channels with different values of parameter w_{gg}. These values are distributed in the range [a, b]. Figure 8 shows the one-to-one correspondence between the coupling parameter w_{gg} of an oscillatory channel and its frequency of oscillation. The horizontal axis shows the number of the oscillatory channel, with the channels ordered such that the coupling parameter w_{gg} increases

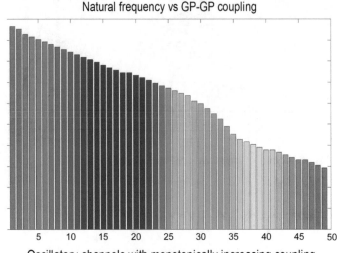

Fig. 8 Frequency distribution

with the channel number. The vertical axis shows the frequency of oscillations, which monotonically decreases.

Figure 9a–d shows results of SLNN simulations. Each panel of this figure corresponds to some particular frequency of CO and a group of oscillatory channel is selected to be in synchrony with the CO. The partial synchronization mode means that a selected group of oscillatory channels has the same frequency as the CO but all other channels oscillate with different frequencies. Figure 9a–d show the effect of increasing the natural frequency of the CO, such that a group of oscillatory channels with frequencies close to the frequency of the CO become synchronous with it.

Conclusions

Our study of the dynamics of oscillatory neuronal network comprised of excitatory and inhibitory populations demonstrated that an oscillatory regime is stable and exists for a broad range of parameter values, particularly under Parkinsonian conditions. Bifurcation theory allows us to study dynamical modes under parameter variation and provides a valuable correspondence between regions in parameter space and the rhythmic activity of some particular frequency. In fact, the parameter space can be partitioned into zones of rhythmic activities in a particular band: a region of alpha-rhythm, a region of beta-rhythm, etc. These results provide the possibility of controlling both the dynamical mode and the frequency of oscillations, as well as finding the parameter values that correspond to desirable dynamics. Our study shows that the regime of partial synchronization could be of interest for

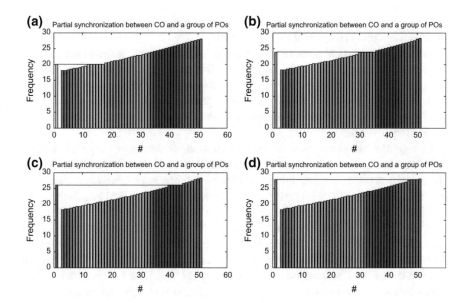

Fig. 9 Partial synchronization

modelling motor control and action selection. Further work should be done on frequency-phase action coding.

The model presented in this section was based on population-level modelling of the nuclei within the basal ganglia, where the underlying rhythms relating to action selection were generated within the BG themselves. In the next section we will demonstrate another model for action selection via partial synchronization in the basal ganglia that is instead based on spiking neurons. In this model, the oscillations that select actions are assumed to arise in the cortex, and act to synchronize neurons in the basal ganglia to select particular output channels.

Action Selection via Partial Synchronization in a Spiking Model of the Basal Ganglia

Phase Locking and Arnold Tongues

Much of the present study in physiology focuses on oscillatory activity at different frequencies that arises in various regions in the brain. According to Smeal et al. [82], a large body of evidence suggests that such rhythms are generated by synchronized neural activity that results from functional connectivity between neurons. According to Timmermann et al. [83], "it is widely agreed that excessive synchronizations are linked to the motor symptoms of PD". Moreover, "partial synchrony in cortical networks is believed to generate various brain oscillations, such as the alpha and gamma rhythms" [84].

From a mathematical point of view we can model a single neuron in the absence of noise as non-linear oscillator. This assumption may seem unrealistic but there is experimental evidence to support it [82]. Furthermore, Izhikevich [84] was able to reproduce a huge variety of different firing patterns from a system of only two non-linear ordinary differential equations, simply by changing the parameters that govern the equations.

One of the simplest cases of synchronization in oscillatory regimes can be observed in a self-sustained oscillator driven by an external force. Such mechanisms can describe many biological phenomena that can be observed in nature, such as synchronization of clocks that govern the circadian rhythm [85] or simultaneous flashing of fireflies [86]. Assuming the frequency of forcing is close enough to the oscillator's intrinsic frequency, the steady state solution of a periodically-forced oscillator is synchronous with the periodic input. Under this condition the phase difference of the two oscillators approaches a constant value—this is a stable fixed point of the system. In dynamical system theory the oscillator in this state is called **phase-locked**, because its frequency is locked to the forcing frequency.

We can extend the concept of phase-locking to two neurons coupled by single or multiple synapses. It is important to note that phase-locking does not necessarily mean that the two cells are firing at the same time, but rather they may fire with some constant delay between spikes. However, when a subset of spiking neurons in a population have similar phases and are phase-locked to the same periodic force they will fire in unison. We will call such behavior **partial synchronization**.

Synchronization in a periodically forced oscillator may also appear at other frequencies besides the one close to the driving frequency. In fact, we can generalize the concept of phase-locking when the ratio between the period of the force and the period of the oscillator is a rational number. We will say that a forced oscillator with period T is **p:q—phase-locked** to the force with period T_f if $pT \approx qT_f$ where p and q are positive integers. If the period of forcing (T_f) is kept constant then two parameters affect the ability of an oscillator to phase-lock: the strength of the forcing input (k) and the oscillator's intrinsic period of oscillation (T_i). For each pair p, q we may thus calculate the region in the (k, T) parameter space where p:q phase locking occurs; these regions are called **Arnold tongues**.

Phase locking can appear also in maps that act as non-forced oscillators. Let us consider a simple example: the circle-map. In dynamical systems a map is defined as an equation with discrete times, and it can be represented through an evolving sequence $\theta = \{\theta_n\}_{n=1}^{+\infty}$. The circle map is a particular θ that solves:

$$\theta_{n+1} = \theta_n + \Omega - \frac{K}{2\pi} \cdot \sin(2\pi\theta_n) \quad \mod 1$$

Starting from a fixed initial value θ_0, mod is the standard modulo operator, Ω and K are system parameters. The sequence θ represents the angle variation over a circle. We define rotational number:

$$r = \lim_{n \to +\infty} \frac{\sum_{i=1}^{n} \theta_i}{n}$$

Thus p:q—phase-locked Arnold tongues are defined as a region in the space of parameters where there the system solution has is locally constant rotational number $r = p/q$. In Fig. 10 the different coloured regions indicate the different rotational numbers obtained by varying $(\Omega, K) \in [0, 1] \times [0, 2\pi]$ are shown. Some values of the rotational number in Fig. 10 are rational, and eventually are surrounded by other equal values of r, in this case, they will form an Arnold tongue region in the parameter space.

Cellular Model

The basal ganglia are a group of subcortical brain areas thought to play a role in action selection. Here we investigate the role of interplay between neurons in the *subthalamic nucleus* (STN) and those in the internal segment of the *globus pallidus* (GPi). We used a simple Izhikevich model of spiking neurons with class 1 excitability [42] for all neurons. This model is computationally tractable yet still reproduces two important features of basal ganglia neurons: excitability and spiking behaviour. We will not discuss in detail the validity of the model in the description of the neural dynamics (for a deeper insight see [42]).

The model consists of two populations of neurons representing the STN and GPi, containing $N_s = 200$ and $N_g = 25$ neurons respectively. Each unit is modelled according to the two dimensional Izhikevich model, where the equations governing the dynamics of neuron i are:

$$V_i' = 0.04V_i^2 + 5V_i + 140 - u_i + I_i + I_{ext} + I_{syn} + \eta_i$$
$$u_i' = a(bV_i - u_i)$$
$$V_i > 30: V_i \leftarrow c; u_i \leftarrow u_i + d$$

We used standard values for all the parameters $(a = 0.02, b = 0.2, c = -65, d = 6)$, which correspond to neurons that are normally quiescent and have

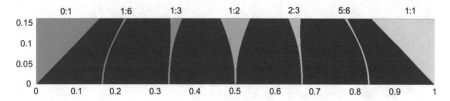

Fig. 10 Colored plot of the rotational number r for circle map at varying Ω in $[0, 1]$ (horizontal axis) and K in $[0, 1/2\pi]$ (vertical axis). Not every value of r corresponds to a rational number, and thus to an Arnold tongue

class 1 excitability according to Izhikevich's classification. The term I_i corresponds to a constant external current that is different for each neuron and has the effect of giving each neuron a different intrinsic spiking frequency. The term I_{ext}, which is defined more fully below, is the oscillatory forcing input that is identical for each neuron in a population. I_{syn} represents the sum of all the synaptic currents flowing in neuron i. Finally, η_i is a Brownian input of white noise. Initially we set $\eta_i = 0$ to consider the simpler deterministic case as this makes it easier to determine the different partial synchronization regimes, but we will later add noise to obtain more realistic results.

In the model, we choose values of I_i for the STN population uniformly from the range 4–7, so that without external input the STN units all spike independently, with a linear range of frequencies from 7–18 Hz that is similar to the range of spiking frequencies seen experimentally in monkeys [13]. In this model we consider the STN neurons as being arranged on a line, with the injected current (and therefore frequency of spiking) varying monotonically along the line. For GPi neurons we set $I_i = 0$, so that the neurons in the GPi do not intrinsically spike. Figure 11 shows the intrinsic spiking activity of the STN population.

In order to simulate oscillatory cortical input to the STN we apply an identical external current comprising multiple frequency components to all STN neurons:

$$I_{ext} = \sum_j a_j \sin(2\pi\omega_j t)$$

where a_j and ω_j are the amplitude and frequency of oscillatory component j, respectively. For the GPi neurons we set $I_{ext} = 0$.

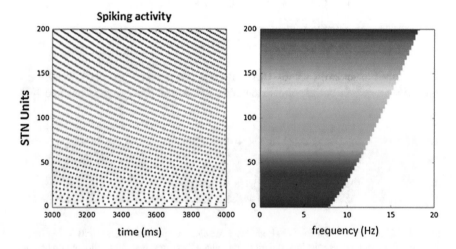

Fig. 11 STN spiking activity without external input. *Left* Raster plot showing each neuron's spike train. *Right* The intrinsic spiking frequency of each STN neuron

Synaptic Transmission

Each STN neuron makes an excitatory synapse onto one GPi neuron in a uniform manner, such that each GPi neuron receives $\frac{N_s}{N_g} = 8$ synapses. Like in the STN, we also consider the GPi units to be arranged on a line, with the STN $\rightarrow$ GPi projection organised in a topographical fashion, such that each GPi neuron receives input from a group of STN neurons that are adjacent to each other (and therefore have similar intrinsic spiking frequencies). Additionally, since there is biological evidence for inhibitory synaptic connections between nearby neurons in the globus pallidus [44], model GPi neurons receive inhibitory connections from their $N_{gg} = 4$ neighbouring neurons on each side (Fig. 12).

For both excitatory and inhibitory connections we used standard exponential chemical synapses. The total synaptic current flowing at each time t in neuron i is given by:

$$I_{syn}(t) = I_E + I_I = g_E(t)(E_{rev} - v(t)) - g_I(t)(I_{rev} - v(t))$$

where $E_{rev} = 0$ mV and $I_{rev} = -80$ mV are the reversal potentials, and determine if the synapse is inhibitory (I) or excitatory (E). We initially set $g_E = g_I = 0$. If two cells are connected through a synapse of type x (where x is E or I), the postsynaptic response in neuron j in following each presynaptic action potential from neuron i arises by an increment of the conductance $g_x \leftarrow g_x + w_x$, where w_x is a parameter. The synaptic conductance follows an exponential decay determined by equation:

$$\begin{cases} g'_E = -\frac{g_E}{\tau_E} \\ g'_I = -\frac{g_I}{\tau_I} \end{cases}$$

Fig. 12 Overview of the populations in the model and their connections

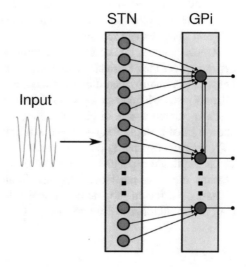

Here τ_x is the decay time of each synapse. Each STN neuron i can form excitatory synapses onto the output GPi units. We scaled the excitatory impact of STN
units according to their intrinsic frequency using the following procedure: We fixed
the maximal weight w_x^1 and decay time τ_x^1 for the highest frequency STN unit and
we scaled these values down uniformly with decreasing frequency, reaching minimum weight $w_x^2 = w_x^1 - 0.15$ and decay time $\tau_x^2 = \tau_x^1 - 0.015$ for the lowest frequency STN unit.

For the moment we do not specify the values of $\bar{w}_x$ and $\bar{\tau}_x$ because we will
discuss them later in more detail, and show that they play a very important role in
the switching time between selected actions.

Software

Simulations of unconnected periodically forced Izhikevich oscillators were performed in MATLAB (MathWorks, Inc), while simulations of connected neural
networks were developed in NEURON [78] interfaced with Python [87].

Results

We first studied the effects of a cortical input with a single sinusoidal oscillatory
component $I_{ext} = a_0 sin(2\pi\omega_0 t)$ on the STN population under a deterministic regime.
With this forcing input applied to each of the STN neural oscillators, the system
produces regions of partial synchronization with a set of p:q—phase-locked regions
for any fixed input frequency. Each STN neuron remains an oscillator under forcing
but its period of oscillation may be different to its intrinsic period. When the period
of oscillation of an STN neuron is such that it completes p cycles for every q cycles
of the forcing signal, we describe that neuron as being p:q phase-locked. In Fig. 13a
we show three phase-locked regions found with fixed $\omega = 16.5$ and $a_0 = 0.1$. Figure 13b shows plots against time and phase portraits from STN neurons in the
different phase-locked regions.

Arnold tongues illustrate the multiple synchronizations found in STN cells
depending on their intrinsic frequencies. The Arnold tongue diagrams in Fig. 14
helps us to visualize the different areas of partial synchronization obtained for three
different external frequencies in STN neurons at varying external amplitude.

Each colored region in Fig. 14 corresponds to a partial synchronization region,
showing that a single frequency cortical input can synchronize multiple groups of
STN neurons depending on their intrinsic frequencies. Increasing the input oscillation amplitude expands each 1:1 phase-locked region (orange colour), squeezing
together the other p:q locked regions. Changing the forcing frequency shifts the
phase-locked regions up and down, whilst largely preserving the total area of each
region. Arnold tongues provide a useful tool for determining the correct parameters

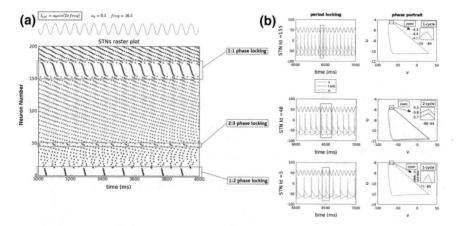

Fig. 13 a Activity in STNs using a single sinusoidal input with parameters $\omega = 16.5$, $A = 0.1$. Three regions of partial synchronization appear, corresponding to different p and q values. **b** A member of each locked region is represented with the plot of v, u and I_{ext}. In order to make u and I_{ext} visible in the plot they were multiplied by values 5 and 120, respectively. A small *rectangle* points out the period of q input oscillations and the number of p spikes in such period. The phase portrait of the two solutions u and v shows the correspondence between the number of p-cycles and the value p in the phase-locking

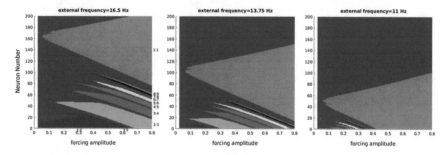

Fig. 14 Phase-locked regions of Arnold tongues. Each colour represents a single p:q pair for three different values of external frequencies: 16.5, 13.75 and 11 Hz. The *horizontal axis* shows the strength of external forcing and the *vertical axis* shows the STN number (arranged in order of increasing intrinsic frequencies). The biggest colored region (*orange*) of partial synchronization corresponds to 1:1 locking. From the plot with external frequency of 11 Hz we notice that no Arnold tongues are present above the 1:1 locking region. Decreasing the external frequency doesn't change the area of the 1:1 region, but shifts the synchronized regions downwards while slightly squeezing together the other phase locked regions

to obtain a desired number of synchronized cells. The Arnold tongues suggest that neurons with intrinsic frequencies above that of the forcing input cannot become synchronized, and this appears to be a universal property.

Fig. 15 Raster plot of 2000
STN units forced by a
16.5 Hz single-frequency
sinusoidal force with
amplitude of 0.12. The STN
have frequencies ranging
from 0 Hz (*bottom*) to 40 Hz
(*top*). No phase locked units
appear in cells with frequency
above 16.5 Hz. This
important property is not
maintained when we have
inputs with multiple
frequency components

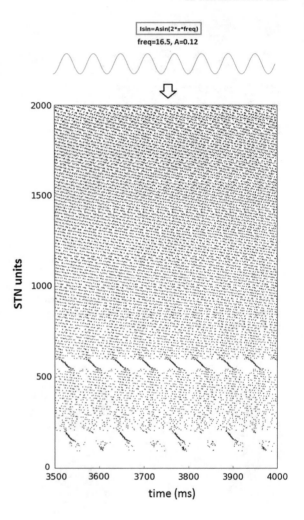

For example, Fig. 15 shows a simulation with a large number of neurons that
have intrinsic frequencies above that of the input, yet none of these neurons become
synchronized. As we will show shortly, this feature is no longer true when the
forcing input contains multiple frequency components.

The phenomena of partial synchronization may lead to the selection of different
channels via activation of GPi units. We now add the GPi units into the network, by
defining non-zero weights for the excitatory and inhibitory connections. We chose
maximum weights $w_I^1 = -2$, $w_E^1 = 1.2$ and decay times $\tau_I^1 = \tau_E^1 = 0.1$. We also added
the white noise component $\eta_i \sim N(0, \sigma^2)$ to STN cells, where $N(0, \sigma^2)$ represents a
Brownian random variable. We choose the value of σ^2 ranging from 0.002 to 0.01,
such that lower-frequency neurons receive weaker noise than higher frequency ones.

In our model actions are selected via activation of groups of GPi neurons, which we assume belong to different "channels" of information flow through the basal ganglia. Figure 16 shows the spiking activity of the same network of STN and GPi neurons in response to a single-frequency (16.5 Hz) oscillatory input of varying amplitude. When the amplitude is low (bottom panel), two small groups of partially synchronized STN neurons appear via 1:1 and 1:2 phase-locking, and their synchronized firing activates two groups of GPi neurons. GPi neurons with above-average firing rates are considered "activated", and are indicated by the solid black bar on the far right of the figure. However, at higher amplitudes a single-frequency oscillatory input can give rise to other selected channels' combinations, as shown in Fig. 16 (top panel). This follows from the result of the Arnold tongue diagrams shown in Fig. 14, as these showed that multiple phase-locked groups of STN neurons can appear for a single oscillatory input if the input amplitude is big enough.

We now consider external inputs that contain multiple frequency components. Figure 17 shows how an input containing two, three or four frequency components can select groups of GPi neurons. The strength of the oscillatory input in Fig. 17 was chosen such that in each case the selected neurons in the GPi are activated by groups of STN neurons that are in 1:1 phase-locked synchronization. The ability to select multiple combinations of output channels in response to mixed-frequency input may be the basis of action selection in the basal ganglia.

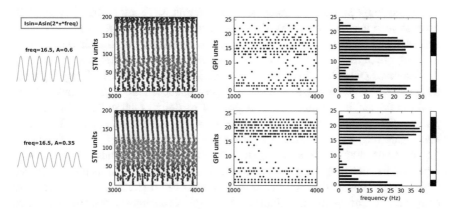

Fig. 16 A single-frequency (16.5 Hz in this case) oscillatory input can give raise to multiple synchronization regions, in agreement with the Arnold tongues. The *bars* on the far *right* of the figure show which GPi neurons are considered selected, which is defined as those that have an above-average firing rate. An input amplitude of 0.6 selects only two clear output channels despite the high number of different p:q phase-locked regions in the Arnold tongue diagram (Fig. 14), while an amplitude of 0.35 selects three channels. Unsurprisingly, the size of the channels (number of GPi neurons recruited) increases with the amplitude of forcing

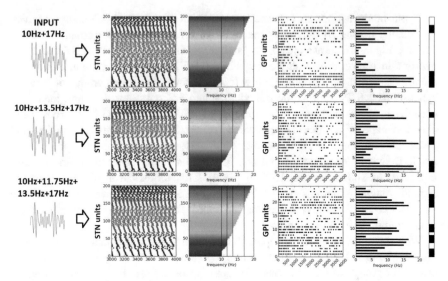

Fig. 17 A single forcing input with multiple frequency components having the same amplitude of 0.12 activates multiple GPi "channels" via partial synchronization. This figure shows, from *top* to *bottom*, the activation of two, three or four output channels by an oscillatory input containing two, three or four frequency components. The number of active channels is in one to one correspondence with the number of forcing frequencies

Arnold tongues are determined using a deterministic regime, but they can still give us an idea of how many different channels may be open in a stochastic approach and predict the amplitude required for synchronizing a desired amount of cells. A small amount of white noise preserves the mean number of partially synchronized cells identified by the tongues. If this number is sufficiently high, the STN excitatory strength will activate a single channel in the GPi neurons.

As in the single frequency case, regions of partially synchronized STN units in phase locked regimes other than 1:1 can cause additional channels to be selected. To investigate this, we calculated Arnold tongue diagrams for the case with multiple forcing frequency components. Figure 18 shows Arnold tongues for an external input that has two frequency components: 16.5 and 13.75 Hz. In this figure there are 400 STN units with intrinsic firing rates from 0 to 40 Hz. Arnold tongues were computed according to the definition of p:q–phase locking, considering the two input periods separately to produce two diagrams. The shapes of the Arnold tongues in Fig. 18 are the same for the two frequencies, since each neuron is forced by both frequency inputs, thus its period is a multiple of both the two-forcing periods. However, the values of p and q for any given region differ between the two plots. For example, the biggest region of partial synchronization (green on the left, dark blue on the right) corresponds to 1:1 phase–locking for the 16.5 Hz component of the input and to 6:5 phase–locking for the 13.75 Hz component 13.75. The second biggest region (cyan on the left, sea green on the right) correspond to 1:1 phase–locking for the 13.75 Hz component and 5:6 phase–locking for the 16.5 Hz component.

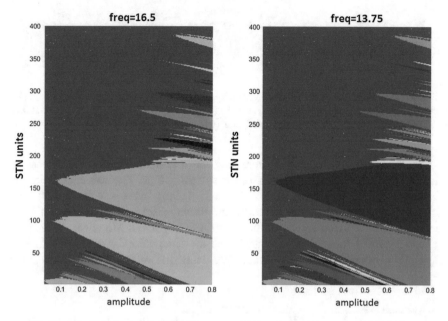

Fig. 18 Two forcing frequencies Arnold tongues. Values for p and q are computed from the definition of p:q phase-locked regions using each of the two frequency components. The *left figure* shows Arnold tongues calculated in relation to the 16.5 Hz component of the input, and the *right figure* shows Arnold tongues for the 13.75. The shapes are equal, but the values of p and q differ

The two main regions of partial synchronization thus correspond to 1:1 phase-locking in the case of two forcing frequencies, which is similar to the results in Fig. 14. However, these regions have decreased area compared to the single-frequency case and they squeeze together, forming more tongues in between them. Also, in contrast to the single-frequency component case, other Arnold tongues appear for intrinsic frequencies higher than the biggest phase-locked regions, leading to more complex situations. The input signal coming from the cortex is likely composed of many frequency components and amplitudes, thus our consideration of only single and double frequency input is a simplification. Nevertheless we discovered that big regions of 1:1 phase-locking are preserved for each frequency, and their impact on GPi units is maintained even a noisy regime.

Finally, we consider the time taken to select different channels in the output GPi population. Physiologically, the switching time between different actions is thought to be in the order of hundreds of milliseconds [88].

We investigated the effects of synaptic weights and decay time constants on the time taken to switch between sets of activated channels. Specifically, we varied the maximal weights and decay times for the excitatory synapses, and looked for changes in the action selection time. Figure 19 shows four different activated channel switches and each switch takes approximately 400 ms, here we use the fixed coupling strength and decay time: $\left(w_E^1; \tau_E^1\right) = (0.5, 0.16)$.

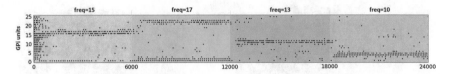

Fig. 19 Changing selected channels by switching the forcing input. Each *coloured region* corresponds to a different input frequency, which changes every 6 s. The amplitude is fixed at 0.12 for the entire simulation time. In this simulation $\left(\mathbf{w}_E^1; \tau_E^1\right) = \left(\mathbf{0.5}, \mathbf{0.16}\right)$

Figure 20 demonstrates that increasing the maximum excitatory conductance and decay time of synapses decreases the switching time. The switching time was calculated as the first firing of the GPi units (having indexes in the set Ω) with closest intrinsic frequency to the switching forcing frequency in the deterministic case, so that we avoid sporadic situations caused by noise. Moreover, in order to obtain a correct comparison we maintained the same noise for different values of synaptic conductance and decay time. We define the vector of spike timings for

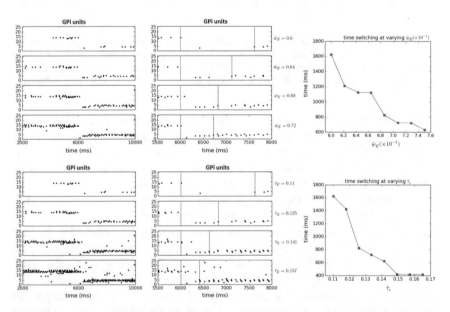

Fig. 20 Decrease in action-selection switching time at varying excitatory strength (*top*) and decay time (*bottom*). The values of conductance $\mathbf{w}_E^1$ vary in the interval [0.6, 0.75] with fixed maximal decay time 0.12. The decay times τ_E^1 vary in the interval [0.11, 0.165] with fixed maximal conductance 0.65. The simulation lasts 12 s. The switch takes longer than a second with low $\mathbf{w}_E^1$ and τ_E^1, but we can achieve realistic results increasing $\mathbf{g_e}$ and $\tau_\mathbf{e}$. The amplitude of the forcing oscillator is of 0.12 for the entire simulation, while the frequency switches from 14 to 10 Hz at 6 s. The set $\Omega = \{\mathbf{3}\}$ contains the index of the GPi unit excited by most STN units with intrinsic frequency closer to 10 Hz used in the calculation of the switching time

each unit i (i.e. spike train) as spk_i, so that the formula used to calculate the switching time has the form:

$$t_{switch} = \min \left\{ \min_t \left\{ spk_i(t) : t \geq \frac{tstop}{2} \right\} : i \in \Omega \right\}$$

Discussion

We are interested in oscillations arising in the firing rate neuronal activity of the basal ganglia, since experimental evidence has shown that excessive oscillatory synchronization is positively correlated with the symptoms of Parkinson's disease [63]. Moreover, oscillations are thought to play a fundamental role in the functional physiology of the basal ganglia [22].

In the past years several computational models were developed that reproduced pathological oscillations in the basal ganglia. Different approaches showed how oscillations can be generated either intrinsically within the basal ganglia circuitry [44, 89] or extrinsically by an oscillatory input coming from an external source, such as the cortex. Experiments on rats by [90] suggest that increase low-frequency oscillations in the STN and GP are due to inappropriate processing of rhythmic cortical input. Ebert et al. [89] included STN self-excitation in their model, but experimental evidence [91] shows that STN axonal arborisation is not self-targeted, but rather it projects into the globus pallidus (either external or internal part), the substantia nigra pars reticulata and the striatum. Terman et al. [44] focused on the interplay between STN excitation on the GP units and the back-inhibition on the STN from the GP combined with STN strong hyperpolarization during back-inhibition.

Here we have presented a model in which an oscillatory cortical input partially synchronizes unconnected STN neurons, and consequently selects different combinations of output channels in the GPi through excitation. Our results support the idea that the input coming from the cortex could be separated into multiple frequency components, where the dominant components (higher amplitudes) would influence the STN by creating regions of partial synchronization, and consequently activating combinations of different channels on the output GP. We assumed the existence of inhibitory connections between GPi units, because there is evidence for self-inhibition in the GPe [49] but similar data for the GPi is lacking. However, this feature is not a critical part of the model, and similar results can be achieved without it.

Switching time of new actions is strongly dependent on the synaptic strength and decay time, and we were able to find a proper set of parameters that fit the realistic switching time [88]. Future development of the model could study how dopamine input influences the switching time of new actions and compare parkinsonian (lack of dopamine) conditions to healthy ones. Experimental evidences suggest that dopamine increases the activity in the inhibitory GPe projection on the STN [92], and the activity in the STN neurons is consequently reduced [90]. According to Nevado-Holgado [63], advanced Parkinson's disease increases specific synaptic weights in the basal ganglia, in particular the increase in the STN excitation towards the GP. According to this predictions our results would demonstrate that switching time of new actions would much faster during the Parkinsonian state. Drastic increases of synaptic conductance and decay time lead to almost instantaneous switching. If we further increase the synaptic strengths (g_E^1 and τ_E^1) becomes too high and we do not clearly recognize which channels are selected. Thus Parkinsonian conditions may lead to critical values of synaptic strengths in which the selection of new actions are hardly processed by the brain. This may be related to the motor symptoms of the disease.

Take Home Message for Neurologists

Although oscillations may play a physiological role in the basal ganglia in terms of action selection, the oscillatory nature of basal ganglia activity can reveal many pertinent features of Parkinson's disease pathology, as shown by electrophysiologcal recordings and discussed in section "The Basal Ganglia and Parkinson's Disease". Using biologically realistic computational models to understand the dynamics of the pathological activity and to visualise the effects of treatment has the capacity to directly aid the clinician. Such studies can then allow one to explore how to minimise the pathological activity and promote healthy activity (section "Results: Modelling of Action Selection via Partial Synchronization in the Basal Ganglia"). In particular, computational modelling can visualise and clarify the effect of deep brain stimulation on neurons in the basal ganglia and compare how different stimulation settings change the neuronal firing patterns. Using such models in the clinic would highlight for clinicians what stimulation paradigms can maximise therapeutic benefit while minimising side effects, energy use and damage at the interface (section "Modelling of DBS"). Such work is already underway, with patient-specific models being built into DBS planning software, and therefore directly integrated with the hardware. The use of DBS extends to the treatment of an increasing number of neurological and psychological disorders, for example, recent work from our group has looked into the thalamocortical cerebellar network and examined the oscillations of the network in essential tremor. In the future, such computational models could be used routinely, pre-, intra- and post-operatively to aid the entire clinical process. Aside from this clearly very practical benefit, the results presented in section "Results: Modelling of Action Selection via Partial Synchronization in the Basal Ganglia" show how computational modelling allows hypotheses how about activity (both physiological and pathological) in the basal ganglia might arise. In the longer term, the understanding of the basal ganglia and PD that such models give us might permit even more advanced novel treatments to be developed.

Take Home Message for Computationalists

Both experimental and theoretical studies strongly support a concept of oscillatory neural activity in the basal ganglia. Irregular and regular oscillations of different frequencies, synchronised, coherent and not synchronised can be observed both in healthy and pathological cases (section "The Basal Ganglia and Parkinson's Disease"). The goal of computational modelling in this respect is to clarify the correspondence between oscillatory dynamics and the functional state, be it healthy or pathological. These computational studies aim to reveal the neuronal mechanisms of movement and action selection both in the healthy brain, to try to understand how these mechanisms go wrong in disease (section "Results: Modelling of Action Selection via Partial Synchronization in the Basal Ganglia"). Our modelling of oscillatory dynamics shows that partial synchronization is a powerful theoretical approach and can be used for formulation of new theories on brain functioning. In particular, we demonstrate how partial synchronization can be applied to model the action coding and movement selection in the healthy and parkinsonian basal ganglia. Furthermore, we have discussed the important role of patient-specific models, taking into account individual differences in anatomy, or oscillation frequency or for DBS in electrode location (section "Modelling of DBS"). These subtle differences can have an important impact if we are to use such models to optimize therapy on a patient-to-patient basis.

Acknowledgments RB and RM-H acknowledge financial support from BBSRC UK (Grant BB/L000814/1).

References

1. Joundi R, Jenkinson N, Brittain JS, Aziz TZ, Brown P (2012) Driving oscillatory activity in the human cortex enhances motor performance. Curr Biol 22:403–7
2. de Lau LM, Breteler MM (2006) Epidemiology of parkinson's disease. Lancet Neurol 5 (6):525–535
3. Jankovic J (2008) Parkinson's disease: clinical features and diagnosis. J of Neurol Neurosurg Psychiatry, 79(4):368–376
4. Dauer W, and Przedborski S (2003) Parkinson's disease: mechanisms and models. Neuron 39 (6):889–909
5. National Institute for Health and Clinical Excellence (2006) NICE. Parkinson's disease: diagnosis and management in primary and secondary care. London
6. Rascol O, Goetz C, Koller W, Poewe W, Sampaio C (2002) Treatment interventions for parkinson's disease: an evidence based assessment. The Lancet 359(9317): 1589–1598
7. Bain P, Aziz T, Liu X, Nandi (2009) Deep Brain Stimulation. Oxford University Press, Oxford, UK
8. Kringelbach LM, Green AL, Owen SL, Schweder PM, Aziz TZ (2010) Sing the mind electric - principles of deep brain stimulation. Eur J Neurosci 32(7):1070–1079
9. DeLong MR (1990) Primate models of movement disorders of basal ganglia origin. Trends in Neurosciences 13(7):281–285

10. Filion M, Le´on Tremblay (1991) Abnormal spontaneous activity of globus pallidus neurons in monkeys with MPTP-induced parkinsonism. Brain Research 547(1):140–144

11. Miller WC, DeLong MR (1987) Altered tonic activity of neurons in the globus pallidus and subthalamic nucleus in the primate mptp model of parkinsonism. Advances in Behavioral Biology 32:415–427

12. Bergman H, Wichmann T, DeLong MR (1990) Reversal of experimental parkinsonism by lesions of the subthalamic nucleus. Science, 249(4975):1436–1438

13. Bergman H, Wichmann T, Karmon B, DeLong MR (1994) The primate subthalamic nucleus II. neuronal activity in the MPTP model of parkinsonism. Journal of Neurophysiology 72 (2):507–520

14. Calabresi P, Picconi B, Tozzi A, Ghiglieri V, Di Filippo M (2014) Direct and indirect pathways of basal ganglia: a critical reappraisal. Nature Neurosc 17:1022–30

15. Wei W., Rubin, J, Wang X-J (2015) Role of the indirect pathway of the basal ganglia in perceptual decision making. J Neurosc 35(9):4052–64

16. Kleiner-Fisman G, Herzog J, Fisman DN, Tamma F, Lyons KE, Pahwa R, Lang AE, Deuschl G (2006) Subthalamic nucleus deep brain stimulation: Summary and meta-analysis of outcomes. Movement Disorders 21(S14):S290–S304

17. Kumar R, Lozano AM, Kim YJ, Hutchison WD, Sime E, Halket E, Lang AE (1998) Double-blind evaluation of subthalamic nucleus deep brain stimulation in advanced parkinson's disease. Neurology 51(3):850–855

18. Patricia L, Krack P, Pollak P, Benazzouz A, Ardouin C, Hoffmann D, Benabid AL (1998) Electrical stimulation of the subthalamic nucleus in advanced parkinson's disease. N Engl J Med 339(16):1105–1111

19. McIntyre CC, Mori S, Sherman DL, Thakor NV, Vitek JL (2004a) Electric field and stimulating influence generated by deep brain stimulation of the subthalamic nucleus. Clinical Neurophysiology 115:589–595

20. McIntyre CC, Savasta M, Kerkerian-Le Goff L, Vitek JL (2004b) Uncovering the mechanism (s) of action of deep brain stimulation: activation, inhibition, or both. Clinical Neurophysiology 115:1239–1248

21. Mallet N, Pogosyan A, Márton LF, Bolam JP, Brown P, Magill P (2008) Parkinsonian Beta Oscillations in the External Globus Pallidus and Their Relationship with Subthalamic Nucleus Activity. The Journal of Neuroscience 28(52):14245–14258

22. Jenkinson N, Brown P (2011) New insight into the relationship between dopamine, beta oscillations and motor function. Trends Neurosci 34(12):611–618

23. Hutchison W, Dostrovsky J, Walters J, Courtemanche R, Boraud T, Goldberg J, Brown P (2004) Neuronal oscillations in the basal ganglia and movement disorders: evidence from whole animal and human recordings. J Neurosc 24:9240–9243

24. Bar-Gad I, Elias S, Vaadia E, Bergman H (2004) Complex locking rather than complete cessation of neuronal activity in the globus pallidus of a 1-methyl-4-phenyl-1, 2, 3, 6-tetrahydropyridine-treated primate in response to pallidal microstimulation. J Neurosci 24:7410–7419

25. Betarbet R, Sherer TB, Greenamyre T (2002) Animal models of parkinson's disease. BioEssays 24(4):308–318

26. Benabid AL, Pollak P, Louveau A, Henry S, de Rougemont J (1987) Combined (thalamotomy and stimulation) stereotactic surgery of the VIM thalamic nucleus for bilateral Parkinson disease. Appl Neurophysiol 50:344–6

27. Medtronics DBS Therapy for Parkinson's Disease and Essential Tremor Clinical Summary (2013) http://manuals.medtronic.com/wcm/groups/mdtcom_sg/@emanuals/@era/@neuro/documents/documents/contrib_181407.pdf

28. Borisyuk R, Yousif N (2007) International symposium: theory and neuroinformatics in research related to deep brain stimulation. Expert Rev Med Devices 4(5):587–9

29. Borisyuk R (2002) Oscillatory activity in the neural networks of spiking elements. BioSystems 67(1–3):3-16

30. Borisyuk R, Chik D, Kazanovich Y (2009) Visual perception of ambiguous figures: synchronization based neural models. Biological Cybernetics 100(6):491–504

31. Borisyuk R, Kazanovich Y, Chik D, Tikhanoff V and Cangelosi A (2009) A neural model of selective attention and object segmentation in the visual scene: An approach based on partial synchronization and star-like architecture of connections. Neural Networks 22(5–6):707–19

32. Kazanovich Y, Borisyuk R (2006) An oscillatory model of multiple object tracking. Neural Computation 18:1413–1440

33. Yousif N, Liu X (2009) Investigating the depth electrode–brain interface in deep brain stimulation using finite element models with graded complexity in structure and solution. J Neurosci Methods 184:142–151

34. Merrison-Hort R, Borisyuk R (2013) The emergence of two anti-phase oscillatory neural populations in a computational model of the Parkinsonian globus pallidus. Front Comput Neurosci 7:173

35. Merrison-Hort R, Yousif N, Njap F, Hofmann UG, Burylko O, Borisyuk R. (2013) An interactive channel model of the Basal Ganglia: bifurcation analysis under healthy and parkinsonian conditions. J Math Neurosci 3(1):14

36. Tass P, Majtanik M (2006) Long-term anti-kindling effects of desynchronizing brain stimulation: a theoretical study. Biological Cybernetics 94:58–66

37. Hauptmann C and Tass P (2007) Therapeutic rewiring by means of desynchronizing brain stimulation. Biosystems 89:173–181

38. Rosenblum M, Pikovsky A (2004) Delayed feedback of collective synchrony: An approach to pathological brain rhythms. Physical Review E70, 041904

39. Nevado-Holgado AJ, Mallet N, Magill PJ, Bogacz R (2014) Effective connectivity of the subthalamic nucleus-globus pallidus network during Parkinsonian oscillations. J of Physiology 592: 1429–1455

40. Morris C, Lecar H (1981) Voltage oscillations in the barnacle giant muscle fiber. Biophysical journal 35(1):193

41. FitzHugh, R (1961) Fitzhugh-nagumo simplified cardiac action potential model. Biophys J 1:445-466

42. Izhikevich EM (2003) Simple Model of Spiking Neurons. IEEE Transactions on neural networks.14(6):1569–1572

43. Brette R, Gerstner W (2005) Adaptive exponential integrate-and-fire model as an effective description of neuronal activity. Journal of neurophysiology 94(5):3637-3642

44. Terman D, Rubin J, Yew A, Wilson C (2002) Activity patterns in a model for the subthalamopallidal network of the basal ganglia. J Neurosci 22:2963–76

45. Rubin J, Terman D (2004) High Frequency Stimulation of the Subthalamic Nucleus Eliminates Pathological Thalamic Rhythmicity in a Computational Model. J of Computational Neurosci, 16:211–235

46. Hahn PJ, McIntyre CC (2010) Modeling shifts in the rate and pattern of subthalamopallidal network activity during deep brain stimulation. J Computational Neurosci 28(3):425–41

47. Humphries M, Stewart R, Gurney K (2006) A Physiologically Plausible Model of Action Selection and Oscillatory Activity in the Basal Ganglia. J Neurosci 26:12921–12942

48. Chersi F, Mirolli M, Pezzulo G, Baldassarre G (2013) A spiking neuron model of the cortico-basal ganglia circuits for goal-directed and habitual action learning. Neural Networks 41: 212–224

49. Goldberg J, Bergman H (2011) Computational physiology of the neural networks of the primate globus pallidus: function and dysfunction. Neuroscience 198: 171–192

50. Günay C, Edgerton J, Jaeger D (2008) Channel Density Distributions Explain Spiking Variability in the Globus Pallidus: A Combined Physiology and Computer Simulation Database Approach. J Neurosc 28(30): 7476–7491

51. Deister C, Chan C, Surmeier D, Wilson C (2009) Calcium-Activated SK Channels Influence Voltage-Gated Ion Channels to Determine the Precision of Firing in Globus Pallidus Neurons. J Neurosci 29(26): 8452–8461

52. Gillies A, Willshaw D (2006) Membrane channel interactions underlying rat subthalamic projection neuron rhythmic and bursting activity. J Neurophysiol 95(4): 2352–2365
53. Gillies A, Willshaw, D (2006) Membrane channel interactions underlying rat subthalamic projection neuron rhythmic and bursting activity. J Neurophysiol 95(4):2352–2365
54. Miocinovic S, Parent M, Butson C, Hahn P, Russo G, Vitek J, McIntyre C (2006) Computational Analysis of Subthalamic Nucleus and Lenticular Fasciculus Activation During Therapeutic Deep Brain Stimulation. J of Neurophysiology 96: 1569–80
55. Yousif N, Borisyuk R, Pavese N, Nandi D, Bain P (2012) Spatiotemporal visualisation of deep brain stimulation induced effects in the subthalamic nucleus. Eur J of Neurosci 36(2):2252–2259
56. Wolf J, Moyer J, Lazarewicz M, Contreras D, Benoit-Marand M, O'Donnell P, Finkel L (2005) NMDA/AMPA Ratio Impacts State Transitions and Entrainment to Oscillations in a Computational Model of the Nucleus Accumbens Medium Spiny Projection Neuron. J Neurosci 25:9080–95
57. Wilson H, Cowan J (1992) Excitatory and inhibitory interactions in localized populations of model neurons. Biophys J 12:1–24
58. Amari SI (1977) Dynamics of pattern formation in lateral-inhibition type neural fields. Biological cybernetics 27(2):77-87
59. Gestner W, Kistler W (2002) Spiking neuron models. Cambridge University Press, Cambridge, UK
60. Stephen Coombes (2006), Scholarpedia, 1(6):1373
61. Borisyuk RM, Kirillov AB (1992) Bifurcation analysis of a neural network model. Biological Cybernetics 66:319–325
62. Gillies A, Willshaw D, Li Z (2002) Subthalamic-pallidal interactions are critical in determining normal and abnormal functioning of the basal ganglia. Proc Biol Sci 269(1491):545–551
63. Nevado-Holgado AJ, Terry J, Bogacz R (2010) Conditions for the Generation of Beta Oscillations in the Subthalamic Nucleus-Globus Pallidus Network. J Neurosci 30(37):12340–12352
64. Eusebio A, Thevathasan W, Doyle Gaynor L, Pogosyan A, Bye E, Foltynie T, Zrinzo L, Ashkan K, Aziz T, Brown P (2011) Deep brain stimulation can suppress pathological synchronisation in parkinsonian patients. J Neurol Neurosurg Psychiatry 82:569–573
65. Vitek JL (2002) Mechanisms of deep brain stimulation: excitation or inhibition. Mov Disord 17(S3):S69–72
66. Butson CR, Cooper SE, Henderson JM, Wolgamuth B, McIntyre CC (2011) Probabilistic analysis of activation volumes generated during deep brain stimulation. Neuroimage 54(3): 2096–2104
67. McIntyre CC, Savasta M, Kerkerian-Le Goff L, Vitek J (2004) Uncovering the mechanism(s) of deep brain stimulation: activation, inhibition, or both. Clinical Neurophysiology, 115 (6):1239–1248
68. Vasques X, Cif L, Hess O, Gavarini S, Mennessier G, Coubes P (2008) Stereotactic model of the electrical distribution within the internal globus pallidus during deep brain stimulation. J of Comput Neurosci 26:109–118
69. Yousif N, Bayford R, Bain PG, Liu X (2007) The peri-electrode space is a significant element of the electrode–brain interface in deep brain stimulation: A computational study. Brain Research Bulletin 74:361–368
70. Yousif N, Bayford R, Liu X (2008a) The influence of reactivity of the electrode–brain interface on the crossing electric current in therapeutic deep brain stimulation. Neuroscience 156:597–606
71. Yousif N, Bayford R, Wang S, Liu X (2008b) Quantifying the effects of the electrode–brain interface on the crossing electric currents in deep brain recording and stimulation. Neuroscience 152: 683–691
72. Butson CR, McIntyre CC (2005) Tissue and electrode capacitance reduce neural activation volumes during deep brain stimulation. Clinical Neurophysiology(116) 2490–2500
73. Hofmann L, Ebert M, Tass PA, Hauptmann C (2011) Modified pulse shapes for effective neural stimulation. Front Neuroeng 4:9

74. Sotiropoulos SN, Steinmetz PN (2007) Assessing the direct effects of deep brain stimulation using embedded axon models. J Neural Eng 4:107–119
75. http://www.vercise.com/vercise-and-guide-dbs-systems/guide-dbs/
76. Ranck JB Jr (1975) Which elements are excited in electrical stimulation of mammalian central nervous system: a review. Brain Res 98:417–440
77. Yousif N, Purswani N, Bayford R, Nandi D, Bain P, Liu X (2010) Evaluating the impact of the deep brain stimulation induced electric field on subthalamic neurons: A computational modelling study. J Neurosci Methods 188:105–112
78. Carnevale NT, Hines ML (2006) The NEURON Book. Cambridge University Press
79. Wilson HR, Cowan JD (1973) A mathematical theory of the functional dynamics of cortical and thalamic nervous tissue. Kybernetik 13(2):55-80
80. Hoover JE, Strick PL (1993) Multiple output channels in the basal ganglia. Science 259: 819–21
81. Prescott TJ, Gonzalez FMM, Gurney K, Humphries MD, Redgrave P (2006) A robot model of the basal ganglia: Behavior and intrinsic processing. Neural Networks 19 31–61
82. Smeal M, Ermentrout B, White J (2010) Phase-response curves and synchronized neural networks. Philos Trans R Soc Lond B Biol Sci 2407–2422
83. Timmermann L, Florin E (2012) Parkinson's disease and pathological oscillatory activity: is the beta band the bad guy? - New lessons learned from low-frequency deep brain stimulation. Exp Neurol 233(1):123–5
84. Izhikevich EM (2007) Dynamical Systems in Neuroscience: The Geometry of Excitability and Bursting, The MIT Press, Cambridge, MA
85. Pikovsky A, Rosenblum M, Kurths J (2003) Synchronization, A universal concept in nonlinear sciences. Cambridge University Press, Cambridge, UK
86. Strogatz S (1994) Nonlinear Dynamics and Chaos. Westview Press
87. Hines ML, Davison AP, Muller E (2009) NEURON and Python. Front in Neuroinf 3:1
88. Anderson JR, Labiere C (2003) The Newell Test for a theory of cognition. Behavioural and Brain Sciences 26:587–640
89. Ebert M, Hauptmann C, Tass PA (2014) Coordinated reset stimulation in a large-scale model of the STN-GPe circuit. Front in Comp Neurosci 8:154
90. Magill PJ, Bolam P, Bevan M D (2001) Dopamine regulates the impact of the cerebral cortex on the subthalamic nucleus-globus pallidus network. Neurosci 106(2):313–330
91. Parent M, Parent A (2007) The microcircuitry of primate subthalamic nucleus. Parkinson Relat Disord 13(S3):S292–295
92. Kultas K, Iliknsky I (2001) Basal Ganglia and Thalamus in Health and Movement Disorders. Springer US

Mathematical Models of Neuromodulation and Implications for Neurology and Psychiatry

Janet A. Best, H. Frederik Nijhout and Michael C. Reed

Introduction

Mathematical models can test hypotheses of brain function in health and disease and can be used to investigate how different systems in the brain affect each other. We are particularly interested in how the electrophysiology affects the pharmacology and how the pharmacology affects the electrophysiology of the brain. Thus, many of the questions we address are on the interface between the electrophysiological and pharmacological views of the brain.

The human brain contains approximately 10^{11} neurons, each of which makes on the average several thousand connections to other neurons [66]. It is natural and comforting to think of the brain as a computational device, much like the computers that we build and understand. In this electrophysiological view of the brain, the neurons are the elementary devices and the way they are connected determines the functioning of the brain. But the brain is a much more flexible, adaptive, and complicated organ than this point of view suggests. The brain is made up of cells and there are several times as many glial cells as there are neurons. Glial cells protect neurons from oxidative stress by helping them synthesize glutathione [64], and astrocytes store glucose as glycogen [9]. Neurons synthesize and release more than 50 different kinds of neurotransmitters and myriad different receptor types allow neurons to influence each other's electrophysiology by volume transmission in which cells in

J.A. Best (✉)
The Ohio State University, Columbus, OH 43210, USA
e-mail: best.82@osu.edu

H. Frederik Nijhout · M.C. Reed
Duke University, Durham, NC 27708, USA
e-mail: hfn@duke.edu

M.C. Reed
e-mail: reed@math.duke.edu

© Springer International Publishing AG 2017
P. Érdi et al. (eds.), *Computational Neurology and Psychiatry*,
Springer Series in Bio-/Neuroinformatics 6,
DOI 10.1007/978-3-319-49959-8_8

one nucleus change the local biochemistry in a distant nucleus. This is the pharmacological view of the brain.

This is just the beginning of the full complexity of the problem. The functioning of neurons and glial cells is affected by an individual's genotype and dynamic changes of gene expression levels, on short and long time scales. These dynamic changes are influenced by the endocrine system, because the brain is an endocrine organ and is influenced by other endocrine organs like the gonads and the adrenal glands. And, although we think of the brain as producing behavior, in fact our behavior influences the electrophysiology, the pharmacology, and endocrine status of the brain, and therefore the gene expression levels. This is true both in the short term and in the long term. Individuals who exercise in their 30 and 40 s are 30 % less likely to get Parkinson's disease [3, 29] and the progression of Parkinson's symptoms is slower in those who exercise [38]. Thus the functioning of an individual brain depends on the history of environmental inputs and behavior throughout the individual's lifetime. And, we haven't even mentioned the complicated and changing anatomy, by which we mean the morphology of individual cell types, the connection patterns of neurons, and the proprioceptive feedback to the brain from the body [31].

Mathematical models are an important tool for understanding complicated biological systems. A model gives voice to our assumptions about how something works. Every biological experiment or psychology experiment is designed within the context of a conceptual model and its results cause us to confirm, reject, or alter that model. Conceptual models are always incomplete because biological systems are very complex and incompletely understood. Moreover, and as a purely practical matter, experiments tend to be guided by small conceptual models of only a very small part of a system, with the assumption (or hope) that the remaining details and context do not matter or can be adequately controlled.

Mathematical models are formal statements of conceptual models. Like conceptual models, they are typically incomplete and tend to simplify some details of the system. But what they do have, which experimental systems do not, is that they are completely explicit about what is in the model, and what is not. Having a completely defined system has the virtue of allowing one to test whether the assumptions and structure of the model are sufficient to explain the observed, or desired, results.

The Scientific Problem—Volume Transmission

The Electrophysiological View of the Brain

It is natural for us to think of the brain as a large computational device that processes information analogously to a computer. In this view, which we like to call the electrophysiological point of view, the basic elements are the neurons that receive inputs from other neurons and, via action potentials, send information to other neurons. There are then two fundamental classes of biological (and mathematical) questions.

How do individual neurons receive and process their inputs and decide when to fire? How do connected sets of neurons perform information processing functions that individual neurons cannot do? The electrophysiological point of view is natural for two reasons. First, we have had great success in building computational machines and we understand completely how they work. If brains are like our computational devices then we can use computational algorithms as metaphors and examples of what must be going on in the brain. Secondly, the electrophysiological point of view fits well with our modern scientific method of trying to understand complex behavior in the large as merely the interaction of many fundamental parts (the neurons) whose behavior we understand very well. The electrophysiological point of view is perfect for mathematical analysis and computation. One need not deal with the messy details of cell biology, the existence of more than 50 identified neurotransmitters, changing gene expression levels, the influence of the endocrine system, or the fact that neurons come in a bewildering variety of morphological and physiological types. All of these things appear, if they appear at all, as parameters in models of neurons, or as parameters in local or global network simulations. In particular, the chemistry of neurotransmitters themselves is not very important, since their only role is to help the electrophysiological brain transmit information from one neuron to the next.

The Pharmacological View of the Brain

There is a different point of view that we call the pharmacological view of the brain. It has been known for a long time that not all neurons are engaged in the one-to-one transfer of information to other neurons [39]. Instead, groups of neurons that have the same neurotransmitter can project densely to a distant volume (a nucleus or part of a nucleus) in the brain and when they fire they increase the concentration of the neurotransmitter in the extracellular space in the distant volume. This increased concentration modulates neural transmission in the distant region by binding to receptors on the cells in the target region. This kind of neural activity is called **volume transmission**. It is also called **neuromodulation** because the effect of the neurotransmitter is not one-to-one neural transmission but instead the modulation of other transmitters that are involved in one-to-one transmission. Two examples of volume transmission are the dopaminergic projection to the striatum in the basal ganglia from cells of the substantia nigra pars compacta and the serotonergic projection to the striatum from the dorsal raphe nucleus (DRN); these are discussed in more detail below. The locus coeruleus projects widely throughout the brain and spinal cord, with thin varicose norepinephrine (NE) axonal networks of low to moderate densities [45]. Projections of NE neurons from the locus coeruleus to the cortex play an important role in initiating and maintaining wakefulness [112].

There are many pieces of evidence that suggest that volume transmission plays a fundamental role in the functioning of the brain. Dopamine (DA) has been linked to fundamental brain functions such as motivation, pleasure, cognition, memory, learning, and fine motor control, as well as social phobia, Tourette's syndrome,

Parkinson's disease, schizophrenia, and attention deficit hyperactivity disorder [39]. In most experiments it is the concentration of dopamine in a particular nucleus that is important. Similarly, serotonin (5-HT) has been linked to feeding and body weight regulation, aggression and suicidality, social hierarchies, obsessive compulsive disorder, alcoholism, anxiety disorders, and affective disorders such as depression [39]. Many pharmaceutical and recreational drugs have been shown to act by binding to certain receptors and thus changing the local concentrations of various neurotransmitters in regions of the brain. For example, the immediate effect of selective serotonin reuptake inhibitors (SSRIs) is to inhibit the reuptake of 5-HT after it has been released thus increasing its concentration in the extracellular space in certain brain regions. Adenosine is an important neuromodulator that protects the brain from continuous neuronal activation: adenosine concentrations increase with neuronal activity and in turn inhibit individual neurons while also facilitating a transition into sleep, thereby promoting rest at a systemic level [34, 45]. Caffeine binds to adenosine receptors promotes wakefulness. Cocaine blocks the reuptake of DA, 5-HT, and norepinephrine [39] and has strong psychological effects.

Furthermore, various morphological and physiological features of the brain are consistent with the idea that the purpose of some neurons is to change the local biochemistry at distant regions of the brain. Often the projections are dense in the target volume suggesting that the idea is to change the local concentration at all parts of the target region simultaneously by the same amount. There are more than a dozen subtypes of receptors for 5-HT in the brain [91], suggesting that this great variety allows the concentration of 5-HT to modulate neurons in different ways depending on what receptors they express. As illustrated conceptually in Fig. 1, the 5-HT neurons in the dorsal raphe nucleus (DRN) have very thin unmyelinated axons and release 5-HT from many small varicosities rather than synapses [61], suggesting that their purpose is not one-to-one neural transmission. 5-HT neurons in different parts of the DRN project to many different brain regions that frequently project back, suggesting that the DRN is differentially changing the local biochemistry in many distinct regions [82], in particular the DRN sends a dense projection to the striatum

Fig. 1 Volume transmission and axonal varicosities

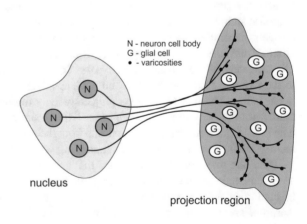

N - neuron cell body
G - glial cell
• - varicosities

nucleus

projection region

[106, 111]. The higher the concentration of 5-HT in the striatum, the more DA is released from the DA neurons projecting from the SNc per action potential [19, 22, 37]. Thus the neurotransmitters affect each other. There are also multiple DA receptor types in the striatum [6] that enrich neurotransmitter interactions with important functional consequences for the basal ganglia as we describe below.

Notice that what is important in volume transmission is that groups of neurons project to distant nuclei and change the local biochemistry there. That is, they project changes in biochemistry over long distances [97]. Of course they do this by firing action potentials. But the action potentials do not carry information in the usual sense; their only purpose is to allow the neurons to project biochemistry over long distances. This is the pharmacological view of the brain. To understand the brain one must understand both the electrophysiology and the pharmacology, and how they affect each other. For excellent reviews of volume transmission with a historical perspective and many examples, see [45, 117].

Volume Transmission and Balance in the Basal Ganglia

Here we discuss in more detail the dopaminergic volume transmission in the basal ganglia that motivates much of the computational work described in this chapter.

The basal ganglia (BG) are a group of subcortical nuclei including the striatum, subthalamic nucleus, internal and external globus pallidus, and substantia nigra. Cortical-BG-thalamic circuits are critically involved in many functions including sensorimotor, emotion, cognition [51, 75]. Multiple paths and subcircuits within BG have been identified. In some cases the different circuits perform different functions; for instance the striatum, the input nucleus of the BG, has anatomic and functional subdivisions including sensorimotor and associative. In other cases, pathways may compete, as has been postulated for action selection.

Two of the most studied pathways through the basal ganglia are the direct and indirect pathways through the striatum. The names reflect the fact that the direct pathway proceeds from the striatum directly to either the internal portion of the globus pallidus (GPi) or the Substantia Nigra pars reticulata (SNr), the two output nuclei of the BG. The indirect pathway, on the other hand, also involves a subcircuit that includes the external portion of the globus pallidus (GPe) and the subthalamic nucleus (STN) before reaching the output nuclei. The two pathways have opposing effects on the thalamus: the indirect pathway has an inhibitory effect, while the direct pathway has an excitatory effect [48, 103].

Albin and DeLong [4, 35] proposed that the balance of these opposing pathways is important for healthy function. Dopaminergic cells in the SNc project to the striatum and inhibit medium spiny neurons (MSNs) in the indirect pathway by binding to D2 receptors while the same neurons excite MSNs in the direct pathway by binding to D1 receptors [103]. Albin and DeLong noted that, during PD, as cells in the SNc die, less DA is released in the striatum; the result is that the direct pathway is less excited and the indirect pathway is less inhibited, so the thalamus receives more inhibition and less excitation. Thus the loss of dopaminergic cells in the SNc has the effect of shifting the balance in favor of the indirect pathway, and they reasoned that the increased inhibitory output from BG to the thalamus might account for some of the motor symptoms of PD, such as bradykinesia and difficulty in initiating movement. This view later lost favor in the face of new experimental observations that appeared to contradict the Albin-DeLong theory. The fact that pallidotomy—lesioning the GPi—alleviates some PD motor symptoms fit well with the theory, but, paradoxically, it emerged that high frequency stimulation of GPi was equally effective therapeutically. The solution to this conundrum seemed to be that the pattern of neuronal firing in the BG was as important for symptoms as the rate of firing, which led some to dismiss the Albin-DeLong theory. Interestingly, as PD progresses, firing patterns in the GPi become bursty and cells become more synchronized [12, 62, 92]. Note that synchronous bursting and pausing result in higher amplitude variation in the GPi output compared to the uncorrelated, irregular firing observed in the healthy GPi and thus constitutes an effectively stronger signal. This observation allows the possibility that the Albin-DeLong theory retains merit but the notion of balance needs to be interpreted more generally, recognizing that not only firing rate but also firing patterns and correlation among cells can contribute to the strength of the signal. With this more general notion of balance, it is again widely hypothesized that many of the motor symptoms of PD are due to an imbalance between the direct and indirect pathways [48, 71, 115].

The projection from the SNc to the striatum is very dense and the evidence is strong that it is the DA concentration in the extracellular space that is important for keeping the balance between the direct and indirect pathways, not one-to-one neural transmission. In particular, DA agonists given to Parkinson's patients are somewhat successful in restoring function [24]. Thus, DA, projected from the SNc, is acting as a neuromodulator of the direct and indirect pathways.

Mathematical Modeling of Volume Transmission

It is clear that understanding the interaction between the electrophysiology of the brain and the pharmacology of the brain is fundamental to brain function in health and disease. Because of the myriad receptor types and the complicated anatomy of the brain, it is unlikely that there are simple recipes that specify how the phamacological-electrophysiological interactions work in different local brain regions and different functional systems. In this chapter, we review some of our investigations into those interactions, concentrating on serotonin and dopamine, especially in the basal ganglia. In section "Computational Methods" we indicate how we go about constructing our mathematical models by sketching our dopamine model. In section "A Serotonin Model", we show three applications of our 5-HT model. We show how the serotonin autoreceptors stabilize extracellular 5-HT in the face of genetic polymorphisms. We investigate how substrate inhibition of the enzymes tyrosine hydroxylase and tryptophan hydroxylase determines how sensitive the brain concentrations of DA and 5-HT are to the content of meals. And, in section "Homeostasis of Dopamine" we propose a new mechanism of action for selective serotonin reuptake inhibitors. In section "Serotonin and Levodopa" we explain why levodopa is taken up by 5-HT neurons and is used in those neurons to make DA. A mathematical model is used to investigate the consequences for levodopa therapy for Parkinson's disease. Finally, in section "Homeostasis of Dopamine", we investigate various homeostatic mechanisms that stabilize the extracellular concentration of DA in the striatum.

Computational Methods

In 2009, we constructed a mathematical model of dopamine (DA) terminal [15] so that we could study synthesis, release, and reuptake and the homeostatic mechanisms that control the concentration of DA in the extracellular space. We investigated the substrate inhibition of tyrosine hydroxylase (TH) by tyrosine, the consequences of the rapid uptake of extracellular dopamine by the dopamine transporters, and the effects of the autoreceptors on dopaminergic function. The main focus was to understand the regulation and control of synthesis and release and to explicate and interpret experimental findings. We started with a model of a DA terminal because dopamine is known to play an important role in many brain functions. Dopamine affects the sleep-wake cycle, it is critical for goal-directed behaviors and reward learning, and it modulates the control of movement via the basal ganglia. Cognitive processing, such as executive function and other prefrontal cortex activities, are known to involve dopamine. Finally, dopamine contributes to synaptic plasticity in brain regions such as the striatum and the prefrontal cortex.

Dysfunction in various dopaminergic systems is known to be associated with a number of disorders. Reduced dopamine in the prefrontal cortex and disinhibited striatal dopamine release is seen in schizophrenic patients. Loss of dopamine in the striatum is a cause of the loss of motor control seen in Parkinson's patients. Studies have indicated that there is abnormal regulation of dopamine release and reuptake in Tourette's syndrome and dopamine appears to be essential in mediating sexual responses. Furthermore, microdialysis studies have shown that addictive drugs increase extracellular dopamine and brain imaging has shown a correlation between euphoria and psycho-stimulant-induced increases in extracellular dopamine. These consequences of dopamine dysfunction indicate the importance of maintaining dopamine functionality through homeostatic mechanisms that have been attributed to the delicate balance between synthesis, storage, release, metabolism, and reuptake. It is likely that these mechanisms exist both at the level of cell populations and at the level of individual neurons.

A schematic diagram of the mathematical model is given in the Fig. 2 that represents a DA terminal or varicosity. The boxes contain the acronyms of substrates and the ellipses the acronyms of enzymes and transporters. For convenience in the equations below and in the diagram we denote the concnetrations in the

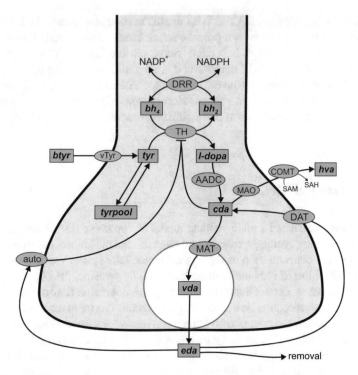

Fig. 2 Dopamine metabolism

mathematical model by lower case letters and we omit the brackets for concentration. Thus, *cda*, *vda*, and *eda* represent the concentrations of cytosolic DA, vesicular DA, and extracellular DA, respectively. Each arrow indicates a biochemical reaction, a transport velocity or an influence. Dopamine is synthesized in the nerve terminal from tyrosine *tyr* which is transported across the blood brain barrier. We include exchange between tyrosine and a tyrosine pool that represents all the other uses and sources of tyrosine in the terminal. Tyrosine is converted into L-3,4-dihydroxyphenylalanine, $l-dopa$, by tyrosine hydroxylase, *TH*, and $l-dopa$ is converted into cytosolic dopamine, *cda*, by aromatic amino acid decarboxylase, *AADC*. *cda* inhibits *TH* and is transported into the vesicular compartment by the monoamine transporter, *MAT*, and vesicular dopamine, *vda*, is released from the vesicular compartment into the extracellular space at a rate proportional to the firing rate of the neuron. In the extracellular space, extracellular dopamine, *eda*, affects the autoreceptors, is taken up into the terminal by the dopamine transporters, *DAT*, and is removed from the system by uptake into glial cells and the blood and by diffusion. Dopamine is also catabolized in the terminal by monoamine oxidase, *MAO*.

The variables in the mathematical model are the concentrations of the 10 boxed substrates. Each differential equation simply reflects mass balance: the rate of change of the concentration of the substrate is the sum of the rates by which it is being made minus the sum of the rates by which it is lost. So, for example, the differential equation for cda is

$$\frac{dcda}{dt} = V_{\text{AADC}}(l - dopa) + V_{\text{DAT}}(eda) - V_{\text{MAO}}(cda) - V_{\text{MAT}}(cda, vda).$$

Each V is a velocity (a rate) and the subscript indicates which rate. These velocities depend on the current concentrations of one or more of the variables, as indicated. So, for example, V_{MAT} depends on both cda and vda because there is leakage out of the vesicles back into the cytosol. Similarly the differential equation for eda is

$$\frac{deda}{dt} = auto(eda)fire(t)(vda) - V_{\text{DAT}}(eda) - k(eda).$$

The first term on the right is release of dopamine into the extracellular space, which depends on the current state of the autoreceptors, $auto(eda)$, the current firing rate, $fire(t)$, and the vesicular concentration, vda. The second term is the uptake back into the terminal cytosol and the third term is removal in the extracellular space by uptake into glial cells and blood vessels.

Determination of the functional forms of the velocities. Each of the velocities, such as V_{TH}, V_{AADC}, or V_{DAT} depends on the current state of one or more of the variables. How do we determine the functional form of the dependence? If we can, we assume simple Michaelis-Menten kinetics. For example,

$$V_{\text{DAT}}(eda) = \frac{V_{max}(eda)}{(K_m + eda)}.$$

In other cases the formula might be much more complicated depending on what is known about about how enzymes or transporters are activated or inhibited by molecules that are or are not its substrates. For example the formula for VTH is:

$$V_{\text{TH}}(tyr, bh4, cda, eda) =$$

$$\left(\frac{V_{max}(tyr)(bh4)}{(tyr)(bh4) + K_{tyr}(bh4) + K_{tyr}K_{bh4}(1 + \frac{(cda)}{K_{i(cda)}})} \right) \cdot \left(\frac{.56}{1 + \frac{(tyr)}{K_{i(tyr)}}} \right) \cdot \left(\frac{4.5}{8(\frac{eda}{.002024})^4 + 1} + .5 \right)$$

The velocity V_{TH} depends on the current concentrations of its substrates, tyr and $bh4$, and on cda because cda inhibits the enzyme TH. The first term on the right is standard Michaelis-Menten kinetics with the additional inhibition by cda. The second term expresses the fact that TH is inhibited by its own substrate, tyr. The last term on the right is the effect of the autoreceptors and that depends on eda. Not so much is known about the mechanism of this effect, so we took a functional form that was consistent with in vitro experiments in the literature.

How are the parameters determined? The answer is, alas, with difficulty. There are measurements of K_m values in the literature. Sometimes they vary over two orders of magnitude, which is not surprising because often they are measured in test tubes or in vitro or measured in different cell types under different conditions. There are few V_{max} values in the literature because fluxes are hard to measure, especially in vivo. Typically we adjust the V_{max} values so that the concentrations at steady state in the model are in the ranges measured experimentally. We take the choice of parameters seriously, and we do the best we can.

But if we don't know the exact, correct values of the parameters, is the model right? This question (that we frequently get) is based on two misunderstandings. First, there are no exact, correct values of the parameters. In each of us, the parameters are somewhat different, because of genotype, because of environmental inputs, and because of changing gene expression levels. If we do our job well, we can hope to have a dopamine terminal system for an "average" person, and it is of course very relevant to ask how the behavior of the system depends on the parameters. Secondly, there's no "right" model. Every model is a simplification of a very complicated biological situation. We don't regard our models as fixed objects, or "right" or "wrong", but as growing and changing as we learn more from experiments and from our computations. The purpose of the model is to have a platform for in silico experiments that will increase our understanding of the biology.

Results

A Serotonin Model

We created a similar model for a serotonin (5-HT) terminal or varicosity [16] and have used the DA and 5-HT models to study various questions in brain physiology including depression and Parkinson's disease. In this section we provide three examples.

Homeostatic Effects of 5-HT Autoreceptors

The $5\text{-}HT_{1B}$ autoreceptors on terminals and varicosities provide a kind of end product inhibition for the extracellular 5-HT concentration. When the extracellular concentration rises, the autoreceptors inhibit synthesis (a long term effect) by inhibiting the enzyme TPH (the first step in the 5-HT synthesis pathway) and they also inhibit release of 5-HT from vesicles (a short term effect). Figure 3 shows some of the consequences autoreceptor regulation. In Panel A the firing rate is varied, in Panel B the activity of the serotonin transporter (SERT) is varied, and in Panel C the activity of tryptophan hydroxylase (TPH) is varied. TPH is the rate-limiting enzyme for the synthesis of 5-HT. In Panels B and C, common polymorphisms in the population are

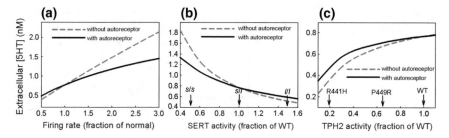

Fig. 3 Homeostatic effects of the 5-HT autoreceptors

shown on the x axis. In each panel, the y axis shows the concentration of 5-HT in the extracellular space at steady state in the model. The black curves show the steady state values in the model with the autoreceptors present and the dashed grey curves show the steady state values when the autoreceptors are turned off. In each case, the variation in extracellular 5-HT is much less in the presence of the autoreceptors. This shows how the autoreceptors buffer the extracellular concentration of 5-HT against changes in firing rate and genetic polymorphisms in two key proteins.

The Effect of Substrate Inhibition of TH and TPH

It is interesting that TH, the key enzyme for creating DA, and TPH, the key enzyme for creating 5-HT both show substrate inhibition. One can see this in the velocity curves in Fig. 4. That is, the substrate of the enzyme inhibits the enzyme itself. Substrate inhibition was emphasized by Haldane in the 1930s [55], but has been regarded as a curiosity although many enzymes exhibit this property. In all the cases that we have examined, substrate inhibition has a biological purpose [94]. In substrate inhibition, instead of the normal Michaelis-Menten shape where the velocity curve

Fig. 4 Substrate inhibition of tyrosine and tryptophan hydroxylase

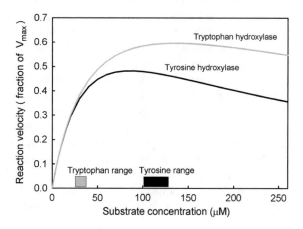

Fig. 5 5-HT and DA
changes due to meals

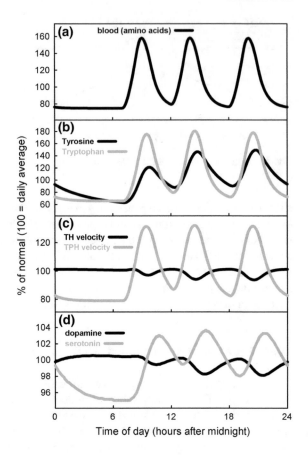

saturates as the concentration of substrate gets large, the curves reach a maximum
and then begin to descend as one can see in Fig. 4. The effect is much stronger for
TH than for TPH. Does this matter? Well, that is something one can try out using
the models. Panel A in Fig. 5 shows (assumed) amino acid curves in the blood for
tyrosine and tryptophan due to three daily meals. Panel B shows the tyrosine and
tryptophan concentrations in the cytosols of the terminals. Panel C shows the velocities of the TH and TPH reactions. Notice that the TH velocity varies little but the
TPH velocity varies a lot. Why is that? The normal fasting concentration of tyrosine
in the DA cells is 100–125 µM which puts it in the flat part of the TH velocity curve
in Fig. 4, so changes in tyrosine in the cell don't change the synthesis rate of DA
very much. In contrast, the synthesis rate of 5-HT varies quite a bit with the changes
in blood and cytosolic tryptophan because the fasting concentration of tryptophan
in 5-HT cells is on the sharply rising part of the velocity curve. As a consequence,
the concentration of 5-HT in the vesicles (Panel 4) and the extracellular space (not
shown) varies modestly while the concentration of DA varies very little. In fact, it is
known that brain DA is quite insensitive to the protein content of meals [40, 41], but

that the brain content of 5-HT does vary with meals [39] and these simulations show why. For more information on substrate inhibition, see [94]. These simulations show that sometimes the details of the biochemistry, in this case the Michaelis-Menten constants of TH and TPH and the normal concentrations of tyrosine and tryptophan in cells, really do matter.

How Do SSRIs Work?

Depression, which is characterized by feelings of worthlessness and lack of motivation, is a major health problem and has important economic consequences (treatment and lost productivity) as well [21, 49]. The antidepressants used to treat depression are among the most widely prescribed drugs, but unfortunately most have unwanted side effects and limited therapeutic effects for most patients [47, 104]. In fact, none of the drugs are efficacious for a majority of the patients for whom it is prescribed [32, 80, 110]. Most of the commonly used antidepressants are selective serotonin reuptake inhibitors (SSRIs). It is not known what their mechanism of action is, or why they work on some patients and not on others. In this section we discuss these issues in the context of volume transmission.

Early evidence indicated that low 5-HT levels in the brain are linked to depression [39, 101]. 5-HT is synthesized in terminals and varicosities from the amino acid tryptophan and there is evidence that acute tryptophan depletion causes depression in humans [10, 114] and a decrease in 5-HT release in the rat hippocampus [107]. Thus it was natural to use SSRIs as antidepressants. Since SSRIs block the reuptake of 5-HT into the cytosol by the serotonin transporters (SERTs), it was expected that the SSRIs would raise the level of 5-HT in the extracellular space. Figure 6 shows a 5-HT neuron in the DRN sending its axon to a projection region where 5-HT is

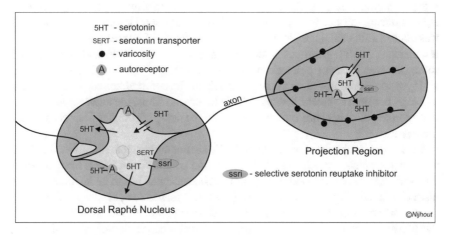

Fig. 6 SSRIs block the serotonin transporters

released from varicosities when the action potential arrives. Since the SSRIs block some of the SERTs, the effects of the SSRIs should be to raise the 5-HT level in the extracellular space in projection region. However, there is a complication. When a DRN neuron fires an action potential, 5-HT is also released from the cell body in the DRN [46]. The released 5-HT binds to 5-HT_{1A} autoreceptors on the cell bodies and the autoreceptors decrease cell firing when stimulated. Thus, when one gives an SSRI one blocks some SERTs in the projection region which should tend to make 5-HT go up there. However, the SSRIs also block SERTs in the DRN and the resulting increase of 5-HT in the extracellular space there will decrease firing and this will tend to *lower* extracellular 5-HT in the projection region. So, what will happen? The answer is that it depends on the balance between these two effects, and so it is not surprising that experimentalists found that 5-HT goes up in some projection regions and down in others and the magnitudes of the changes are dose dependent [1, 8, 57, 77]. Even at the beginning, it was clear that other effects besides the 5-HT concentration in the extracellular space must be important, because changes in concentration will happen in minutes and hours after an SSRI dose, but patients usually feel beneficial effects only after 3–6 weeks of treatment [39].

Attention focused next on the 5-HT_{1A} autoreceptors on the raphe nucleus cell bodies. It was shown in numerous studies (for example, [28]) that giving 5-HT_{1A} antagonists potentiates the SSRI-induced increase of 5-HT in projection regions. Similarly, 5-HT_{1A} knockout mice show increased release in projection regions [70]. Both types of studies confirm the role of the 5-HT_{1A} autoreceptors in decreasing tonic firing of 5-HT neurons in raphe in the presence of SSRIs. Furthermore, a number of studies showed that chronic treatment with SSRIs desensitizes the 5-HT_{1A} autoreceptors [20, 28, 58, 63]. And, thus, one could explain the improvements of patients on the time scale of 3–6 weeks by the slow desensitization of autoreceptors. Consistent with this hypothesis were several studies that showed that 5-HT levels in projection regions are higher after chronic treatment as compared to acute treatment [72, 100, 109]. These studies did not measure e5-HT in projections regions during the full course of chronic SSRI treatment. Unfortunately, when this was done, Smith et al. [105] found that extracellular 5-HT concentrations in neocortex, caudate, and hippocampus of awake monkeys went up initially and then declined somewhat over the course of treatment. Similar findings were found by Anderson et al. [5] who saw an initial quick rise in 5-HT in the cerebral spinal fluid of rhesus monkeys but then a plateau during chronic treatment. Thus, the autoreceptor desensitization hypothesis seems unlikely to explain the delay of beneficial effects of SSRI treatments.

In fact, the mechanisms of action of SSRIs are not understood, nor is it understood why some patients are helped and others not and why different SSRIs have different efficacies. The problem is extremely difficult because one has to understand mechanism and function on 4 different levels, genomic, biochemical, electrophysiological, and behavioral, but changes on each level affect function on the other 3 levels, and this makes the interpretation of experimental and clinical results very difficult. As we indicated above, the release of 5-HT affects dopamine signaling. 5-HT may activate the hypothalamic-pituitary-adrenal axis by stimulating production of corticotropin-releasing hormone [56] and the endocrine system affects the 5-HT system [18, 105]. This may be the basis of gender differences in depression and response to SSRIs. And, finally, both gene expression and neuronal morphology are changing in time. In this circumstance, it is not surprising that many variables on all 4 levels are correlated with depression or to the efficacy of the SSRIs. All such correlations are candidates for causal mechanisms, so sorting out which mechanisms are causal is extremely difficult.

We used our mathematical model of serotonin varicosities and terminals to propose and investigate a new hypothesis about the action of SSRIs. The serotonin neurons in the DRN fire tonically at about 1 Hz but some of the individual spikes are replaced by short bursts. In a series of pioneering studies [42, 59, 65, 109], Jacobs, Fornal and coworkers studied the relationship between the electrophysiology of the 5-HT system and various behaviors in nonanaesthetized cats. They showed that the firing rate and pattern of some DRN 5-HT neurons differ in active waking, quiet waking, slow-wave sleep, and REM sleep [65]. And Hajos showed that some DRN neurons have bursts and others do not [53, 54]. Thus, it is plausible that the purpose of tonic firing is to maintain a basal 5-HT concentration in projection regions, but that it is bursts that contain incoming sensory input and stimulate specific behavioral responses.

If depression is caused by low tissue levels of 5-HT then vesicular 5-HT must be low in depressed patients since the normal concentrations in the cytosol and extracellular space are extremely low. So we assumed that our model depressed patient had low vesicular 5-HT, about 20–25 % of normal. For example, this could be caused by low tryptophan input. As a result, the model depressed patient had low extracellular 5-HT in projection regions. We modeled chronic treatment by SSRIs by assuming that the SSRIs block 2/3 of the SERTs. As we expected, this does not change extracellular 5-HT in projection regions very much because of the two competing effects discussed above. We then included the result of Benmansour et al. [11] that the expression level of SERTs declines considerably in rats during a 21 day treatment by SSRIs. Here are the results of our modeling. After 21 days, the average levels of 5-HT in projection regions of the model depressed patient did *not* return to normal. However, the response to bursts did return to normal. The intuitive reason behind this is that as the available SERTs decline considerably, reuptake of 5-HT becomes must slower. This has a much greater effect on bursts than on tonic firing because during bursts the extracellular 5-HT in projection regions is still high when the next action potential arrives. Thus, our proposed hypothesis is that it is burst firing that is connected to behavior and that, in depressed patients, the response to burst firing is brought back to normal by SSRIs because after 21 days the number of available SERTs is further depressed. It is interesting that Zoli et al. [116] emphasized that neurons that communicate via one-to-one neural transmission during tonic firing may contribute to volume transmission during burst firing. During bursts, the large amount of neurotransmitter in the extracellular space cannot be taken up in the synapse but spills out into the rest of the extracellular space.

Serotonin and Levodopa

Parkinson's disease has been traditionally thought of as a dopaminergic disease due to the death of dopaminergic cells in the substantia nigra pars compacta (SNc). These dopaminergic cells project to the striatum where low levels of DA cause dysfunction in the motor system. DA does not cross the blood-brain barrier because it doesn't have a carboxyl group and is not recognized as an amino acid. However, its precursor, L-DOPA, still has the carboxyl group and does cross the blood-brain barrier. Thus, the idea of levodopa therapy is to fill the remaining DA terminals in the striatum with L-DOPA so that these terminals will release more DA into the extracellular space when action potentials arrive, compensating for the DA terminal loss caused by cell death in the SNc. However, accumulating evidence implies an important role for the serotonergic system in Parkinson's disease in general and in physiological responses to levodopa therapy. We used a mathematical model [95] to investigate the consequences of levodopa therapy on the serotonergic system and on the pulsatile release of dopamine (DA) from dopaminergic and serotonergic terminals in the striatum.

Levodopa Makes DA in 5-HT Neurons

The key idea is the recognition of the similarities in the synthesis pathways of 5-HT in 5-HT neurons and DA in DA neurons. DA is synthesized from the amino acid tyrosine (tyr) that crosses the blood-brain barrier and is taken up into DA nerve terminals by the L-transporter. In the DA terminal, the enzyme tyrosine hydroxylase (TH) adds an OH group to tyr making levodopa (L-DOPA). We will abbreviate levodopa by L-DOPA and by LD. The amino acid decarboxylase (AADC) cuts off the carboxyl group to make cytosolic DA. The monoamine transporter (MAT) packages DA into vesicles. When the action potential arrives a sequence of events, including Ca^{++} influx, causes some vesicles to move to the boundary of the terminal and release their contents into the extracellular space. The extracellular DA is taken back up into the cytosol by the dopamine transporter (DAT). Extracellular DA also binds to DA autoreceptors (A-DA) that inhibit synthesis and release. This control mechanism stabilizes the concentration of DA. Of course, the actual situation is more complicated, for example, cytosolic DA itself inhibits TH and extracellular DA can be taken up by glial cells.

The situation for 5-HT is remarkably similar. 5-HT is synthesized from the amino acid tryptophan (tryp) that crosses the blood-brain barrier and is taken up into 5-HT nerve terminals by the L-transporter. In the 5-HT terminal, the enzyme tryptophan hydroxylase (TPH) adds an OH group to tryp making 5-HTP. The enzyme amino acid decarboxylase (AADC) cuts off the carboxyl group to make cytosolic 5-HT. The monoamine transporter (MAT) packages 5-HT into vesicles. When the action potential arrives, some vesicles to move to the boundary of the terminal and release their contents into the extracellular space. The extracellular 5-HT is taken back up into the cytosol by the serotonin transporter (SERT). Extracellular 5-HT also binds to 5-HT autoreceptors (A-5HT) that inhibit synthesis and release.

The main difference between DA neurons and 5-HT neurons is that DA neurons express the enzyme TH and thus make DA, and 5-HT neurons express TPH and thus make 5-HT. As we will explain, this distinction is eliminated in 5-HT neurons when one gives a dose of levodopa.

L-DOPA is taken up into all cells by the L-transporter, just like tyr and tryp. When L-DOPA is taken up into 5-HT terminals, the enzyme AADC cuts off the carboxyl group to make DA, which is then packaged into vesicles by MAT. Thus vesicles in the 5-HT neurons are filled with both 5-HT and DA, and when the action potential arrives, both are released into the extracellular space. There is a large dense projection of 5-HT neurons from the dorsal raphe nucleus (DRN) to the striatum. So, during a dose of levodopa, the 5-HT neurons release a large pulse of DA into the striatum.

All the aspects of this story have been verified experimentally in the last 15 years. Experiments have verified that 5-HT neurons can store and release DA in vivo and in vitro [88]. In levodopa treatment of a hemiparkinsonian rat, striatal extracellular DA decreased substantially when the serotonergic system was lesioned [108]. Glial cells also express AADC and so could contribute to the conversion of LD to DA, but experiments using reserpine to block vesicular packaging showed a great reduction of extracellular DA, suggesting that most of the levodopa-derived DA is released by exocytosis of vesicles rather than by glia, at least at physiological levels of levodopa administration [67]. It has also been shown that 5-HT_{1A} autoreceptor agonists (that decrease DRN firing) and 5-HT_{1B} autoreceptor agonists (that decrease release at 5-HT terminals) both lower extracellular DA in the striatum in a dose-dependent manner after an LD dose [76].

The new understanding of 5-HT neurons in levodopa therapy has helped to explain a serious side effect of levodopa therapy. Within 5 years of chronic LD treatment, many patients experience a variety of complications [84, 85]. For instance, the length of the therapeutic time window in which a given LD dose relieves PD symptoms gradually shortens and approaches the plasma half-life of LD (wearing-off). Rapid variations in efficacy may occur (on-off fluctuations). Another, particularly troubling, complication of chronic LD therapy is the appearance of involuntary movements (levodopa-induced dyskinesia, LID). These complications increase patients disability substantially, pose a therapeutic dilemma, and limit the use of LD.

There is good evidence that large pulses of DA in the striatum are the proximal cause of LID that are seen in long-term dosing [44]. And there is conclusive evidence that these large pulses result from DA release from 5-HT neurons in the striatum. Lesioning the 5-HT system or giving selective serotonin autoreceptor (5-HT_{1A} and 5-HT_{1B}) agonists results in a nearly complete elimination of LID [26].

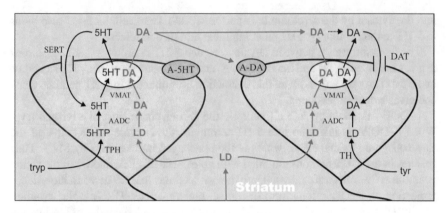

Fig. 7 Levodopa makes DA in 5-HT neurons

Mathematical Modeling

In order to investigate these phenomena, we created a mathematical model that corresponds to Fig. 7 [95]. What we discovered was that the size of these large pulses of DA coming from 5-HT neurons depends critically on the fraction, f, of SNc cells left alive, which is why there are more and more dyskinesias as Parkinson's disease progresses. Here is the intuitive explanation. As long as there are lots of SNc cells alive, there will be lots of DA terminals in the striatum with DATs and DA autoreceptors. The DATs take up a lot of the excess DA that comes from the 5-HT neurons and it is stored in DA terminals, and the DA autoreceptors restrict DA release from the DA terminals when the extracellular DA concentration is high. These effects keep the DA concentration in the extracellular space from going too high. However, as the fraction of SNc cells left alive gets smaller and smaller these two control mechanisms have less and less effect. The DA released from 5-HT neurons causes high pulses of DA in the striatum because it is not taken up quickly by the remaining DATs and it cannot be taken up into 5-HT terminals by the SERTs. It is these high pulses of DA in the striatum that lead to the aforementioned dyskinesias. In addition, the extra DA created by the levodopa dose and the 5-HT neurons is used up much faster because it cannot be stored in the small number of remaining DA terminals. It diffuses away or is taken up by glial cells, thus shortening the period of efficacy of the LD dose.

Figure 8, Panel A, shows model calculations of the time course of extracellular DA in the striatum for different values of f, the fraction of SNc cells left alive. Each curve is labeled with the corresponding value of f. As f declines from 1 (normal) to 0.2 and then 0.1, the level of extracellular DA gets higher because there are fewer DATs to take up the DA released by 5-HT neurons. However, when f is very small (f = 0.05 or 0.01), the peaks decline because removal mechanisms such as catabolism, diffusion, and uptake into glial cells become more important. The dashed black horizontal line in (A) represents the level of extracellular DA needed in the striatum for anti-Parkinsonian effects. In Panel B, the two solid curves reproduce the simulations

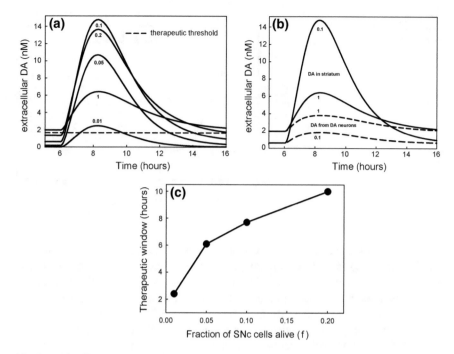

Fig. 8 5-HT effects as SNc cells die

from A for f = 1 and f = 0.1. The corresponding dashed curves in show the amount of extracellular DA in the striatum that comes from the DA neurons. For a normal individual (f = 1) the DA neurons contribute approximately 60 %, but as SNc cells die (f = 0.1) most of the DA comes from the 5-HT neurons.

Panel C shows that the amount of time that extracellular DA stays above the therapeutic level (the dashed black line in (A)) declines as PD progresses and f gets smaller until the therapeutic window becomes approximately 2 h.

How does an LD dose affect the 5-HT system? We will describe verbally what we saw in our modeling; details and figures can be found in [95]. First of all, LD competes with tyrosine and tryptophan for the L-transporter at the blood-brain barrier, so during an LD dose there is less tryptophan in the extracellular space in the brain. The tryptophan and LD compete again to be transported into 5-HT neurons, resulting in lowered tryptophan in 5-HT neurons. TPH turns some tryptophan into 5-HTP, but 5-HTP then has to compete with LD for the enzyme AADC that cuts off carboxyl groups and makes DA and 5-HT. Then, DA and 5-HT compete for the monoamine transporter, MAT, that packages them into vesicles. Thus it is not surprising that the concentration of 5-HT in the vesicles and in the extracellular space goes down approximately 50 % during an LD dose; Fig. 9 shows the extracellular 5-HT curve computed by the model. This drop in extracellular 5-HT is consistent with exper-

Fig. 9 LD dosing lowers
5-HT in the striatum

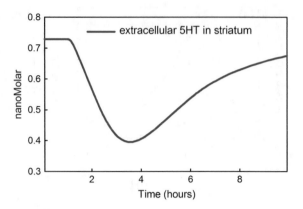

imental findings in animals. Extracellular 5-HT was found to decrease 50–90 % in
different brain regions [23]. Carta et al. [26] found that tissue 5-HT decreased 48 % in
the striatum during an LD dose, and Navailles et al. [86] showed that 5-HT decreases
30 % in the striatum and 53 % in motor cortex after chronic LD dosing. All this can
be summed up by saying that dosing with LD turns 5-HT neurons partially into DA
neurons, which is good for relieving the symptoms of Parkinson's disease but has
the side effect of compromising the 5-HT system in the brain.

This raises the interesting question of whether levodopa doses could be one of the
reasons for depression in Parkinson's patients. Decreased serotonergic signaling has
been linked to depression [78]. As we pointed out above, acute tryptophan depletion
is known to lower 5-HT brain levels in various animals [83], and results in lowered
mood in humans [114]. While depression is frequently described as the most common psychiatric complication in PD [74], reported rates vary widely, from 2.7 %
to greater than 90 % [98], due to factors including whether both major and minor
depression are included and how subjects are selected for inclusion in the study.
Moreover, many complicating factors make it difficult to draw conclusions about the
possible connections between LD therapy and depression [43, 90, 95]. Nevertheless, the case of LD therapy and its effect of the 5-HT system is a cautionary tale.
We create drugs and prescribe them because we expect them to have a specific effect
in a specific brain location (in this case more DA in the striatum). However, the drug
may have many other effects throughout the brain (in this case by lowering serotonin
in all brain nuclei to which the raphe nuclei project).

Homeostasis of Dopamine

Neurotransmitters provide the mechanism by which electrical signals are communicated from one neuron to the next. However, as we discussed above, there is strong
evidence that in many cases it is the concentration of a neurotransmitter in the extra-

cellular space of a nucleus that affects electrophysiological neurotransmission by other neurotransmitters. This volume transmission raises several natural questions. What are the mechanisms by which the extracellular concentrations of neurotransmitters are controlled? How do neurotransmitters in the extracellular space affect synaptic transmission by other neurotransmitters? How robust are these mechanisms in the face of polymorphisms in the enzymes affecting synthesis, release, and reuptake of neurotransmitters? How are dysfunctions in these control mechanisms related to neurological and neuropsychiatric diseases? In this section we briefly describe our work on several of these questions.

Passive Stabilization of DA in the Striatum

As discussed above, progressive cell loss from the substantia nigra pars compacta (SNc) is the proximal cause of the symptoms of Parkinson's disease [39]. The dopaminergic cells of the SNc send projections to the striatum where the loss of dopaminergic tone is thought to be the main cause of tremor and other motor symptoms of parkinsonism [25, 33]. An interesting and important feature of the disease is that symptoms do not appear until a very large percentage (75–90 %) of SNc cells have died and therefore this feature has been the focus of much experimental and clinical investigation [2, 118]. Experiments with animal models [13, 17, 36] have shown that although striatal tissue content of dopamine declines more or less proportionally to cell death in the SNc, the extracellular concentration of dopamine in the striatum remains near normal until more than 85 % of SNc neurons have died. This is widely believed to be the reason that symptoms do not appear until very late in the degeneration of the SNc.

What is the basis of this remarkable homeostasis of striatal extracellular DA in the face of progressive cell death in the SNc? Some researchers proposed that the nigrostriatal system adapts to cell death to maintain extracellular DA level by increasing DA synthesis in the living terminals or by sprouting more terminals. But in 2003, Bergstrom and Garris proposed a very simple explanation that they called "passive stabilization" and provided experimental evidence for it [13]. The extracellular concentration of DA depends on the balance between release of DA and reuptake of DA by the dopamine transporters (DATs). If half of the SNc cells die, there will be only half as much release, but there will also be only half as much reuptake, so the concentration of DA in the extracellular space should remain the same.

We used our mathematical model of a DA terminal to investigate the proposal of Bergstrom and Garris [13]. Notice that their hypothesis does not explain why passive stabilization breaks down when f, the fraction of SNc cells left alive, gets small. We believe that passive stabilization breaks down at small f because there is always some removal of DA from the system in the extracellular space by uptake into glial cells and blood vessels or simply diffusion out of the tissue. As the number of DA terminals in the striatum gets smaller, these removal effects get proportionally larger because the reuptake DATs become sparser and sparser. This hypothesis was confirmed and explained by our mathematical modeling. Figure 10 shows the con-

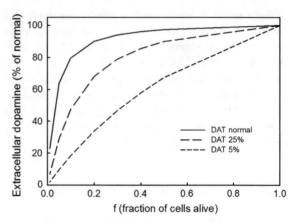

Fig. 10 DA concentration in the striatum as the fraction of SNc cells alive changes

centration of DA in the striatum in the model as a function of f, the fraction of SNc cells left alive. One can see that the passive stabilization effect of Bergstrom and Garris keeps the extracellular DA concentration quite constant until approximately 80 % of the SNc cells have died. As even more cells die the concentration drops to zero because the removal effects dominate more and more. The dashed curves show that the passive stabilization depends on the dopamine transporters.

Homeostasis of DA and Cryptic Genetic Variation

In our 2009 DA model [15] that included synthesis, packaging into vesicles, release, and reuptake via the DATs, we also included the effects of the DA autoreceptors that sense the DA concentration in the extracellular space. When extracellular DA gets higher than normal, the autoreceptors inhibit synthesis and release of DA, and when extracellular DA gets lower than normal this inhibition is withdrawn stimulating synthesis and release. Thus the autoreceptors act to modulate extracellular DA against both long term and short term perturbations such as changes in the supply of tyrosine or changes in firing rate. The mechanisms by which extracellular DA affects synthesis and release via the autoreceptors are mostly unknown and an important topic of current research that involves difficult questions in cell biology. The control of DA in the extracellular space is also affected by other neurotransmitters. For example, there is a dense serotonergic projection to the striatum from the dorsal raphe nucleus (DRN). The released 5-HT binds to 5-HT receptors on DA terminals and increases DA release when the SNc cells fire.

An important field of study in the past 15 years has been to quantify the effects of gene polymorphisms on the proteins that are important for the dopaminergic system, for example, tyrosine hydroxylase (TH) or the dopamine transporter (DAT). Typically, these experiments are done in test tubes or in vitro and the polymorphisms often have large quantitative effects on the function of the proteins. And, it is very tempting to conclude that the polymorphisms are therefore the causes of various

Table 1 Polymorphisms in TH and DAT

Gene	Mutation	Relative activity (%)	Citation
TH	T245P	150	[99]
TH	T283M	24	[99]
TH	T463M	116	[99]
TH	Q381K	15	[69]
DAT	V382A	48	[79]
DAT	VNTR10	75	[79]
DAT	hNET	65	[52]

neurological or neuropsychiatric diseases. Some of these polymorphisms that have large in vitro effects are shown in Table 1.

However, in vivo there are many control mechanisms (we've discussed two above) that buffer the DA concentration in the extracellular space against perturbations in the DA system. We pointed this out already in our 2009 paper [15], but the point is made dramatically by the two dimensional surface taken from [89]. The surface shows the extracellular DA concentration (z-axis) at steady state compared to normal, as a function of the activity of tyrosine hydroxylase and the efficacy of the dopamine transporter computed from our model. In both cases, 1 indicates normal activity for TH and DAT. The large white dot on the surface is wild type, the concentration of extracellular DA when TH and DAT have their normal activities. The smaller white dots on the surface indicate points that correspond to common polymorphisms (homozygotes and heterozygotes) in the human population taken from the table. Notice that all the white dots lie on the flat part of the surface where the polymorphisms cause only very modest changes in extracellular DA despite the fact that they cause large changes in protein activity. This is the effect of the autoreceptors. It is quite amazing that these polymorphisms all lie on the flat part of the surface. Presumably, if they didn't, they would have been selected against and would not be common in the human population. This example shows why one has to be very careful about jumping to physiological conclusions from in vitro experiments. The polymorphisms in Table 1 have large effects on the activity of the proteins but homeostatic mechanisms ensure that the effect on the phenotype (the extracellular DA concentration) is very small.

The surface in Fig. 11 is a perfect example of cryptic genetic variation in which large variation in genes (gene products), that is, TH and DAT, produce very little variation in a phenotypic variable, the extracellular DA concentration. It should be kept in mind that the actual situation is much more complicated than this two-dimensional surface in three-dimensional space would lead one to believe. There are many other variables, both genetic variables (for example a polymorphism in the monoamine transporter) or phenotypic variables (for example the 5-HT concentration, see below) that could affect the shape of this surface. The "real" surface is a high dimensional surface in a high dimensional space. Nevertheless this surface does tell us a lot, and

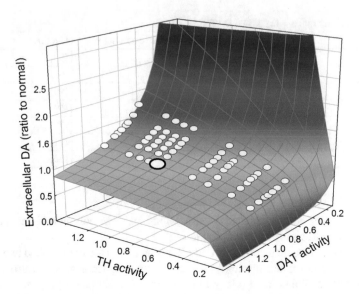

Fig. 11 DA homeostasis in the striatium. On each axis, 1 corresponds to normal

it is interesting to think about the people who have the 15 % mutation in TH. They are the ones to the right sitting at the edge of the cliff where DA drops to zero. Interestingly, these genotypes sometimes show a dystonia, involuntary muscle contractions that affect posture, brought about by low levels of extracellular DA that can be alleviated by levodopa [69]. So, one could say that their position at the edge of the cliff (that is, having the 15 % TH polymorphism) predisposes them to the dystonia. Some of them are pushed over the cliff by other variables not pictured and thus show the dystonia. The job of a precision medicine provider would be to advise a patient with the 15 % TH polymorphism how to flatten the region around where they lie on the surface and thus to avoid being pushed over the cliff by other variables. In our simulations, the region around these individuals gets flatter if one increases the strength of the autoreceptor effect.

Escape from Homeostasis and Neurological and Neuropsychiatric Diseases

In Parkinson's disease, many motor symptoms are caused by very low DA in the striatum of the basal ganglia, which, in turn, is caused by cell death of the dopaminergic neurons in the substantia nigra pars compacta (SNc) [39]. By contrast, the dyskinesias that may result from levodopa therapy are known to be caused by unusually high concentrations of extracellular DA in the striatum [44]. The chorea of Huntington's disease is also associated with high concentrations of extracellular DA in the striatum, that in turn may be caused by degeneration of inhibitory projections from the striatum to the SNc [30]. There is a hypothesis that hyperactivity of dopamine transmission underlies many of the symptoms of schizophrenia [68]. The fact that amphetamines, cocaine, and other drugs that increase levels of extracellular DA cause similar symptoms to schizophrenia supports this hypothesis. These three diseases illustrate the idea that when one departs from the homeostatic region, either above or below, disease processes may occur.

Homeostasis does not mean that a system is rigid. In means that an important biological variable (in this case the extracellular DA concentration in the striatum) is kept within a narrow range by a host of homeostatic mechanisms. Note that not all variables are homeostatic; on the contrary, some variables change dramatically so that other variables can remain homeostatic [89]. Each of the homeostatic mechanisms works well only over a certain range of biological variation. If inputs or other biological parameters leave this range then the biological variable is no longer homeostatic and departs from its normal range and neurological and neuropsychiatric symptoms appear.

Does 5-HT Stabilize DA in the Striatum?

The 5-HT neurons in the dorsal raphe nucleus (DRN) send a dense projection to the basal ganglia, in particular to the striatum and the substantia nigra pars compacta [111]. We have discussed above that the projection to the striatum plays an important role in levodopa therapy for Parkinson's disease because much of the newly created DA during a dose comes from 5-HT terminals (see section "Serotonin and Levodopa"). The 5-HT projection to the striatum is an example of volume transmission. The released 5-HT binds to receptors on DA terminals, such as the ones from the SNc, and increases the release of DA in the striatum when the DA neurons fire [19, 22, 37]. Thus, increased firing in the DRN causes increased release of 5-HT in the striatum, which in turn causes inhibition of the indirect pathway and excitation of the direct pathway from the cortex through the striatum to the thalamus; see Fig. 12. This circuitry is even more complicated because there are excitatory projections from the thalamus to the cortex [102]. And the DRN sends projections to many regions in the brain and most of those regions project back to the DRN [82]. One of those returning projections is an inhibitory projection from the medial prefrontal cortex [27] pictured in Fig. 12.

Here is a plausible mechanism by which 5-HT release from DRN neurons could partially compensate for cell loss in the SNc. When SNc cells die, then some inhibition of the indirect pathway is withdrawn and some excitation of the direct pathway is withdrawn. Since the indirect pathway inhibits the thalamus and the direct pathway excites the thalamus, the effect of cell loss in the SNc is greater net inhibition of the thalamus. Since projections from the thalamus excite cortical neurons there will be less stimulation of the cortex. But the inhibitory projections from the medial prefrontal cortex will fire less, removing inhibition from the DRN. Thus the DRN will fire more, which will increase the release of DA from DA terminals in the striatum partially compensating for cell loss in the SNc. Thus the "purpose" of the 5-HT

Fig. 12 A 5-HT circuit that could stabilize DA in the striatum

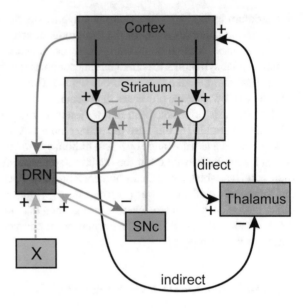

projection from the DRN to the striatum might be to stabilize the balance between the direct and indirect pathways against cell loss in the SNc. That this idea would work is supported by a simple mathematical model [96], but not enough is known about the details of projections from the thalamus to the cortex and from the cortex to the DRN to be sure of the anatomy.

"Take Home" Message for Neurologists, Psychiatrists

The complicated electrophysiological, pharmacological, and anatomical structure of the brain makes the design and delivery of drugs to achieve specific ends a very daunting challenge. There are several difficulties that are implicit in the examples that we've given in this chapter, but it is useful to make them explicit.

1. The brain contains many homeostatic mechanisms (in the electrophysiology, the pharmacology, the endocrinology, etc.) that tend to act to counter the intent of specific interventions. For example, as discussed in section "How Do SSRIs Work?", the original idea of SSRI development was to block the reuptake of 5-HT in projection regions and thus raise the 5-HT concentration in the extracellular space of the projection region. However, the SSRI will also block reuptake of the 5-HT released in the DRN when 5-HT cells fire. The released 5-HT stimulates the 5-HT$_{1A}$ autoreceptors on DRN cell bodies lowering the firing rate of the DRN 5-HT neurons, and this would tend to *lower* the 5-HT concentrations in projection regions.

2. A drug will not only affect the cells that you want it to affect, but, may also affect many other cells in the brain. A perfect example is the use of levodopa for Parkinson's patients where the intent is to increase the production of DA in the remaining SNc neurons. But levodopa is taken up by all cells of the body and in 5-HT varicosities and cell bodies it is used to manufacture and store DA in 5-HT neurons. This not only causes the large pulses of DA in the striatum that have been implicated in the development of dyskinesias, but also severely impairs the 5-HT system during the dose.

3. As emphasized by Fuxe et al. [45], brain cells express a myriad of different receptors. Often receptor populations are at locations distant from the endogenous sources of neurotransmitter and in fact may never be reached by endogenous transmitter [116]. When one gives an exogenous drug, one may stimulate receptors that are not normally stimulated under physiological conditions, and therefore the consequences are difficult to predict.

4. The brain is not homogeneous, receptors are not spread out evenly, and local consequences in one nucleus can differentially project to other brain regions. For example, suppose that one designs an antagonist for the $5\text{-}HT_{1A}$ receptors on the cell bodies of 5-HT neurons with the intent of raising DRN cell firing and the release of 5-HT in the striatum. Then, depending on the strength of projections, the 5-HT concentration will likely change differentially in all regions to which the DRN projects. Moreover, a number of studies have identified roles for $5\text{-}HT_{1A}$ receptors in processes such as thermoregulation, immune function, and memory [91], where side effects might be anticipated.

The examples that we have given all involve volume transmission. Mathematical modeling of volume transmission can help us to understand the differential effects of drugs in local brain regions as well as side effects caused by projections to other regions. And, thus, mathematical modeling is an important tool for a better and more rational design of drug interventions.

There is another way in which the study of volume transmission can help us understand the brain. There are biophysical models of individual neurons and models of small and large networks of neurons. On the other hand, there are top-down models of behavior created by cognitive scientists in which different brain regions or nuclei are treated as black boxes and one studies how the regions influence each other and cause behavior. Connecting the models at these two very different levels is a difficult but important problem in brain science. Here the study of volume transmission can help because volume transmission operates at intermediate levels between these two types of models [50, 81, 87]. For example, studies of cholinergic modulation of neural circuits (volume transmission) have helped bridge understanding from the cellular effects of acetylcholine to its role in cognitive performance [87].

"Take Home" Message for Computationalists

Most of the mathematical modeling that has been done in computational and mathematical neuroscience addresses the electrophysiological view of the brain. All mathematical models are simplifications of the real biology, of course, and a natural simplification is that the brain consists of its electrical circuitry. If so, one should study the fundamental unit of the circuitry, the individual neuron, how neurons influence each other, and the behavior of small, medium, and large networks of interacting neurons. Because of the sheer size of the circuitry and the biological variation in individual neurons and connection patterns, these problems have been a fertile source of interesting biological and mathematical questions since the time of Hodgkin and Huxley [60]. All along it was understood that neurons are cells and that they are influenced by glia, local biochemistry, diet, the endocrine system, behavior, and changing gene expression levels, but it was hoped that those other details could be ignored because function arises mainly from the electrical circuitry. If function arises from the coordination of all those systems, then understanding function in the brain is a daunting challenge, indeed.

We believe, and we have been making the case in this chapter, that volume transmission is an important new area for computational and mathematical modelers who study the brain. By volume transmission, a local nucleus, for example the DRN or the SNc, can change the local biochemistry in the extracellular space in a distant nucleus. And, as we have indicated, there are 5-HT receptors on DA neurons that change the amount of DA released in the striatum, when the concentration of 5-HT goes up. So the different volume transmission systems are not independent, but affect each other. Unraveling these long distance biochemical networks and their interactions will be fundamental to understanding the brain in health and disease. In addition, the study of volume transmission raises interesting mathematical questions. We mention three such questions here.

If the purpose of a projection is to keep the neurotransmitter in the extracellular space within close upper and lower bounds, how precise does the placement of varicosities or terminals have to be to accomplish the goal? This is an interesting mathematical question because the neurons are firing stochastically, and the varicosities and terminals are both the sources of the neurotransmitter and the sinks into which it is absorbed. What is the spatial dependence in the extracellular space of the long term (average) neurotransmitter concentration? A natural first assumption would be to assume that the glial cells do not take up neurotransmitter, so in this case it is just a question of release, diffusion, and reuptake in the tortuous extracellular space. However, glial cells do take up neurotransmitters, and this means that the boundaries of the extracellular space are weakly absorbing. We have some preliminary results on these questions [73].

Secondly, though it has been known for years that autoreceptors play an important role in controlling the extracellular concentrations of neurotransmitters, not so much is known about the intracellular mechanisms involved in inhibiting synthesis and release or in the strengths of the inhibitions or the ranges over which they operate. In most of our models, we suppose that the inhibition of synthesis and release depends on the current extracellular concentration of the neurotransmitter. However, recent experimental and computational evidence [113] shows that autoreceptor effects are long-lived. The autoreceptor effect can last 30–60 s after the concentration in the extracellular space has returned to normal. Thus models will have to take into account the dynamics of autoreceptor effects inside of cells.

How should volume transmission and pharmacology appear in electrophysiological network models? For firing rate models, the firing rate of a neural population might depend on the concentration of a neurotransmitter released from a presynaptic population. Behn and Booth [7] used such an approach in modeling the rat sleep-wake regulatory network. Thus they were able to simulate experiments in which neurotransmitter agonists and antagonists were microinjected into the locus coeruleus to see the effects on the structure of the sleep-wake cycle. In conductance-based models, such as those based on the Hodgkin-Huxley formalism, modulation can be accounted for through effects on model parameters [50]. Researchers often have found bifurcations in synaptic conductance parameters that dramatically change the dynamics of the system. On inspection, these conductance strengths would depend on local neurotransmitter concentrations through volume transmission. For example, in [14], the transition from normal to pathological (parkinsonian) neural activity may be due in part to an increased level of inhibition from the striatum; as discussed above, this is expected to result from decreased volume transmission of DA from the SNc. This is just one example of a neural circuit that would exhibit very different dynamics at different concentration levels of a neurotransmitter. This possibility shows that there will be very interesting dynamical systems questions in the interactions between the volume transmission network and local electrophysiological systems.

Acknowledgements The authors would like to thank Professor Parry Hashemi for stimulating and useful discussions. This work was supported in part by NSF grant EF-1038593 (HFN, MCR), the Mathematical Biosciences Institute and the NSF under grant DMS-0931642 (JAB, MCR), and an NSF CAREER Award DMS-0956057 (JAB).

References

1. Adell, A., Celada, P., Abella, M.T., Artigasa, F.: Origin and functional role of the extracellular serotonin in the midbrain raphe nuclei. Brain Res Rev **39**, 154–180 (2002)
2. Agid, Y.: Parkinson's disease: pathophysiology. Lancet **337**, 1321–1324 (1991)
3. Ahlskog, J.E.: Does vigorous exercise have a neuroprotective effect in Parkinson disease? Neurology **77**, 288–294 (2011)
4. Albin, R.L., Young, A.B., Penney, J.B.: The functional anatomy of basal ganglia disorders. Trends Neurosci. **12**, 366–375 (1989)
5. Anderson, G.M., Barr, C.S., Lindell, S., Durham, A.C., Shifrovich, I., Higley, J.D.: Time course of the effects of the serotonin-selective reuptake inhibitor sertraline on central and peripheral serotonin neurochemistry in the rhesus monkey. Psychopharmacology **178**, 339–346 (2005)

6. Beaulieu, J.M., Gainetdinov, R.R.: The physiology, signaling and pharmacology of dopamine receptors. Pharmacological Reviews **63**(1), 182–217 (2011)

7. Behn, C.D., Booth, V.: Simulating microinjection experiments in a novel model of the rat sleep-wake regulatory network. J Neurophysiol **103**, 1937–1953 (2010)

8. Bel, N., Artigas, F.: Fluoxetine preferentially increases extracellular 5-hydroxytryptamine in the raphe nuclei: an in vivo microdialysis study. Eur. J. Pharmacol. **229**, 101–103 (1992)

9. Belanger, M., Allaman, I., Magistretti, P.J.: Brain energy metabolism: focus on astrocyte-neuron metabolic coooperation. Cell Metabolism **14**, 724–738 (2011)

10. Benkelfat, C., Ellenbogen, M.A., Dean, P., Palmour, R.M., Young, S.N.: Mood-lowering effect of tryptophan depletion. Arch. Gen. Psych. **51**, 687–697 (1994)

11. Benmansour, S., Owens, W.A., Cecchi, M., Morilak, D., Frazer, A.: Serotonin clearance in vivo is altered to a greater extent by antidepressant-induced downregulation of the serotonin transporter than by acute blockade of the transporter. J. Neurosci. **22**(15), 6766–6772 (2002)

12. Bergman, H., Wichmann, T., Karmon, B., DeLong, M.: The primate subthalamic nucleus. II. neuronal activity in the MPTP model of parkinsonism. J Neurophysiol **72**, 507–520 (1994)

13. Bergstrom, B., Garris, P.: "Passive stabilization" of striatal extracellular dopamine across the lesion spectrum encompassing the presymptomatic phase of Parkinson's disease: a voltam-metric study in the 6-OHDA-lesioned rat. J. Neurochem. **87**, 1224–1236 (2003)

14. Best, J., Park, C., Terman, D., Wilson, C.: Transitions between irregular and rhythmic firing patterns in an excitatory-inhibitory neuronal network. J Comp Neurosci **23**, 217–235 (2007)

15. Best, J.A., Nijhout, H.F., Reed, M.C.: Homeostatic mechanisms in dopamine synthesis and release: a mathematical model. Theor Biol Med Model **6**, 21 (2009)

16. Best, J.A., Nijhout, H.F., Reed, M.C.: Serotonin synthesis, release and reuptake in terminals: a mathematical model. Theor Biol Med Model **7**, 34– (2010)

17. Bezard, E., Dovero, S., C, C.P., Ravenscroft, P., Chalon, S., Guilloteau, D., Crossman, A.R., Bioulac, B., Brotchie, J.M., Gross, C.E.: Relationship between the appearance of symptoms and the level of nigrostriatal degeneration in a progressive 1-methyl-4-phenyl-1,2,3,6tetrahydropyridine-lesioned macaque model of Parkinson's disease. J Neurosci **21**, 6853–6861 (2001)

18. Birzniece, V., Johansson, I.M., Wang, M.D., Secki, J.R., Backstrom, T., Olsson, T.: Serotonin 5-HT1A receptor mRNA expression in dorsal hippocampus and raphe nuclei after gonadal hormone manipulation in female rats. Neuroendocrinology **74**(2), 135–142 (2001)

19. Blandina, P., Goldfarb, J., Craddock-Royal, B., Green, J.P.: Release of endogenous dopamine by stimulation of 5-hydroxytryptamine3 receptors in rat striatum. J. Pharmacol. Exper. Therap. **251**, 803–809 (1989)

20. Blier, P., de Montigny, C., Chaput, Y.: Modifications of the serotonin system by antidepressant treatment: implications for the therapeutic response in major depression. J. Clin. Psychohar-macol. **7**, 24S–35S (1987)

21. Bloom, D.E., Cafiero, E.T., Jané-Llopis, E., Abrahams-Gessel, S., Bloom, L.R., Fathima, S., Feigl, A.B., Gaziano, T., Mowafi, M., Pandya, A., Prettner, K., Rosenberg, L., Seligman, B., Stein, A.Z., Weinstein, C.: The global economic burden of noncommunicable diseases. Harvard School of Public Health, World Economic Forum (2011)

22. Bonhomme, N., Duerwaerdère, P., Moal, M., Spampinato, U.: Evidence for 5-HT4 receptor subtype involvement in the enhancement of striatal dopamine release induced by serotonin: a microdialysis study in the halothane-anesthetized rat. Neuropharmacology **34**, 269–279 (1995)

23. Borah, A., Mohanakumar, K.P.: Long-term L-DOPA treatmeant causes indiscriminate increase in dopamine levels at the cost of serotonin synthesis in discrete brain regions of rats. Cell. Mol. Neurobiol. **27**, 985–996 (2007)

24. Brooks, D.J.: Dopamine agonists: their role in the treatment of Parkinson's disease. J. Neurol. Neurosurg Psychiatry **68**, 685–689 (2000)

25. Carlsson, A.: Perspectives on the discovery of central monoaminergic neurotransmission. Annu. Rev. Neurosci. **10**, 19–40 (1987)

26. Carta, M., Carlsson, T., Kirik, D., Björklund, A.: Dopamine released from 5-HT terminals is the cause of L-DOPA-induced dyskinesia in parkinsonian rats. Brain **130**, 1819–1833 (2007)

27. Celada, P., Puig, M.V., Casanovas, J.M., Guillazo, G., Artigas, F.: Control of dorsal raphe serotonergic neurons by the medial prefrontal cortex: Involvement of serotonin-1A, GABA$_A$, and glutamate receptors. J. Neurosci **15**, 9917–9929 (2001)

28. Chaput, Y., Blier, P., de Montigny, C.: In vivo electrophysiological evidence for the regulatory role of autoreceptors on serotonergic terminals. J. Neurosci **6**(10), 2796–2801 (1986)

29. Chen, H., Zhang, S.M., Schwarzschild, M.A., Hernan, M.A., Ascherio, A.: Physical activity and the risk of Parkinson disease. Neurology **64**, 664–669 (2005)

30. Chen, J., Wang, E., Cepeda, C., Levine, M.: Dopamine imbalance in Huntington's disease: a mechanism for the lack of behavioral flexibility. Front. Neurosci. **7** (2013)

31. Chiel, H.J., Beer, R.D.: The brain has a body: adaptive behavior emerges from interactions of nervous system, body, and environment. Trends Neuroscience **20**, 553–557 (1997)

32. Cipriani, A., Furukawa, T., Salanti, G., Geddes, J., Higgins, J., Churchill, R., Watanabe, N., Nakagawa, A., Omori, I., McGuire, H., Tansella, M., Barbui, C.: Comparative efficacy and acceptability of 12 new-generation antidepressants: a multiple-treatments meta-analysis. Lancet **373**, 746–758 (2009)

33. Cooper, J., Bloom, F., Roth, R.: The Biochemical Basis of Neuropharmacology. Oxford U. Press, New York, NY (2003)

34. Cunha, R.A.: Different cellular sources and different roles of adenosine: A$_1$ receptor-mediated inhibition through astrocytic-driven volume transmission and synapse-restricted A$_{2A}$ receptor-mediated facilitation of plasticity. Neurochem. Int. **52**, 65–72 (2008)

35. DeLong, M.: Primate models of movement disorders of basal ganglia origin. TINS **13**, 281–285 (1990)

36. Dentresangle, C., Cavorsin, M.L., Savasta, M., Leviel, V.: Increased extracellular DA and normal evoked DA release in the rat striatum after a partial lesion of the substantia nigra. Brain Res **893**(178–185) (2001)

37. Deurwaerdère, P., Bonhomme, N., Lucas, G., Moal, M., Spampinato, U.: Serotonin enhances striatal overflow in vivo through dopamine uptake sites. J. neurochem. **66**, 210–215 (1996)

38. Duncan, R.P., M.Earhart, G.: Randomized controlled trial of community-based dancing to modify disease progression in Parkinson disease. Neurorehabilitation and Neural Repair **26**(2), 132–143 (2012)

39. Feldman, R., Meyer, J., Quenzer, L.: Principles of Neuropharmacology. Sinauer Associates, Inc, Sunderland, MA. (1997)

40. Fernstrom, J.: Role of precursor availability in control of monoamine biosynthesis in brain. Physiol. Rev. **63**, 484–546 (1983)

41. Fernstrom, J., Wurtman, R.D.: Brain serotonin content: physiological dependence on plasma tryptophan levels. Science **173**, 149–152 (1971)

42. Fornal, C., Litto, W., Metzler, C.: Single-unit responses of serotonergic dorsal raphe neurons to 5-ht1a agonist and antagonist drug administration in behaving cats. J. Pharmacol. Exper. Therap. **270**, 1345–1358 (1994)

43. Frisina, P.G., Haroutunian, V., Libow, L.S.: The neuropathological basis for depression in Parkinson's disease. Parkinsonism Relat Disord **15**(2), 144–148 (2009)

44. de la Fuente-Fernandez, R., Lu, J.Q., Sossi, V., Jivan, S., Schulzer, M., Holden, J.E., Lee, C.S., Ruth, T.J., Calne, D.B., Stoessl, A.J.: Biochemical variations in the synaptic level of dopamine precede motor fluctuations in Parkinson's disease: PET evidence of increased dopamine turnover. Annals of Neurology **49**(3), 298–303 (2001)

45. Fuxe, K., Dahlstrom, A.B., Jonsson, G., Marcellino, D., Guescini, M., Dam, M., Manger, P., Agnati, L.: The discovery of central monoamine neurons gave volume transmission to the wired brain. Prog. Neurobiol. **90**, 82–100 (2010)

46. Gartside, S.E., Umbers, V., Hajos, M., Sharp, T.: Interaction between a selective 5-HT1A receptor antagonist and an SSRI in vivo: effects on 5-HT cell firing and extracellular 5-HT. Br. J. Pharmacol. **115**, 1064–1070 (1995)

47. Gelenberg, A., Chesen, C.: How fast are antidepressants? J. Clin. Psychiatry **61**, 712–721 (2000)
48. Gerfen, C.R., Surmeier, D.J.: Modulation of striatal projection systems by dopamine. Annu. Rev. Neurosci. **34**, 441–466 (2011)
49. Gonzalez, O., Berry, J., McKight-Eily, L., Strine, T., Edwards, K., Kroft, J.: Current depression among adults - United States, 2006, 2008. Center for Disease Control and Prevention (2010)
50. Graupner, M., Gutkin, B.: Modeling nicotinic neuromodulation from global functional and network levels to nAChR based mechanisms. Acta Pharmacologica Sinica **30**, 681–693 (2009)
51. Haber, S.N., Calzavara, R.: The cortico-basal gangia integrative network: the role of the thalamus. Brain Research Bulletin **78**, 69–74 (2009)
52. Hahn, M., Mazei-Robison, M., Blakely, R.: Single nucleotide polymorphisms in the human norepinephrine transporter gene affect expression, trafficking, antidepressant interaction, and protein kinase C regulation. Mol. Pharmacol. **68**, 457–466 (2005)
53. Hajos, M., Gartside, S.E., Villa, A.E.P., Sharp, T.: Evidence for a repetitive (burst) firing pattern in a sub-population of 5-hydroxytryptamine neurons in the dorsal and median raphe nuclei of the rat. Neuroscience **69**, 189–197 (1995)
54. Hajos, M., Sharp, T.: Burst-firing activity of presumed 5-ht neurones of the rat dorsal raphe nucleus: electrophysiological analysis by antidromic stimulation. Brain Res. **740**, 162–168 (1996)
55. Haldane, J.: Enzymes. Longmans, Green and Co, New York (1930)
56. Heisler, L.K., Pronchuk, N., Nonogaki, K., Zhou, L., Raber, J., Tung, L., Yeo, G.S.H., O'Rahilly, S., Colmers, W.F., Elmquist, J.K., Tecott, L.H.: Serotonin activates the hypothalamic–pituitary–adrenal axis via serotonin 2C receptor stimulation. J. Neurosci. **27**(26), 6956–6964 (2007)
57. Hervás, I., Artigas, F.: Effect of fluoxetine on extracellular 5-hydroxytryptamine in rat brain. role of 5-HT autoreceptors. Eur. J. Pharmacol. **358**, 9–18 (1998)
58. Hervás, I., Vilaró, M.T., Romero, L., Scorza, M., Mengod, G., Artigas, F.: Desensitization of 5-HT$_{1A}$ autoreceptors by a low chronic fluoxetine dose. effect of the concurrent administration of WAY-100635. Neuropsychopharmacology **24**, 11–20 (2001)
59. Heyn, J., Steinfels, G., Jacobs, B.: Activity of serotonin-containing neurons in the nucleus raphe pallidus of freely moving cats. Brain Res. **251**, 259–276 (1982)
60. Hodgkin, A.L., Huxley, A.F.: A quantitative description of membrane current and its application to conduction and excitation in nerve. The Journal of Physiology **117**(4), 500–544. (1952)
61. Hornung, J.P.: The human raphe nuclei and the serotonergic system. J. Chem. Neuroanat. **26**, 331–343 (2003)
62. Hurtado, J., Gray, C., Tamas, L., Sigvardt, K.: Dynamics of tremor-related oscillations in the human globus pallidus: a single case study. Proc. Nat. Acad. Sci. **96**, 1674–1679 (1999)
63. Invernizzi, R., Bramante, M., Samanin, R.: Citalopram's ability to increase the extracellular concentrations of serotonin in the dorsal raphe prevents the drug's effect in the frontal cortex. Brain Res. **260**, 322–324 (1992)
64. Iwata-Ichikawa, E., Kondo, Y., Miyazaki, I., Asanuma, M., Ogawa, N.: Glial cells protect neurons against oxidative stress via transcriptional up-regulation of the glutathione synthesis. J Neurochem **72**(6), 2334–2344 (1999)
65. Jacobs, B.L., Fornal, C.A.: 5-ht and motor control: a hypothesis. TINS **16**, 346–352 (1993)
66. Kandel, E., Schwartz, J., Jessell, T., Siegelbaum, S., Hudspeth, A.: Principles of Neural Science, 5th edn. McGraw-Hill Education/Medical (2012)
67. Kannari, K., Tanaka, H., Maeda, T., Tomiyama, M., Suda, T., Matsunaga, M.: Reserpine pretreatment prevents increases in extracellular striatal dopamine following L-DOPA administration in rats with nigrostriatal denervation. J Neurochem **74**, 263–269 (2000)
68. Kegeles, L., Abi-Dargham, A., Frankle, W., Gil, R., Cooper, T., Slifstein, M., Hwang, D.R., Huang, Y., Haber, S., Laruelle, M.: Increased synaptic dopamine function in associative regions of the striatum in schizophrenia. Arch. Gen Psychiat. **67**, 231–239 (2010)

69. Knappskog, P., Flatmark, T., Mallet, J., Lüdecke, B., Bartholomé, K.: Recessively inherited L-DOPA-responsive dystonia caused by a point mutation (Q381K) in the tyrosine hydroxylase gene. Hum. Mol. Genet. **4**, 1209–1212 (1995)

70. Knobelman, D.A., Hen, R., Lucki, I.: Genetic regulation of extracellular serotonin by 5-hydroxytryptamine$_{1A}$ and 5-hydroxytryptamine$_{1B}$ autoreceptors in different brain regions of the mouse. J. Pharmacol. Exp. Ther. **298**, 1083–1091 (2001)

71. Kravitz, A.V., Freeze, B.S., Parker, P.R.L., Kay, K., Thwin, M.T., Deisseroth, K., Kreitzer, A.C.: Regulation of parkinsonian motor behaviors by optogenetic control of basal ganglia circuitry. Nature letters **466**, 622–626 (2010)

72. Kreiss, D.S., Lucki, I.: Effects of acute and repeated administration of antidepressant drugs on extracellular levels of 5-HT measured in vivo. J. Pharmacol. Exper. Therap. **274**, 866–876 (1995)

73. Lawley, S., Best, J., Reed, M.: Neurotransmitter concentrations in the presence of neural switching in one dimension. AIM's J. **(under revision)** (2016)

74. Lemke, M.R.: Depressive symptoms in Parkinson's disease. European Journal of Neurology **15**(Suppl. 1), 21–25 (2008)

75. Lincoln, C.M., Bello, J.A., Lui, Y.W.: Decoding the deep gray: A review of the anatomy, function, and imaging patterns affecting the basal ganglia. Neurographics **2**, 92–102 (2012)

76. Lindgren, H.S., Andersson, D.R., Lagerkvist, S., Nissbrandt, H., Cenci, M.A.: L-DOPA-induced dopamine efflux in the striatum and the substantia nigra in a rat model of Parkinson's disease: temporal and quantitative relationship to the expression of dyskinesia. J. Neurochem. **112**, 1465–1476 (2010)

77. Malagie, I., Trillat, A.C., Jacquot, C., Gardier, A.M.: Effects of acute fluoxetine on extracellular serotonin levels in the raphe: an in vivo microdialysis study. Eur. J. Pharmacol. **286**, 213–217 (1995)

78. Mann, J.J.: Role of the serotonergic system in the pathogenesis of major depression and suicidal behavior. Neuropsychopharmacology **21**(2S), 99S–105S (1999)

79. Miller, G., Madras, B.: Polymorphisms in the 3'-untranslated region of the human and monkey dopamine transporter genes effect reporter gene expression. Mol. Psychiat. **7**, 44–55 (2002)

80. Moncrieff, J., Kirsch, I.: Efficacy of antidepressants in adults. Brit. Med. J. **331**, 155 (2005)

81. Montague, P.R., Dolan, R.J., Friston, K.J., Dayan, P.: Compuational psychiatry. Trends in Cognitive Sciences **16**(1), 72–80 (2012)

82. Monti, J.M.: The structure of the dorsal raphe nucleus and its relevance to the regulation of sleep and wakefulness. Sleep Med. Rev. **14**, 307–317 (2010)

83. Moore, P., Landolt, H.P., Seifritz, E., Clark, C., Bhatti, T., Kelsoe, J., Rapaport, M., Gillim, C.: Clinical and physiological consequences of rapid tryptophan depletion. Neuropsychopharmacology **23**(6), 601–622 (2000)

84. Mouradian, M.M., Juncos, J.L., Fabbrini, G., Chase, T.N.: Motor fluctutations in Parkinson's disease: pathogenetic and therapeutic studies. Annals of Neurology **22**, 475–479 (1987)

85. Mouradian, M.M., Juncos, J.L., Fabbrini, G., Schlegel, J., J.Bartko, J., Chase, T.N.: Motor fluctuations in Parkinson's disease: Central pathophysiological mechanisms, part ii. Annals of Neurology **24**, 372–378 (1988)

86. Navailles, S., Bioulac, B., Gross, C., Deurwaerdère, P.D.: Chronic L-DOPA therapy alters central serotonergic function and L-DOPA-induced dopamine release in a region-dependent manner in a rat model of Parkinson's disease. Neurobiol. Dis. **41**, 585–590 (2011)

87. Newman, E.L., Gupta, K., Climer, J.R., Monaghan, C.K., Hasselmo, M.E.: Cholinergic modulation of cognitive processing: insights drawn from computational models. Frontiers in Behavioral Neuroscience **6**(24), 1–19 (2012)

88. Nicholson, S.L., Brotchie, J.M.: 5-hydroxytryptamine (5-HT, serotonin) and Parkinson's disease - opportunities for novel therapeutics to reduce problems of levodopa therapy. European Journal of Neurology **9**(Suppl. 3), 1–6 (2002)

89. Nijhout, H.F., Best, J., Reed, M.: Escape from homeostasis. Mathematical Biosciences **257**, 104–110 (2014)

90. Pålhagen, S.E., Carlsson, M., Curman, E., Wålinder, J., Granérus, A.K.: Depressive illness in Parkinson's disease – indication of a more advanced and widespread neurodegenerative process? Acta Neurol Scand **117**, 295–304 (2008)
91. Raymond, J.R., Mukhin, Y., Gelasco, A., Turner, J., Collinsworth, G., Gettys, T., Grewal, J., Garnovskaya, M.N.: Multiplicity of mechanisms of serotonin receptor signal transduction. Pharmaco. Therap. **92**, 179–212 (2001)
92. Raz, A., Vaadia, E., Bergman, H.: Firing patterns and correlations of spontaneous discharge of pallidal neurons in the normal and tremulous 1-methyl-4-phenyl-1,2,3,6 tetra-hydropyridine vervet model of parkinsonism. J Neurosci. **20**, 8559–8571 (2000)
93. Reed, M., Best, J., Nijhout, H.: Passive and active stabilization of dopamine in the striatum. BioScience Hypotheses **2**, 240–244 (2009)
94. Reed, M., Lieb, A., Nijhout, H.: The biological significance of substrate inhibition: a mechanism with diverse functions. BioEssays **32**, 422–429 (2010)
95. Reed, M., Nijhout, H.F., Best, J.: Mathematical insights into the effects of levodopa. Frontiers Integrative Neuroscience **6**, 1–24 (2012)
96. Reed, M., Nijhout, H.F., Best, J.: Computational studies of the role of serotonin in the basal ganglia. Frontiers Integrative Neuroscience **7**, 1–8 (2013)
97. Reed, M., Nijhout, H.F., Best, J.: Projecting biochemistry over long distances. Math. Model. Nat. Phenom. **9**(1), 133–138 (2014)
98. Reijnders, J.S.A.M., Ehrt, U., Weber, W.E.J., Aarsland, D., Leentjens, A.F.G.: A systematic review of prevalence studies of depression in Parkinson's disease. Movement Disorders **23**(2), 183–189 (2008)
99. Royo, M., Daubner, S., Fitzpatrick, P.: Effects of mutations in tyrosine hydroxylase associated with progressive dystonia on the activity and stability of the protein. Proteins **58**, 14–21 (2005)
100. Rutter, J.J., Gundlah, C., Auerbach, S.B.: Increase in extracellular serotonin produced by uptake inhibitors is enhanced after chronic treatment with fluoxetine. Neurosci. Lett. **171**, 183–186 (1994)
101. Schildkraut, J.J.: The catecholamine hypothesis of affective disorders: a review of supporting evidence. Amer. J. Psych. **122**, 509–522 (1965)
102. Shepherd, G.M. (ed.): The Synaptic Organization of the Brain, 5th edn. Oxford U. Press, (2004)
103. Smith, Bevan, M., Shink, E., Bolam, J.P.: Microcircuitry of the direct and indirect pathways of the basal ganglia. Neuroscience **86**, 353–387 (1998)
104. Smith, D., Dempster, C., Glanville, J., Freemantle, N., Anderson, I.: Comparative efficacy and acceptability of 12 new-generation antidepressants: a multiple-treatments meta-analysis. Brit. J. Physchiatry **180**, 396– (2002)
105. Smith, T., Kuczenski, R., George-Friedman, K., Malley, J.D., Foote, S.L.: In vivo microdialysis assessment of extracellular serotonin and dopamine levels in awake monkeys during sustained fluoxetine administration. Synapse **38**, 460–470 (2000)
106. Soghomonian, J.J., Doucet, G., Descarries, L.: Serotonin innervation in adult rat neostriatum i. quantified regional distribution. Brain Research **425**, 85–100 (1987)
107. Stancampiano, R., Melis, F., Sarais, L., Cocco, S., Cugusi, C., Fadda, F.: Acute administration of a tryptophan-free amino acid mixture decreases 5-HT release in rat hippocampus in vivo. Am. J. Physiol. **272**, R991–R994 (1997)
108. Tanaka, H., Kannari, K., Maeda, T., Tomiyama, M., Suda, T., Matsunaga, M.: Role of serotonergic neurons in L-DOPA- derived extracellular dopamine in the striatum of 6-OHDA-lesioned rats. NeuroReport **10**, 631–634 (1999)
109. Tanda, G., Frau, R., Chiara, G.D.: Chronic desipramine and fluoxetine differentially affect extracellular dopamine in the rat pre-frontal cortex. Psychopharmacology **127**, 83–87 (1996)
110. Turner, E., Rosenthal, R.: Efficacy of antidepressants. Brit. Med. J. **336**, 516–517 (2008)
111. Vertes, R.P.: A PHA-L analysis of ascending projections of the dorsal raphe nucleus in the rat. J. Comp. Neurol. **313**, 643–668 (1991)
112. W., B.C., E., S.B., A., E.R.: Noradrenergic modulation of wakefulness/arousal. Sleep Med. Rev. **16**(2), 187–197 (2012)

113. Wood, K.M., Zeqja, A., Nijhout, H.F., Reed, M.C., Best, J.A., Hashemi, P.: Voltammetric and mathematical evidence for dual transport mediation of serotonin clearance in vivo. J. Neurochem. **130**, 351–359 (2014)

114. Young, S.N., Smith, S.E., Pihl, R., Ervin, F.R.: Tryptophan depletion causes a rapid lowering of mood in normal males. Psychopharmacology **87**, 173–177 (1985)

115. Zold, C.L., Kasanetz, F., Pomata, P.E., Belluscio, M.A., Escande, M.V., Galinanes, G.L., Riquelme, L.A., Murer, M.G.: Striatal gating through up states and oscillations in the basal ganglia: Implications for Parkinson's disease. J. Physiol-Paris **106**, 40–46 (2012)

116. Zoli, M., Jansson, A., Syková, E., Agnati, L., Fuxe, K.: Volume transmission in the CNS and its relevance for neuropsychopharmacology. Trends in Pharmacological Sciences **20**, 142–150 (1999)

117. Zoli, M., Torri, C., Ferrari, R., Jansson, A., Zini, I., Fuxe, K., Agnati, L.: The emergence of the volume transmission concept. Brain Res. Rev. **28**, 136–147 (1998)

118. Zygmond, M., Abercrombie, E.D., Berger, T.W., Grace, A.A., Stricker, E.M.: Compensation after lesions of central dopaminergic neurons: some clinical and basic implications. TINS **13**, 290–296 (1990)

Attachment Modelling: From Observations to Scenarios to Designs

Dean Petters and Luc Beaudoin

Introduction

Computational psychiatry is an emerging field of study at the intersection of research on psychiatric illness and computational modelling. Using a biological approach to understanding psychiatric disorders has enabled researchers to make great progress in understanding the causes of disorders like schizophrenia and depression. A computational approach helps bridge an explanatory gap from biological details to observable behaviours and reportable symptoms at the psychological level [60]. Connectionist modelling, and the use of reinforcement learning algorithms have been applied to computational psychiatry [60]. For example, connectionist models have been used to simulate schizophrenia [22] and reinforcement learning algorithms have been used to model anxiety and mood disorders [25, 26].

This paper describes an agent-based modelling approach to attachment relationships, and attachment disorders. Agent-based models can incorporate connectionist and reinforcement learning components within complete cognitive architectures [68]. In addition, they may also include other representations and algorithms that exist within complex cognitive architectures that are 'deep' in the sense of providing detailed implementations of particular subsystems, such as planning, reasoning or learning [5, 46, 67]. A key benefit of an agent-based approach in computational psychiatry is that it allows multiple agents to represent multiple people interacting in an 'online' dynamic fashion [76]. Agent-based attachment models are therefore complementary to attachment models based upon neural networks that do not exist in dynamically changing 'online' virtual environments [20, 29, 32].

D. Petters (✉)
Birmingham City University, Birmingham, UK
e-mail: d.d.petters@cs.bham.ac.uk

L. Beaudoin
Simon Fraser University, Burnaby, Canada
e-mail: LPB@sfu.ca

© Springer International Publishing AG 2017
P. Érdi et al. (eds.), *Computational Neurology and Psychiatry*,
Springer Series in Bio-/Neuroinformatics 6,
DOI 10.1007/978-3-319-49959-8_9

227

Attachment Theory describes how our closest relationships develop and function across the life span [19]. Attachment bonds are formed early in infancy and can reform and develop through the life-span [13]. Attachment relationships are realised in different ways in different contexts and different developmental stages. From early in infancy, humans demonstrate a strong innate predisposition to emotionally attach to familiar people around them who provide physical or emotional security. Then in developing towards adulthood, humans also show attachment responses to romantic partners and act as caregivers [30, 97].

Attachment Theory also provides biological, comparative, evolutionary, cognitive, cross-cultural and psychopathological perspectives [9, 13, 27, 31, 67, 77, 94]. It does not merely explain moment to moment interactions between infants and carers. Attachment interactions can also be observed longitudinally through human lifespan development from infancy to old age, and in adult romantic relationships. Because Attachment Theory explains phenomena over a range of timescales from moment to moment interactions to ontogenetic and phylogenetic development a complete modelling approach to attachment phenomena needs to be capable of simulating temporally nested scenarios. Sometimes the modeller will just want to explain a few minutes of interaction, but on other occasions modelling attachment development over a lifespan or over evolutionary trajectories may be desired [73].

The early sensori-motor/behavioural core of the relationship (support for exploration and independence and haven of safety when needed) continues for the duration of the relationship. Then with development of language and representational skills, mental representations of relationship history and expectations about availability and responsiveness in relationships become a significant factors in attachment related influences on psychological problems.

Attachment Theory provides detailed descriptions of many phenomena of interest to a computational modeller that range from: normative attachment development through the lifespan [13, 67]; classifications of secure, insecure-avoidant, insecure-ambivalent and insecure-disorganised behaviour patterns in infancy [1, 52, 54, 68]; to measurement of analogous categories in adolescents [3], adult caregivers [37] and adults in romantic relationships [30]. Given that infant classifications are made in response to mostly non-verbal behaviour, infant attachment behaviour patterns can sometimes be compared to phenomena described in comparative psychology and ethology [13, 39]. In disorganised attachment classification, infants neither approach nor avoid their caregivers in a consistent and organised way, resolving apparent conflict by producing behaviours that seem out of place. These disorganised responses in human infants have been compared to ethological displacement behaviour. For example, in studies of birds [39], where birds in confrontations neither fight or flee but instead preen.

In the Adult Attachment Interview (AAI), classification typically occurs through interpretation of verbal behaviour. It is the quality of discourse in terms of appropriate quantity, relevance and coherence, rather than its content, that allows patterns in verbal behaviour to be interpreted as autonomous, dismissing, preoccupied, or unresolved/disorganised [37]. Whilst avoidance in infancy has been described as an organised response that involves managed disengagement from a caregiver, avoidance in dismissing adults, as observed in the AAI, may be better presented as a form of psychodynamic defense, with dismissing adults producing verbal responses that minimise discussion or description of emotionally uncomfortable content.

One of the challenges of modelling and one of the ways in which modelling can be helpful is in making explicit the parallels between the behavioural and representational facets of attachment relationships. How parallel are they? Can we develop testable hypotheses? And can we develop ideas about how early experience and problems in the interactive domain are related to the representational realm? It is a significant shortcoming that, so far, attachment theorists have not managed to be very explicit about these issues or proposed realistic mechanisms.

The same cognitive architecture may simulate normal attachment interactions and pathological attachment relationships and internal states, depending on the environment and experiences included in the model [67]. In addition, a simulation of pathological processing should be able to engage with therapeutic interventions, such as interventions to change behavioural patterns, or modelling a response to talking therapies. Taken together these benefits reinforce each other so that whilst it is not practical to model the whole brain or whole mind, designing relatively broad complete architectures for attachment suggests a promising approach for computational psychiatry.

The Scientific Problem

The scientific problem focused upon in this paper involves updating the information processing framework for attachment theory, originally set out by John Bowlby between 1958 and 1982 [12–16]. As reviewed in the next section, Bowlby's theoretical approach to explaining attachment phenomena evolved from psychoanalytic, to ethological, control systems and finally to including Artificial Intelligence structures and mechanisms. However, he did not develop a systematic Artificial Intelligence perspective, and never became aware of the kinds of developments that are routinely used in contemporary research in cognitive architectures, machine learning or agent based modelling. So the purpose of the research programme detailed in this paper is to update the attachment control system framework that Bowlby set out by reconceptualising it as a cognitive architecture that can operate within multi-agent simulations [67].

The scientific contributions made by this paper are two-fold. Firstly, it frames descriptions of attachment behaviour patterns as directing and evaluating computational modelling efforts. Secondly, it demonstrates the progress that has been made in producing running simulations of the behaviour patterns of interest. To fulfil the first scientific contribution, behavioural patterns will be expressed in scenarios (specification of requirements for the modelling effort), to guide the modelling effort and to allow models to be validated when the simulations are constructed and 'run'. The important characteristics of the behaviours that we want to capture may involve numerical quantities, such as the frequencies of occurrence of particular behaviours.

However there may be aspects that cannot be quantified, such as capturing rule based patterns of behaviour. In this case structural descriptions may provide an effective method of assessment of what has and has not been achieved. This paper will therefore use a scenario-based method of specifying and evaluating requirements that can provide precise metrics for elements of the behaviours we want to capture that are not easily represented in quantitative ways [65, 67]. Scenarios can capture abstract patterns of behaviour and allow researchers to recast them at a level of concreteness and appropriate detail. For example, attachment measures involve taking detailed observations and coming up with generalisations such as discourse patterns being more or less coherent. These generalisations can then be modelled in simulations with precise metrics. The second part of the scientific problem is therefore creating agent-based simulations that can match the requirements set out by the scenarios. But the process is iterative with scenarios and simulations both deepening over design cycles [65, 67].

The particular attachment behaviour patterns this paper will consider have been organised into four specific scenarios. These are the patterns in empirical observations ranging from secure base behaviour of infants to the discourse patterns produced by adults, and ultimately, simulating causal links between adult states of mind, caregiving patterns and infant attachment patterns:

- after describing secure-base behaviour in infancy the scientific problem is posed as explaining infant exploration and proximity seeking behaviour patterns that might be demonstrated when an infant explores a park. In terms of the modelling techniques used, this is similar to an artificial life foraging scenario (using simple 'animat' autonomous agents) [49, 67], but with two types of agent: those to be protected and those that do the protecting [17, 62, 63].
- after describing how attachment studies using the Strange Situation procedure involve year long intensive observations of home behaviour, and then short observations of behaviour in controlled laboratory conditions, the scientific problem for attachment modelling is posed as analogous to a machine learning study. In this scenario, different infant home experiences result in different attachment classifications in the laboratory. Thus an overall simulation (the infant and carer agents interacting within the virtual environment) has a training stage and a test stage. The scientific problem is therefore to show how 'test stage' behaviour can be produced by learning from the experiences in the training stage. This second scenario develops the first scenario by including infant agent learning about responsiveness and sensitivity over a simulated year, and by describing behaviour at one year of age in minute by minute detail.

Two scenarios are presented which have not until now been implemented in simulations:

- after analysing the Strange Situation in more detail, the scientific problem in this scenario focuses more on the least frequently found individual differences category, the 'D' disorganised/disoriented classification. So the scenario is posed as how the observable 'displacement' behaviour of 'D' infants is caused. The development of 'D' pattern behaviour in infants is linked in high risk populations to maltreatment, and in low risk samples to their caregiver experiencing unresolved loss. It develops the first two scenarios because it focuses on the scheduling of behaviour units and how behaviour units might be constructed 'on the fly'.

- after describing response patterns in the Adult Attachment Interview (AAI), the scientific problem is posed as how coherent discourse about attachment by caregivers, and different types of failure to produce coherent discourse, might be modelled. The AAI involves questions and answers about the same broad subject over the course of an hour. So the agent will therefore be able to engage in language processing with a simple grammar and attachment-focused lexicon. The agent will possess an ability to make moment to moment working memory retrievals about what was previously communicated in the interview, and about its previous experiences in the virtual world [47]. The agent will also produce sentences that do not need to be as complex as those produced by real humans in the actual AAI, Just complex enough to allow variations between agents that present balanced recollections of their attachment history, and those that present either responses that lack key details or that include redundant information. The agent will therefore possess control processes that subserve both language comprehension and production. In addition to how the agents engage in language, this scenario will also specify how agents will engage in caregiving, in terms of responsiveness, sensitivity, and effectiveness. So the structures and mechanisms that support conversation need to be integrated with a broader architecture that supports selection of actions that manipulate the world, deliberation about the consequences of such actions, self-reflection, and failures in self-reflection. The challenge is therefore that the same agent will be required to act in a simulated environment as a caregiver in a particular manner and then converse about its actions and experiences in that environment in an associated communicational manner. Since attachment studies also describe longitudinal data on trajectories from infant to adult classifications, this scenario might also be extended to a single 'lifespan' scenario where agents experience attachment interactions from infancy to adulthood, and at each point in their 'lifespan' possess an attachment classification that reflects their experiences.

Bowlby's Development of Attachment Theory

For thirty years after the Second World War, Attachment Theory [1, 12–15] was formulated and developed by John Bowlby and co-workers. Briefly reviewing how Attachment Theory developed in this long time period will help show that this theory is made of behavioural and cognitive components [66]. The behavioural component in Attachment Theory is comprised of many different systematic ways to measure and observe attachment-related phenomena. Attachment models need to produce these systematically observed behaviour patterns in simulations. There are also many cognitive components in Attachment Theory explaining different ways in which individuals process information relevant to attachment phenomena. Cognitive components in Attachment Theory are useful for attachment modellers as a starting point for designing the internal information processing structures and mechanisms that will produce the required behaviour patterns when incorporated in simulated agents. In each stage in its theoretical development new layers of theoretical cognitive constructs and empirical measurement tools have been added to what is now a rich and broad approach to social and emotional development. So a recounting of this theoretical development will be used to structure the development of attachment modelling scenarios and attachment models which have been implemented as running simulations.

While a medical student, Bowlby became interested in personality development and the key role played by an individual's early caregiving environment [43]. He trained as a psychoanalyst and used psychoanalytic ideas to explain how infants and their mothers emotionally bond [2, p. 333]. During the early 1950s he brought together and integrated a significant amount of empirical data regarding the effects of separation and loss on emotional development and attachment in a landmark report to the World Health Organization [11]. However, during this period he increasingly found psychoanalytic theory inadequate as an explanation for phenomena related to social and emotional attachment, separation, and loss [12, 13]. For Bowlby, this approach to motivation required revision because it was rooted in a drive theory which suggested infants were primarily focused on their inner drives and drive representations, and little interested in the social or physical environment per se. This focus inwards was in part driven by the psychoanalytic retrospective case study method which Bowlby had rejected. Critiques from psychology and philosophy of science also made clear that the drive theory of motivation was not tenable [90]. It was not well supported by their own evidence, which itself was problematic, and seemed inaccessible to ordinary standards of empirical analysis and falsification [95].

Although Bowlby rejected drive theory, psychoanalysis did possess a number of key insights into early experiences and relationships which Bowlby valued and wanted to maintain in his own approach [95]. Bowlby wanted to reform psychoanalytic theory rather than replace it, to conserve theoretical insights related to the importance of the inner-life [95], such as:

- human infants lead a complex emotional and cognitive early life
- the strength of attachment between mother and infant is not related to overt behaviours like separation protest
- early close relationship attachments form prototypes for relationships through the lifespan

Through contact with Robert Hinde, Bowlby became aware that Ethology offered an alternative motivational model upon which to base his explanations for attachment phenomena. Ethology was a more scientifically respectable approach than psychoanalysis, with instincts and fixed action patterns that were observable in behaviour, as opposed to the empirically inaccessible sexually based mental energy hypothesized by psychoanalysts. This led to his 1958 presentation of Ethological and Evolutionary Attachment Theory [12]. In this theory, the attachment behaviour of human infants was explained in terms of ethological behaviour systems. Within the attachment control system, the goals related to attachment were organized according to four behaviour systems, the attachment, fear, sociability and exploration systems [13, 19]. According to Bowlby, what defines the attachment control system is not a set behaviour repertoire but the outcomes that predictably follow from these behaviours. Similar behaviours may be produced by different behaviour systems. The three main ethological claims [12, p. 366] incorporated in this version of his theory were that:

- attachment behaviours are species-specific behaviour patterns—following a similar and predictable pattern in nearly all members of a species. This is theorised as occurring because each behavioural system was evolutionarily designed to increase the likelihood of survival and adaptation to environmental demands, in the environment of evolutionary adaptedness (EEA) [13].
- these behaviour patterns are activated and terminated by various external and internal stimuli. Each behavioural system involves a set of contextual activating triggers; a set of interchangeable, functionally equivalent behaviours that constitute the primary strategy of the system for attaining its particular goal state; and a specific set-goal (a state of the person-environment relationship that terminates the systems activation).
- the actions of a behaviour system are not always a simple response to a single stimulus but can be a behavioural sequence with a predictable course. Simple sequences of behaviour can be integrated into more complex behavioural patterns.

Although Bowlby's use of Ethology provided a rigorous observational methodology and a rich comparative/evolutionary framework, it did so at the cost of some of the descriptive richness of the inner life that the previous psychoanalytic approach provided [43]. So in the 1960s Bowlby again searched for a motivational framework, that was scientifically rigorous but also rich and complex enough to explain attachment phenomena. He introduced a much broader and richer set of constructs to Attachment Theory as he formulated the concept of the Attachment Control System. This explanatory framework integrated ideas from Piagetian developmental psychology, Cybernetics, Systems Theory, and Artificial Intelligence, as well as retaining ideas from Ethology.

Where Bowlby's 1958 version of a behaviour system was representationally and mechanistically simple, in the 1969 version of the behaviour system this construct had become an organising component in the representationally and algorithmically richer attachment control system. In this new framing, a behaviour system might be supported by different information processing mechanisms and structures at different stages of development. So extra claims incorporated by Bowlby into how behaviour systems operate in this updated control system approach to attachment included that:

- behaviour systems are a product of the interaction of genetic endowment with environment, and so moulded by social encounters so that the person's behavioural

capacities fit better with important relationship partners and other relational constraints. So each behaviour system has its own ontogenetic development, initially
producing reflex actions and later in infancy producing fixed action patterns which
increase in the complexity of their organisation in sequences and chains.

- each behavioural system also includes cognitive-behavioural mechanisms, such as
  monitoring and appraising the effectiveness of behaviours enacted in a particular
  context, which allow flexible, goal-corrected adjustment of the primary strategy
  whenever necessary to get back on the track of goal attainment.
- representations guide behaviour, operating partly unconsciously but also partly at
  the level of conscious thoughts and intentions,
- later in development, the operation of behaviour systems is increasingly mediated
  through high level representations, such as consciously accessible internal working models and natural language

Up till this point in this historical review, we have strongly emphasised the organising focus provided by Bowlby's conceptualisation of behaviour systems. This theoretical framework has a central place in the cognitive component of Attachment Theory. This is of great value for computational psychiatrists taking their first steps to
modelling in this domain. However, there are several reasons for an attachment modeller to look beyond modelling of attachment behaviours like secure-base and safe-
haven responses. These include finding modelling approaches that capture the complexities of ontogenetic development, and capture the complexities of adult attachment behaviour.

Regarding ontogenetic development, the continuity of exploration, fear, attachment, and socialisation behavioural systems through lifespan development is linked
to continuity of the goals that are held by these systems through developmental
stages. In the field of Situated Cognition, Clark has introduced the concept of 'soft
assembly' to explain the flexible manner in which systems can be constructed when
they are focused on achieving particular set goals rather than using particular means
or actions [21, p. 44]. Part of the scientific problem set out in this paper is how
to computationally model the soft assembly of attachment behaviour systems that
use reflexes and goal corrected mechanisms early in infancy but use more advanced
chaining of actions, simple plans, hierarchical plans, internal working models, and
mediate actions using natural language, at increasing mature stages in development.
Individuals plan future actions with particular outcomes in mind, and can discuss
these plans using natural language. In terms of 'soft assembly' it is the organism's
progress to achieving whole-organism goals that helps shape the components of its
cognitive architecture as they develop. To understand ontogenetic development and
the complexities of adult attachment behaviour, we need to explain the ability to
revise, extend, and check models of self and other in attachment relationships [13,
p. 82].

Bowlby formulated the attachment control system as relying on internal working
models that allowed predictions to be made about the results of possible actions,
as well as including planning representations and processes, and feedback mechanisms such as homeostatic control [13]. Internal working models of self and others

are mental representations which store the residues of person-environment transactions. They organise memories of behavioural-system functioning and guide future attempts to attain the overall set goal of the system. Bowlby invokes internal working models at early stages in development and later on, when linguistic skills and conscious reflection can enable models to become more adequate [13, p. 84]. Internal working models transmit, store and manipulate information and allow the individual to conduct small scale experiments within the head [13, p. 81]. For Bowlby, their function was to replace the internal worlds of traditional psychoanalytic theory. Bowlby emphasises the requirements for internal working models to be updated. For a computational psychiatrist, a particularly important aspect of Bowlby's internal working model construct is that pathological sequelae of separation and bereavement can be understood as outdated or insufficiently updated models that may contain inconsistencies and confusions [13, p. 82]. So internal working models are very germane to psychopathology in the attachment domain.

Bowlby had observed that people often fail to bring to conscious awareness some aspects of their previous attachment experiences. Psychoanalytic theory suggested that this occurs to minimise an individual's psychic pain and distress. Freud viewed the mind as involving conflict between conscious thoughts and unconscious emotions 'trying' to become conscious and be discharged, like a *"boil or abscess which, being unable to find a path to the surface of the body, cannot discharge the toxic matter it contains"* [90, p. 13]. Freud theorised that unconscious psychically toxic thoughts were held below consciousness by an active process of defensive control-repression:

> How had it come about that the patients had forgotten so many of the facts of their external and internal lives but could nevertheless recollect them if a particular technique was applied? Observation supplied an exhaustive answer to these questions. Everything that had been forgotten had in some way or other been distressing; it had been either alarming or painful or shameful by the standards of the subject's personality. It was impossible not to conclude that was precisely why it had been forgotten - that is, why it had not remained conscious. In order to make it conscious again in spite of this, it was necessary to overcome something that fought against one in the patient; it was necessary to make efforts on one's own part so as to urge and compel him to remember. The amount of effort required of the physician varied in different cases; it increased in direct proportion to the difficulty of what had to be remembered. The expenditure of force on the part of the physician was evidently the measure of a **resistance** on the part of the patient. It was only necessary to translate into words what I myself had observed, and I was in possession of the theory of **repression**. [33, p. 18]

Bowlby accepted that there was a phenomenon to be explained regarding differential awareness of past aversive events. However, he rejected the psychoanalytic account for how such a failure in awareness might arise. When comparing psychoanalytic defense processes to his own conceptualisation, Bowlby emphasised that inaccessible memories and desires were *"due to certain information of significance to the individual being systematically excluded from further processing"* [15, p. 65]. He therefore proposed inaccessibility occurring due to a person cognitively redefining or re-categorising previous experiences to exclude them [28]. So the concept of defensive exclusion in the attachment domain [15] captured important clinical phenomena that psychoanalysts had previously noted but which Bowlby realized were not dependent on their theory of libidinal (sexually based) drives. Therefore implementations of internal working models in computational psychiatry should allow for differences in accessibility of memories of past events and biases in the information provided about the consequences of possible future actions.

Bowlby used the term 'defensive exclusion' to describe an informational process with an affective dimension, in which internal and external inputs are excluded *"together with the thoughts and feelings to which such inflows give rise"* [15, p. 65]. He suggested that individuals can selectively exclude available information relevant to attachment interactions and focus instead on salient but less distressing thoughts. Defensive processes may be necessary and adaptive if they stop disorganising negative affects. Or they might lack the subtlety and flexibility necessary to serve adaptive ends and result in restricted, overgeneralized, or distorted versions of reality that are inconsistent with good adjustment. Defensive exclusion is an important phenomena for computational psychiatry because it is concerned with limits on the updating of internal working models of the self and other, and how these limitations can consolidate over time to result in major differences in self-image and personality [28]. For example, Bowlby suggested multiple internal working models can be formed with incompatible information. Some primitive models may be formed early and become dominant and exert influence either consciously or unconsciously over later forming radically incompatible subordinate models that the individuals is actually aware of [14, p. 238]

Without using such terminology, he also portrays kinds of self-reflective metaprocessing on mental life which occurs in therapy in explicitly computational terms:

> The psychological state may then be likened to that of a computer that, once programmed, produces its results automatically when activated. Provided the programme is the one required, all is well. [When] representational models and programmes are well adapted, the fact that they are drawn on automatically and without awareness is a great advantage. When however, they are not well adapted, for whatever reason, the disadvantages of the arrangement become serious.

> For the task of changing an over-learned programme of action and/or of appraisal is enormously exacerbated when rules long implemented by the evaluative system forbid its being reviewed.[...] A psychological state of this kind in which a ban on reviewing models and action systems is effected outside awareness is one encountered frequently during psychotherapy. It indicates the existence of another stage of processing at which defensive exclusion can also take place, different to the stage at which perceptual defence takes place. [15, pp. 55–56]

Work has been undertaken in Artificial Intelligence on computational systems which can undertake limited forms of self-examination [7, 45], though this remains an under-explored area of AI.

In conclusion, Bowlby has provided the computational psychiatrist with a rich description of information processing structures and mechanisms, but these are not set out in enough detail to be readily incorporated using contemporary modelling techniques. A process of interpretation and 're-imagining' is required to translate Bowlby's attachment control system and its components into a design for a contemporary cognitive architecture [65, 67, 69, 71–73]. This section has shown that scenarios for attachment modelling need to include behavioural descriptions ranging from infant exploration and security seeking to adult caregiver behaviour patterns, including complex adult thought and verbal action patterns.

A Scenario for Secure-Base Behaviour

Bowlby's descriptions of infant and childhood attachment responses highlight several hallmarks of attachment [74], and of these, using the carer as a secure-base and haven of safety are central behavioural characteristics of an attached child. Secure-base behaviour is typically observed in infant-carer attachment relationships from infancy to middle childhood. This involves infants using their carers as secure-bases from which to explore, and havens of safety to return to when tired or anxious. Infants will use a flexible repertoire of behaviours when pursuing currently active attachment goals. The outcomes that reliably follow from activating these behaviours define the attachment control system. For example, if an infant is anxious and its current goal is to increase its proximity to a carer the infant may cry (which predictably brings the carer closer), or crawl towards the carer.

A secure-base scenario at one year would require that when in an open area like a park, infants move away from their carers to explore, but stay within a line of sight to the caregiver, and periodically 'check-in' by gaining the attention of their carers or by moving back to closer proximity [4, 13], switching between approach to the caregiver and exploratory behaviour [67, pp. 51–78].

The secure-base scenario needs to capture the temporal contingencies of the real infants and carers whose behaviours we want to explain, thus capturing interactions that in reality occur within a continuing stream of time and affect. Moreover, people involved in social interactions will often control the nature of the sensations they receive from their external and internal environments. In a secure-based scenario, if an infant turns the direction of their head, they can then see a completely different view from before that may activate different desires, goals or intentions. So, for example, when an infant turns their face away from the person they are interacting with because they are concerned about that person's response, this may have the result of provoking that person less, so containing the situation. In a simulation focused on adults in conversation about attachment the same types of contingencies through time exist—an agent will respond at any point in time to the previous statements made to it, and the previous responses it has already made back. So to capture the nature of the rich contingencies and dependencies that exist over short and long term timescales a modelling methodology requires the following capabilities:

- a simulated environment that can support dynamic interactions that unfold over time and are contingent on the immediately previous context (a world with entities that respond in an appropriate timely fashion to events).
- agents which exist within the simulated environment that can engage in extended exchanges or 'dialogs' with their environments, including other agents, where each new step in the dialog can be influenced by all the preceding steps (agents within

the world that have an encoded memory or adaptive structure that 'records' previous experiences).

- a way of populating and running simulations that allows for inclusion of a wide variety of different types of agent cognitive architectures.
- being able to combine data from experimental studies (that treat data points as an independent sequence of discrete training exemplars) with data from naturalistic studies (that see each data point as part of a linked series of events) [76].

In summary, what is needed to model social interactions ranging from secure-base behaviour to linguistically mediated dialogues are agents which exist in a virtual environment over multiple time-slices, and scenarios with goals so that the effectiveness of the attachment system can be evaluated as an aid or impediment to achieving the goals. To validate such models the running simulations need to be matched against scenarios based upon results from standardised empirical procedures carried out with many participants. Fortunately, such studies were completed by colleagues of Bowlby, most importantly, Mary Ainsworth and colleagues' development of the Strange Situation procedure [1]; and Mary Main and colleagues' development of the AAI [37, 53].

An Individual Differences Approach to Attachment in Infancy—The Strange Situation

Whilst Bowlby was setting out the information processing underpinnings for Attachment Theory, Mary Ainsworth and co-workers [1] studied how differences in infant-care interactions can affect the course of emotional and social development. The focus on individual differences in attachment status and development led to an empirically productive new direction for attachment research. Much of the contemporary attachment research on mental health issues and psychopathology is linked to attachment categories derived from the Strange Situation procedure [1]. This is not an experiment where infant-carer dyads are randomly assigned to different conditions in the laboratory. Rather it is a standardised laboratory procedure where all infants are presented with the same controlled and replicable set of experiences.

To capture infant responses to changes in context, the Strange Situation procedure consists of 8 three minute episodes which are designed to activate, intensify or relax the one-year-olds attachment behaviour in a moderate and controlled manner. The infant and carer enter the laboratory setting together, but then undergo a separation. The carer leaves the room, before a reunion in a subsequent episode. As the first reunion episode ends the infant meets an unfamiliar stranger in the laboratory, before a further separation. In each episode infant behaviour is recorded behind a two-way mirror. In the final episode the mother is reunited with her one-year-old infant after the infant has been left alone for three minutes in the unfamiliar setting.

In the Strange Situation infants show normative trends in how they respond to context changes that occur in the transitions between the eight episodes. For example, infants, irrespective of home environment, typically exhibit increased distress when their carer leaves the room. When the infant is left with a stranger (in episode four) or completely alone (in episode six) they typically gain some support from interaction with the 'stranger'. Even though the stranger is an unfamiliar, interacting with a warm and friendly adult is typically preferable to being alone. Nested within the normative trends are several patterns of response reflecting the infants confidence in the caregivers response patterns. The infant's response in the reunion episodes correlates strongly with patterns of maternal behaviour and infant responses intensively observed throughout the previous year. Therefore a key finding of the Strange Situation, and which makes it such a valuable research tool, is that infant behavioural patterns observed when the carer returns to the infant after a separation (infant-carer reunions occur in episodes five and eight) provide the best short-hand classification for the attachment behavioural patterns of infant and carer observed at length in the home environment.

Individual differences in the Strange Situation cluster into four patterns:

- Secure (type B) infants form the largest proportion in non-clinical populations and secure behaviour is the reference pattern against which the other classifications are evaluated. Infant responses in reunion episodes in the Strange Situation include approaching their mothers in a positive manner and then returning to play and exploration in the room quickly. They receive care at home which is consistently sensitive, more emotionally expressive and provides less contact of an unpleasant nature; at home these infants are less angry and they cry less.
- Avoidant (type A) infants typically make up the second largest proportion of non-clinical populations. Infants responses in reunion episodes in the Strange Situation include not seeking contact or avoiding their carer's gaze or avoiding physical contact with her. These children return quickly to play and exploration but do so with less concentration than secure children. Whilst playing they stay close to and keep an eye on their carer. It may seem that they are not distressed or anxious in the Strange Situation. However, research employing telemetered autonomic data and salivary hormone assays has demonstrated that, despite their relative lack of crying, avoidant infants are at least as stressed by the procedure as secure and resistant infants. Their care at home is consistently less sensitive to infant signals and less skilled in holding the baby during routine care and interaction. At home these infants are more angry and cry more than secure infants.
- Ambivalent (type C) infants typically make up a small but measurable proportion of non-clinical populations. Infants responses in reunion episodes in the Strange Situation include not being comforted and being overly passive or showing anger towards their carers. These infants do not return quickly to exploration and play. Their care at home is less sensitive and particularly inconsistent. In comparison with average levels across all groups, C type carers are observed at home being more emotionally expressive; they provide physical contact which is unpleasant at a level intermediate between A and B carers and leave infants crying for longer durations. At home these infants are more angry and cry more than secure infants.

- Disorganised (type D). This last attachment pattern has been more recently categorised, is the not as well characterised or understood, and forms a very small proportion of infants in the general population [34, p. 26]. Disorganised attachment
  is typically conceptualised as a lack of a coherent, organised attachment strategy,
  and is linked to frightened or frightening behaviour on the part of the caregiver
  (we detail disorganised attachment more fully in the next section).
- Cannot classify—another recently defined category is comprised of the small
  number of infants who do not fit into any of the other four classifications.

The major impact of the Strange Situation is in part because numerous additional
attachment measures have been developed which show that the individual difference
categories exist at one year of age are found later in childhood [56, 97] and in adulthood [36]. Attachment Theory describes how our closest relationships develop and
function across the life span. In a risk factors approach to psychopathology, secure
status has been suggested as a protective factor whereas the three insecure attachment
patterns have been suggested as risk factors for various subsequent psychopathologies [27]. Whilst the evidence linking secure attachment to improved social competence is relatively clear, the relationship of insecure attachment to mental health is
not so straightforward. There are several possibilities for simple causal relationships
between insecure attachment status and psychopathology which can be discounted.
Firstly, insecure avoidant (type A) and ambivalent/resistant (type C) attachments are
not a form of psychopathology warranting clinical attention, and are often adaptive responses to particular caregiving environments [34]. Secondly, empirical data
show that these two insecure attachment patterns do not have a direct causal role
in the later development of psychological disturbance. The effort to find the 'Holy
Grail' of main effects of infant attachment on later psychopathology has so far been
characterised as a "fruitless search" [27, p. 638]. Sroufe et al. [89] and Mikulincer and Shaver [59] both note that statistical relationships found linking attachment
insecurity and psychopathology may in part be caused by insecurity arising or being
increased in a context of pre-existing psychological problems. Thirdly, the two main
insecure attachment patterns are not even strongly linked to specific threats to mental well-being. Rather, there are multiple pathways to any given disorder. A single
disorder might be reached from a combination of other risk factors. So in addition
to early attachment relationships, other risk factors are: child characteristics such as
temperament; parental socialisation and management strategies; and family ecology
(family life stress and trauma, family resources and social support). Also, insecure
attachment may contribute, along with these other risk factors, to multiple disorders
[27].

A commonly accepted view is that early attachment is just one factor among
many that either add protection or increase vulnerabilities to subsequent disorders. Attachment relationship dysfunction can give rise to serious psychopathology. For example, 'reactive attachment disorder' is one of a small number of psychopathological disorders diagnosable in young children [27, 34]. However, such
psychopathologies are linked to significant abuse and negligence by carergivers.
Whilst avoidant and ambivalent/resistant attachment have a complex and indirect

relationship with psychopathology, the less frequent D disorganised pattern shows a different relationship—elaborated in section "More Detail on Disorganised/Disoriented Attachment in Infancy".

A Scenario Based Upon the Original Strange Situation Studies

From the perspective of computational psychiatry the original Strange Situation study (that defined the A, B and C categories of infant attachment) is extremely valuable because it provides a year long 'training' phase for a simulation, where identical infant agents can be placed with a mother-agent that act towards them with caregiving patterns that are abstractions of the codings and rating actually observed by Ainsworth et al. [1] in year long home observations. The infant agents can then be placed in an abstraction of the 24 min Strange Situation procedure to 'test' if their agent architectures have adapted to their experiences and respond in the same way as real infants do. Training and test phases therefore support the same kinds of qualitative and quantitative evaluation that can occur in machine learning [66, 75]. A key justification for this approach is that A, B and C infant attachment patterns are learned, in the sense of a response to experience, and not innate temperamental traits [67, 93]. More details of the Strange Situation scenario for A, B and C behaviour patterns is found in Petters [67, 68]. In brief, the Strange Situation scenario needs to include a description of how infant and carer agents interact with each other to reproduce the different home and laboratory behaviour patterns observed in the year long home experiences and short test experiences of the Strange Situation procedure [1]. This includes a task that carer agents engage with and have to break off from (an abstraction of all the non-childcare tasks a childcarer does), thus allowing the programmer to create carer agents which are more or less responsive to caregiving. In addition, the scenario needs to include objects that the infant agent can explore ('toys'). So the infant agent is carrying out an exploration activity from which it can break off to seek proximity. Mini-scenarios for the Strange Situation component of the simulation need to set out requirements for how separation and reunion behaviours should result from learning about past experiences over ontogeny.

More Detail on Disorganised/Disoriented Attachment in Infancy

The A, B and C classification criteria were gained from a relatively small sample, 24 infant-carer dyads in the first study, and 105 infant-carer dyads overall in the four studies combined in the 1978 "*Patterns of Attachment*" book [1]. Main and Solomon [54] recount that after the Strange Situation was developed researchers studying middle-class populations typically classified every infant participant to whichever of the A, B, or C attachment classification fitted them best. Most infants fitted well into one of these three categories. But some infants needed to be 'forced' into a classification they did not fit well. This raised the question of whether some new categories might be found. Perhaps a more fine grained analysis would add extra categories—D, E, F and G—to the already existing A, B and C attachment patterns. Main and Solomon [54] review a number of studies where infants whose separation and reunion responses failed to make a good fit within A, B or C classifications. However, they did not find any clear new categories because these 'non-fitting' infants did not resemble each other in coherent or organised ways. What they shared in common was *"bouts or sequences of behavior which seemed to lack a readily observable goal, intention, or explanation."* [54, p. 122]. The term selected to describe these diverse behaviour patterns was 'disorganized/disoriented' but such infants were disorganised in different ways. Some were disorganised with respect to temporal behaviour patterns, and acted in an unusual and unpredictable manner given the context. For others, disorganisation went beyond merely seeming to behave out of context. These infants showed more obvious odd behaviours, such as *"approaching with head averted, stilling or freezing of movements for thirty seconds with a dazed expression"* [54, p. 122].

In discussing the D behaviour pattern in the Strange Situation, Main and Hesse [52] emphasise several issues: the contradiction of action as it is carried out, often at the moment it is initiated, and specifically related to attachment and to fear of the caregiver. For example, they describe one infant: *"immediately upon the parent's appearance in the doorway, the infant orients, then places hand to mouth; or rises, then falls prone, or cries, calls, and eagerly approaches the parent, then turns about and avoids the parent with a dazed expression. Later, in the same episode, the infant may approach the parent obliquely or with head averted; cling to the parent while averting gaze; cry while moving away from the parent, make hesitant, stop-start movements on approach to the parent, or undertake movements of approach which have a slow, limp, "underwater" quality as though being simultaneously inhibited or contradicted"* [52, p. 173]. A classification scheme for both forms of disorganisation includes the following criteria:

- disordering of expected temporal sequences
- simultaneous displays of contradictory behaviour
- incomplete, undirected, misdirected or interrupted movements and expressions (including stereotypes, mistimed movements and anomalous postures)
- behavioural and expressive slowing, stilling, and freezing
- direct indices of confusion and apprehension towards parents
- direct indices of disorganisation and disorientation

Context is key for a disorganised classification. It would not be uncommon for an infant to show many of these behaviours in particularly stressful situations—for example, when the parent is absent in the separation episodes of the Strange Situation. It is when such behaviours are seen in reunion episodes that they indicate disorganisation [54, p. 164]. A key characteristic of disorganised and disoriented behaviour patterns is that actions are contradicted or inhibited as they are actually being undertaken. So actions can be undermined almost as soon as they are initiated.

The inadequacy of the original classification scheme without the D category is shown by the fact that in this scheme many infants showing disorganised or disoriented behaviour patterns would be classified as B—the secure pattern. Despite their behaviour being unusual, they would not be showing pronounced positive instances of avoidance or ambivalence, as these are both organised strategies and D category infants seemed to lack a readily observable goal, intention or other explanation. For example, they might follow full proximity seeking by then turning away sharply and showing odd responses in their motion or expression [54]. In addition, some best fit 'A's were avoidant on reunions but also very distressed by each separation, and so seemed to not maintain a fully avoidant strategy of minimising the public exhibition of attachment behaviour whilst diverting attention to exploration. Some best fit C infants would show strong distress, but also strong avoidance. All these behaviour patterns are badly fitting to an A, B, C classification system but are readily classifiable with the addition of the D category [54, p. 130].

Studies that have specifically investigated the nature of parenting for D category infants show a difference between high and low risk samples that included 'D' category infants. In high risk samples a large majority of the infants of maltreating parents have been judged disorganised/disoriented in Strange Situation studies. However, the mere presence of a 'D' infant classification does not indicate an infant has been maltreated as this pattern is also found in 'low risk' groups, with carers who have been found to have unresolved loss of an attachment figure. In medically normal samples it is likely to reflect aspects of the child's interaction or experience of the parents' displays of anxiety, distress, conflicting signals, including tone of voice, facial expressions, visual attention, and bodily postures [52]. The mediating factor for high and low risk samples is hypothesised as being frightened/frightening behaviour. This hypothesis explains Strange Situation behaviour patterns where an observer cannot determine what is causing things—like screaming for the carer in a separation episode and then moving silently away on reunion, or crying loudly to gain proximity at the start of a reunion episode, but then freezing in an odd posture for several seconds. Main and Hesse [52, p. 178] suggest that infants produce D behaviour because they are experiencing anxiety and distress that is so intense that it cannot be deactivated by a shift in attention (the avoidant strategy) or approach to the carer (B and C strategy). They emphasize the 'dilemma' faced by infants with frightening or frightened caregivers. Normally, in stressful situations, a caregiver is the 'solution' and helps reduce anxiety and alarm. Infants with frightening or frightened caregivers may view their carer as such a solution, but they are paradoxically also the source of anxiety. The D pattern indicates a momentary state where the infant experiences an irresolvable conflict. So according to Main and Hesse [52], fear is the likely cause of inhibition or contradictory movements, and more direct markers of fear such as fearful expressions, extremely tentative approaches, and tense, vigilant body postures [52, p. 173].

Disorganised/Disoriented Behaviour Patterns as Displacement Behaviour

Main and Solomon [54] infer from the observed contradiction in movement patterns in 'D' type behaviour a contradiction in intention or plan, and go onto explain the phenomena in terms of the ethological displacement concept. Ethologists studying animal behaviour have found that a range of different behaviour selection strategies are used by animals in situations where different possible behaviours might be activated so are in conflict with each other. Behaviour conflicts occur when an animal is in a situation where either of two types of behaviour might be activated and it is impossible to produce both behaviours in a complete form. Conflict resolution can occur by: alternating the two competing behaviours; redirection activities (the same action with a different target); and the combination of the two behaviours in ambivalent or 'compromise' actions showing intention movement contributions from both [38]. Sometimes conflict resolution behaviours produced by an animal give a hint of a behaviour that is being inhibited, for example, showing intention movements of one before switching to the other. Alternatively, displacement behaviours can occur where two contextually obvious competing behaviours can be entirely replaced by a seemingly out of place behaviour. A classic example of behavioural conflict resolution through displacement behaviour is where a bird is in an approach/avoidance conflict with a conspecific. If the conspecific is much larger and likely to win a conflict, the best strategy is flight. But if the other bird is smaller and will likely withdraw the best strategy is to fight and force the flight of the other bird. But what to do if the opponent is in between the limit which trigger unambiguous decisions, leaving the bird without a clear decision to either fight or flee? In this case, the bird may produce a preening behaviour which appears to be irrelevant to the tendencies in conflict [38, p. 406].

Displacement behaviours involve irrelevant behaviour substitutes, in the sense of occurring through motivational factors other than those which normally activate them. However 'irrelevance' is a relative term, and the main issue is that they occur at a point of balance between two other behavioural tendencies which would dominate the scene if they did not counter-balance each other [38, p. 414]. Examples from comparative behavioural biology range from organised to disorganised. Preening in passerine birds is organised in that it is frequently observed and whilst not directly related to the aggressive or sexual behaviour it displaces, may have some function as a way of not committing to another behaviour and therefore maintaining proximity and engagement, keeping options open [38]. Disorganised examples involve broken or piecemeal behaviour patterns, for example, with just intention movements observable. So the comparative literature on displacement show examples ranging from more to less relevant and from more to less organised. These dimensions are pertinent to theories of infant attachment. For example Main and Weston [55] suggest displacement is the ethological mechanism for A type avoidant behaviour. In this view, infants are not really exploring when avoiding their caregivers but are displacement-exploring as an alternative to showing distress and proximity seeking. But this is an organised displacement that is either an evolved predisposition or a learned adaptive response [67, pp. 122–123]. D type displacement seems to be a different phenomenon, which may not be an evolved predisposition or learned behavioural pattern. The displacement behaviours seem less relevant and occur when no organised strategy fits.

A Scenario that Refines the Scientific Problem in Light of the 'D' Strange Situation Category

The discovery of the 'D' category makes things more complicated for computational models of the Strange Situation in several important ways. Firstly, we have to rethink how we interpret the 'C' category of infant attachment. Main and Weston [54] note that the Strange Situation behaviour of C infants seems less well organised than A and B infants, and when studies classified just in terms of A, B and C were reclassified with A, B, C and D alternatives more C infants were ultimately judged D than the A and B infants. Secondly, computational experiments run with simulations of the Strange Situation A, B, and C categories are simple in the sense of starting with identical infant agents. As these simulations will show how A, B and C differences emerge in the same agent architecture from experience. The 'D' category is less well studied and it may be that many infants who demonstrate disorganised or disoriented behaviour are temperamentally (innately) predisposed to insecurity in a way that A and C infants are not. Lastly, there are implications for how computational modellers implement behaviour systems. If all infants produce organised action sequences directed towards clear goals, then the modeller does not have to show how actions emerge from behaviour systems—these elements of the simulation can be 'hand-coded' at a basic level of behaviour, with sequences of behaviour emerging from such 'hand-coded' behaviour units. If the simulation needs to show how and why the units of action production fail then a greater level of granularity needs to be incorporated in the simulation. Perhaps emergence of both units of behaviour, and behaviour sequences needs to be explained.

The observation that there exist two kinds of disorganisation suggests two levels of behaviour construction. One level of behaviour construction is where units of behaviour that are put together as blocks—the organisation of behaviour sequences. A, B, and C behaviour patterns arise from organised predictable patterns for units of behaviour; so in disorganised behaviour patterns actions units may be scheduled in an odd order or in a way that fails to reach an appropriate goal. If we view the Strange Situation procedure as a 'stress test' then for A, B and C infants the heightened anxiety is evidence for the nature of their internal working models but still leaves them free to organise behaviour in pursuit of their goals in light of these internal working models. In the 'stress test' context of the Strange Situation, a disorganised/disoriented infant finds their action scheduler does not work effectively. A second level of behaviour construction is that multi-element behavioural schema also need to be constructed, and when not constructed the result is stereotypies or stilling. What is failing in these cases is the basic organisation of motor movements.

Regarding setting out an abstract description of 'home' behaviour for the 'training' phase of a 'D' category scenario: Infant agents will be required to not only be able to differentiate more and less responsive and sensitive care, but also respond to acts that model maltreatment—such as aversive communication and holding, and extreme neglect in responsiveness. Sensory and motor systems will also be required to simulate synchrony of micro-behaviours between a carer agent and infant agent, with some scenarios requiring that these synchronous interactions are not supported.

The Adult Attachment Interview (AAI)

The introduction of the Strange Situation procedure into attachment research had a major effect on direction and nature of attachment research. The Strange Situation relies on observation of overt behaviour. In attachment measures looking to map the individual difference patterns found in the Strange Situation to older individuals, alternative assessments strategies have been developed. For example, the rationale for the Adult Attachment Interview (AAI) is that language is a better indicator of adult attachment assessment. The AAI is a pre-specified interview format of 20 questions in a fixed order, but with additional specific follow up probes to these main set questions. The questioning and following arrangement must only highlight but not alter the participant's natural response tendencies. This is because the AAI is designed to elucidate structural variations in how life history is presented that allow reliable inferences about the participants internal state with regard to attachment [37]. It opens with a question asking for a general description of family relationships in the speaker's childhood. Further questions are asked about separations and experiences of rejection; and the effects of these experiences on adult personality. A key section probes experiences of bereavement. Experience of abuse is also asked about. The AAI ends with the speaker being invited to express wishes for his or her real or imagined child in the future [37].

Comparing the Strange Situation and the AAI, both procedures uncover 'inner workings' related to attachment which might otherwise not be so readily observable. Hesse [37] emphasises another aspect of the AAI that it has in common with the Strange Situation, namely, that the procedure acts as a mild 'stress test'—though neither procedure is presented as such in the attachment literature:

> The central task the interview presents to participant is that of (1) producing and reflecting on memories related to attachment, while simultaneously (2) maintaining coherent, collaborative discourse with the interviewer (Hesse 1996). This is not as easy as it might appear, and George and colleagues (1985, 1996) have remarked upon the potential of the protocol to "surprise the unconscious". As indicated above, the interview requires the speaker to reflect on and answer a multitude of complex questions regarding his or her life history, the great majority of which the speaker will never have been asked before. In contrast to ordinary conversations, where the interviewee has time for planning, the AAI moves at a relatively rapid pace, and usually all questions and probes have been presented within an hour's time. Ample opportunities are thereby provided for speakers to contradict themselves, to find themselves unable to answer clearly, and/or to be drawn into excessively lengthy or digressive discussions of particular topics [37, p. 555]

One of the earliest studies that used the AAI was carried out by Main et al. [53], who found a predictive relationship between a parent's hour long discussion of his or her own attachment history and his or her child's Strange Situation behaviour that was observed five years previously. Interviewees could be placed into three classification categories:

- secure autonomous adults value attachment relationships and experiences, and give apparently objective responses when asked about any particular relationship experience. When reporting specific experiences they provide confirming detailed memories and ability to reflect on those experiences with an understanding of why they, and others, behaved the way they did—and this is the case for happy and troubled experiences. So a secure autonomous adult might describe episodes of rejection but recognise the limitations of attachment figures in a balanced way, as well as including positive aspects of inadequate attachment figures [34]. This adult response pattern is associated with the infant Strange Situation secure behaviour pattern. However, there is also evidence that individuals who were insecure in childhood have gained 'earned secure' status as adults [37].
- dismissing adults devalue, or are emotionally cut off from attachment relationships and experiences. These individuals provide very short transcripts, with little to say about specific incidents and attachment experiences from their childhood in general. Responses are not only short but minimise the importance of relationships in general. They may idealize relationships as 'loving' but not provide detailed examples to justify such positive summary statements. This adult response pattern is associated with the infant Strange Situation insecure-avoidant behaviour pattern [37].
- preoccupied/enmeshed adults are preoccupied with (enmeshed by) early attachments or attachment related experiences. When reporting experiences these adults give plenty of detail and long transcripts but fail to provide a good overview because they become so entangled in the details. They may seem to be still engaged in emotional struggles related to attachment relationships [34]. Priming experiments that vary primes between attachment related primes, such as 'mother', and non attachment related primes, such as 'basket', showed that adults with this behaviour pattern, but not dismissing or secure adults, took much longer in a lexical decision task, thus demonstrating an example of specific attachment related interference to cognitive processing (Blount-Matthews cited in [37]). This adult response pattern is associated with the infant Strange Situation insecure-resistant/ambivalent behaviour pattern [37].

That the AAI produces a single classification when adults may have had different attachment relationships with either parent, and significant other carers, is based on the assumption that different attachment patterns found with each parent will have coalesced in adulthood [37]. Over the last 25 years, methods of analysing AAI transcripts have become more sophisticated. In addition, to the three classifications found in early AAI studies, two further classifications have been found in more recent studies [37]:

- unresolved/disorganised adults speak in unusual ways about loss experiences, demonstrating either chronic or failed mourning [37]. Unresolved/disorganised parents are more frightened or frightening and dissociative [37]. Such 'unre-solved' caregivers exhibit a range of behavioural attributes characteristic of loss and bereavement, including: interruptions to cognitive processes, particularly in contexts associated with the lost person; disbelief that loss has occurred or is per-manent; unfounded fear of death. So the AAI is an interview based method that can assess unresolved trauma which relies upon apparent lapses in metacognitive monitoring of reasoning and discourse [52].
- cannot classify adults produce responses that do not fit easily into any of the other categories. This pattern emerged in the early 1990s as a result of Main and Hesse noticing that a very small number of transcripts showed contradictions that stopped them fitting well into any of the other four classifications, for example, responses that showed both idealisation (suggestive of avoidant status) and highly preoccupied angry speech (suggestive of preoccupied/enmeshed status) [37].

Hesse [37] reviews the strong evidence of associations between AAI classification of an adult and the current Strange Situation categorisation of an infant who is cared for by that adult. As noted in the AAI descriptions above, secure infants are associ-ated with secure-autonomous caregivers; insecure-avoidant infants with dismissing caregivers; insecure-resistant/ambivalent infants with preoccupied caregivers; and disorganised/disoriented infants with unresolved/disorganised caregivers. An early finding by Main and Goldwyn [51] showed a correspondence between infant A, B and C Strange Situation and classifications and AAI classifications. This finding has subsequently been replicated by a number of studies. For example, van Ijzen-doorn [92] carried out a meta-analysis of 18 AAI studies comprising 854 parent-infant pairs from six countries and found that 82 % of secure-autonomous mothers had secure offspring; 64 % dismissing mothers had insecure-avoidant offspring; but just 35 % preoccupied mothers has insecure resistant/ambivalent offspring. In more recent studies evidence for the association between disorganised/disoriented infants with unresolved/disorganised caregivers has also been demonstrated. Hughes et al. [44] found that mothers who had previously experienced a stillbirth with their first child were more likely to be subsequently classified unresolved in the AAI, and for their later second-born infant to be classified disorganised in the Strange Situation. Studies have also investigated the relation between an adult's AAI classification and current caregiving behaviour—showing strong correlations [98]. Longitudinal stud-ies have looked at the relation between current AAI classification for adults and their previous Strange Situation categorisation of 20 years previously, when that adult was an infant, showing that 72 % had the same secure or insecure classification in infancy and adulthood [96]. For the computational psychiatrist, these associations between individual difference categories provide valuable constraints to evaluate and validate attachment models. The modelling effort can start by re-producing response patterns from the AAI, designing architectures to produce these linguistically mediated interview responses. Modelling can then go on to show how the same architectures can produce different caregiving response patterns, and hence demonstrate empirically observed patterns of intergenerational transfer.

Explanation in Terms of Attentional Flexibility in the AAI

The three organised responses in the AAI can be related to the three organised responses in the Strange Situation [37]. Main [50] highlighted how Strange Situation patterns show differences in attentional flexibility/inflexibility of the infant's responses to their carer and the environment. Secure babies showed greatest attentional flexibility—switching between proximity seeking and exploration quite easily. Insecure avoidant infants can be seen as attentionally inflexible with attention focused away from the carer. Insecure resistant/ambivalent infants can be seen as attentionally inflexible with attention focused strongly and singularly towards the carer [37]. The same attentional flexibility patterns were later related to the three organised responses in the AAI in terms of their discourse strategy. Adults in the AAI avoid thoughts or are unable to avoid thoughts the way that infants avoid behaviours or cannot avoid behaviours. Secure autonomous adults fluidly and flexibly shift between describing attachment related experiences and then responding to requests to evaluate those experiences. Dismissing adults show an inflexible attentional focus away from both description and evaluation. Preoccupied adults focus so strongly on describing attachment experiences that they do not break away to appropriately respond to requests for evaluation [37]. Parents of disorganised infants have limits and lapses to metacognitive monitoring of their own reasoning and shared discourse when discussing experiences which would be expected to involve trauma.

Explanation in Terms of Grice's Maxims of Discourse in the AAI

Coherence in the AAI has been related to the idea of linguistic coherence as defined by the linguistic philosopher Grice 1975 cited in [37], who proposed that coherent discourse is co-operative, and identified as adhering to four maxims: being true and accurate (maxim of quality); being succinct without being terse (maxim of quantity); being relevant (maxim of relationship); and being clear and well-ordered (maxim of manner) [34].

The AAI uses adherence to (or violation of) these maxims as a proxy for how an individual thinks and feels about their attachment experience. Transcripts with a lack of overall coherence end up being categorised as such due to major contradictions or inconsistencies, passages that are exceptionally short, long, irrelevant or difficult to understand and follow. It is not what actually happened to an individual in their past that is important for predicting an adult's attachment approach, but the coherence of the attachment narrative that the adult produces in the constrained AAI. So adults of all AAI classifications may publicly state the same kinds of values. The adult's internal 'state of mind' as indicated by coherence with respect to past attachment relationships is the best predictor of how they will conduct future attachment interactions, not the actual nature of their previous attachment interactions or their explicitly professed values [34]. So the coding for the AAI considers the use of language rather than making retrospective inferences about the person's actual attachment history [37].

A Scenario Based on Modelling Adult Coherent Discourse; and Failures in Coherence

This scenario involves refining the scientific problem to modelling adult meta-processing and executive control in four AAI classifications: secure autonomous; dismissing; preoccupied/enmeshed; and unresolved disorganised. A computational model that can simulate all variants of adult attachment as measured by the AAI would include adult conversation that was: (1) coherent; (2) partial or incomplete in how it described past events or current mental status with regard to attachment; (3) uncontrolled in how it was too long and repeated itself; and (4) showed significant disorganisation.

Simulating AAI behaviour patterns is significantly more challenging than simulating infant behaviour patterns observed in Strange Situation studies. Firstly, the AAI involved natural language mediated comprehension and production, whereas infant behaviour patterns in Strange Situation studies can be abstracted to changes of location and orientation and non-linguistic communications. Secondly, in contrast with infant cognition, much richer internal processes need to be included. Once a question is comprehended, this question refers to some memory or value. So memory retrieval is required. Then planning a response needs to occur before an utterance is produced. Lastly, an adult scenario needs to take account of a richer social-cultural mileu than for infants. So the scenario needs to set out how social-cultural values and standards should moderate adult responses. However, a scenario based upon the AAI can make helpful abstractions. Whilst the AAI is a face-to-face interview its analysis is completed on transcripts—which do give timing information but do not give information about details of bodily movements and facial expressions. So these embodied details can be omitted from the scenario.

Whilst an implemented simulation of the AAI is highly likely to include rich internal states a scenario that acts as a specification of requirements for the simulation only needs to set-out the externally observable behaviour of agents. So details of the linguistic interchanges are required but details of internal processing of language, values and standards are not required in the scenario. These details are required in the design. Multiple designs might be constructed to simulate one scenario. Linguistic interaction can also be constrained to be a subset of natural language.

Since a scenario for language mediated attachment phenomena will focus on public linguistic utterances it will need to specify an attachment related lexicon and grammar that is rich enough to capture the range of response patterns in the AAI—but no more rich than this. The intention of modelling the AAI is not to simulate the full sophistication of language processing. Rather it is to provide a medium in which agents can explain, and fail to explain, past attachment related events.

AAI response patterns have been explained in terms of attentional flexibility and discourse properties. The secure, dismissing and preoccupied enmeshed patterns can be explained as organised responses. Only secure autonomous agents will be able to access all memories and respond to queries about those memories in a controlled and appropriate manner. Dismissing agents will show attenuated accessibility to memories, bias in the memories that are provided, and an avoidance of the conversational subject of attachment. Preoccupied/enmeshed agents will have access to past memories, but will show bias in the memories that are reported, and show a lack of control in how these memories are reported. To simulate the unresolved/disorganised AAI response will involve modelling adaptive and pathological responses to bereavement. Healthy mourning involves accepting change, some mental reorganisation and reorientation. The unresolved category of the AAI includes adults with incomplete mental and behavioural search processes, disorientation in contexts linked to the lost person, vulnerability to disbelief that actual loss has really occurred or is permanent, and an enhanced and unfounded fear of death. So at an information processing level this will involve lapses in metacognitive monitoring of reasoning and discourse processes.

The secure-autonomous, dismissing, preoccupied/enmeshed and unresolved/ disorganised AAI classifications predict the attachment status of their children, their responsiveness to those children, and their emotional/clinical status [37]. For example, the status of mind with respect to attachment as measured in the AAI has been found to moderate the relation between maternal postnatal depression and infant security as classified in the Strange Situation [57]. In particular, unresolved/ disorganised AAI classification for a person has been used to predict D infant attachment status for that person's infant offspring [37, 52]. So whilst this scenario starts by requiring simulation of AAI linguistic behaviour patterns, it will be extended to include behaviour typical of AAI classifications in naturalistic contexts, including caregiving.

Whilst starting by narrowly simulating AAI question and responses, this scenario may develop by extending the adult agents requirements include undertaking caregiving. The link is that both simulated interview and simulated caregiving responses will be guided by the same 'state of mind' with regard to attachment, and the scenarios for both kinds of interaction are constrained by empirical evidence. An extension to this scenario might also consider lifespan development. Infants around one year of age maintaining physical proximity to their carer and exploring their environments using their caregiver as a secure-base. Older children may be satisfied by more indirect distant contact. Furthermore, children of four or five years of age will represent their attachment status pictorially, as well as linguistically [53]. As an individual develops, attachment changes from a sensorimotor to higher-order representational phenomenon. And longitudinal studies have shown significant associations between infant and 20-year later AAI classifications for that individual—suggestive of more complex scenarios [96].

The Design-Based Approach to Studying Minds

In this section, we describe the design-based research methodology followed by The Cognition and Affect project, led by Aaron Sloman and co-workers [84]. The implemented attachment models which are described in section "Architectural Designs Which Have Been Implemented as Simulations" were created as part of the Cognition and Affect Project, and greatly benefited from the prior conceptual and technical work in this project. The Cognition and Affect project is a broad project, possessing many elements that are beyond the specifics of attachment modelling but that are very relevant to computational psychiatry. The Cognition and Affect project uses two contrasting research methods: a broad designed-based approach to exploring information processing architectures that may be implemented in running software or hardware simulations [88]; and a philosophical technique of conceptual analysis [78]. Both research approaches consider what cognitive and affective phenomena might be produced by what kind of architecture. The design-based approach involves exploring the space of possible requirements, designs, and simulation implementations for the cognitive or affective states being studied. For example, within this paper, requirements for attachment modelling are expounded in the scenarios related to empirical observations from Strange Situation and AAI studies. The design-based approach and conceptual analysis methods are broad explorations of their respective domains (designs and possible meanings).

Detailed findings of the Cognition and Affect project include that the mind is an information processing virtual machine, which contrasts with the brain, which is a physical machine. Both mind and brain have been produced by evolution. The mind is a virtual machine which is implemented (realised) by the brain. Some mental disorders arise because of physical damage or disease in the brain. Other disorders are akin to a computer software 'bug' in a running program. So some therapies for psychological disorders are analogous to run-time de-bugging of a virtual machine [85]. Therefore, in order to understand pathological mental states and processes and understand how to treat them we need to understand the information processing architectures that support normal and healthy cognitive, volitional, motivational, emotional and other affective states and processes.

Evolution, in designing humans, has created a hybrid computational architecture containing a highly reactive layer whose mechanisms are tightly coupled with the environment (in monitoring and or acting upon them), a higher level deliberative (management) layer, and self-reflective (meta-management) processes. The reactive layer of control and cognition was produced earlier in phylogenetic development and emerges earlier in ontogenetic development, but remains active throughout the lifespan. The deliberative and meta-management layers both develop later in individual development and emerged later in evolution. These proposed divisions, however, are not simple. Layers may share mental resources [64], are tightly interwoven and all their mechanisms are to some extent reactive [83]. Figure 1 shows the CogAff schema and the H-CogAff architecture. The CogAff schema (Fig. 1a) is a systematic architectural framework which is a useful conceptual tool because it makes high

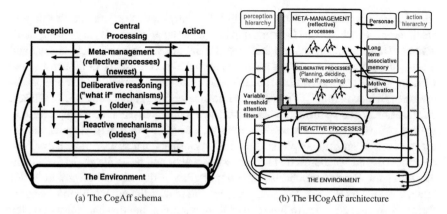

(a) The CogAff schema   (b) The HCogAff architecture

Fig. 1 Diagrammatic representations of the Cognition and Affect approach to understanding mind: the CogAff schema sets out the kinds of processes which can occur in different kinds of architecture so highlighting distinctions between different possible architectures in a broader architectural design-space; and the HCogAff architecture is a high level description of structures and processes which exist in the human mind. So this is one architecture from the broader design-space illustrated by the CogAff schema (thanks to Aaron Sloman for permission to use these graphics)

level distinctions that are helpful in situating attachment models within a shared integrative framework [67, 70, 81]. The CogAff schema organises information processing by overlaying three layers (reactive; deliberative; and meta-management) and three columns (perception, central processing and action). The HCogAff architecture (Fig. 1b) is a special case (or subclass) of CogAff which has not been implemented in its entirety though the production of some subsets have been accomplished [67, 99].

When designing cognitive models to simulate a broad range of affective processes the models are better thought of as control systems rather than computational or symbol processing systems (though they may carry out computations and symbol processing). Within control systems affective states can be cascading control states, which can involve dispositions to produce evaluations, which tend to produce or activate motives, which tend to drive behaviour [80]. They can vary in duration, malleability and information processing features. Neural and physical events will be very short term control states. Intermediate duration control states include intentions, plans and desires, as well as emotions such as joy or fear and moods and preferences. Longer term control states include personality types, temperament and attachment patterns, as well as skills and attitudes and emotions such as love or grief [67, p. 86]. All affective states may engender further affective states. When one focuses on the stability and relative longevity of a state, one tends to refer to it as a trait; but the distinction is contextual [79]. We can view attachment classifications as arising from longer term control states which engender shorter term affective states in temporary episodes of emotions like happiness or anxiety [100].

The Cognition and Affect proposes a rich set of structures, mechanisms and processes that are very germane to modelling attachment. Architectures that support complex phenomena like attachment will have many motivators, and motivator generactivators (which are mechanisms that can generate new or activate existing motivators). These mechanisms are suffused throughout the reactive, deliberative and meta-management layers. Goals are motivators that specify states of affairs towards which agents have a motivational attitude, such as to cause, prevent, accelerate or otherwise influence them [7]. They are the information processing substrate of wishes, wants, desires, and so forth. Various assessments of a goal can be created according to the perceived costs and benefits of satisfying or failing to satisfy the goal, and its urgency. An agent often explores and elaborates plans that specify how a goal can be achieved, and schedules, that specify when a goal will be achieved. In the attachment domain, Bowlby emphasised the importance of the set-goal of proximity for infants. Adults have more complex goals related to attachment, taking into account their self-image, attitudes to interpersonal interactions, and epistemically complex motivators like social standards (with normative values representing what goals and other motives they ought to have) [8].

A very important distinction in the HCogAff architecture is between non-attentive reactive or perceptual processes and the attentive processes which occur within the variable attention filter in Fig. 1b. Many reactive processes can operate in parallel whilst deliberative management and reflective meta-management processes are attention-requiring and serial. Since deliberative management and reflective meta-management processes are resource-bound there is a possible cost to even considering a goal. For instance, this might interrupt more important processes. So reactive motive generactivators can operate in parallel in the non-attentive reactive component of HCogAff and act to generate and activate motives. They are triggered by internal and external events. They can be thought of as 'scouring' the world for their firing conditions [100]. When these conditions are met a motivator is constructed which may 'surface' above the attentional filter. This filter prevents many motivators from disrupting attention and being operated upon by processes in the resource/attention bound deliberative or meta-management levels. Amongst the information processes generated by motivators are: evaluations, prioritisation, selection, expansion into plans, plan execution, plan suspension, prediction, and conflict detection. Management processes can form and then operate upon explicit representations of options before selecting options for further deliberation [100]. Representations in the deliberative layer can be formed with compositional semantics that allow a first order ontology to be formed which includes future possible actions. The meta-level can operate using meta-semantics which allows a second order ontology to be formed which refers to the first order ontology used within the deliberative level [86].

The Cognition and Affect approach also sets out how different classes of emotions can be distinguished in architectural terms. Emotions can be distinguished by whether they are interrupts to reactive or deliberative processes—so the fear of a child when their parent has unexpectedly left a room is different from the fear of a child in close proximity from their carer who they think may leave them in the future. Being startled, terrified or delighted, involves global interrupts in reactive processes. Interruptions in deliberative processes give rise to attachment emotions like being anxious, apprehensive or relieved [82]. A particularly important class of emotions, termed perturbance, emerge as a result of the interplay between several components of the architecture when meta-management processes fail to control management processes. This can be due to insistent motivators and also because meta-management processes are quite limited. So effective meta-management, such as demonstrated through coherent dialogues in the AAI, is a skilled process that takes practice to perfect. Perturbant emotions involve uncontrolled affective content, negative or positive, tending to interrupt and capture management processing, making it difficult for other concerns to gather or maintain attention [100]. Perturbance is likely to be important in explaining attachment phenomena like AAI preoccupied/enmeshed, dismissing and disorganised classifications. Modelling emotions like infatuated love and grief, that involve a loss of control requires first showing how information processing architectures maintain control by the meta-management layer in typical unemotional circumstances.

The design-based modelling approach taken by the Cognition and Affect project is facilitated by using the SIM-AGENT toolkit [87, 88]. This is open source software which facilitates simulating very different kinds of architectures. Allowing a modeller to compare and contrast examples from across a wide design space of information processing architectures. These examples might or might not include various combinations of a reactive layer, a deliberative layer, self-reflective (meta-management) abilities, different kinds of short term and long-term memory, and different kinds of motive generators. The toolkit can therefore be used to simulate different models of various branches in phylogenetic and ontogenetic development, and the development of different kinds of psychopathology. The SIM-AGENT toolkit helps developers overcome common technical challenges in implementing, simulating and debugging broad models of cognition and affect. It also enables the use of a wide variety of programming paradigms and facilities helpful to agent programmers, such as rule-based systems, neural nets, reflective systems, automatic memory management, incremental compilation, and extendable syntax and semantics. An advantage of the SIM-AGENT toolkit for capturing different kinds of self-reflection, compared with other architectures [5, 46, 58] is that it is more general than a single architecture, and so might easily incorporate diverse structures, mechanisms and subsystems. The SIM-AGENT toolkit can provide a coherent and systematic approach to comparing very different architectures that might otherwise reduce to a disordered and unprogrammatic mix of simulations created with different tools that cannot be easily compared.

Architectural Designs Which Have Been Implemented as Simulations

This section describes architectural design intended to fulfil the scenarios previously described in sections "A Scenario for Secure-Base Behaviour" and

"A Scenario Based Upon the Original Strange Situation Studies" which have been implemented as running autonomous agent simulations.

A Reactive Architectural Model of Basic Secure Base Behaviour

The secure-base scenario described in section "A Scenario for Secure-Base Behaviour" has been simulated with multiple software agents in a 2D virtual environment implemented using the sim-agent toolkit. Figure 2 shows a screenshot of the simulation in graphical mode. It has been implemented in autonomous agents using the sim-agent toolkit. In this simulation, infant, caregiver and stranger agents interact with objects representing food, furniture and toys. The infant's actions can include changing its physical location and signalling with an affective tone from positive (like smiling) through neutral to negative (like crying) to very negative (like 'hard crying'). Figure 3a illustrates the four Behaviour Systems the architecture possesses: attachment-security, exploration, wariness, and socialisation; interaction of these behaviour systems produces behaviour patterns that match the requirements set out in the scenarios. In the infant control architecture the highest activated behaviour system controls the actions that are carried out by the agent. A pattern of alternating behaviours occur because, when goals are satisfied, activation levels in behaviour systems with this goal are reset. Consequently over-time activation levels for safety, exploration, social interaction, and physical contact, rise and fall as they are each satisfied and reset. The activation levels vary because inputs to component behaviour systems vary according to the current context—for example, placing a novel toy right in front of an infant agent leads to higher activation of the exploration behaviour system irrespective of other activation levels. If several behaviours are activated, only the one with the highest activation level generates a goal. So this is a 'winner-take-all' action selection mechanism. Secure-base behaviour is produced by this architecture because when the infant is close to the caregiver security producing behaviours have low activation and exploration behaviours are higher—leading the infant to move towards exploration goals (toys). However, the further the infant agent moves from the caregivers the security producing goals are increasingly activated. At some point, as the infant explores away from the caregiver, the winner-take-all goal switches from exploration to security, and the infant signals to the caregivers and starts moving towards the caregiver. When it gets closer, and receives signals from the caregiver, the activation of security goals decreases, and the cycle of exploration and proximity seeking starts again.

A number of other computational models simulate movement patterns similar to secure-base behaviour patterns in human infants and other species that possess attachment or imprinting processes [6, 10, 18, 40, 48]. Multi-generational evolutionary simulations explore the evolutionary pressures linked to care-giving and feeding [62, 63] and lifetime learning and protection [17]. However, these models

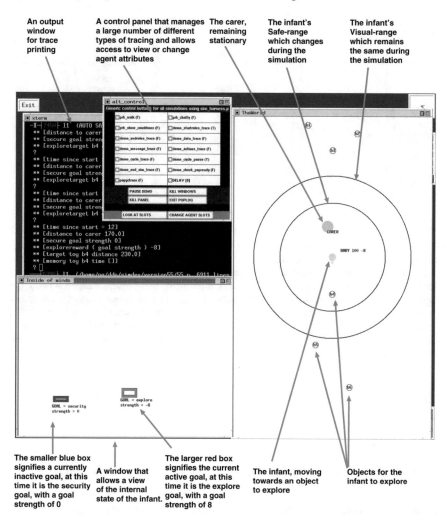

An output window for trace printing

A control panel that manages a large number of different types of tracing and allows access to view or change agent attributes

The carer, remaining stationary

The infant's Safe-range which changes during the simulation

The infant's Visual-range which remains the same during the simulation

The smaller blue box signifies a currently inactive goal, at this time it is the security goal, with a goal strength of 0

A window that allows a view of the internal state of the infant.

The larger red box signifies the current active goal, at this time it is the explore goal, with a goal strength of 8

The infant, moving towards an object to explore

Objects for the infant to explore

Fig. 2 A screenshot of the simulation of secure-base behaviour in graphical mode. The large window down the *right hand side* of the screen is the main window, and shows the positions of all the agents and objects present in the 2D virtual world in this experiment. These are a BABY, a CARER and six objects named b1 through to b6 which represent objects that the BABY agent can explore. The activity in the main window shows the BABY moving away from the CARER towards an object. Of the three windows on the *left hand* of the screen the *top* window is a control panel, which allows different types of trace printing to be sent to the output window. The output window is below and partially hidden by the control panel. The window at the *bottom* of the *left hand side* of the screen represent elements of the internal state of the BABY agent. In this particular experiment only the security and exploration goals are represented. The security goal is relatively low because the carer agent is within the baby agents secure-range limit. The exploration goal is active and the baby is moving towards a toy and away from its carer

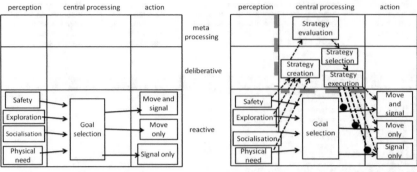

(a) A reactive attachment architecture

(b) A hybrid attachment architecture with reactive, deliberative and meta-management subsystems

Fig. 3 A reactive and a hybrid architecture that both simulate secure-base behaviour. In both architectures safety, exploration, socialisation and physical need 'behaviour systems' gain activation levels as a result of internal and external factors and compete to control actions for movement and signalling in a 'winner-take-all' selection mechanism. In the hybrid architecture only, the resource-constrained deliberative subsystem takes input from the reactive-subsystem, carries out 'look ahead' reasoning, and can inhibit the reactive subsystem and execute alternative actions. The *green dashed line* represents the fact that in the human attachment system deliberative and meta-management processes require attention and so are resource bound, which limits the number that can be concurrently active

are not closely constrained by the details of actual psychological data like the Strange Situation procedure studies.

A Hybrid Architectural Model of Basic Secure Base Behaviour

Even when modelling just infant behaviour there will be a range of possible answers to how to model particular actions. In the study of cognitive development there is a dispute which Munakata et al. [61] describes as between "*Rich interpretation vs. deflationary accounts*" of infant behavioural phenomena. Rich interpretation involves "*casting simple responses in terms of overly complex abilities and cognitive processes*", whereas deflationary accounts involve "*underestimating the abilities of preverbal and non-verbal populations ... mistakenly casting thoughtful behaviours in terms of simplistic processes*" p. B43, [61]. Bowlby's (1969) account of how the information processing related to attachment behaviour develops predates this dispute and therefore does not take a position in it. However, contemporary computational modellers need to construct modelling frameworks that can support both rich and deflationary accounts. Figure 3b illustrates a hybrid architecture that is a 'rich interpretation' in contrast to the reactive architectural 'deflationary account' presented in Fig. 3a. The hybrid architecture situates reactive subsystems alongside a deliberative planning subsystem (that allows 'look-ahead' reasoning) and a simple meta-management subsystem (where cognitive meta-processes operate on other cognitive processes) [67, pp. 103–151]. In this hybrid architecture, the attentive processes that occur are those not stopped by a resource limited variable attention filter. These resource bound serial deliberative processes take input from non-attentive reactive or perceptual processes which operate in parallel. Reactive motive gener-activators are triggered and activated by any possibly relevant internal and external events. In the attachment domain there will be possible threats but also possible exploratory and social opportunities. When these conditions are met a motivator is constructed which may 'surface' through the attentional filter and be operated upon by processes in the deliberative or meta-management levels. Amongst the deliberative attachment processes generated by motivators are the creation, selection, and execution of action plans. Deliberative processes that evaluate other processes occur in the meta-management layer. This hybrid architecture can produce the same external behaviour patterns as the reactive architecture which it extends. What it adds is a reconceptualisation of the attachment control system in light of insights from dual process theories in psychology and artificial intelligence [71], including the findings of the Cognition and Affect project described above in section "The Design-Based Approach to Studying Minds".

Both architectures in Fig. 3a, b possess reactive learning mechanisms so that toy objects and 'strangers' become familiar. But neither architecture possesses learning mechanisms than change the long term relative tendencies for particular behaviours to become activated. That is, neither architecture changes its predisposition to explore or seek security according to evidence from the level of responsiveness or sensitivity of the caregiving it receives. This deficiency is remedied by the architecture described in the next section "A Reactive Architectural Model of the Development of Individual Differences in Infant-Carer Attachment Dyads".

A Reactive Architectural Model of the Development of Individual Differences in Infant-Carer Attachment Dyads

The infant agent architecture created to fulfil the requirements of the Strange Situation scenario experiences very many episodes where the infant agent signals for a response from the carer agent [67, pp. 79–102], [66, 68]. The architecture includes reactive level learning mechanisms that allow the infant agents to infer a measure of implicit trust from the responsiveness and sensitivity of their carer agents from the results of these episodes. In long run simulations that model the first year of life there are repeated instances where the carer agent goes beyond the infant's Safe-range limit, and is called back by the infant agent. In episodes of infant attachment distress, the carer agent responses vary from very prompt to tardy. If an infant experiences a series of very prompt responses it will learn to have a level of trust in the carer agent's responsiveness. This is implemented as a large safe-range distance.

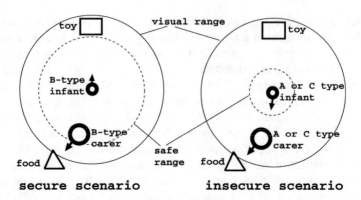

Fig. 4 An illustration of the difference that having a high or low value for the safe-range variable has on the behaviour of an infant agent. On the *left*, a secure infant agent with a high safe-range value moves towards a toy as its carer moves away towards food. On the *right*, an insecure infant agent with a low safe-range value experiences the same external event (carer agent and toy objects in the same positions) but it behaves differently, signalling to, and moving towards, its carer agent. This is because its behaviour system for safety is highly activated because its carer agent is outside its safe range

When the carer is within this distance the infant's security goal is not activated. If an infant agent experiences tardy responses the learned secure-range is small, and when the carer is outside this small range the security goal gains activation, until it becomes the active goal and controls behaviour and results in the infant moving towards the carer and signalling to the carer to come closer.

Figure 4 shows an illustration of how infant agent behaviour differs when contrasting levels of trust are held by two infant agents. The infant on the left is trusting in the responsiveness of its carer agent and has secure status. Whilst the infant on the right shows the large safe-range of a infant that does not trust its carer agent. Figure 5a illustrates how a reactive level reinforcement learning subsystem is connected to a more basic reactive level non-learning architecture, and allows infant agent trust in the carer agent to adapt to the carer agent's responsiveness. Figure 5b illustrates the results of a computational experiment where ten identical infant agents (all initialised with the same level of trust) paired with carer agents with identical responsiveness and sensitivity. The figure shows that over time these averagely trusting infants with their averagely responsive carers diverge into two classes of infant-carer dyads (one group of five dyads with high trust and one group with low trust). This occurs because of positive feedback that operates on small differences in the initial conditions in the randomly located agents and toy and food objects. The positive feedback arises because when infants receive a timely response their trust increases, allowing the carer to move farther away before calling them back. That makes subsequent responses more likely to be rated timely. Conversely, tardy responses make the infant agent request a response sooner in any episode where the carer is moving away. Since the carer agent's policy does not change, it then seems less responsive to

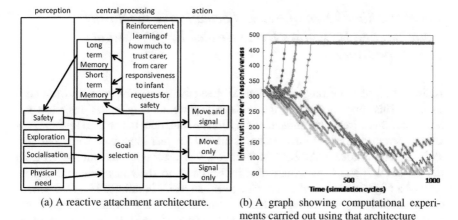

(a) A reactive attachment architecture.

(b) A graph showing computational experiments carried out using that architecture

Fig. 5 A reactive architecture that includes 'winner-take-all' selection mechanism and reinforcement learning mechanism for learning a trust level from repeated carer responses to infant requests for proximity and communication (empty deliberative and meta-management mechanisms layers have been omitted from the diagram) which learns about trust. The graph shows evidence that positive feedback can drive taxon formation in infant security classification. Small random differences in locations were acted upon by positive feedback so that 10 initially identical infant-carer dyads developed into 5 secure relationships, and 5 insecure relationships

this infant agent. This positive feedback mechanism, operating over a long training period, may be what drives the infant-carer pairs into the Secure/Insecure clustering seen in the Strange Situation studies, and so makes a novel contribution to the taxonicity debate in attachment categorisation [67, p. 97]. A carer whose performance is initially intermediate between Secure and Insecure may come to be perceived as at either extreme of caregiving responsiveness. Figure 5b shows an asymmetry between the shape of the positive and negative updates to the Safe-range limit because two different novel discount rules for positive and negative immediate rewards are implemented [73].

Future Work

This section describes architectural design intended to fulfil the scenarios previously described in sections "A Scenario that Refines the Scientific Problem in Light of the 'D' Strange Situation Category", and "A Scenario Based on Modelling Adult Coherent Discourse; and Failures in Coherence" which have not yet been specified in enough detail to be implemented in running programs.

Towards a Simulation of Action Formation in the Strange Situation; and Its Failure Demonstrated by 'D' Category Behaviour Patterns

A simulation of the Strange Situation 'D' response pattern has not yet been constructed. This simulation needs to be able to produce not only the 'D' response pattern in the short 'test' environment of the Strange Situation, but be able to show how all four Strange Situation infant patterns arise from year long experience and how all four patterns arise in the Strange Situation. What sets this simulation apart from models of A, B and C patterns is that it also needs to show how previous experience of a frightened or frightening caregiver can result in failures in action formation and scheduling in a Strange Situation reunion episode. A number of attachment models have been simulated with Artificial Neural Networks [29, 32]. Robotic models inspired by Attachment Theory have used neural network controllers to produce exploratory behaviour and arousal regulation [41, 42]. Future work might combine online agent-based models with offline neural network models by incorporating the latter within the former, or adapt existing robot models to more closely simulate detailed empirical observations.

Towards a Simulation of Language Using Agents that Demonstrate Executive Function and Its Failure in the AAI

A simulation of the language mediated response patterns that arise in the AAI has not yet been constructed. One of the psychological phenomena that a simulation of the AAI will have to reproduce is defensive exclusion. This concept is a reframing of psychoanalytic repression in contemporary scientific terms. The opposite of defensive exclusion will also need to be simulated, as the AAI includes response patterns where individuals do the opposite of holding back or avoiding emotionality. Instead they seem to excessively dwell on previous negative events. This section will first present two computational models of psychodynamic defense produced very early on in the development of cognitive modelling by Colby [23, 24]. Listing the limitations of Colby's model of neurosis will show the design elements that need to be included in a contemporary cognitive architecture that can simulate the response patterns that are seen in the AAI.

Corby's first model was a simulation of the defensive processes that occur in neurosis [23]. Slightly later he produced a cognitive model of paranoia based upon similar defensive processes [24]. These simulations were advanced for their day, but compared to contemporary agent based simulations they are rather simplistic. Colby's model of neurosis substitutes 'active cognitions' which are 'unacceptable and painful' with other more acceptable cognitions through defensive processes such as: deflection, substitution, reversal and projection. An example of such an 'unacceptable cognition' might be that 'I am defective', or 'mother is helpless'. Colby's simulation searches for beliefs that conflict with these cognitions, and when conflicting beliefs are found that have high emotional charge the unacceptable and painful cognition is transformed. 'I am defective' may be transformed by deflection, shifting the object of the belief to view someone else as defective. However, Colby's model has a number of limitations:

- it is not grounded in a model of a virtual microworld in which it interacts;
- the defensive mechanisms are all 'programmed in' rather than emerging from the operation of a psychologically plausible model of language processing, memory operation, and motivation (with appropriate processing, performance and resource constraints);
- there is no learning or emergence of behaviour patterns over time in interaction with an interactive environment;
- there is no opportunity for meta-processing in the model that might lead to recognition of its own neurotic and defended behaviour patterns.

The hybrid model of secure-base behaviour described in section "A Hybrid Architectural Model of Basic Secure Base Behaviour" and the reactive learning model described in section "A Reactive Architectural Model of the Development of Individual Differences in Infant-Carer Attachment Dyads" rectify some of the omissions of Colby's models. They are grounded in a model of a virtual microworld where an infant and carer agent interact over multiple timeslices. The hybrid architecture demonstrates how an agent can remember past events and plan ahead to future events representing attachment episodes using symbolic propositions. This model interprets events and stores memories of events in symbolic chunks which can vary in complexity but are all able to be retrieved as single items into working memory. All cognitive processing in the model is carried out using production rules. From their internal model the agents can then create plans which consider about the possible effects of the actions they are able to take. However, this attachment model as currently implemented is too simple to address questions about these memories and models—it has no linguistic ability for agent to agent communication.

Since the attachment domain is helpfully circumscribed, the capability to discuss attachment memories, attitudes and predictions would require only a limited lexicon and grammar–just complex enough to express questions and responses about objects and events and processes relevant to attachment interaction. In addition to a simplified lexicon and grammar a conversational attachment agent would also need a processing model of language use that is constrained in its processing as normal human conversation is. A key issue is that even with a simple language, parsing and understanding messages is a subtle and tricky business where much may go wrong, resulting in lapses in coherence. So a challenge for the attachment modeller is to produce systems that possess coherence through appropriate cognitive control. Then failures of coherence may be modelled through different biases: in memory processes (with defensively excluded memories failing to be encoded or retrieved and enmeshed memories being more strongly encoded, rehearsed, and retrieved); in processes of language comprehension, deliberation, and language production; and in the activation of motivators responding to attitudes and standards. The structures and mechanisms used by a wide variety of contemporary cognitive architectures might be integrated in agents to produce the required linguistic competences [5, 46, 58, 91].

In the same way that Bowlby borrowed from the currently popular cybernetic and A.I. concepts of the 1960s to 1980s, the hybrid model can be augmented by incorporating current modelling techniques and approaches from diverse areas. Because a model of attachment will require many capabilities that are general rather than specific to the attachment domain an attachment modeller can look to integrate a broad range of computational techniques from diverse aspects of cognition including language and memory processing, problem solving, and reasoning. As we have noted above, a model of AAI discourse can be presented at a high level of abstraction and so not need to include low level language details of syntax and parsing. To give an example for a starting point in modelling AAI discourse, Lewis and Vasishth [47] present one element for a language processing extension to the hybrid attachment model. Lewis and Vasishth [47] used the ACT-R architecture to model sentence comprehension using general cognitive principles to model moment to moment parsing rather than language specific mechanisms. In particular, the key principles that underly their model of long 'garden path' sentences include similar declarative memory, procedural production rule processing and working memory mechanisms that already exist in the hybrid model. These features act alongside a memory retrieval activation function that fluctuates according to frequency of usage and decay, and is subject to interference. These latter features are not possessed in the hybrid model but if added could provide an explanation for both dismissing and preoccupied behaviour patterns. This is because when memory retrieval relies upon frequency and decay, if you do not frequently retrieve memories, they will become less accessible. So in an extreme, memories may seem defensively excluded as with dismissing adults in the AAI. In contrast, preoccupied adults may be stuck in an opposite cycle of frequent retrievals driving a predisposition for ever more retrievals and enmeshed attentionally inflexible deliberation. These general cognitive mechanisms might be further augmented with affective interference, capturing the psychoanalytic idea that memories that are painful may be less easily accessed. In addition, as Eagle [28] notes, psychodynamic defense is not just concerned with memories, but also a range of cognitive and affective contents, including desires, plans, evaluative schemes, and predispositions that in part constitute one's overall personality. Simulating discourse in the AAI can therefore be based on contemporary cognitive science rather than psychoanalytic theory. With skilled activation and retrieval of control processes and appropriate memories giving rise to coherent discourse. Dismissing, preoccupied or disorganised discourse will arise from a breakdown of cognitive control and inaccessibility of memories and other control processes.

Clearly simulating the AAI is more complex than simulating the comprehension or production of a single sentence. Instead of relating single pieces of information into a serial message over seconds, the AAI is relating pieces of information to an internal working model of attachment over minutes and dozens of minutes. However, a simulation of the AAI only needs to include a simple lexicon and grammar, not the richer linguistic constructs Lawis and Vasishth [47] simulate. A lexicon and grammar for the AAI merely needs to be complex enough to support the types of frequency of usage, decay over time and interference that Lawis and Vasishth [47] implement. At the longer timescale of the AAI self-reflection and failures of self-reflection will also come into play. Architectures to simulate the AAI need to show failures in executive function of just the right type. And failures of the right type will only arise from control of the right type. Gray [35] reviews the different type of control of production selection:

> The architectural approaches, the mechanisms imputed to the central controller vary greatly. Both ACT-R and Soar postulate a number of special mechanisms to handle the selection and firing of production rules. In contrast EPIC's central controller plays a very minor role in selecting which productions to fire and when to fire them. Instead, EPIC (executive processing interactive control)[uses] production rules to program explicitly different control strategies for different tasks [35, p. 5]

So there are different extant cognitive architectures providing alternative inspirations for meta-cognitive processing. Lastly, from the CLARION cognitive architecture we can borrow a symbolic-subsymbolic interface [91]. As we saw in the discussion of 'D' category behaviour, we may need to consider how individuals acts are scheduled and formed at a lower level of granularity than productions. To validate the AAI simulation the computational psychiatrist should also be able to run the simulation in a slower, less demanding interview than the AAI. This could show that dismissing, preoccupied and disorganised agents can demonstrate a greater level of control and coherence in a different context removed from the 'stress test' environment of the AAI. Scaffolding for meta-processes leading to self-awareness and self-control might be provided by giving the agent more time, and repeated questions that may involve revisiting prior responses.

Conclusion

The major 'take home' message from this paper is that computational psychiatry can be directed at a high cognitive level. This paper has demonstrated that models of individual people can be simulated as autonomous agents in simulated virtual microworld where they can interact together with other agents over multiple moments in time. Perhaps capturing minutes of interaction. Or sometimes even simulating many years of interaction. Agents in these microworlds can learn from their experiences. In future simulations they may be designed to reflect upon their experiences, show failures in their reflections, and even come to recognise those failures in their reflection. Interaction between agents can occur by acting on virtual objects, but also through linguistically mediated communication. Agents would communicate with microworld languages that are simpler than natural language but expressive enough to capture the external and internal events and processes and inter-agent communication that occur in the microworld. Computational experiments can be run where agents experience events which lead to maladaptive control states forming. Or systems may be designed so that human users can carry out 'therapy' with autonomous agents, acting upon agents or engaging them in discourse to brings about change in maladaptive states. The simulations that have until now been implemented only show a subset of the phenomena described in scenarios. Using attachment phenomena as a candidate domain to illustrate computational psychiatry provides several benefits. Attachment phenomena are broad in scope and range from healthy everyday interactions to psychopathology. They are also very helpfully circumscribed for the computational modeller because attachment relations are just one aspect of a person's social relations and broader experience. So attachment models can represent attachment interactions at a level of abstraction above much of the details of a person's day to day experiences. This is because the essence of attachment interactions revolve around expectations of responsiveness, sensitivity and effectiveness, not the concrete details of perceptions, actions, or low level details of plans and learning. Developmental trajectories from infancy to adulthood define the boundaries of the attachment domain. In addition, through lifespan attachment development, goal structures are comparable because there is continuity of motivational systems. Also, behavioural and cognitive manifestations are analogous. At later stages in development, we represent what we enacted in sensorimotor interactions at earlier stages in development.

Bowlby presented his theory of defensive exclusion as an updating of Freudian repression. As with repression, defensive exclusion can be seen as a theory of personality [28]. This is because what we have conscious access to is used by us to make up our conscious sense of self or identity. Other unconscious aspects of our cognitive processes, may direct or influence our behaviour. However, such unconscious aspects may not seem to derive from 'ourselves' when they do drive or influence behaviour—seeming 'ego-alien' rather than 'ego-syntonic' [28]. So the AAI scenario presented in this paper show that models that are based on how behaviour arises from the differential accessibility of memories can also act as the basis for models of other personality constructs beyond attachment phenomena that are relevant and of interest to psychiatry. This is because they are exploring how a person's self-image is composed from the balance of the parts of their outer and inner experiences that are excluded from, and included in, access and awareness. This approach might be extended to modelling therapy as different processes of self and other relation. For example, therapy in these models could be conceived as opening inaccessible ego-alien processes to awareness and accepting them. Or therapy may be learning to reject ego-alien processes as not part of the self, despite rising from within the agent's broader information processing architecture. In other words, perceptual or memory mechanisms can exclude information from awareness and what is excluded from awareness defines what remains as accessible to an agent's self-created self-image [28]. Although autonomous agent architectures are not conscious they can be a natural tool to use to examine the interactions of accessible thought processes with inaccessible, processes that still influence behaviour. A computational model of AAI behaviour patterns that demonstrates a systematic exclusion of some types of memories, attributions, desires, or intentions is therefore a model of personality, in this sense of personality as deriving from consciously accessible ego-syntonic structures and processes. These types of interactions also occur beyond the attachment domain and so are of broader interest in computational psychiatry [71].

In conclusion, whilst the research presented in this paper is clearly in progress it demonstrates how a class of emotions and other psychological and psychiatric phenomena may be computationally modelled. This is important for theoretical developments but also for translational research into designing technology for diagnosis and intervention.

References

1. Ainsworth, M., Blehar, M., Waters, E., Wall, S., 1978. Patterns of Attachment: A psychological study of the strange situation. Erlbaum, Hillsdale, NJ.
2. Ainsworth, M., Bowlby, J., 1991. An Ethological Approach to Personality Development. American Psychologist 46, 333–341.
3. Allen, J., 2008. The attachment system in adolescence. In: Handbook of Attachment, (Second edition, eds. J. Cassidy & P.R. Shaver. Guilford Press, London, pp. 419–435.
4. Anderson, J., 1972. Attachment behaviour out of doors. In: Blurton-Jones, N. (Ed.), Ethological Studies of Child Behaviour. Cambridge University Press, Cambridge, UK, pp. 199–215.
5. Anderson, J., 2009. How Can the Human Mind Occur in the Physical Universe? OUP, New York.
6. Arkin, R., 2005. Moving up the food chain: Motivation and emotion in behaviour base robots. In: Fellous, J., Arbib, M. (Eds.), Who Needs Emotions? The brain meets the Robot. Oxford University Press, Oxford, UK, pp. 245–279.
7. Beaudoin, L., 1994. Goal processing in autonomous agents. Ph.D. thesis, School of Computer Science, The University of Birmingham, (Available at http://www.cs.bham.ac.uk/research/cogaff/)
8. Beaudoin, L., Sloman, A., 1993. A study of motive processing and attention. In: Sloman, A., Hogg, D., Humphreys, G., Partridge, D., Ramsay, A. (Eds.), Prospects for Artificial Intelligence. IOS Press, Amsterdam, pp. 229–238.
9. Belsky, J., 1999. Modern evolutionary theory and patterns of attachment. In: Handbook of Attachment, eds. J. Cassidy & P.R. Shaver. Guilford Press, London, pp. 141–162.
10. Bischof, N., 1977. A systems approach toward the functional connections of attachment and fear. Child Development 48 (4), 1167–1183.
11. Bowlby, J., 1952. Maternal care and mental health. World Health Organization, Geneva.
12. Bowlby, J., 1958. The nature of a child's tie to his mother. International Journal of Psychoanalysis 39, 350–373.
13. Bowlby, J., 1969. Attachment and loss, Volume 1: Attachment. Basic books, New York.
14. Bowlby, J., 1973. Attachment and loss, Volume 2: Separation, anxiety and anger. Basic books, New York.
15. Bowlby, J., 1980. Attachment and loss, volume 3: Loss, sadness and depression. Basic books, New York.
16. Bowlby, J., 1982. Attachment and loss, Volume 2: Attachment. Basic books, New York, (Second edition 1982).
17. Bullinaria, J., 2009. Lifetime learning as a factor in life history evolution. Artificial Life 15, 389–409.
18. Canamero, L., Blanchard, A., Nadel, J., 2006. Attachment bonds for human-like robots. International Journal of Humanoid Robotics 3, 301–320, 3.
19. Cassidy, J., 1999. The nature of the child's ties. In: Handbook of Attachment, eds. J. Cassidy & P.R. Shaver. Guilford Press, London, pp. 3–20.

20. Cittern, D., Edalat, A., 2015. Towards a neural model of bonding in self-attachment. In: Proceedings of International Joint Conference on Neural Networks, 2015. IJCNN, Killarney, pp. 1–8.
21. Clark, A., 1998. Being There: Putting Brain, Body and World Together Again. MIT Press, Boston.
22. Cohen, J., Servan-Schreiber, D., 1992. Context, cortex and dopamine: a connectionist approach to behavior and biology in schizophrenia. Psychological Review 99, 45–77.
23. Colby, K., 1963. Computer simulation of a neurotic process. In: Computer Simulation of Personality: Frontiers of psychological research', eds. S.S. Tomkins and S. Messick. Wiley, New York, pp. 165–179.
24. Colby, K., 1981. Modeling a paranoid mind. Behavioural and Brain Sciences 4, 515–560.
25. Daw, N., Niv, Y., Dayan, P., 2005. Uncertainty-based competition between prefrontal and dorsolateral striatal systems for behavioral control. Nature Neuroscience 8, 1704–1711.
26. Dayan, P., Seymour, B., 2008. Values and actions in aversion'., In: 'Neuroeconomics: Decision making and the brain', eds. P. Glimcher and C. Camerer and R. Poldrack and E. Fehr. Academic Press, New York, pp. 175–191.
27. Deklyen, M., Greenberg, M., 2008. Attachment and psychopathology in childhood. In: Handbook of Attachment: : Theory, research, and clinical applications, 2nd Edition, eds. J. Cassidy & P.R. Shaver. Guilford Press, London, pp. 637–665.
28. Eagle, M., 1987. The Psychoanalytic and the Cognitive Unconscious. In: Stern, R. (Ed.), Theories of the Unconscious and Theories of the Self. The Analytic Press, Hillsdale, N, pp. 155–189.
29. Edalat, A., Mancinelli, F., 2013. Strong attractors of Hopfield neural networks to model attachment types and behavioural patterns. In: Proceedings of IJCNN 2013. IEEE. http://dx.doi.org/10.1109/IJCNN.2013.6706924
30. Feeney, J., 2008. Adult romantic attachment: Developments in the study of couple relationships. In: Handbook of Attachment, (Second edition, eds. J. Cassidy & P.R. Shaver. Guilford Press, London, pp. 456–481.
31. Fox, N. A., Card, J., 1999. Psychophysiological measures in the study of attachment. In: Handbook of Attachment, eds. J. Cassidy & P.R. Shaver. Guilford Press, London, pp. 226–248.
32. Fraley, R., 2007. A connectionist approach to the organization and continuity of working models of attachment. Personality and Social Psychology Review 6, 1157–80.
33. Freud, S., 1925|1995. An autobiographical study. In: Gay, P. (Ed.), The Freud Reader. Vintage, London.
34. Goldberg, S., 2000. Attachment and Development. Arnold, London.
35. Gray, W., 2007. Composition and control of integrated cognitive systems. In: Integrated Models of Cognitive Systems, ed. W. Gray. Oxford University Press, New York, pp. 3–12.
36. Hesse, E., 1999. The adult attachment interview, historical and current perspectives. In: Handbook of Attachment, eds. J. Cassidy & P.R. Shaver. Guilford Press, London, pp. 395–433.
37. Hesse, E., 2008. The adult attachment interview, protocol, method, of analysis, and empirical studies. In: Handbook of Attachment, (Second edition, eds. J. Cassidy & P.R. Shaver. Guilford Press, London, pp. 552–598.
38. Hinde, R., 1970. Animal Behaviour: A Synthesis of Ethology and Comparative Psychology. McGraw-Hill, London, 2nd Edition.
39. Hinde, R., 1983. Ethology and child development. In: Hand book of child psychology, eds. P.H. Mussen. J. Wiley and Sons, New York, pp. 27–93.
40. Hiolle, A., Canamero, L., 2007. Developing sensorimotor associations through attachment bonds. In: Proc. 7th International Conference on Epigenetic Robotics (EpiRob 2007), eds. C. Prince, C. Balkenius, L. Berthouze, H. Kozima, M. Littman. Lund University Cognitive Studies, Lund, pp. 45–52.
41. Hiolle, A., Canamero, L., Davila-Ross, M., Bard, K., 2012. Eliciting caregiving behavior in dyadic human-robot attachment-like interactions. ACM Trans. Interact. Intell. Syst 2, 3.

42. Hiolle, A., Lewis, M., Canamero, L., 2014. Arousal regulation and affective adaptation to human responsiveness by a robot that explores a novel environment. Frontiers in neurorobotics 8, 17.

43. Holmes, J., 1993. John Bowlby and Attachment Theory. Routledge, (revised edition).

44. Hughes, P., Turton, P., Hopper, E., McGauley, G., Fonagy, P., 2001. Disorganised attachment behaviour among infants born subsequent to stillbirth. Journal of Child Psychology and Psychiatry 42, 791–801.

45. Kennedy, C., Sloman, A., 2003. Autonomous recovery from hostile code insertion using distributed reflection. Cognitive Systems research 4, 89–117.

46. Laird, J., 2012. The SOAR Cognitive Architecture. MIT Press, Cambridge, Mass.

47. Lewis, R., Vasishth, S., 2005. An activation-based model of sentence processing and skilled memory retrieval. Cognitive Science 29, 375–419.

48. Likhachev, M., Arkin, R., 2000. Robotic comfort zones. In: Proceedings of SPIE: Sensor Fusion and Decentralized Control in Robotic Systems. pp. 27–41.

49. Maes, P., 1994. Modeling adaptive autonomous agents. Artificial Life 1 (1), 135–162.

50. Main, M., 1990. Cross cultural studies of attachment organisation; recent studies, changing methodologies, and the concept of conditional strategies. Human Development 33, 48–61.

51. Main, M., Goldwyn, R., 1984. Predicting rejection of her infant from mother's representation of her own experience: Implications for the abused-abusing intergenerational cycle. International Journal of Child Abuse and Neglect 8, 203–217.

52. Main, M., Hesse, E., 1990. Parents' unresolved traumatic experiences are related to infant disorganized attachment: Is frightened and/or frightening parental behavior the linking mechanism? In: Attachment in the Preschool Years, eds. M.T. Greenberg and D. Cicchetti and E.M. Cummings. The University of Chicago Press, Chicago, pp. 161–184.

53. Main, M., Kaplan, N., Cassidy, J., 1985. Security in infancy, childhood and adulthood: A move to the level if representation. In: Growing points of attachment theory and research, eds. I. Bretherton and E. Waters. Monographs of the society for research in child development, 50(1–2 serial no 209), London, pp. 66–107.

54. Main, M., Solomon, J., 1990. Procedures for identifying infants as disorganized/disoriented during the ainsworth strange situation. In: Attachment in the Preschool Years, eds. M.T. Greenberg and D. Cicchetti and E.M. Cummings. The University of Chicago Press, Chicago, pp. 121–160.

55. Main, M., Weston, D., 1982. Avoidance of the attachment figure in infancy. In: The place of attachment in human behavior, eds. M. Parkes & J. Stevenson-Hinde. Basic Books, New York, pp. 31–59.

56. Marvin, R., Britner, P., 1999. Normative development: The ontogeny of attachment. In: Handbook of Attachment, eds. J. Cassidy & P.R. Shaver. Guilford Press, London, pp. 44–67.

57. McMahon, C., Barnett, B., Kowalenko, N., Tennant, C., 2006. Maternal attachment state of mind moderates the impact of post-natal depression on infant attachment. Journal of Child Psychiatry and Psychology 147, 660–669.

58. Meyer, D. E., Kieras, D. E., 1997. A computational theory of executive control processes and human multiple-task performance: Part 1. Basic Mechanisms. Psychological Review 104, (1), 3–65.

59. Mikulincer, M., Shaver, P., 2012. An attachment perspective on psychopathology. World Psychiatry 11, 11–15.

60. Montague, M., Dolan, R., Friston, K., Dayan, P., January 2012. Computational Psychiatry. Trends in cognitive sciences 16, 72–81.

61. Munakata, Y., Bauer, D., Stackhouse, T., Landgraf, L., Huddleston, J., 2002. Rich interpretation vs. deflationary accounts in cognitive development: the case of means-end skills in 7-month-old infants. Cognition 83, B43–B53.

62. Parisi, D., Cecconi, F., Cerini, A., 1995. Kin-directed altruism and attachment behaviour in an evolving population of neural networks. In: Gilbert, N., Conte, R. (Eds.), Artificial societies. The computer simulation of social life. UCL Press, London, pp. 238–251.

63. Parisi, D., Nolfi, S., 2006. Sociality in embodied neural agents. In: Sun, R. (Ed.), Cognition and multi-agent interaction: from cognitive modelling to social simulation. Cambridge University Press, Cambridge, pp. 328–354.
64. Pessoa, L., 2015. Prcis on the cognitive-emotional brain. Behavioral and Brain Sciences 38, 71–97.
65. Petters, D., 2004. Simulating infant-carer relationship dynamics. In: Proc AAAI Spring Symposium 2004: Architectures for Modeling Emotion - Cross-Disciplinary Foundations. No. SS-04-02 in AAAI Technical reports. Menlo Park, CA, pp. 114–122.
66. Petters, D., 2005. Building agents to understand infant attachment behaviour. In: Bryson, J., Prescott, T., Seth, A. (Eds.), Proceedings of Modelling Natural Action Selection. AISB Press, School of Science and Technology, University of Sussex, Brighton, pp. 158–165.
67. Petters, D., 2006a. Designing agents to understand infants. Ph.D. thesis, School of Computer Science, The University of Birmingham, (Available online at http://www.cs.bham.ac.uk/research/cogaff/)
68. Petters, D., 2006b. Implementing a theory of attachment: A simulation of the strange situation with autonomous agents. In: Proceedings of the Seventh International Conference on Cognitive Modelling. Edizioni Golardiche, Trieste, pp. 226–231.
69. Petters, D., 2014a. Towards an Enactivist Approach to Social and Emotional Attachment. In: ABSTRACTS. AISB50. The 50th annual convention of the AISB. Goldsmiths University of London. AISB, Goldsmiths College, London, pp. 70–71.
70. Petters, D., 2014b. Losing control within the HCogaff architecture. In: From Animals to Robots and Back: reflections on hard problems in the study of cognition, eds. J. Wyatt & D. Petters & D. Hogg. Springer, London, pp. 31–50.
71. Petters, D., Waters, E., 2010. A.I., Attachment Theory, and simulating secure base behaviour: Dr. Bowlby meet the Reverend Bayes. In: Proceedings of the International Symposium on 'AI-Inspired Biology', AISB Convention 2010. AISB Press, University of Sussex, Brighton, pp. 51–58.
72. Petters, D., Waters, E., 2014. From internal working models to embodied working models. In: Proceedings of 'Re-conceptualizing Mental Illness: Enactivist Philosophy and Cognitive Science - An Ongoing Debate', AISB Convention 2014. AISB, Goldsmiths College, London.
73. Petters, D., Waters, E., 2015. Modelling emotional attachment: an integrative framework for architectures and scenarios. In: Proceedings of International Joint Conference on Neural Networks, 2015. IJCNN, Killarney, pp. 1006–1013.
74. Petters, D., Waters, E., Schönbrodt, F., 2010. Strange carers: Robots as attachment figures and aids to parenting. Interaction Studies: Social Behaviour and Communication in Biological and Artificial Systems 11 (2), 246–252.
75. Russell, S., Norvig, P., 2013. Artificial Intelligence, A Modern Approach. Prentice Hall, (Third edn.).
76. Schlesinger, M., Parisi, D., 2001. The agent-based approach: A new direction for computational models of development. Developmental Review 21, 121–146.
77. Simpson, J., 1999. Attachment theory in modern evolutionary perspective. In: Handbook of Attachment, eds. J. Cassidy & P.R. Shaver. Guilford Press, London, pp. 115–141.
78. Sloman, A., 1982. Towards a grammar of emotions. New Universities Quarterly 36 (3), 230–238.
79. Sloman, A., 1992. Prolegomena to a theory of communication and affect. In: Ortony, A., Slack, J., Stock, O. (Eds.), Communication from an Artificial Intelligence Perspective: Theoretical and Applied Issues. Springer, Heidelberg, Germany, pp. 229–260.
80. Sloman, A., 1993. The mind as a control system. In: Hookway, C., Peterson, D. (Eds.), Philosophy and the Cognitive Sciences. Cambridge University Press, Cambridge, UK, pp. 69–110.
81. Sloman, A., 2000. Architectural requirements for human-like agents both natural and artificial. (what sorts of machines can love?). In: Dautenhahn, K. (Ed.), Human Cognition And Social Agent Technology. Advances in Consciousness Research. John Benjamins, Amsterdam, pp. 163–195.

82. Sloman, A., 2001. Beyond Shallow Models of Emotion. Cognitive Processing: International Quarterly of Cognitive Science 2 (1), 177–198.

83. Sloman, A., 2002. How many separately evolved emotional beasties live within us? In: Trappl, R., Petta, P., Payr, S. (Eds.), Emotions in Humans and Artifacts. MIT Press, Cambridge, MA, pp. 29–96.

84. Sloman, A., 2008. The cognition and affect project: Architectures, architecture-schemas, and the new science of mind. Tech. rep., University of Birmingham, (Available at http://www.cs.bham.ac.uk/research/cogaff/)

85. Sloman, A., 2009. Machines in the Ghost. In: Dietrich, D., Fodor, G., Zucker, G., Bruckner, D. (Eds.), Simulating the Mind: A Technical Neuropsychoanalytical Approach. Springer, Vienna, pp. 124–177.

86. Sloman, A., Chrisley, R. L., June 2005. More things than are dreamt of in your biology: Information-processing in biologically-inspired robots. Cognitive Systems Research 6 (2), 145–174.

87. Sloman, A., Logan, B., March 1999. Building cognitively rich agents using the Sim_agent toolkit. Communications of the Association for Computing Machinery 42 (3), 71–77.

88. Sloman, A., Poli, R., 1996. Sim_agent: A toolkit for exploring agent designs. In: Wooldridge, M., Mueller, J., Tambe, M. (Eds.), Intelligent Agents Vol II (ATAL-95). Springer-Verlag, pp. 392–407.

89. Sroufe, L., Carlson, E., Levy, A., Egeland, B., 1999. Implications of attachment theory for developmental psychopathology. Development and Psychopathology 11, 1–13.

90. Storr, A., 1989. Freud. OUP, Oxford.

91. Sun, R., 2007. The motivational and metacognitive control in clarion. In: Integrated Models of Cognitive Systems, ed. W. Gray. Oxford University Press, New York, pp. 63–76.

92. van Ijzendoorn, M., 1995. Adult Attachment representations, parental responsiveness, and infant attachment: A meta-analysis of the predictive validity of the Adult Attachment Interview. Psychological Bulletin 117, 387–403.

93. van Ijzendoorn, M., Bakermans-Kranenburg, M., 2004. Maternal sensitivity and infant temperament. In: Theories of Infant Development, eds. G. Bremner & A. Slater. Blackwell Press, Oxford, pp. 233–257.

94. van Ijzendoorn, M., Sagi, A., 1999. Cross-cultural patterns of attachment: Universal and contextual dimensions. In: Handbook of Attachment, eds. J. Cassidy & P.R. Shaver. Guilford Press, London, pp. 713–734.

95. Waters, E., Kondo-Ikemura, K., Posada, G., Richters, J., 1991. Learning to love: Mechanisms and milestones. In: Minnesota Symposium on Child Psychology (Vol. 23: Self Processes and Development), eds. M. Gunner & Alan Sroufe. Psychology Press, Florence, KY, pp. 217–255.

96. Waters, E., Merrick, S., Treboux, D., Crowell, J., Albersheim, L., 2000. Attachment stability in infancy and early adulthood: A 20-year longitudinal study. Child Development 71, 684–689.

97. Weinfield, N., Sroufe, L., Egeland, B., Carlson, E., 1999. The nature of individual differences in infant-caregiver attachment. In: Handbook of Attachment, eds. J. Cassidy & P.R. Shaver. Guilford Press, London, pp. 68–88.

98. Wolff, M. D., van IJzendoorn, M., 1997. Sensitivity and attachment: A meta-analysis on parental antecedents of infant attachment. Child Development 68, 571–591, 2.

99. Wright, I., 1997. Emotional agents. Ph.D. thesis, School of Computer Science, The University of Birmingham.

100. Wright, I., Sloman, A., Beaudoin, L., 1996. Towards a design-based analysis of emotional episodes. Philosophy Psychiatry and Psychology 3 (2), 101–126.

Self-attachment: A Holistic Approach to Computational Psychiatry

Abbas Edalat

Introduction

In the past few years, Computational Psychiatry has grown as an emergent subject following remarkable advances in Computational Neuroscience in general and in Reinforcement Learning in particular. Work in Computational Psychiatry so far has focused on computational models to explicate cognitive impairment and deficits, including sub-optimal decision making, that are ubiquitous in all varieties of mental disorder. This has led to a number of interesting results which use Reinforcement Learning or Game Theory to model depression, borderline personality disorder and other psychiatric illnesses [1, 2]. The underlying approach is reductionist and consists of constructing models that describe a given form of normal or healthy decision making or behaviour which are then modified to understand the abnormal or impaired form of the same process in mental illness. It is hoped that these models can help us better understand mental illness and can aid us in developing treatments and in particular pharmacological interventions for psychiatric diseases.

In this article, we propose a holistic approach to Computational Psychiatry, called Self-attachment that was first introduced in [3, 4], and focuses on early attachment insecurities in childhood. There has been increasing evidence over the past decades to show that early attachment experiences of infants with their primary care-givers play a crucial role in the development of their capacity for affect regulation, which is vital for enhancing resilience in the face of stress in life and averting mental illness. It is now believed that early negative attachment interactions have an adverse impact on the development of this capacity in children, making them vulnerable to psychological disorders later in life [5, 6]. This view has been supported by Attachment Theory, a scientific paradigm in developmental psychology introduced by John

A. Edalat (✉)
Algorithmic Human Development Group, Department of Computing,
Imperial College London, London, UK
e-mail: a.edalat@ic.ac.uk

© Springer International Publishing AG 2017
P. Érdi et al. (eds.), *Computational Neurology and Psychiatry*,
Springer Series in Bio-/Neuroinformatics 6,
DOI 10.1007/978-3-319-49959-8_10

Bowlby in 1960's [7], which has also had an impact on psychotherapy in the past few decades [8]. This will be described in detail in the next section.

Whereas attachment insecurities create vulnerability to mental illness, a number of experiments–using a technique called "security priming"–have been able to artificially activate mental representations of supportive attachment figures and thereby improve the mental health of individuals suffering from various mental disorders [5]. Individuals are also capable of using a constructive marriage or a therapeutic framework to "earn their secure attachment" [5]. In particular, in schema therapy "limited reparenting" is used to redress early maladaptive schemas in the parent-child interactions, help the individual find experiences missed in childhood, and establish a secure attachment through the therapist [9].

Many studies in the past two decades have provided compelling evidence that for thousands of years human beings have used religion to mirror a caring parent as an "attachment object" to redress their attachment insecurities and regulate their emotions, what has been called the "compensatory pathway" to religious practice [10]. Attachment objects were originally studied in the context of children's development by Donald Winnicott under the name transitional objects [11] and were later used as cloth-covered surrogate dummy mothers by Harry Harlow in his experiments with infant monkeys [12]. Bowlby referred to them as "substitute objects" or "attachment figures" that acted as a substitute for the mother [7]. It has also been argued that religious practice in human beings resembles the use of transitional objects in children's development as described by Winnicott [13].

In addition, functional Magnetic Resonance Imaging (fMRI) studies since 2000 have confirmed that the activation of the dopaminergic reward systems of the brain is a common feature of romantic love, maternal love and religious praying, all of which create an affectional bond. From these findings, we infer that creating an affectional bond, be it in the context of parental love, romantic love or religious practice, provides a basis for attachment interactions that in turn are aimed at emotion self-regulation [14–16].

Self-attachment therapy, as described in [4], proposes that early attachment insecurities of adults who have become victims of psychiatric disorders can be addressed and tackled with self-administered protocols that emulate secure child-parent attachment interactions by mental representation. The individual pro-actively takes up the role of a nurturing parent for reparenting a mental representation of the "inner child", the emotionally damaged child that the individual was in early life. For this purpose, an internal affectional bond is created by the "adult self" with the inner child. The next stage is based on two fundamental paradigms in neuroscience: (i) neuroplasticity, i.e., our brain's capacity to reorganise its neural circuits by forming new neural connections throughout life, and (ii) long term potentiation, i.e., a persistent strengthening of synapses based on recent and repeated patterns of activity [17, Chap. 24]. The goal of Self-attachment is to produce in a systematic and repeated way virtual secure attachment experiences in the individual that by neuroplasticity and long term potentiation would create neural circuits that provide a sense of secure attachment in the individual. With this sense of secure attachment the individual then revisits traumatic experiences in connection with the mental image of the inner child in order to reprocess these episodes with a supporting parent represented by the adult self of the individual. The goal of the exercise is to help the individuals to create their own secure attachment objects that can support them in self-regulating their emotions.

In [4], a number of promising case studies of practicing Self-attachment were reported. In these case studies, individuals with chronic depression and anxiety, resistant to various forms of previous psychotherapy, had significant clinical improvement in their symptoms following the practice of Self-attachment. We previously proposed the use of virtual reality as an alternative to imagery techniques for simulating the interaction of the "inner child" with the "adult self" [3]. In the two related virtual reality experiments reported in [18, 19], individuals with either excess self-criticism or with depression embodied a virtual child and then a virtual adult to receive self-compassion, which is one of the first interventions in Self-attachment. In both experiments, the embodied interactions between the virtual child and the virtual adult resulted in an improvement in their self-criticising or depressive symptoms. These compassionate interactions can be regarded as interactions between the "inner child" and the "adult self", even though the individuals had not made a conscious effort to perceive the interactions in this way. We therefore submit that the above virtual reality experiments have provided additional proof of concept for Self-attachment.

The rest of this article is organised as follows. In section "Attachment Theory", we briefly review the basic tenets of attachment theory, the neurobiology of secure and insecure attachment, the vulnerability to mental illness caused by insecure attachment and the impact of attachment theory on psychotherapy. In section "Attachment Objects", we describe the role of attachment objects for emotion self-regulation in ethology, in children's development and in religious practice. In section "fMRI Studies on Bond Making", we explain how fMRI experiments on romantic love, maternal love and religious prayers since 2000 indicate a common overarching paradigm for rewarding affectional bonds in these different contexts. In section "Self-attachment", first the nature of Self-attachment intervention is explained and then the Self-attachment protocol is outlined. In sections "A Game-Theoretic Model" and "Self-attachment and Strong Patterns", a game-theoretic model for Self-attachment and a model based on strong patterns that are learned deeply in an associative artificial neural network are presented. Finally, in section "Neural Models of Self-attachment", we describe several neural models of the human brain for Self-attachment.

Attachment Theory

Attachment Theory was first formulated by Bowlby [7], based on his studies of separation anxiety of infants from their mothers, and was later developed along with Mary Ainsworth. In the past decades, it has been regarded as a main scientific paradigm in developmental psychology [20]. According to Bowlby, human beings, as well as higher primates, are born with an innate tendency to seek proximity to their primary care-givers to feel secure and safe in particular in times of need and stress. He argued that if the infant's needs for attachment are met by a sensitive parent capable of quick and appropriate response when the infant is in distress, then a stable and secure sense of self is developed in the infant which leads to positive association with the self and others in life. However, when the primary care-giver is unreliable, insensitive and thus unable to respond quickly and appropriately to the needs of the infant, an insecure sense of self and negative models of self and others are developed in the infant which would be the source of later emotional problems.

Bowlby [8] proposed that the type of emotional attachment children develop with their primary caregivers will become the foundation of their "internal working model", a cognitive framework with which the child and later the adult interprets the social world and predicts the outcome of her/his behaviour in view of past memory. Thus, through the internal working model, the early attachment types of children have a great impact on their future emotional and personality developments as adults engaging in social and intimate relationships.

In its present form, Attachment Theory classifies the quality and dynamics of the relationship of a child with his/her parent into four types of attachment: secure attachment and three kinds of insecure attachments (avoidant, anxious/ambivalent, and disorganised) [21]. It is proposed and vindicated by longitudinal studies that attachment types strongly impact on the emotional, cognitive and social development of the child into adulthood by determining the individuals working model of relationships [22].

The theory has been corroborated by the Strange Situation experiment [23] developed in the 1970s by Mary Ainsworth who was inspired by Harry Harlow's experiments with monkeys. The Strange Situation experiment has become the standard measure of eliciting attachment in infants [21]. This procedure, involving a mother, her one year old toddler, and a stranger in an unfamiliar room, has been repeated across many different cultures, with results indicating similarities but also differences in the distribution of attachment types among the toddlers in different societies [24, 25].

It is hypothesised that the particular type of attachment developed in a child depends crucially on the kind of response by the parent to the child's emotional needs, which is repeated over and over again during the formative stage of the infants development. The primary care-giver of a securely attached child responds quickly and appropriately to the distress signals of the child and in due course the child learns to use the primary care-giver both as a safe haven and as a secure base in order to explore the world, while feeling assured that whenever a stress situation arises the primary care-giver is available to provide support and reassurance. The primary care-givers of the insecurely attached children, however, often respond to the emotional needs of the child by rejection, in the case of avoidant insecure child, or by inconsistent behaviour, in the case of anxious insecure child, or by frightening the child, in the case of a disorganised insecure child [26, pp. 19–24].

Regulation Theory: Neurobiology of Secure Attachment

John Bowlby's Attachment theory was influenced by a number of disciplines including control theory and cybernetics. In fact, Attachment theory has been considered as a regulation theory in particular in the work of Allan Schore in the past two decades [27]. His work explains how secure attachment leads to the capacity for self-regulation of emotions and how this is mapped in the infant's developing brain. We will briefly explain the essential components of Schore's regulation theory.

Schore quotes the following assertion from developmental psychoanalysis by Emde [28]:

> It is the emotional availability of the caregiver in intimacy which seems to be the most central growth-promoting feature of the early rearing experience.

Visual experience is thought to play an important role in the emotional development of the infant and its emotional attachment to the parent in the first, so-called practicing, year. Through intense mutual gaze, the infant-parent dyad is established, creating a framework for transmission of mutual influence and inter-communication. The parent is attuned to and resonates with the internal state of the infant at rest and is able to contingently fine tune her affective stimulation with the infant's dynamically activated, deactivated or hyperactivated state. The infant's right hemisphere is intensely involved in the visual interactions with the mother and is used as a template for imprinting what it receives from the mother's emotion-regulatory right cortex. This leads to entrenched circuits in the infant's right brain that reflect the mother's right brain activities and will be used in turn to regulate the infant's emotions. The mutual and intense gaze, facial expressions, mirroring and resonance between the infant and the mother release endogenous opiates inducing pleasure and release of dopamine from the ventral tegmental area resulting in dopaminergic-driven arousal and dopamine-mediated elation. High levels of dopamine secretion results in a rapid growth rate of the infant's brain by accelerating the transcription of genes and DNA synthesis and by regulating dendritic growth [6, p. 14].

The affective homeostasis in the child-parent dyad is used by the primary caregiver to minimise the child's negative affects, maximise its positive affects and moderate its arousal level. This drive for attunement and homeostasis with the child is particularly paramount in the first year of life.

It is proposed, based on a large volume of research in neuroscience, that the resulting emotional regulation in the child is rooted in the development of several brain regions, in particular the Orbital Frontal Cortex (OFC) which as part of the prefrontal cortex is densely connected to the limbic system and thus acts as its "executive director". The orbitofrontal cortex develops on the basis of the type of interaction infants have with their primary care-givers and is critically involved in attachment processes that occur during the first two years of life [6, p. 14]. The optimal growth of the OFC in the securely attached toddler allows delayed response based on stored representation rather than immediate information in the environment. The child develops a mental image or a schema of the parent and a primitive capacity for self-reflection which allow the child to withstand short and temporary absence of the parent. The infant thus acquires a learning brain able to explore the world.

Schore summarises the significance of these early infant-mother interactions [6, p. 12]:

> This mutual regulatory system is adaptive in that it allows for the arousal level of the child's developing nervous system to be rapidly recalibrated against the reference standard of the mother's.

In the second, so-called socialising year of life, the child is able to walk and run, and so a key role of the primary care-giver is to restrain the child from dangerous or anti-social behaviour. This for the first time results in a break-up in the homeostasis between the child and the primary care-giver and thereby to the painful experience of shame by the child. By being sensitive, responsive, emotionally approachable and staying with the child, the parent of a securely attached child is able to re-enter into synchrony and mutual gaze with the child. It is in the repeated process of misattunement and reattunement with the primary care-giver that the child gradually learns that the break-up of homoeostasis with the parent can always be restored. The child and the parent then develop the capacity to move from positive to negative and back to positive affect and the child learns that negative affects can be endured. By staying in with the child the "good enough" parent [29], a notion popularised by the British psychologist and paediatrician Donald Winnicott, allows the repair of the misattunement in the child-parent dyad.

Schore evaluates the maturation of the infant's brain as a result of these disruption-repair transactions between the infant and the mother as follows [6, p. 21]:

> These experiences trigger specific psychobiological patterns of hormones and neurotransmitters, and the resultant biochemical alterations of the brain chemistry influence the experience-dependent final maturation of the orbitofrontal cortex.

In this way, the child's brain becomes focused on exploring the world: in the words of Julian Ford, a "learning brain" develops [30], a concept that fits with the idea of "knowledge instinct" [31]. The OFC is also involved in internal regulation of states by its connections to the hypothalamus, which allows cortical containment of the autonomic sympathetic and parasympathetic somatic reactions induced by emotions [6].

Neurobiology of Insecure Attachment

The development of a securely attached child should be compared with that of insecure attached children. Here, we will briefly explain the situation for the avoidant insecure attachment, which can show the discrepancy in acquiring a capacity for emotional regulation [6, pp. 27–28]. In contrast to the parent of a securely attached child, the parent of an avoidant insecure child expresses low levels of affection and warmth, is averse to physical intimacy and actively blocks proximity seeking approaches of the child. In reaction to a long history of dealing with such a primary care-giver, the avoidant insecure child shows no interest in engaging with an adult who tries to interact with him/her and exhibits little interest to get close to the adult. The normal process of attachment bonding is severely disrupted, which leads to failure of the regulatory mechanism and disturbance in the limbic activity and hypothalamic dysfunction. It is suggested that this negative attitude reflects suppressed anger in the infant because of past frustration in attempts to seek proximity with the primary care-giver. A psychobiological disequilibrium based on dysregulated brain chemistry is developed when the primary care-giver does not regularly engage in repairing misattunement to establish homoeostasis. Negative social interactions in the critical early years of development lead to permanent change in opiate, dopamine, noradrenaline and serotonin receptors. The primary caregiver's failure to engage quickly and appropriately in repairing the homoeostasis disruption traumatises the infant and results in defects in the development of the OFC and in biochemical dysregulation and toxic brain chemistry of the infant. In this way, the child's main focus will be how to avoid further trauma: in the words of Julian Ford, a "survival brain" develops [30].

It is thought that in avoidant insecure attachment the sympathetic nervous system, in anxious insecure attachment the parasympathetic nervous system, and in disorganised attachment both the sympathetic and the parasympathetic nervous system are impaired and dysregulated. See [32] for an overview of the subject.

Here, we highlight the results in [33], where the authors examined amygdala activation, feelings of irritation, and the use of excessive force as indicated by grip strength during exposure to infant crying and scrambled control sounds in 21 women without children. Individuals with insecure attachment representations, based on Berkeley Adult Attachment Interview [34], showed heightened amygdala activation when exposed to infant crying compared to individuals with secure attachment representations. In addition, insecure individuals experienced more irritation during infant crying and used more excessive force than individuals with a secure representation.

We can view attachment schemas as a type of implicit memory that has been sculpted in the brain by repeated experience of broadly similar interactions with a primary care giver [32, p. 139]:

> [A]ttachment schemas are a category of implicit social memory that reflects our early experience with care takers. Our best guess is that these schemas reflect the learning histories that shape experience-dependent networks connecting the orbital frontal cortex, the amygdala, and their many connections that regulate arousal, affect and emotion. It is within these neural networks that interactions with caretakers are paired with feelings of safety and warmth or anxiety and fear.

Attachment Insecurity and Vulnerability to Mental Disorder

Since the early attachment schemas are deeply embedded in the brain, their impact on the child's and later the adult's internal working model with which the social environment is processed, interpreted and responded to throughout life can hardly be overestimated. In this section, we will see that insecure attachments and the resulting incapacity to regulate strong emotions will have lasting influence on the psychological conditions of the individual. In the past few decades, there has been increasing evidence to suggest that the root cause of much mental illness lies in a sub-optimal capacity for affect regulation [27]. The adverse impact of early attachment insecurity on the capacity to regulate strong emotions and thus to fend off psychological disorders and mental illness goes very wide.

Bowlby himself emphasised the relationship between attachment problems and many psychiatric disorders [8, p. 152]:

> Advocates of attachment theory argue that many forms of psychiatric disturbance can be attributed either to deviations in the development of attachment behaviour or, more rarely, to failure of its development; and also that the theory casts light on both the origin and the treatment of these conditions.

We now highlight the main findings of Mikulinceri and Shaver in their 2012 review article in World Psychiatry [5], which provides a recent update on Attachment Theory. First and foremost, the authors report that

> attachment insecurity is a major contributor to mental disorders.

Early interactions with inconsistent, unreliable, or insensitive primary care-givers disrupt the development of a secure, self-regulated psychological foundation, which undermines building an increasing capacity to cope with stress and thus predisposes an individual to mental breakdown in crisis situations in adult life. The authors argue that attachment insecurity in childhood therefore creates an overall vulnerability to psychological and mental disorders, with the particular symptomatology depending on genetic, developmental, and environmental factors. It is, however, the combined impact of attachment insecurity with childhood trauma, neglect or abuse that is a predictor of psychological and mental disorders in later life. In their earlier work [35], these authors reviewed hundreds of cross-sectional, longitudinal, and prospective studies of both clinical and non-clinical samples and concluded:

> [A]ttachment insecurity was common among people with a wide variety of mental disorders, ranging from mild distress to severe personality disorders and even schizophrenia.

They have also reported in their review article more recent findings to show that

> [attachment insecurities are] associated with depression....., clinically significant anxiety...., obsessive-compulsive disorder....., post-traumatic stress disorder (PTSD)....., suicidal tendencies......, and eating disorders.......

Attachment insecurity, it is asserted, also plays a key role in various personality disorders with specific attachment insecurity correlating and corresponding to different disorders [5]:

> Anxious attachment is associated with dependent, histrionic, and borderline disorders, whereas avoidant attachment is associated with schizoid and avoidant disorders.

In addition, according to the review article, apart from the problem of emotion regulation, two other basic pathways, mediate between attachment insecurity and psychopathology, namely those involved in (i) problems of self-representation, such as

> lack of self-cohesion, doubts about one's internal coherence and continuity over time, unstable self-esteem, and over-dependence on other peoples approval,...

and (ii) interpersonal problems, so that:

> ...avoidant people generally had problems with nurturance (being cold, introverted, or competitive), and anxious people had problems with emotionality (e.g., being overly expressive). These problems seem to underlie insecure individuals self-reported loneliness and social isolation.... and their relatively low relationship satisfaction, more frequent relationship breakups, and more frequent conflicts and violence....

Impact on Psychotherapy

Attachment Theory has influenced nearly all forms of psychotherapy including psychoanalysis and psycho-dynamic therapy [36], and its impact on Cognitive Behavioural Therapy, as the most widely used type of therapy today, has led to Schema Therapy [9]. In general, clinicians have used the concepts and findings of Attachment Theory to understand, address and resolve the attachment issues that their clients bring into therapy. The therapeutic framework, according to Attachment Theory, should be perceived by the client as a safe environment in which early attachment insecurities could in principle be dealt with and replaced with a secure attachment based on the working relationship with the therapist [37, 38]. In [26], after an extensive review of Attachment Theory, a model of treatment based on this theory is proposed in which the therapist's interventions are tailored to the attachment needs of the client. The aim is to help the client to use the verbal and non-verbal relationship with the therapist to internalise a sense of secure base that was not created as a result of early child-parent interactions.

The Adult Attachment Interview (AAI) scoring system was developed by George et al. [34] and is designed to retrospectively classify the childhood attachment type of adults. It consists of a number of questions to assess the person's understanding of their early childhood relationships with parents. The pioneering work of Pierson et al. [39] examined the notion of earned security using the Adult Attachment Interview scoring system. Earned-security classifies adults who in the AAI describe difficult, early relationships with parents, but who have been able to reflect into their experience through psychotherapy or a constructive marriage and as a result have developed secure working models in their adulthood as shown by their high coherency scores. Mikulinceri and Shaver [5] also provide evidence that a sense of security provided by a psychotherapist improves a client's mental health and can lead to earned secure attachment. According to the study of a group of adults by Pearsonal et al.

in [39], however, earned securers have comparable depressive symptomatology as insecurers.

Another main focus of the review article [5] is what they call the healing effects of attachment security, which is directly related to Self-attachment protocols. Whereas attachment insecurities create vulnerability to mental illness, the review argues that

> the creation, maintenance, or restoration of a sense of attachment security should increase resilience and improve mental health.

The authors then report on studies of experiments published in 2001 and 2007 on so-called "security priming", which artificially activate mental representations of supportive attachment figures, for example, by

> subliminal pictures suggesting attachment-figure availability, subliminal names of people designated by participants as security-enhancing attachment figures, guided imagery highlighting the availability and supportiveness of an attachment figure, and visualization of the faces of security-enhancing attachment figures.

The authors indicate that security priming improves participants' moods even in threatening contexts and eliminates the detrimental effects of threats on positive moods, and found that subliminal priming with security-related words mitigated cognitive symptoms of PTSD (heightened accessibility of trauma-related words in a Stroop-colour naming task) in a non-clinical sample.

Attachment Objects

When secure attachments are not available in normal relationships, attachment figures or attachment objects have been used by human beings to regulate their emotions and create a sense of felt security for themselves. Such attachment substitutes are often used by securely attached individuals to cope with extreme forms of distress such as loss of a loved one, war atrocities, human inflicted traumas, and horrific accidents [10]. While these attachment substitutes are often external objects, the meaning individuals attribute to such an object and the relationship, interactions and contract agreements they establish with it are highly personal, subjective and are created often meticulously by the individuals themselves who may nevertheless use and copy ideas from others. These attachment objects are employed by individuals with the aim of attaining the kind of inner-felt security that can be observed in securely attached children or adults in dyadic relationships. In fact, experiments with monkeys show that higher primates are able to use such attachment objects, which we will describe next.

Evidence in Ethology

By late 1950's, Bowlby had for several years studied separation anxiety in children who were separated from their mothers but he could not explain his clinical findings using his psychoanalytic training. A number of experiments with monkeys by the leading ethologist Harry Harlow however attracted Bowlby's attention and had a profound impact on his ideas. Bowlby wrote at the time [40]:

> The longer I contemplated the diverse clinical evidence the more dissatisfied I became with the views current in psychoanalytical and psychological literature and the more I found myself turning to the ethologists for help. The extent to which I have drawn on concepts of ethology will be apparent.

Motivated by Bowlby's work on separation anxiety, Harlow experimented on infant rhesus monkeys with surrogate dummy mothers that were either bare-wired or cloth-covered. He found that the infant monkeys had an overwhelming preference for cloth-covered mothers and would spend their time clinging to the cloth mother [12]. These experimental studies provided further support for attachment theory [41].

A series of later experiments by Harlow in the 1960's [42, p. 487], regarded as unethical today, showed that clinging to the cloth-covered surrogate mother served as a way of regulating anxiety. Two groups of infant rhesus monkeys were removed by Harlow from their mothers, and given a choice between either a cloth-covered or a bare-wired surrogate mother. In the first group, the cloth-covered mother provided no food, while the wire mother did. In the second group, the cloth-covered mother provided food while the bare-wired mother did not. As expected the infant monkeys would cling to the cloth-covered mother whether it provided food or not and the infant monkeys went to the wire surrogate only when it provided food. Whenever frightened, the infant monkeys would run to the cloth mother for protection and comfort, notwithstanding which mother provided them with food. Placed in an unfamiliar room with their cloth-covered surrogates, the monkeys would cling to them until they felt secure enough to explore and then would occasionally return to the cloth mother for comfort. Monkeys who were placed in an unfamiliar room without their cloth mothers, however, would freeze in fear and cry, crouch down, or suck their thumbs. Some of the monkeys would even run from object to object, apparently searching for the cloth mother as they cried and screamed. Monkeys placed in this situation with their wire mothers exhibited the same behavior as the monkeys with no mother.

The cloth-covered surrogate mother thus provided for the infant monkeys an attachment object for affect regulation. Bowlby in fact described this kind of attachment object both for the infant monkeys and the human child as we will see next.

Children's Attachment Objects

While the word attachment has a broad meaning in every day life, Attachment Theory as conceived by John Bowlby has a narrow domain of discourse which is focused on the relationship between two individuals of the same species, in particular the child-parent and the adult-adult relationships. However, according to Bowlby, when normal attachment relations are unavailable an inanimate object can play an important role as an "attachment figure". In his first seminal work on Attachment Theory, after providing an example of such an inanimate attachment object for an infant chimpanzee, who had been cared for by a human foster-mother, he writes [7, p. 313]:

> Many other examples of such behaviour in primate infants brought up in atypical surroundings could be given.
>
> Thus it seems clear that, whether in human infants or monkey infants, whenever the "natural" object of attachment behaviour is unavailable, the behaviour can become directed towards some substitute object. Even though it is inanimate, such an object frequently appears capable of filling the role of an important, though subsidiary, attachment-"figure". Like a principal attachment-figure, the inanimate substitute is sought especially when a child is tired, ill or distressed.

Bowlby's "attachment figure" or "substitute object" described above had been previously studied by Winnicott [29], the renown British paediatrician and child-psychoanalyst, who had coined the term "transitional objects" for it. The concept of transitional object was introduced by him in 1953 to describe "comfort objects" such as pillows, blankets and soft toys that a child becomes intensely and passionately attached to. According to Winnicott, for toddlers with good enough mothers, these attachments to transitional objects play a key role in ego development: the child projects the comforting properties of a good enough mother to the inanimate object which, unlike the mother who can temporarily disappear, is always under the control of the child whether the mother is present or not. By practicing and interacting with the transitional object, a mother substitute, the child then acquires the capacity for self-soothing by internalising the good enough mother and is able to withstand increasingly longer absences of the mother. Later researchers have formulated a number of key functions of transitional objects, including separation-individuation, libidinal object constancy, capacity for object relation and empathy and symbolisation and creativity [43].

In 1958, Winnicott also introduced another concept, namely the "capacity to be alone" [44]. His article starts by asserting that

> I wish to make an examination of the capacity of the individual to be alone, acting on the assumption that this capacity is one of the most important signs of maturity in emotional development.

He theorised that this capacity can be developed in children who have good enough mothers and have thus, in psychoanalytical terminology, introjected or internalised a good object in their inner psychic world. In Winnicott's view this capacity can however only take shape by the experience of

> being alone, as an infant and a small child, in the presence of the mother.

According to Winnicott, only by acquiring the capacity to be alone the child can discover its "true self" in contrast to "a false life based on reactions to external stimuli.

It is based on this self-discovery that the child's capacity to be alone has been proposed as the foundation of independence and creativity later in life, and is regarded as "an aspect of emotional maturity" [45, p. 18].

The notion of a "self-object" in Heinz Kohut's self-psychology, an established school of object-relation psychoanalysis, also has many parallels with an attachment object. Self-objects are, according to Kohut, external persons (including parents and therapists), objects or activities that [46, p. 220]

> support the cohesion, vigor, and harmony of the adult self... ranging from cultural self-objects (the writers, artists, and political leaders of the group - the nation, for example - to which a person feels he belongs.....) to the mutual self-object functions that the partners in a good marriage provide for each other.

According to Kohut, it is through empathic responsiveness that self-objects support the individual's developmental needs of mirroring, which leads to regulation of a cohesive sense of self and is the most vital part of cure [46, pp. 65–66]:

> [H]ow does self psychology perceive the process of cure? The answer is: as a three-step movement, the first two steps of which may be described as defense analysis and unfolding of the transferences, while the third step–the essential one because it defines the aim and the result of the cure–is the opening of a path of empathy between self and self-objects, specifically, the establishment of empathic in-tunes between self and self-object on mature adult levels.

Religion as Attachment Object

John Bowlby believed that attachment continues in one way or another later in adulthood [47, p. 588]:

> Probably in all normal people [attachment] continues in one form or another throughout life and, although in many ways transformed, underlies many of our attachments to country, sovereign, or church.

Attachment theorists have in later years objected to the use of the term "attachment", as meant in the context of attachment theory, for any type of bond that human beings create [48, p. 846]. In particular the use of the term "attachment" for patriotism, as in the above quotation, has been received with scepticism [49, p. 803]. However, there is now a consensus among attachment theorists for the notion of "religion as an attachment object", which we will elaborate in this section.

In his book "Theological Imagination: Constructing the Concept of God", Gordon Kaufman adopts the tenets of Attachment Theory and quotes John Bowlby to argue that human beings are at their happiest when they feel secure in some trusted relationships and then provides this quotation by him [50, p. 59]:

> Since in the lives of all of us our most trusted companions are our attachment figures, it follows that the degree to which each of us is susceptible to fear turns in great part on whether our attachment figures are present or absent.

Later Kaufman goes on to argue that in the Christian tradition God is above all a loving and caring father and therefore concludes [50, p. 67]:

> The idea of God is the idea of an absolutely adequate attachment-figure.

While Freud had regarded the notion of God as projection of father and considered it as a delusion [51], attachment theorists began to study the different aspects of the relationship individuals perceive to have with God in the context of attachment. The idea that God, or any deity or religion, can be considered as an attachment object has been investigated since 1990s by a group of social psychologists and there is now a considerable amount of studies which provide evidence for this hypothesis in both Christian and Jewish religions; see the comprehensive review [10] by Granvist, Mikuliner and Shaver [10], which we will summarise below.

The review [10] provides a systematic summary of the work of the authors and that of other researchers on the notion of "Religion as Attachment". First, there is the phenotypic resemblance between parental attachment and believers' relationships with God. Asked which of the following best describes their view of faith–(i) a set of beliefs; (ii) membership in a church or synagogue; (iii) finding meaning in life; or (iv) a relationship with God– most Americans by far chose the last description. Central in the perceived relationship of the believer with God is the experience of "love", which is closely akin to the relationship of a child with an adult attachment figure. In addition, the images of God in religious texts and in believers' description of God's traits are also similar to attributes of parenting, and the main underlying factors are "availability" and "benevolence". Second, there are similarities between the criteria of attachment relationships such as proximity seeking and the believers' perceived relationship with God such as his perceived omnipresence, which is felt by the believer when they visit a place of worship and particularly when they pray.

Third, one of the main functions of secure attachment, i.e., to provide a safe haven when there is danger or distress has its direct parallel in the notion of God as a safe haven. In times of extreme stress and trauma, believers resort to praying to God to alleviate their fear and anxiety. Several studies have confirmed that after loss of a loved one, religious practice and prayer increases among believers and these correlate with successful coping at these critical times [52]. Interestingly, in regard to separation anxiety, one study suggests that even subliminal perception of threat can activate the attachment system of a believer to increased access to the concept of God and supports correspondence between internal working models of parents and God [53].

Fourth, the other main function of secure attachment, i.e., to furnish a secure base for exploring the world and taking up challenges in life, has its parallel in the notion of "God as a wiser and stronger secure base", which is in line with Bowlby's assertion that children consider their parents as stronger and wiser than themselves [7]. In an extensive work on empirical research on religious practice [54, pp. 158–164], it is shown that having an intrinsic orientation in religious practice, i.e., considering it as " an end in itself–a final, not instrumental, good", is correlated with two forms of mental health, namely "freedom from worry and guilt" and "a sense of competence and control". Studies on the psychological effect of religious conversion shows that there is generally a significant decrease in negative affects and a notable increase in well-being in individuals who go through religious conversion; see [54, 55, pp. 101–106]. The review [10] also summarises several studies to confirm that particular aspects of religious beliefs that correlate more strongly with psychological well-being, i.e., "divine relationships" and "praying", are precisely those that are in line with the model of religion as an attachment object. In addition, Kirkpatrick et al. in [56] concluded in their study on the effect of religious practice and loneliness that having a personal relationship with God predicted reduced loneliness despite controlling other factors of interpersonal support.

Furthermore, the review describes two hypotheses that describe two different pathways to God as an attachment object for individuals, which are related to the individual differences in the output of the attachment system. The first is the "compensatory pathway" to reduce distress generally chosen by individuals who have had insensitive primary care-givers resulting in attachment insecurities. In particular, sudden religious conversions are correlated with insensitive parenting [57] and a number of studies have indicated that increase in religiousness among individuals with insensitive parenting is precipitated with severe emotional upheavals [58]. This is consistent with the findings of William James in his classic book "Varieties of Religious Experience" who called these individuals "second born" after having a sick soul with a great amount of anguish and pain [59]. The second hypothesis is the "correspondence pathway", generally chosen by individuals who have had caring and religious parents. This pathway expresses a continuity in secure attachment with religious parents in children who grow up to hold the religious orientation of their parents. In his book, William James called these people "once born", i.e., individuals who have a rectilinear life with a happy mind.

Finally, the authors point out the limitations of the concept of religion-as-attachment model and suggest a more inclusive framework for spiritual attachment objects that includes notions such as mindfulness from non-theistic religions like Buddhism and New Age spirituality.

In more recent years, a similar study on Islamic scripture and spirituality has been undertaken to examine Allah as an attachment figure in the Islamic faith. In [60], the authors investigate whether Muslims seek proximity with a loving God as a safe haven in times of distress and a secure base for living in a challenging world. They consider five different types of Islamic texts, namely, (i) the divine names or attributes of Allah, (ii) stories in the Qur'an that represent attachment relations between Allah, His prophets and people, (iii) verses of the Qur'an with an emphasis on the caring and supportive relationship between Allah and His people, (iv) divine sayings and prophetic inspirations that project Allah as a caring and supportive attachment figure, and (v) supplications which describe the believers' relationship with Allah. On all these themes, the relationship of a Muslim believer with Allah is consistent with the relationship with an attachment figure and the paper comes to similar conclusions as in the Christian and Jewish faiths.

Closely related to the concept of God as an attachment object is the notion of God as a transitional object in a sense used by Winnicott. According to him, a transitional object for the child has an illusory aspect and he postulates a general transitional space which has some illusory aspect [11, p. 3]:

> I am staking a claim for an intermediate state between a baby's inability and his growing ability to recognise and accept reality. I am therefore studying the substance of *illusion* that which is allowed to the infant, and which in adult life is inherent in art and religion....

William Meissner, a Jesuit and a psychoanalyst, has used the notion of a transitional object to explain the psychology of religion and, in particular, the psychology of praying. After elaborating on Winnicott's view of transitional objects with their illusory aspect, he writes [13, p. 177]:

> Illusion, therefore, becomes in Winnicott's view a developmental form of transition to reality, in the sense that without the capacity to utilize transitional objects and to generate transitional forms of experience the child's attempt to gain a foothold in reality will inevitably be frustrated. Illusion in this view is not an obstruction to experiencing reality but a vehicle for gaining access to it.

In fact, Winnicott had claimed that the area of illusory experience is a vital potential space for psychological development [11, p. 110]:

> It is useful, then, to think of a third area of human living, one neither inside the individual nor outside in the world of shared reality. This intermediate living can be thought of as occupying a potential space, negating the idea of space and separation between the baby and the mother, and all developments derived from this phenomenon. This potential space varies greatly from individual to individual, and its foundation is the baby's trust in the mother *experienced* over a long-enough period at the critical stage of the separation of the not-me from me, when the establishment of an autonomous self is at the initial stage.

Meissner then asserts [13, p. 183]:

> Within this potential space, then, man must revive the roots of his capacity for creative living and for faith experience.

In this relation, Erik Erikson, a renown developmental psychologist and psychoanalyst, writes the following on religious experience with an implicit reference to Freud's view of religion as regression to childhood [61, p. 176]:

> But must we call it regression if man thus seeks again the earliest encounters of his trustful past in his efforts to reach a hoped-for and eternal future? Or do religions partake of man's ability, even as he regresses, to recover creatively? At their creative best, religions retrace our earliest inner experiences, giving tangible form to vague evils and reaching back to the earliest individual sources of trust; at the same time they keep alive the common symbols of integrity distilled by the generations. If this is partial regression, it is a regression which, in retracing firmly established pathways, returns to the present amplified and clarified.

Later in his book, Meissner writes the following about the individual believer's praying [13, p. 182]:

> It is here that the qualities of the God-representation and their relationship to the believer's representation become immediate. The God he prays to is not ultimately the God of the theologians or of the philosophers, nor is this God likely to be in any sense directly reconcilable with the God of Scripture. Rather, the individual believer prays to a God who is represented by the highly personalised transitional object representation in his inner, private personally idiosyncratic belief system.

He then goes on to write:

One might say that in prayer the individual figuratively enters the transitional space where he meets his God-representation. Prayer thus becomes a channel for expressing what is most unique, profound, and personal in individual psychology.

We note that in the three cases of attachment objects considered in this section, i.e., surrogate cloth-covered monkeys used by infant monkeys in Harlow's experiments, comfort objects used universally by human children and deities used widely in religious practice by adults, the attachment object actually is or is perceived to be an *external* object. We will see how an internal object is used in Self-attachment later in the article. Next, however, we see how the neural activation in the brain corresponding to a relationship with an external attachment object like God overlaps with that of adult love and maternal love.

fMRI Studies on Bond Making

Since 2000, there have been three different types of fMRI studies on bond-making with respect to adult love, maternal love and religious praying which show that these three different forms of bonding in human beings share a common denominator in terms of activation of the neural pathways in the reward system of the human brain.

Bartles and Zeki in [14] reported on fMRI studies of passionate/romantic adult love. In their experiment, six men and eleven women, who were passionately in love with their partners, stared at photos of their partners or at photos of their friends for about 17 s. The conclusion was that looking staringly at the photo of a beloved partner increased activation of the dopaminergic- related brain areas such as the caudate nucleus and putamen. These findings were reinforced by fMRI experiments reported in [62] by Aron et al. on partners in the early stage of passionate love, which showed increased activity in dopamine- rich subcortical brain areas, the ventral tegmental area and caudate nucleus.

In [15], Bartles and Zeki conducted an fMRI experiment, similar to that in [14], on twenty mothers when each stared at the photos of their own child, compared to another child of the same age with whom they were acquainted with, their best friend, and photographs of another person they were acquainted with. There was increased activity in the dopaminergic-rich sub- cortical brain areas (caudate nucleus, putamen, subthalamic nucleus, periaqueductal gray, substantia nigra, and lateral thalamus). There were specific differences in activation patterns in romantic love and maternal love, in particular the activation of the periaqueductal (central) gray matter (PAG) was observed in maternal but not passionate love. However, the conclusion was that there are neural correlates common to both maternal and romantic love, which are based on increased activation of the dopaminergic rich subcortical regions of caudate nucleus and putamen.

The above fMRI studies were related to bond making between human beings. In a completely new type of experiment, Schjoedt et al. in [16] investigated how performing religious prayers changed the neural activity in a group of Danish Christians. The devout believer had five tasks to perform that included two prayers, the Lords Prayer, as a highly formalized prayer, and a personal prayer as an improvised prayer. The participants all reported that they were strong believers in God's existence and regularly prayed. The result was the activation of the caudate nucleus in the participants when praying, which supports the hypothesis that religious prayer is capable of stimulating the dopaminergic system of the dorsal striatum in practicing individuals. This conclusion is consistent with research on the human striatum indicating that repeated behaviours which are expected to elicit future rewards evoke activity in the dorsal striatum. Furthermore, regarding a related study, the authors say:

> we found no significant caudate activation of self-reported religious persons, who did not pray regularly. While one can argue that prayer involves interaction with an abstract idea of some deity in one form or the other, as far as the brain activity is concerned, it is no different than normal interpersonal interaction.

On this subject, Uffe Schjoedt, the lead author of the article, writes [63]:

> Importantly and somewhat contrary to the widespread assumption that communicating with God constitutes a unique experience reserved for believers, our findings suggest that praying to God is comparable to normal interpersonal interaction, at least in terms of brain function. Praying, it seems, is subserved by the basic processing of our biologically evolved dispositions like other complex cultural phenomena, in this case the evolved human capacity for social cognition.

These findings give further support to the concept of God as an attachment object for believers as described in the previous section. The devotional personal relationships believers have with their God have neural correlates with the passionate relationships in adult love and maternal love. Being in love, whether with your child, your partner or a deity, has a common denominator in that in all these cases the reward system of the brain is activated in the anticipation of some reward which gives incentive, energy and hope to the individual to maintain and strengthen their relationship with the beloved object by carrying out an appropriate set of tasks.

Self-attachment

As we have seen in section "Attachment Insecurity and Vulnerability to Mental Disorder", insecure attachment in childhood in general, and disorganised insecure attachment in particular, makes us vulnerable as adults to psychological disorders and mental illness. From this, it follows that a holistic approach to Computational Psychiatry would seek to examine how individuals might be enabled to earn secure attachment in adult life. We have seen in previous sections how psychotherapy, constructive marriages and the compensatory pathway in religious practice can help individuals earning secure attachment. It was also shown how attachment objects are used in very different contexts by infant monkeys, human children and believers to create a bond with an inanimate or abstract object and use it for emotion self-regulation.

The question now arises as to whether it is possible to develop a self-administrable protocol, based on developmental psychology and neuroscience (as described in section "Attachment Theory") that can help individuals use neuroplasticity and long term potentiation to create their own attachment objects in order to earn secure attachment.

It is proposed in [4] that this is a feasible task. The dyadic child-parent interactions of a good enough parent and a child can be self-administered by an individual who is considered to consist of an inner child and an adult self. The inner child, representing the emotional self, rooted mostly in the right brain and the limbic system, becomes dominant under stress, whereas the adult self corresponding to the logical self, rooted mostly in the left brain and the prefrontal cortex, is dominant in the absence of stress. The adult self connects to and imaginatively creates an affectional bond with the inner child taking up the role of a new primary carer who retrains the inner child to acquire the capacity for emotion self-regulation. In the process of these interactions, the aim is for the inner child to be raised to emotional maturity while the adult self is transformed to a secure attachment object for the inner child.

The creation of the internal affectional bond with the inner child is proposed to activate the dopaminergic pathways of the reward system in the brain. This activation, we argue, provides the energy, incentive and hope to persevere with the protocol and plays the same role that the primary care-giver's love for the child has in the healthy emotional growth of the child. This bond-making is the distinguishing feature of Self-attachment and is in line with fMRI studies on romantic and maternal love as well as religious practice described in section "Religion as Attachment Object". It is also consistent with Bowlby's description of how bonds are created and maintained [8, p. 155]:

> Thus, many of the most intense emotions arise during the formation, the maintenance, the disruption, and the renewal of attachment relationships. The formation of a bond is described as falling in love, maintaining a bond as loving someone, and losing a partner as grieving over someone. Similarly, the threat of loss arouses anxiety and actual loss gives rise to sorrow; whilst each of these situations is likely to arouse anger. The unchallenged maintenance of a bond is experienced as a source of joy.

Self-attachment aims to create an unchallenged maintenance of the bond between the adult self and the inner child so that it becomes a source of joy for the individual. Subsequent to the formation of the Self-attachment bond, the training practices that are at the basis of the adult self and inner child interactions emulate those of "good enough" primary care givers, as described in section "Regulation theory: Neurobiology of Secure Attachment", to minimise negative and maximise positive affects and modulate the inner child's arousal level. It is hypothesised that based on neuroplasticity and long term potentiation, these practices lead to neural circuits corresponding to secure attachment that will increasingly challenge the sub-optimal circuits produced as a result of insecure attachment in childhood as described in section "Neurobiology of InSecure Attachment". While the sub-optimal circuits cannot be wiped off and under high stress can become dominant again for a while, the new optical circuits will gradually counter them effectively and reduce the severity and the duration of the resulting symptoms.

It is on this basis that the notion of Self-attachment has been proposed in [4], which, in a broad sense of the term, can be regarded as a type of self-directed behaviour and interaction that employs a real or imaginative object of devotion and affection and is practiced regularly in order to regulate emotions and harmonise social interactions. We submit that the use of attachment objects as described in section "Attachment Objects", whether by infant monkeys, human children or by adults in religious practice, has many parallels with the notion of Self-attachment. In these cases, the attachment object is external or is perceived to be external and the emotion regulation by the individual is therefore mediated by an externally perceived object.

Self-attachment is essentially a self-help technique which, depending on the individual, may initially need the support of a psychotherapist for a few sessions. It aims to use neuroplasticity and long term potentiation to help individuals to create their own secure attachment objects by the direct intervention of their "adult self", representing their rational self, in order to reparent the "inner child", representing the emotional self. The attachment object to be created can be regarded as a comfort object (cf. Bowlby), transitional object (cf. Winnicott) or empathetic self-object (cf. Kohut). Self-attachment intervention seeks to closely emulate the dyadic interactions of a "good enough" primary care giver and a child by first taking a compassionate attitude to the inner child and then developing an internal affectional bond with the "inner child". This internal bonding, it is hypothesised, activates the dopaminergic reward system of the brain inducing hormones and neurotransmitters including dopamine, serotonin, oxytocin and vasopressin that provide the incentive, energy, hope, tranquillity and caring attitude required for persevering with the reparenting protocol to achieve emotion self-regulation as in secure attachment of children [6, p. 14].

Self-attachment can be regarded as an extension of attachment theory but it is also related to and incorporates ideas from a range of psychotherapeutic methods. This includes the notion of "inner child" from transactional analysis [64], Mentalisation [65]—defined as the capacity to understand the emotions and mental states of the other people as well as those of oneself—exposure as in behavioural therapy [66], compassionate focused psychotherapy [67], schema therapy and reparenting [9] and object-relation psychodynamic therapy in a wider sense of the term in which objects can be impersonal as well as personal [45, p. 150–152]. Self attachment integrates these techniques into its main focus of intervention, which is the creation of an internal affectional bond to emulate what occurs naturally between an infant and a parent. Self-attachment also employs protocols such as singing, dancing and massage that are known to increase dopamine, serotonin, oxytocin and reduce cortisol levels [68–71]. It can also be combined with any well-established therapeutic technique.

Playing and Role Playing

The internal interactions in Self-attachment resemble role playing, in which the individual plays simultaneously both the role of the child within and that of the adult self. This shows that, while these interactions aim to create new types of cognition and behaviour corresponding to secure attachment, Self-attachment is in a sense a form of playing. It is therefore useful here to highlight the significance of play in general in psychotherapy.

As explained in the previous section, Winnicott considers a potential illusory sphere of play between the inner reality of an individual and the external reality. He also presents the hypothesis that psychotherapy can only be successful when both the therapist and the patient play [11, p. 54]:

> The general principle seems to me to be valid that *psychotherapy is done in the overlap of the two play areas, that of the patient and that of the therapist.* If the therapist cannot play, then he is not suitable for the work. If the patient cannot play, then something needs to be done to enable the patient to become able to play, after which psychotherapy can begin. The reason why playing is essential is that it is in playing that the patient is being creative.

As far as child development is concerned, play has been recognised as vital for the cognitive, physical, social, and emotional well-being of children and for maintaining strong parent-child bonds [72]. There has been a growing body of evidence in children supporting the many connections between cognitive and social competence as well as abstract thinking on the one hand and high-quality pretend play on the other hand. In particular, role playing in children has been linked to cognitive functioning and impulse control. Pretense starts in children age between one and two and plays a vital role in young children's lives through the primary school years [73].

In addition, role playing is an established method in psychotherapy and is defined by Corsini [74, p. 6] as follows:

> Essentially, role playing is a "make believe" process. In therapy, the patient (and if it is to be an interactional situation, the others involved) will act for a limited time "as if the acted-out situation were real".

Asserting that role playing can even be used in self-therapy, Corsini then postulates that role playing has the following basic features [74, pp. 5 and 9]:

- Is a close representation of real life behaviour.
- Involves the individual holistically.
- Presents observers with a picture of how the patient operates in real life situations.
- Because it is dramatic, focuses attention on the problem.
- Permits the individual to see himself while in action in a neutral situation.

The dyadic interactions in Self-attachment between the adult self and the inner child differ in two ways from the usual role playing as in the first item listed above: (i) the behaviour of the inner child is not just a close representation of real life behaviour but rather *the real life behaviour itself,* and (ii) the behaviour of the adult self, who follows the Self-attachment protocol, is the *optimal behaviour of a "good enough" parent in relation to a child.* In the next section, we turn to a more detailed outline of the various stages of Self-attachment.

Self-attachment Protocol

There are four stages in the Self-attachment protocol which are briefly described here [4]:

(1) Introduction to secure Self-attachment therapy. The Self-attachment protocol is challenging and demanding for any volunteer as it requires a great deal of dedication and motivation. For this reason, it is essential to understand why so much time and effort should be invested in this method before it can have an impact. Therefore, in the preliminary stage, the individuals become familiar with the scientific basis and the underlying hypothesis of the proposed Self-attachment therapy. This includes a basic introduction to attachment theory, regulation theory and their neural correlates, fMRI studies on being in love in the case of maternal and romantic love as well as that of bonding with abstract and imaginative objects as in prayer, neuroplasticity and long term potentiation.

(2) Connecting with the "inner child". In this first stage of the protocol, the volunteers start to have a relationship with their inner child with a view to establish empathy and ultimately compassion with the child. While looking at photos of their childhood, they think introspectively and recall their basic childhood environment including their relationships with parents and other care-givers. The aim is to have a feeling for the inner child as a result of these exercises. Since the early attachment type of a child is formed in the pre-verbal years, when visualisation is the main tool for observation and sensing, there is much focus on imagery in this stage. A happy or positive looking photo of childhood that has always been favoured is selected by the volunteer as well as a sad or gloomy looking photo that has been avoided, disliked or less fond of. Several novel exercises are designed to connect to the inner child in various emotional states while the eyes are kept closed: trying to visualise the two chosen childhood photos, to imagine that the child that they were is present and is close to them and that they can touch and hold this child. The objective of the this stage is to conceptualise the inner child as concretely as possible and develop empathy and then compassion toward it.

(3) Making an affectional bond with the inner child. In this stage an imaginative but passionate affectional bond is made with the inner child that is subjectively experienced as falling in love. This resembles passionate devotion to a deity that has neural correlates with maternal and romantic love. It is hypothesised that this step can in principle be taken by all individuals based on their primary narcissism, a notion of "self-love" in children originally introduced by Freud as a defense mechanism that is to protect the child from psychic damage during the formation of the individual self [51]. Bowlby has himself argued that separation anxiety in children from their mothers, which is the root cause of insecure attachment, is a form of injury to primary narcissism [7, p. 11].

The inner child is subsequently adopted imaginative by the volunteers who vow to consistently support and love the inner child to reprocess past traumatic episodes and reparent the child to emotion self-regulation. The imaginative bond making is in practice attained by passionately and repeatedly reciting a favourite happy love song while staring at the positive looking childhood photo trying to rekindle happy memories. The aim here is to bring the inner child into life, excitement and joy again, inducing dopamine release and providing energy and hope for the volunteer who requires constant motivation to keep up the momentum and resolve to persevere with carrying out the protocol. This is in line with latest findings in neuroscience on singing. The study in [69] found increased activation in nucleus accumbens for singing as opposed to speaking words of a familiar song and [70] reported increased activation in regions including caudate nucleus and putamen when professional classical singers imagined singing an aria (love song). In addition, [75] revealed dopamine release in caudate nucleus and nucleus accumbens during anticipation and experience of peak emotional response to passive listening to self-reported pleasurable music.

(4) Developmental retraining and re-parenting the inner child. The last and main stage consists of several types of interactions between the adult self and the inner child that emulate the function of a good enough parent in interacting with a securely attached child in order to minimise the negative emotions, named the Sad-Child protocol, and to maximise the positive affects, named the Happy-Child protocol. We provide one example here on how to reprocess painful and traumatic past events: the volunteers with their eyes closed recall some traumatic episode in their childhood, remembering in as much detail as possible the associated emotions of fear, helplessness, humiliation and rage. Then, they imagine that their inner adult responds quickly to the child in distress by embracing, cuddling, and loudly reassuring the child. Cuddling the inner child is simulated by giving oneself a head, face or neck massage, which is known to reduce cortisol levels and increase oxytocin and endorphin release; see [32, 68, p. 103]. By revisiting the neural circuits of past traumas, these sessions thus induce dopamine, serotonin, oxytocin and vasopressin secretion and are designed to build new optimal neural circuits in relation to the old pathological ones.

A basic undertaking by the volunteers throughout the treatment period is to gradually construct, using their own imagination, a visually potent picture of the protocol that depicts a secure attachment object, for example as a new bright and solid house erected and built in the place of a dark and derelict shelter, which depicts insecure attachment. This visual construction, which either remains completely in the mind or is drawn on paper, symbolises the secure attachment the volunteers would earn themselves in contrast to their past insecure attachment anxieties.

Distinguishing Characteristic and Proof of Concept

The distinguishing characteristic of Self-attachment therapy is the internal affectional bond that is self-administered to emulate the loving relationship of a primary caregiver and a child. The imaginative but passionate relationship between the adult self and the inner child mimics the real interactions of a parent-child dyad that lead to secure attachment in the first years of the child's development. These interactions aim to maximise the positive affects and strengthen the bond and homeostasis between the adult self and inner child, similar to the real interactions of loving parents and their children as described in section "Regulation Theory: Neurobiology of Secure Attachment". They have the combined result of creating more positive affects and establishing an intimate, internal dialogue of adult self with the inner child. This combination is unique to Self-attachment and, in particular, provides a more effective tool to tackle and contain negative affects and the corresponding psychological disorders.

There have been a number of case studies of Self-attachment undertaken by volunteers since 2010 and by clients of professional psychotherapists trained in the protocol since 2014, which were first reported in [4] as mentioned in the Introduction. A detailed report of a number of case studies is now under preparation.

As pointed out in the Introduction, the central idea in the virtual reality experiments reported in [18, 19] is similar to the basic intervention in the first stage of Self-attachment described in (2) above. The difference is that in Self-attachment the child is displayed, imagined and perceived as the "inner child" of the individual, representing the emotionally unregulated childhood of the individual with its own history of attachment insecurities. While this is a crucial distinction that can lead to a much more effective intervention, the results reported in these two papers provide a proof of concept of Self-attachment.

It may be argued here that Self-attachment, as described in section "Self-attachment Protocol", is not a natural undertaking. There are three basic points to be made to counter this argument.

First, looking after yourself or taking care of yourself is a main principle of mental health in all cultures and there are many organisations and manuals dedicated to this principle for the general public. Self-attachment takes this principle further for those who have had traumatic childhood background and are thus vulnerable to mental disorders.

Second, one can argue with the same reasoning that the use of religions or deities as attachment objects for individuals is not natural. Yet, such use of religions and deities have existed for thousands of years and continue to play a significant role in the mental health of human beings.

Third, there seems to be in fact a natural role for Self-attachment in higher primates: There is some evidence for a direct but rudimentary form of Self-attachment in ethology. Experiments, on rhesus monkeys, isolated at birth, show that after six months in isolation they exhibit a type of self-directed behaviour, such as self-orality, clasping a limb or rocking, which are considered soothing and comforting since they are precisely described in ethology as the kind of actions that would have been carried out on the monkey infant by the monkey's mother if she had been present [76]. These self-directed soothing behaviours must have biological underpinnings as they are not learnt and therefore give more evidence that Self-attachment is a particular component of the attachment system discovered by Bowlby.

A Game-Theoretic Model

We have argued that Self-attachment has many parallels with role playing in psychotherapy and that it can be regarded as a type of interactive play between the adult self and the inner child. In this section, a game-theoretic model for the Self-attachment protocol is presented. We first briefly review the notion of a game [77]. A strategic game is a model of decision making by a number of agents called players in which each player has a set of possible actions and in each round of the game, the players simultaneously choose their individual actions independently. A strategy profile is a particular choice of action by each player and a game is fully defined by providing the utility of each player for each strategy profile. The concept of stability in strategic games is captured by the notion of a Nash equilibrium. A strategy profile is said to be a Nash equilibrium for a given game if no player can increase its utility by deviating from its action in the strategy profile when all other players keep their actions intact. A repeated game (or iterated game) is an extensive form game which consists in some number of repetitions of some base game (called a stage game), for example a 2-person game. It captures the idea that a player will have to take into account the impact of its current action on the future actions of other players. Game theory has been applied to various areas in applied mathematics when different decisions are made by competing players; see [77].

In [78], the dynamics of interaction of a child-parent dyad for different attachment types has been modelled using two player and two action games. The two actions of the child are "Go", meaning "go to seek support from the parent" or "Don't Go", whereas the actions of the parent are given by "Attend", meaning respond appropriately to the child's distress, or "Ignore", meaning "ignore the child's distress signal". For example, the game on the left in Fig. 1, which has a Nash equilibrium for the strategy profile (Go, Attend), is proposed for the dynamics of secure attachment, whereas the game on the right in Fig. 1, which has a Nash equilibrium for the strategy profile (Don't Go, Ignore) gives a model of avoidant attachment.

In [79], a framework is introduced which, given a game with an undesirable Nash equilibrium, will generate using reinforcement learning a dynamics which changes

Fig. 1 *Left*: A game with a secure NE. *Right*: A game with an insecure NE

		Parent	
		Attend	Ignore
Child	Go	**4,4**	3,3
	Don't Go	2,1	2,2

		Parent	
		Attend	Ignore
Child	Go	4,2	2,3
	Don't Go	3,1	**3,4**

the utility matrix by providing additional reward for more desirable actions until a new Nash equilibrium at some predetermined desirable strategy profile is established. The change in the utility matrix is brought about by reinforcement learning and additional reward provided to one or both players for taking more desirable actions. Reinforcement learning algorithms model dopamine release (corresponding to a reward-prediction error) [80], which gives the incentive to a player to deviate from its action in the Nash equilibrium in the anticipation of future reward.

Specifically, the framework in [79], employs Q-learning, a particular type of reinforcement learning [81], which is suitable to model dopamine release [82]. As an example, a process of psychotherapeutic intervention to induce change in the behaviour of a parent with an avoidantly attached child has been modelled using attachment games, which offers internal rewards to the parent whenever the action "Attend" is selected and eventually leads to a secure attachment game.

The above framework can be adapted to provide a model of Self-attachment. We assume the individual undertaking the Self-attachment protocol has had an avoidant attachment in childhood and has thus internalised an adult self in the mirror image of his or her primary care-giver. Therefore, we assume that initially the inner child of the individual is avoidantly attached with the individual's adult self and their interactions can be modelled at the start by an avoidant attachment game as on the right in Fig. 1.

The two actions of the Inner Child, namely "Go" and "Don't Go" signify whether the individual seeks help from himself/herself or not, whereas the two actions of the Adult Self, namely "Attend" and "Ignore", represent the two cases when the individual takes action to comfort himself/herself or not. The four strategy profiles then describe the four possible alternatives when the Inner Child is distressed, e.g., when the individual is suffering from some anxiety or depressive symptom. The effect of Self-attachment practice is modelled by the reinforcement learning framework in [83] applied to the avoidant game so as to change it dynamically to the secure attachment game in Fig. 2, which has a Nash equilibrium (Go, Attend) corresponding to secure attachment as well as one (Don't Go, Ignore) corresponding to avoidant attachment.

In reinforcement learning one learns what action to take in order to maximise a reward signal: agents incrementally adapt their estimates of reward associated with state-action pairs based on observed outcomes following each action choice [81]. We give an overview of the adapted reinforcement learning framework for Self-attachment here. A state of the learning procedure is given by a pair of what we call an M-state and its associated Q-state. An M-state is given by a utility matrix $M \in \mathbb{R}_+^{2 \times 2}$ of the Adult Self with its associated Q-state given by the ordi-

		Adult self	
		Attend	Ignore
Inner Child	Go	**4,4**	2,2
	Don't Go	3,1	**3,3**

Fig. 2 A game with a secure and an insecure NE

nal representation $[M] = M/\equiv$, where the equivalence relation $\equiv$ on the set $\mathbb{R}_+^{2\times2}$ is defined by $M \equiv N$ iff for all $i, j, i', j' \in \{1, 2\}$ we have: $M_{ij} < M_{i'j'} \iff N_{ij} < N_{i'j'}$ and $M_{ij} = M_{i'j'} \iff N_{ij} = N_{i'j'}$. Thus, the Q-state can be represented as a matrix in $\{1, 2, 3, 4\}^{2\times2}$ with at least one entry equal to 1. The actions in the Q-learning are the four possible strategy profiles (Go, Attend), (Go, Ignore), (Don't Go, Attend) and (Don't Go, Ignore).

We assume that the utility matrix of the Inner Child is static and does not change throughout the course of Q-learning. The state transitions are defined as follows. If the strategy profile (Go, Attend) is played then $M_{11} \rightarrow rM_{11}$ where $r > 1$ is a fixed multiplicative reward factor. Similarly, if the strategy profile (Don't Go, Attend) is played then $M_{21} \rightarrow rM_{21}$. Otherwise, when the Adult Self plays Ignore, its utility matrix does not change. We note that when the utility matrix does change from M to M' its associated Q-state may or may not change, i.e., we can have $[M] = [M']$ or $[M] \neq [M']$. We assume the Inner Child plays reactively: if the Adult Self plays "Attend" at time t then the Inner Child plays "Go" at time $t + 1$, whereas if the Adult Self plays "Ignore" at time t then the Inner Child Plays "Don't Go" at times $t + 1$. The Q-value $Q([M], \beta)$ for the pair of Q-state and action $([M], \beta)$ where $\beta \in \{$Attend, Ignore$\}$ is updated when $M \rightarrow M'$ according to the standard Q-learning rule:

$$Q([M], \beta) \leftarrow Q([M], \beta) + \ell \left(R(M, \beta) + \delta \max_{\beta' \in S} Q([M'], \beta') - Q([M], \beta) \right)$$

where $S = \{$Attend, Ignore$\}$, $\ell > 0$ is the learning rate, δ is the discount factor with values $0 \leq \delta \leq 1$ and the reward R for a pair of M-state and action given by:

$$R(M, \text{Attend}) = \begin{cases} rM_{11} & \text{if (Attend, Go) is played} \\ rM_{12} & \text{if (Attend, Don't Go) is played} \end{cases}$$

and

$$R(M, \text{Ignore}) = \begin{cases} M_{11} & \text{if (Ignore, Go) is played} \\ M_{12} & \text{if (Ignore, Don't Go) is played} \end{cases}$$

The learning rate has been chosen as $\ell = \ell([M], \beta) = (n([M], \beta))^{-1}$, where $n([M], \beta)$ equals the number of times action β has been selected in Q-state $[M]$. This means that initially $\ell([M], \beta) = 1$ and subsequently $\ell([M], \beta)$ decreases with each new selection of action β in Q-state $[M]$. A softmax action selection rule [81], also called the Boltzmann probabilistic rule, is used to choose the action taken by the Adult Self according to the following probability distribution:

$$\Pr(\beta | [M]) = k^{Q([M], \beta)} / \sum_{\beta \in S} k^{Q([M], \beta)}$$

with exploration parameter $k > 1$. The greater the exploration parameter the more it is likely that, in a given Q-state, actions with higher Q-values would be selected. Thus, the reinforced Adult Self chooses actions according to a path-dependent, non-stationary stochastic process. The initial M-state M_0 is set to be the utility matrix of the Adult Self on the right in Fig. 1. The initial Q-values of a Q-state and action pair is set as follows

$$Q([M], \text{Attend}) = [M]_{21}, \qquad Q([M], \text{Ignore}) = [M]_{22}$$

i.e., it is set to the ordinal value of the Q-state $[M]$ assuming that the Inner Child plays "Don't Go".

The additional reward that reinforces the action "Attend" provides the incentive for the Adult Self to deviate with some probability from the undesirable "Ignore" action in the undesirable Nash equilibrium and to choose "Attend". It is proved mathematically in [83] that in this setting the utility matrix almost surely, i.e., with probability one, converges to the game in Fig. 2 to provide a secure attachment as a Nash equilibrium. The question is how fast is the rate of convergence.

A simulation of this framework in [83] for $r \in \{1.1, 1.3, 1.5\}$ produces Fig. 3 when $k = 1.5$ and Fig. 4 when $k = 2$ for the average number of iterations of the Q-learning algorithm for a range of values of the discount factor δ so that the utility matrix converges to give a secure attachment Nash equilibrium. As the reinforcement parameter $r > 1$ increases, the average number of iterations to yield secure attachment decreases, which can be interpreted by asserting that individuals who

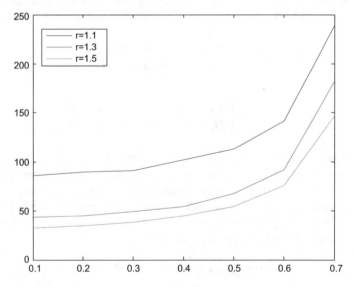

Fig. 3 Average no. of rounds to reach secure attachment for various values of the multiplicative reward factor when $k = 1.5$

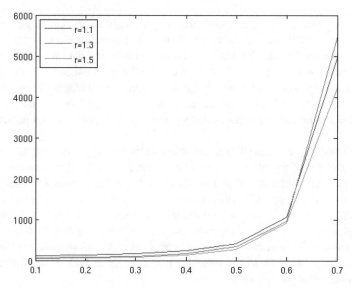

Fig. 4 Average no. of rounds to reach secure attachment for various values of the multiplicative reward factor when $k = 2$

value the Self-attachment protocol at a higher level obtain the desired result quicker. Furthermore, the lower the discount factor δ, the quicker the convergence to secure attachment, which implies that those focused on the present and more immediate reward would see a faster path to the desired result. Finally, as the exploration parameter k increases, the speed of convergence to secure attachment decreases. In fact, for a larger value of k, the choice of "Ignore" dominates the initial rounds of the reinforcement learning because the initial Q-values of the pairs of Q-states and actions are determined by the action "Don't Go" of the Inner Child in the initial Nash equilibrium ($Don'tGo, Ignore$). Therefore, larger values of k capture individuals who show resistance to therapy and are more entrenched in their original behaviour.

Self-attachment and Strong Patterns

In this section, we first review neural models of psychotherapy in the literature. A neural model for personality is constructed in [84], which is based on Cloninger's three-dimensional character cube [85]. It aims to model psychotherapy for individuals who are already functioning quite well but seek to improve their effectiveness. Cloninger's character cube is also used in [86] to develop a neural model for how human beings suppress or enhance certain types of behaviour. Galatzer-Levy [87] presents an outline of how non-linear dynamical systems and attractors can be used qualitatively to model the psychoanalytical process, and includes other related references in this subject. The concept of "working through" in psychoanalysis has been modelled in [88] using a Hopfield network with two weakly connected layers.

The first neural model for Self-attachment we describe here uses the notion of strong patterns in artificial neural networks, in particular Hopfield networks, an early model of energy based associative networks introduced by Hopfield [89]. The patterns stored in such a network under the right conditions, become, with high probability, the minima of the network energy and the fixed points of the network; see below.

Attachment types and cognitive and behavioural prototypes are entrenched in the neural pathways of the brain as a result of some key repeated and strong interactions that an individual undertakes or is exposed to, which become stored patterns in the brain. In [90, pp. 132–144], Lewis, Amini and Lannon emphasise that our attachment types and key emotional attitudes in relation to others are sculpted by limbic attractors as a result of repeated exposure to similar patterns of interactions in childhood, which will then profoundly impact our emotional world for the rest of our lives. They employ artificial neural networks to describe how such patterns of attitude and behaviour are developed.

Similarly, Smith et al. in [91, p. 222] proposed the Hopfield network to model cognitive and behavioural patterns considered as:

> *prototypes*-deeply learned patterns of thought and social activity. In the sense developed by cognitive psychologists, prototypes are cognitive structures that preserve in memory common or typical features of a person's experience. By matching perceptions and thoughts in prototypes stored in memory, persons categorize and identify objects, form inferences and expectations, and construct predictions about the future. Prototypes thus serve an orienting function, since persons use them to guide their behaviour. In general, a person seeks the closest possible match between ongoing experience and these prototype patterns. When confronted with the unfamiliar, a person will search for the closest match to a learned prototype.

The question how to model repeatedly or strongly stored patterns in a Hopfield network has been addressed in [92, 93]. Assume we have a Hopfield network with N neurons $i = 1, \ldots, N$ each taking values ± 1. A configuration of the network is given by $X \in \{-1, 1\}^N$ with components $X_i = \pm 1$ for $1 \leq i \leq N$. Assume we have the deterministic asynchronous updating rule (i.e., with temperature $T = 0$) and zero bias in the local field at each node i.

The updating rule is:

$$\text{If } h_i \geq 0 \text{ then } 1 \leftarrow X_i \text{ otherwise } -1 \leftarrow X_i$$

where $h_i = \sum_{j=1}^{N} w_{ij} X_j$ is the local field of configuration X at i. The energy of the network (assuming zero biases at the nodes) for configuration X is given in terms of the synaptic couplings w_{ij} by

$$E(X) = -\frac{1}{2} \sum_{i,j=1}^{N} w_{ij} X_i X_j.$$

It is easy to check that if the updating rule is applied asynchronously then the energy of the network will not increase. Since there are only a finite number, in fact 2^N, configurations possible for a fixed N, it follows that with the asynchronous updating rule the network will always settle down to one of the minima of its energy landscape, which will be a fixed point of the network.

Assume we have p patterns $X^k \in \{-1, 1\}^N$, with $1 \le k \le p$, each given by its components X_i^k for $i = 1, \ldots, N$, which are to be stored in the memory. The generalized Hebbian rule for the synaptic couplings to store these patterns is defined as follows [92, 93]:

$$w_{ij} = \frac{1}{N} \sum_{k=1}^{p} d_k X_i^k X_j^k, \tag{1}$$

for $i \ne j$ with $w_{ii} = 0$ for $1 \le i, j \le N$. where d_k is the *multiplicity* or *degree*, also called the *strength*, of the pattern X^k. In a standard Hopfield network we have $d_k = 1$, i.e., all patterns are *simple*. In this case, if all patterns are random and $p/N \le 0.138$ then the network behaves like an associative network with a good memory: the p patterns become with high probability fixed points or attractors of the network. If the network is initialised with a configuration X, then by asynchronous updating the network converges with a high probability to one of the patterns X^k for $1 \le k \le p$. The number 0.138 is the *retrieval capacity* of the network which can be determined both experimentally and theoretically.

If $d_k > 1$, then X^k is a *strong pattern* indicating either that the pattern has been multiply stored with the integer d_k as its multiplicity or that the pattern has been deeply stored with a high level of dopamine secretion that has reinforced the learning [94]. The corresponding attractor produced by a strong pattern in the network is called a *strong attractor*. Strong attractors are more stable than those of simple patterns and have a larger basin of attraction and lower energy level [93]. In [92], a square law has been obtained for the retrieval capacity of a single strong pattern in the presence of simple patterns in the Hopfield network. The square law for the stability of a single neuron is deduced by a theorem of Lyapunov which generalises the Central Limit theorem in the case we have independent but non-identically distributed random variables. Assuming that there is only a single strong pattern and all patterns are independent, one obtains for the probability that a single node of the strong pattern with degree $d \ll p$ becomes unstable:

$$\Pr_{error} \approx \frac{1}{2} \left(1 - \operatorname{erf}(\sqrt{Nd^2/2p}) \right)$$

where erf is the error function. This formula reduces to the corresponding error for the standard Hopfield network for $d = 1$ and thus we can see that for $d > 1$, the stability is increased by a factor d^2.

To establish the square law for the retrieval capacity of a whole strong pattern much more work is required. In [92], this is done by moving to the stochastic Hopfield network for a pseudo temperature $T > 0$ in which the updating rule is probabilistic. In this setting, one derives the conditions for the strong pattern to be retrievable with a non-zero probability and then one takes the limit $T \to 0$ which recovers the deterministic network. Assuming that the strong pattern has multiplicity $d > 1$ with $d \ll p$, it is shown analytically that provided p satisfies $p/N \leq 0.138d^2$, the strong pattern can still be retrieved with high probability, showing that the retrieval capacity of a strong pattern grows proportional to the square of its degree.

This square law provides us with a technique to model cognitive and behavioural proto-types as strong patterns that remain stable compared with simple patterns. In fact, it enables us to model the gist of the Self-attachment protocol, which we now describe using a simple example. Assume we store 30 copies of a generic sad face and 100,000 random images in a Hopfield network with $N = 48 \times 48 = 2304$ providing a screen of neurons regarded as pixels with values ±1. Note that in this case we have $0.138Nd^2 = 0.138 \times 2304 \times 30^2 \approx 286, 156$. Since we have a total of $100, 030 < 286, 156$ stored patterns, it follows by the square law that the sad face is retrievable. In fact, a simple experiment shows that with high probability any initial image will be repeatedly updated using the asynchronous rule to converge and retrieve the sad face as a fixed point. This shows that the Hopfield network is now an associative model of a sad brain which with high probability interprets any image as a sad face. We now store additionally 40 copies of a generic happy face in the same Hopfield network. In the resulting network, we have two strong patterns, the old generic sad face with degree 30 and the new generic happy face with degree 40. In the competition between these two strong pattern, the stronger one i.e., the generic happy face, wins [93]. Therefore, after initialising the network with a random pattern, with high probability, one eventually retrieves the happy face. This shows that the associated memory network which had modelled a sad brain–was biased toward retrieving the sad face– is now modelling a happy face– is biased toward retrieving the happy face.

Therefore, the process of Self-attachment therapy works according to this model by constructing, using neuroplasticity and long term potentiation, a strong optimal cognitive and behavioural pattern corresponding to each suboptimal pattern that had previously been learned. Psychotherapy is successful once the strength of the new pattern exceeds that of the old one.

Neural Models of Self-attachment

In this section, we will describe three neural models of the human brain for Self-attachment.

An Energy Based Neural Model with Reinforcement Learning

The first neural model of the human brain for Self-attachment we describe here uses Levine's pathways for emotional-cognitive decision making [95]. Levine's model contains two networks: (i) An energy based competitive *needs* network, which will be modelled here by a Hopfield network whose attractors represent the individual's competing needs that include physiological as well as higher cognitive and emotional needs, motivated by Maslow's hierarchy of needs. (ii) A network of brain regions that make either deliberate or heuristic decisions and is comprised of four connected areas: the amygdala, the OFC, ACC and DLPFC, which account for various decision rules on specific tasks as will be explained below. These four regions comprise a three-layer network, in which the vigilance threshold of the individual, represented by the ACC, determines the status of activation of each layer. The lower the vigilance threshold the quicker the alertness of the individual will be activated. The state of the needs network influences the vigilance threshold so that the winning needs become dominant to implement the corresponding decision rules.

In [4], the above framework is employed to construct a model of Self-attachment by using reinforcement learning in the form of Q-learning. This extends the results in [96] that models psychotherapy based on Mentalisation only. The needs network is represented by strong patterns in a competitive Hopfield network consisting of the two categories of (a) need for cognitive closure, which includes the six basic emotions, and (b) need for cognition, which contains Mentalisation and two sub-protocols of Self-attachment named Happy-Child, aimed at increasing inner joy, and Sad-Child, aimed at reducing the negative affects of the inner child. Strong patterns with different degrees model the six basic emotions, and the three cognitive states Mentalisation, Happy-Child and Sad-Child. These patterns are represented by generic (smiley) faces in the needs network which also contains a large number of random patterns. Three identical Restricted Boltzmann Machines (RBM) [97, 98] with 17 hidden units are pre-trained to recognise the six emotions, the Mentalisation, Happy-Child and Sad-Child patterns. An RBM is a stochastic generative neural network which learns a probability distribution over the inputs it sees. They are used here to model the amygdala, the OFC and the DLPFC. These three regions together with the needs network in the brain account for various decision rules on specific tasks and comprise a three-layer decider network, in which the vigilance threshold of the individual determines the status of activation of each layer. See Fig. 5.

Suppose the Hopfield network receives a random input, modelling a random stimulus to the brain. Then, with high probability one of the strong patterns is retrieved. If the Mentalisation pattern, the Happy-Child or the Sad-Child is recalled, then the Hopfield network will send a low level vigilance threshold to the error detector (ACC), and the DLPFC-OFC circuit is chosen to generate complex decision rules. This includes practicing the protocols for Happy-Child or Sad-Child. As the secondary sensory device, the RBM in the DLPFC receives a Mentalisation signal from the Hopfield network, and categorises it into a 17-unit vector of the hidden layer. In addition, the RBM in the OFC, the primary sensory device, accounts for categorising the input pattern into another 17-unit vector. If the Hamming distance of these two generated vectors (i.e., the number of nodes they differ in) is greater than the vigilance threshold, then a mismatch occurs and the network generates a deliberative rule. Otherwise, a heuristic decision is made. However, because of the low vigilance threshold, the DLPFC- OFC circuit would most likely not make heuristic decisions in this case.

If, on the other hand, the retrieved pattern is one of the six emotion patterns, the Hopfield network will send a high-level vigilance threshold to ACC, and the OFC-amygdala circuit is selected for making decisions. Again, as the secondary sensory device, the RBM representing the OFC classifies the emotion pattern from the Hopfield network, and amygdala classifies the input pattern. The Hamming distance between the two 17 unit output vectors is compared to the high-level vigilance threshold. A heuristic decision is made if the Hamming distance is lower than the vigilance threshold; otherwise, a deliberative decision is taken. Since the vigilance threshold is high, the OFC-amygdala circuit would most likely not make deliberate decisions in this case.

Self-attachment therapy is modelled by a Q-learning process that targets the needs network in the above framework. For simplicity, it is assumed that only six patterns are involved in the learning, namely: the three basic emotions for Angry, Sad, Happy and the three cognitive states for Mentalisation, Happy-Child and Sad-Child. The M-state of the Q-learning is given by the set of degrees or strengths of these patterns at any point in time, whereas the Q-state is given by the ordinal representation of these M-states, i.e., a list of six positive numbers between 1 and 6 that includes 1 as the lowest rank. In addition, it is assumed that there are six actions corresponding to these states. The reward table for Q-leaning is as follows:

Angry : 0   Happy : 0.3   Sad : 0

Mentalisation : 0.4 Happy-Child : 1 Sad-Child : 0.6

This means that whenever any of the six actions Angry, Happy, Sad, Mentalisa-

Fig. 5 The decider network consists of the DLPFC-OFC circuit for deliberate decisions on the *left* and the OFC-amygdala circuit which makes heuristic decisions on the *right*

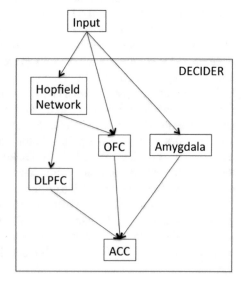

tion, Happy-Child, Sad-Child is selected the degree or strength of the corresponding state is increased respectively by $0, 0.3, 0, 0.4, 1.0, 6$. As can be seen in the table, the highest rewards are received for carrying out the sub-protocols for Happy-Child and Sad-Child. It is hypothesised that these sub-protocols activate the reward system of the brain, inducing dopamine, serotonin, oxytocin and vasopressin: thus these actions are more deeply learned corresponding to a relatively higher increase in their strengths in the M-state compared to Mentalisation or Happy actions.

Initially, we start with an M-state in which the pattern Angry (and/or Sad) are dominant in the needs network. As in the case of the game-theoretic model in section "A Game-Theoretic Model", the Boltzmann probabilistic rule is used for choosing an action in a given Q-state. At each iteration of the algorithm, a random pattern, regarded as input, stimulates the Hopfield network and the two RBM's representing the Amygdala and OFC. The Hamming distance between the hidden layers of these two RBM's provides the measure of discrepancy and either a heuristic or deliberate decision is made. Initially, most decisions will be heuristic as a negative emotion is dominant in the needs network. As the algorithm iteratively progresses, the Q-learning process will gradually increase the strength of the more optimal patterns, i.e., Happy-Child, Sad-Child, Mentalisation and Happy. Eventually, the Happy-Child pattern becomes dominant in the needs network and as a result most decisions will be deliberate. We consider this process as modelling a successful course of Self-attachment.

In Fig. 6, the blue curve shows the average number of iterations required for a successful course of Self-attachment starting with different degrees of the initial Angry pattern. The red curve shows the average number of iterations required for the Mentalisation pattern to become dominant as a result of psychotherapy based on Mentalisation (i.e. without the sub-protocols for Happy-Child and Sad-Child). We see, as expected, that the average number of iterations required when the sub-protocols for Happy-Child and Sad-Child are also used is significantly lower than when they are not included in the algorithm.

A Neural Model of Bonding

In [99], we built on a previous model by Levine [100] concerning how emotional appraisals in the OFC can mediate between activity in neural circuits that drive stress and facilitative responses to social stimuli. Activity in stress circuitry (focused on the central nucleus of the amygdala (CeA), locus coeruleus, and the parvocellular part of the paraventricular nucleus of the hypothalamus (PVN)) results in the release of norepinephrine and corticotropin-releasing hormone (CRH, the precursor to cortisol), while stimulation of facilitative networks (involving the magnocellular part of the PVN along with reward circuitry) leads to the release of dopamine and oxytocin.

Based on this, we hypothesised that a main effect of the Self-attachment bonding protocols is to associate broad classes of social stimuli that have previously been conditioned as being fearful or threatening in nature with new representations of naturally-induced reward in the OFC. These new, additional reward representations are proposed to emerge as a result of the application of various activities such as directed singing with inner-child imagery.

We simulated our model computationally, using a deep belief net to model bi-directional connectivity between the OFC and basolateral amygdala (BLA) and feed-forward networks for the other pathways in the model. The pathway from the BLA to the CeA was assumed to be proportionally strengthened by the magnitude of unexpected punishments, while the pathway from ventral parts of the medial prefrontal cortex (vmPFC) to the CeA (via the intercalated cells) was proportionally strengthened by the magnitude of unexpected rewards.

Using this model we showed how, as the bonding protocols progress, OFC-dorsomedial hypothalamic pathways could increasingly facilitate natural oxytocin release from the magnocellular part of the PVN, and inhibit the release of CRH from the parvocellular part of the PVN. We additionally hypothesised that the bonding protocols would result in dopaminergic reward-prediction errors which would drive a further reduction in activity of stress circuitry via the strengthening of the inhibitory pathway from the vmPFC to the CeA.

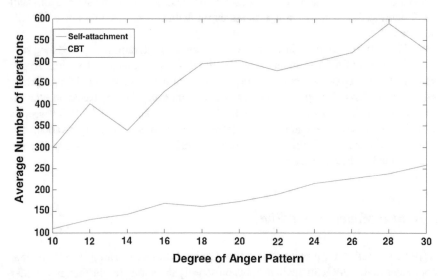

Fig. 6 *Blue curve*: The average number of iterations required so that, starting with the angry dominant pattern with different degrees, the Happy-Child strong pattern becomes dominant. *Red curve*: The average number required when only Mentalisation is used, i.e., without the self attachment protocols

Empathically-Motivated Bonding

The current focus of our work [101] is concerned with how the individual undergoing therapy might further increase motivation to apply these bonding protocols, by taking the perspective of the adult-self and attempting to enter into an empathic state with an inner-child who is conceptualised as being in distress. We build on a model by Numan [102, p. 278] which considers how empathy circuitry (involving the anterior insular, anterior midcingulate cortex and medial prefrontal cortex) might stimulate a mesolimbic-ventral pallidum pathway involved in caregiving behaviour. We additionally consider circuits involved in the perception of pain in self and others, and how sufficient self-other distinction might drive such caregiving behaviour while insufficient distinction might instead result in a distress state within the individual.

Conclusion

After reviewing basic attachment theory and the neural basis of various attachment types, we provided compelling evidence from the literature that insecure attachment is a significant risk factor in the development of mental illness, whereas secure attachment nurtures self-regulation of emotions, vastly reducing the risk for psychological disorders. The wide ranging impact of attachment theory on psychotherapy was also looked at. We then reviewed the work on attachment objects in higher primates, children and in two different ways, namely the corresponding and compensatory pathways, among religious believers. The objective in all these cases is to use the attachment object to earn or retain secure attachment and thereby regulate strong emotions. We showed that fMRI studies indicate an overarching common denominator for bond-making, which activates the reward system in the brain, whether in romantic or maternal love or in praying. Hence, bonding with an abstract attachment object or between two individuals have neural correlates.

Based on these background results and findings, we proposed a holistic approach to Computational Psychiatry by considering an individual as an adult-self, representing the more cognitive aspects, and an inner child, representing the more affective aspects of the individual. Self-attachment uses a self-administrable protocol to create a secure attachment object, represented as the adult-self, for the inner child within an individual. The adult-self is to become a "good enough" parent for the inner child who is to be freed from attachment insecurities and grow emotionally.

We presented several computational models for Self-attachment: (i) a game-theoretic model based on reinforcement learning, (ii) an energy based competitive neural network of needs, which is reinforced for optimal decision making using the Amygdala, OFC and DLPFC represented by RBM's, (iii) a neural model for bond-making in Self-attachment which uses a model for how emotional appraisals in the OFC can mediate between activity in neural circuits that drive stress and facilitative responses to social stimuli, and, (iv) a model for empathically-motivated bonding.

What general result can be deduced from this work for psychiatrists? In a few words, we need to consider attachment theory and early development as a central/starting point in psychiatry and psychotherapy. Self-attachment therapy is an attempt to provide a holistic and attachment-centric approach to psychotherapy.

As for computationalists, we can conclude that computational modelling of Self-attachment (and indeed attachment in general) is fertile ground. Such work can help us to understand and develop more fully the therapy going forwards.

Acknowledgements I would like to thank David Cittern for reading the article and providing me with various suggestions.

References

1. B. King-Casas, C. Sharp, L. Lomax-Bream, T. Lohrenz, P. Fonagy, and P. R. Montague. The rupture and repair of cooperation in borderline personality disorder. *Science*, 321(5890):806–810, 2008.
2. P. R. Montague, R. J. Dolan, K. J. Friston, and P. Dayan. Computational psychiatry. *Trends in Cognitive Sciences*, 16(5):306, 2012.
3. A. Edalat. Self-attachment: A new and integrative psychotherapy. Talk at the Institute of Psychiatry, London, 02-05-2013, 2013.
4. A. Edalat. Introduction to self-attachment and its neural basis. In *2015 International Joint Conference on Neural Networks, IJCNN 2015, Killarney, Ireland, July 12-17, 2015*, pages 1–8, 2015.
5. M. Mikulincer and P. R. Shaver. An attachment perspective on psychopathology. *World Psychiatry*, 11(1):11–15, 2012.
6. A. N. Schore. *Affect Dysregulation and Disorders of the Self*. W. W. Norton, 2003.
7. J. Bowlby. *Attachment: Volume One of the Attachment and Loss Trilogy*. Pimlico, second revised edition, 1997.
8. J. Bowlby. *The Making and Breaking of Affectional Bonds*. Routledge, 1989.
9. J. E. Young, J. S. Klosko, and M. E. Weishaar. *Schema Therapy: A Practitioner's Guide*. Guildford Press, 2006.
10. P. L. Granqvist, M. Mikulincer, and P. R. Shaver. Religion as attachment: normative processes and individual differences. *Pers Soc Psychol Rev*, 2010.
11. D. Winnicott. *Playing and Reality*. Basic Books, 1971.
12. H. F. Harlow. The nature of love. *American Psychologist*, pages 673–685, 1958.
13. W. W. Meissner. *Psychoanalysis and religious experience*. Yale University Press, 1986.
14. A. Bartels and S. Zeki. The neural basis of romantic love. *Neuroreport*, 11(17):3829–3834, 2000.
15. A. Bartels and S. Zeki. The neural correlates of maternal and romantic love. *NeuroImage*, 21:1155–1166, 2004.
16. U. Schjoedt, H. Stjoedkilde-Joergensen, A. W. Geertz, and A. Roepstorff. Rewarding prayers. *Neurosci Lett.*, 443(3):165–8, October 2008.
17. D. Purves, G. Augustine, D. Fitzpatrick, W. Hall, A. LaMantia, J. McNamara, and S. M. Williams, editors. *Neuroscience*. Sinauer Associate, third edition, 2004.
18. C. J. Falconer, M. Slater M, A. Rovira, J. A. King, P. Gilbert, and A. Antley. Embodying compassion: a virtual reality paradigm for overcoming excessive self-criticism. *PLoS ONE*, 9(11), 2014.
19. C. J. Falconer, A. Rovira, J. A. King, P. Gilbert, A. Antley, P. Fearon, N. Ralph, and C. R. Brewin M. Slater. Embodying self-compassion within virtual reality and its effects on patients with depression. *British Journal of Psychiatry*, 2 (1):74–80, 2016.

20. J. Cassidy and P. R. Shaver, editors. *Handbook of Attachment*. Guilford, 1999.
21. M. Main and J. Solomom. Procedures for identifying infants as disorganized/disoriented during the Ainsworth strange situation. In *Attachment in the Preschool Years: Theory, Research, and Intervention*, page 12160. University of Chicago Press, 1990.
22. L. A. Sroufe. Attachment and development: A prospective, longitudinal study from birth to adulthood. *Attachment & Human Development*, 7(4):49 367, 2005.
23. M. D. S. Ainsworth, M. C. Blehar, E. Waters, and S. N. Wall. *Patterns of Attachment: A Psychological Study of the Strange Situation*. Psychology Press, 2015.
24. M. H. Van Ijzendoorn and P. M. Kroonenberg. Cross-cultural patterns of attachment: A meta-analysis of the strange situation. *Child Development*, 1988.
25. M. H. Van Ijzendoorn and A. Sagi. Corss-cultural patterns of attachment: Universal and contextual dimensions. In J. Cassidy and P. R. Shaver, editors, *Handbook of Attachment*, pages 713–734. Guilford, 1999.
26. D. Wallin. *Attachment in Psychotherapy*. Guilford Press, 2007.
27. A. N. Schore. *The Science of the Art of Psychotherapy*. Norton, 2012.
28. R. N. Emde. Development terminable and interminable: II. recent psychoanalytic theory and therapeutic considerations. *The International Journal of Psychoanalysis*, 1988.
29. D. Winnicott. Transitional objects and transitional phenomena. *International Journal of Psycho-Analysis*, 34:89–97, 1953.
30. J. D. Ford. Neurobiological and developmental research. *Treating complex traumatic stress disorders: An evidence-based guide*, pages 31–58, 2009.
31. L. I. Perlovsky. *Neural dynamic logic of consciousness: the knowledge instinct*. Springer, 2007.
32. L. Cozolino. *The Neuroscience of Human Relationships*. W. W. Norton, 2006.
33. M. ME. Riem, M. J. Bakermans-Kranenburg, M. H. van IJzendoorn, D. Out, and S. ARB. Rombouts. Attachment in the brain: adult attachment representations predict amygdala and behavioral responses to infant crying. *Attachment & human development*, 14(6):533–551, 2012.
34. C. George, N. Kaplan, and M. Main. The adult attachment interview. Unpublished manuscript, University of California at Berkeley, 1985.
35. M. Mikulincer and P. R. Shaver. *Attachment in adulthood: structure, dynamics, and change*. Guilford, 2007.
36. P. Fonagy. *Attachment Theory and Psychoanalysis*. Other Press, 2001.
37. L. Cozolino. *The Neuroscience of Psychotherapy: Healing the Social Brain*. W. W. Norton, second edition, 2010.
38. J. H. Obegi and E. Berant, editors. *Attachment Theory and Research in Clinical Work with Adults*. Guilford Press, 2009.
39. J. D L. Pearsona1, D. A. Cohna, P. A. Cowana, and C. P. Cowan. Earned- and continuous-security in adult attachment: Relation to depressive symptomatology and parenting style. *Development and Psychopathology*, 6(2):359–373, 2008.
40. J. Bowlby. The nature of the child's s tie to his mother. *International Journal of Psycho-Analysis*, 39:350–373, 1958.
41. F. CP. Van der Horst, H. A. LeRoy, and R. Van der Veer. "when strangers meet": John Bowlby and Harry Harlow on attachment behavior. *Integrative Psychological and Behavioral Science*, 42(4):370–388, 2008.
42. L. Thims. *Human Chemistry*, volume two. Lulu.com, 2007.
43. C. J. Litt. Theories of transitional object attachment: An overview. *International Journal of Behavioral Development*, 9(3):383–399, 1986.
44. D. Winnicott and M. M. R. Khan. *The maturational processes and the facilitating environment: Studies in the theory of emotional development*. Hogarth Press London, 1965.
45. A. Storr. *Solitude*. Flamingo, 1988.
46. H. Kohut, editor. *How Does Analysis Cure?* University of Chicago Press, 1984.
47. J. Bowlby. The growth of independence in the young child. *Royal Society of Health Journal*, 76:587–591, 1956.

48. M. Main. Attachment theory: Eighteen points. In J. Cassidy and P. R. Shaver, editors, *Handbook of Attachment*, pages 845–887. Guilford, 1999.

49. L. Kirkpatrick. Attachment and religious represenations and behaviour. In J. Cassidy and P. R. Shaver, editors, *Handbook of Attachment*, pages 803–822. Guilford, 1999.

50. G. D. Kaufman. *The Theological Imagination: Constructing the Concept of God*. Westminster John Knox Press, 1981.

51. S. Freud. *The Future of an Illusion*. Martino Fine Books, 2011. First published in 1928.

52. L. A. Kirkpatrick. *Attachment, evolution, and the psychology of religion*. Guilford, 2005.

53. A. Birgegard and P. Granqvist. The correspondence between attachment to parents and God Three experiments using sub- liminal separation cues. *Personality and Social Psychology Bulletin*, 30:1122–1135, 2004.

54. C. D. Batson, P. Schoenrade, and W. L. Ventis. *Religion and the Individual: A Social-Psychological Perspective*. Oxford, 1993.

55. C. Ullman. Change of mind, change of heart: Some cognitive and emotional antecedents of religious conversion. *Journal of Personality and Social Psychology*, 42:183–192, 1982.

56. L. A. Kirkpatrick, D. J. Shillito, and S. L. Kellas. Loneliness, social support, and perceived relationships with God. *Journal of Social and Personal Relationships*, pages 13–22, 1999.

57. L. A. Kirkpatrick and P. R. Shaver. Attachment theory and religion: Childhood attachments, religious beliefs, and conver- sion. *Journal for the Scientific Study of Religion*, 29:315–334, 1990.

58. P. Granqvist, T. Ivarsson, A. G. Broberg, and B. Hagekull. Examining relations between attachment, religiosity, and new age spirituality using the adult attachment interview. *Developmental Psychology*, pages 590–601, 2007.

59. W. James. *The varieties of religious experience*, Harvard University Press, 1985.

60. B. Ghobary Bonab, M. Miner, and M. T. Proctor. Attachment to God in Islamic spirituality. *Journal of Muslim Mental Health*, 7(2), 2013.

61. D. Capps, editor. *Freud and Freudians on Religion: A Reader*. Yale University Press, 2001.

62. A. Aron, H. Fisher, D. J. Mashek, G. Strong, H. Li, and L. L. Brown. Reward, motivation, and emotion systems associated with early-stage intense romantic love. *Journal of neurophysiology*, 94(1):327–337, 2005.

63. U. Schjoedt. It's a brain puzzle, 2012. http://www.theeuropean-magazine.com/649-schj-dt-uffe/650-the-neuroscience-of-prayer. Retrieved on 01-12-2016.

64. I. Stewart and V. Joines. *TA today: A new introduction to transactional analysis*. Lifespace Pub., 1987.

65. J. G. Allen, P. Fonagy, and A. W. Bateman. *Mentalizing in Clinical Practice*. American psychiatric Publishing, 2008.

66. J. S. Abramowitz, B. J. Deacon, and S. PH. Whiteside. *Exposure therapy for anxiety: Principles and practice*. Guilford Press, 2012.

67. P. Gilbert. *Compassion: Conceptualisations, research and use in psychotherapy*. Routledge, 2004.

68. T. Field, M. Hernandez-Reif, M. Diego, S. Schanberg, and C. Kuhn. Cortisol decreases and serotonin and dopamine increase following massage therapy. *International Journal of Neuroscience*, 115(10):1397–1413, 2005.

69. K. J. Jeffries, J. B. Fritz, and A. R. Braun. Words in melody: an H215o pet study of brain activation during singing and speaking. *Neuroreport*, 14(5):749–754, 2003.

70. B. Kleber, N. Birbaumer, R. Veit, T. Trevorrow, and M. Lotze. Overt and imagined singing of an Italian aria. *Neuroimage*, 36(3):889–900, 2007.

71. C. Q. Murcia, S. Bongard, and G. Kreutz. Emotional and neurohumoral responses to dancing tango argentino the effects of music and partner. *Music and Medicine*, 1(1):14–21, 2009.

72. K. R. Ginsburg. The importance of play in promoting healthy child development and maintaining strong parent-child bonds. *Pediatrics*, 119(1):182–191, 2007.

73. D. Bergen. The role of pretend play in children's cognitive development. *Early Childhood Research & Practice*, 4(1), 2002.

74. R. J. Corsini. *Role Playing in Psychotherapy*. AldineTransaction, 2011.

75. V. N. Salimpoor, M. Benovoy, K. Larcher, and A. Dagher andR. J. Zatorre. Anatomically distinct dopamine release during anticipation and experience of peak emotion to music. *Nature neuroscience*, 14(2):257–262, 2011.

76. J. P. Capitanio. Behavioral pathology. In G. Mitchell and J. Erwin, editors, *Comparative Primate Biology*, volume Volume 2A: Behavior, Conservation, and Ecology, pages 411–454. Alan R. L iss, 1986.

77. D. Fudenberg and J. Tirole. *Game Theory*. MIT Press, 1991.

78. L. Buono, R. Chaua, G. Lewis, N. Madras, M. Pugh, L. Rossi, and T. Wi-telski. Mathematical models of mother/child attachment. Problem proposed by L. Atkinson, J. Hunter and B. Lancee at the Fields-MITACS Industrial Problem Solving Workshop August, 2006.

79. D. Cittern and A. Edalat. Reinforcement learning for Nash equilibrium generation. In *Proceedings of Autonomous Agents and Multiagent Systems (AAMAS)*, 2015.

80. F. Wörgötter and B. Porr. Temporal sequence learning, prediction, and control: a review of different models and their relation to biological mechanisms. *Neural Computation*, 17(2):245–319, 2005.

81. R. S. Sutton and A. G. Barto. *Reinforcement Learning: An Introduction*. MIT Press, 1998.

82. M. R. Roesch, D. J. Calu, and G. Schoenbaum. Dopamine neurons encode the better option in rats deciding between differently delayed or sized rewards. *Nature neuroscience*, 10(12):1615–1624, 2007.

83. D. Cittern and A. Edalat. Reinforcement learning for nash equilibrium generation. In *Proceedings of Autonomous Agents and Multiagent Systems (AAMAS)*, 2015. Extended Abstract.

84. A. M. Aleksandrowicz and D. Levine. Neural dynamics of psychotherapy: what modeling might tell us about us. *Neural Netw*, 18(5-6):639–45, 2005.

85. C. R. Cloninger. A new conceptual paradigm from genetics and psychobiology for the science of mental health. *Aust N Z J Psychiatry.*, 32(2):174–86, 1999.

86. D. Levine. Angels, devils, and censors in the brain. *ComPlexus*, 2(1):35–59, 2005.

87. R. M. Galatzer Levy. Good vibration s: Analytic process as coupled oscillations. *The International Journal of Psychoanalysis*, 90(5):983 – 1007, 2009.

88. R. S. Wedemann, R. Donangelo, and L. A. de Carvalh. Generalized memory associativity in a network model for the neuroses. *Chaos: An Interdisciplinary Journal of Nonlinear Science*, 19(1), 2009.

89. J. J. Hopfield. Neural networks and physical systems with emergent collective computational abilities. *Proceedings of the National Academy of Science, USA*, 79:2554–2558, 1982.

90. T. Lewis, F. Amini, and R. Richard. *A General Theory of Love*. Vintage, 2000.

91. T. S. Smith, G. T. Stevens, and S. Caldwell. The familiar and the strange: Hopfield network models for prototype-entrained attachment-mediated neurophysiology. In Thomas S. (Ed) Franks, David D. (Ed); Smith, editor, *Mind, brain, and society: Toward a neurosociology of emotion*, pages 213–245. Elsevier Science/JAI Press, 1999.

92. A. Edalat. Capacity of strong attractor patterns to model behavioural and cognitive prototypes. In C.J.C. Burges, L. Bottou, M. Welling, Z. Ghahramani, and K.Q. Weinberger, editors, *Advances in Neural Information Processing Systems 26*. ACM, 2013.

93. A. Edalat and F. Mancinelli. Strong attractors of Hopfield neural networks to model attachment types and behavioural patterns. In *IJCNN 2013 Conference Proceedings*. IEEE, August 2013.

94. R. A. Wise. Dopamine, learning and motivation. *Nature Reviews Neuroscience*, 5:483–494, 2004.

95. D. Levine. Brain pathways for cognitive-emotional decision making in the human animal. *Neural Networks*, 22:286–293, 2009.

96. A. Edalat and Z. Lin. A neural model of mentalization/mindfulness based psychotherapy. In *Neural Networks (IJCNN), 2014 International Joint Conference on*, pages 2743–2751. IEEE, 2014.

97. A. Fischer and C. Igel. Training restricted boltzmann machines: an introduction. *Pattern Recognition*, 47(1):25–39, 2014.

98. R. Salakhutdinov, A. Mnih, and G. Hinton. Restricted Boltzmann machines for collaborative filtering. In *Proceedings of the 24th international conference on Machine learning*, pages 791–798, 2007.

99. D. Cittern and A. Edalat. Towards a neural model of bonding in self-attachment. In *Neural Networks (IJCNN), 2015 International Joint Conference on*, pages 1–8. IEEE, 2015.

100. D. Levine. Neural networks of human nature and nurture. *Avances en Psicología Latinoamericana, junio, año/vol. 26, número 001 Universidad del Rosario Bogotá, Colombia*, 2008.

101. D. Cittern and A. Edalat. A neural model of empathic states in attachment-based psychotherapy. pre-print, 2016.

102. M. Numan. *Neurobiology of social behavior: toward an understanding of the prosocial and antisocial brain*. Academic Press, 2014.

A Comparison of Mathematical Models of Mood in Bipolar Disorder

Amy L. Cochran, André Schultz, Melvin G. McInnis and Daniel B. Forger

Introduction

Bipolar disorder (BP) is a chronic condition characterized by severe and pathological periods, or episodes, of mania and depression. Over 2.4 % of the worldwide population are diagnosed with BP [1], yet little is known about its causes and/or pathology [2]. Studies have provided new qualitative insights as well as vast amounts of data on mood previously unavailable. These serve as an appeal for new modeling frameworks in BP. Mathematical models can provide insight into the longitudinal course of mood in BP and may suggest underlying biophysical processes or mechanisms not previously known. The challenge is determining the right model for a given application.

The Scientific Problem

Mathematical models are commonly used in medicine to explain how a certain biological system is involved in a certain disease's pathology. However, bipolar disorder (like many psychiatric disorders) has a complicated pathology, spanning multiple

A.L. Cochran (✉) · M.G. McInnis · D.B. Forger
University of Michigan, Ann Arbor, MI 48105, USA
e-mail: cochraam@umich.edu

M.G. McInnis
e-mail: mmcinnis@med.umich.edu

D.B. Forger
e-mail: forger@umich.edu

A. Schultz
Rice University, Houston, TX 77251, USA
e-mail: as86@rice.edu

© Springer International Publishing AG 2017
P. Érdi et al. (eds.), *Computational Neurology and Psychiatry*,
Springer Series in Bio-/Neuroinformatics 6,
DOI 10.1007/978-3-319-49959-8_11

315

spatial levels from genetics to cognition [2]. For example, BP is a highly heritable disorder at over 85 % heritability [3], but BP has shared susceptibility with other affective disorders, such as schizophrenia [4]. Moreover, a number of genes have been implicated, but are involved in different molecular pathways [5]. BP also affects most neural systems, including monoamine systems (e.g. dopamine and serotonin), suprachiasmatic nucleus, and hypothalamic-pituitary-adrenal axis [2]. Even the most prescribed drug in BP, Lithium, has multiple molecular targets [6]. Thus, there is no obvious candidate biological system to model.

The most common approach is to model *mood*, the defining feature of BP. Mood has many features of note. It varies dramatically over time, often unpredictably, and can sustain at extreme levels. It also ranges in severity, duration, and polarity. While these features are helpful starting points for modeling mood, they are not sufficiently specific, so additional assumptions must first be made. Existing models of mood, however, use assumptions that differ widely. Some assume that mood swings periodically between mania and depression [7–11]. Others use multistability, arguing that mood in BP tends to distinct mood states, such as mania and depression, and that it is this tendency that sustains mood in BP at extreme levels [10, 12, 13].

Further complicating the modeling dilemma is that mood is erratic and imprecisely defined (Fig. 1). Consequently, most assumptions can only be supported anecdotally. Biomarkers that pervade a large body of mathematical modeling in medicine do not exist in BP and are replaced by subjective evaluation of a patient. Moreover, current classification systems place a patient's mood into disjunctive categories, such as mania and depression, based on duration and type of symptoms; symptoms relate not only to mood, but also to energy, cognition, and sleep, among others. These categories reflect a common understanding among clinicians, but do not have an empirical basis or any underpinning in a biophysical process. In fact, symptoms insufficient to constitute an episode of mania or depression are associated with disability in BP [14]. Mood can also be measured with specific, longitudinal scores

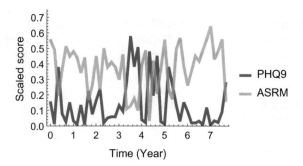

Fig. 1 Bi-monthly samples of mood scores from a bipolar patient in the Prechter Longitudinal Study of Bipolar at the University of Michigan. Manic symptoms are measured using the Altman Self-Rating Mania (ASRM) scale; depressive symptoms are measured using the Patient Health Questionnaire for Depression (PHQ9). Scores were scaled to lie between zero and one

from clinical surveys. Although subjective, clinical surveys quantify mood over a larger range than current classification methods.

Here, we compare recent approaches to modeling mood in BP and discuss how to distinguish between modeling assumptions using only time courses of mood. In what follows, we provide a selective overview of existing models of mood, with an emphasis on random models and/or differential equation models (c.f. [15] for an alternative approach based on iterative maps). For each model we describe, we simulate mood and examine the results from the simulation.

Computational Methods

In this section, we discuss different models of mood in BP and describe their relevance to BP. In addition to existing models, we present two new models and motivate their introduction. For each model we consider, we add random noise to the model formulation (if not already included) in order to capture mood's erratic behavior.

Biological Rhythms Models

BPs characteristic mood swings are often described as cyclic or periodic.[1] A periodic assumption ensures that mood episodes recur and change in polarity between depression and mania. Some modelers assume mood is periodic only during periods of rapid-cycling (defined as a certain number of mood episodes in a year).

A periodic assumption hints at an internal timekeeping mechanism for mood, hypothesized to arise from mitochondrial fluctuations [16] or interactions with circadian rhythms [17]. Certainly, there is a strong connection between circadian and sleep rhythms and mood, although the direction of influence is unclear. BP patients experience diurnal variation in mood and sleep disturbances that depend on the mood state; changes in sleep habits can be therapeutic; and clock genes are associated with BP and with animal models of BP [2, 17]. Even Lithium, the standard treatment for BP, influences circadian pathways by delaying the circadian phase via glycogen synthase kinase 3 (GSK3) [18]. Thus, periodicity is a reasonable assumption given these observations.

For a biological rhythms model, a natural starting point is a van der Pol oscillator. It was first proposed for mood in untreated BP patients by Daugherty et al. in [8] and later extended in [7, 11]. In this model, mood X_t at time t is governed by a

---

[1]Cyclicity/periodicity should not be confused with "cycling" or "rapid-cycling". These latter terms are used in BP to define events in which a patient experiences a certain number of episodes within a certain period of time. Cycling in BP does not require that mood is periodic.

second-order differential equation, which upon applying a suitable transformation (and adding noise to the original formulation) can be written as:

$$dX_t = \left(Y_t + \alpha X_t + \frac{\beta}{3} X_t^3 \right) dt + \sigma dW \tag{1a}$$

$$dY_t = -\omega^2 X_t dt \tag{1b}$$

where W is a Wiener process. When $\sigma = 0$, $\alpha > 0$ and $\beta < 0$, the system has a unique stable limit cycle [8] leading to periodic oscillations in mood. We refer the reader to [8, 11] for extensions of Eq. (1), but do not consider them here.

Understanding that multiple oscillators may be driving mood in BP, [7] built upon the work in [8] by assuming two oscillators contribute to mood X_t:

$$dX_t = c_0 dt + c_1 dY_d^{(1)} + c_2 dY_t^{(2)}, \tag{2}$$

where $Y_t^{(i)}$ ($i = 1, 2$) are the two oscillators. The dynamics of $Y_t^{(i)}$ are based on similar equations to (1) with $\alpha = 1$ and $\beta = -1$ and allowing for coupling between the oscillators:

$$dY_t^{(i)} = \left(Z_i^{(i)} + \left(Y^{(i)} - \frac{(Y^{(i)})^3}{3} \right) + \frac{\eta_i}{1 + \exp(\theta_i Y^{(3-i)})} \right) dt + v(Y^{(i)}) dV_t^{(i)} \tag{3a}$$

$$dZ_t^{(i)} = -\omega_i^2 (y_i + \zeta_i) dt + v(z_i) dW_t^{(i)} \tag{3b}$$

where $V_t^{(i)}, W_t^{(i)}$ are independent Wiener processes ($i = 1, 2$). Ignoring noise and coupling, each oscillator can exhibit a stable limit cycle when $|\zeta_i| < 1$ [7]. We will assume $v(x) = \sigma$ for simplicity.

In each model above, mood is a one-dimensional variable. Goldbeter in [10] proposes another biological rhythms model describing mood as two-dimensional and without van der Pol oscillators. This model uses one variable M to represent manic symptoms and another variable D to represent depressive symptoms. He begins by first formulating a model to capture bistability, where mood is either manic (high M, low D) or depressive (low M, high D) depending on initial conditions and parameters. Bistability is modeled by assuming the variable M is driven by interaction with D and negative self feedback, and similarly, the variable D is driven by interaction with M and negative self feedback. The model is then modified to allow for cycling between mania and depression. This is accomplished by introducing two variables F_m and F_d that delay the interaction between M and D. The result is a system of four differential equations:

$$\frac{dM}{dt} = V_m \frac{K_{i1}^2}{K_{i1}^2 + D^2} \frac{F_d}{K_1 + F_d} - k_m \frac{M}{K_2 + M} \tag{4a}$$

$$\frac{dD}{dt} = V_d \frac{K_{i2}^2}{K_{i2}^2 + M^2} \frac{F_m}{K_3 + F_m} - k_m \frac{D}{K_4 + D} \tag{4b}$$

$$\frac{dF_m}{dt} = k_{c_1} M - k_{c_2} F_m \tag{4c}$$

$$\frac{dF_d}{dt} = k_{c_3} D - k_{c_4} F_d \tag{4d}$$

Their system leads to oscillations in M and D. Another biological rhythms model in BP can be found in [9], focusing on biochemical pathways for mood.

Behavioral Activation System Model

One theory is that BP is caused by dysregulation of the behavioral activation system [19]. The behavioral activation system regulates how goal-relevant cues impact positive affect and approach behavior [20]. Examples of goal-relevant cues include the presence of a goal and its expectations, as well as frustration in not meeting a goal. This theory hypothesizes that the behavior activation system is more sensitive to goal-directed cues in BP patients and that these cues may even trigger a mood episode through activation or inhibition of the behavioral activation system. It explains why BP is associated with certain symptoms ranging from neurobiological, e.g. why the dopamine system (a reward processing system) is dysregulated, to behavioral, e.g. why BP patients are often goal-driven.

The behavioral activation system and its connection to BP were described with a mathematical model in [13]. The model uses one variable X_t to represent mood (i.e. the level of behavioral activation) and another variable R_t to represent goal-directed cues. Behavioral activation X_t is described as a stochastic process with positive and negative feedback loops, linear drift, and goal-directed cues, along with random noise. The goal-directed cues are driven by random events and a drift back to zero. This is all captured with a jump-diffusion model of the form:

$$dX_t = \left(F_p(X_t) + F_n(X_t) + \left(b - k_2 X_t \right) + k_3 R_t \right) dt + \sigma dW_t \tag{5a}$$

$$dR_t = -k_3 R_t dt + \sigma dV_t + dJ_t \tag{5b}$$

where W_t, V_t are Wiener processes; J_t is a jump stochastic process; and

$$F_p(x) = \frac{k_1 x^n}{K^n + x^n} \quad \text{and} \quad F_n(x) = m_h \left(\frac{1}{2} - \frac{1}{1 + \exp(s(x_h - x))} \right)$$

are the positive and negative feedback terms respectively.

Unlike biological rhythms models, the behavioral activation system model does not suppose mood swings are periodic. Rather, swings towards mania or depression are captured with a system that has multi-stability in the absence of noise and goal-directed cues. In dynamical systems, stability describes equilibrium solutions or periodic orbits to which nearby solutions stay close indefinitely. Multistability refers to the existence of multiple stable equilibrium solutions and/or periodic orbits. It has been argued that mania and depression are stable states in the absence of external influences and that these stable states emerge in BP patients due to a parameter bifurcation [10, 12, 13]. For example, the behavioral activation model (ignoring noise and goal-directed cues) can have one, two, or three stable points depending on the exponent n. Their model could capture healthy controls with one stable state for euthymia; recurrent depression with two stable states for depression and euthymia; and BP with three stable points for depression, euthymia, and hypomania/mania [13].

Discrete-Time Random Models

Discrete-time random models describe mood as a stochastic process that changes at discrete time points. They benefit from simple analysis when compared to their continuous-time counterparts and are also sufficient for describing most data, because measurements of mood tend to be scores on clinical mood surveys captured at regular time intervals. Their goal is to model the data directly as opposed to trying to model some biophysical processes that underpins mood.

Discrete-time random models can be categorized based on whether mood is also discrete or continuous. The former type of models are discrete-time Markov chains (DTMC) with finite states. A DTMC formulation has been used in two theses [21, 22], where mood at time step t, which we denote by X_t, takes one of n values in $\{1, \dots, n\}$. The finite values represent n mood states, e.g. depression, hypomania, and mania. Mood X_t has a certain probability of transitioning to another state. A probability transition $n \times n$ matrix P is specified such that P_{ij} defines the probability that $X_{t+1} = j$ conditional on $X_t = i$. The probability distribution of states π_t at time step t is a $1 \times n$ vector of probabilities then satisfies the recurrence relation

$$\pi_{t+1} = \pi_t P, \qquad\qquad t = 0, 1, 2, \dots \qquad (6)$$

These models are Markovian: the future state depends on past states only through the current state.

Models where mood is continuous are based on autoregressive models, standard models for time-series analysis. They were introduced for mood in [23], where mood at time step t, which again we denote by X_t, satisfies the recurrence relation

$$X_{t+1} = f(X_t, X_{t-1}, \dots, X_{t-d+1}) + \varepsilon_t, \qquad\qquad t = d - 1, d, d + 1, \dots \qquad (7)$$

for some function f, random variable ε_t, and an integer d that specifies the time lag in mood states that influence a future mood state. In most applications, ε_t is Gaussian. However, a patient's mood scores is better described as a Gamma random variable, since the scores are non-negative and have a distribution that is positively-skewed. Thus, [23] define ε_t such that X_{t+1} is a Gamma-distributed variable with mean $\mu_t := f(X_t, X_{t-1}, \ldots, X_{t-d+1})$ and variance μ_t/α for some parameter α. That is, the probability density function of X_{t+1} is

$$c_t x^{\alpha-1} \exp\left(-\alpha X_t/\mu_t\right) \tag{8}$$

where c_t is a normalizing constant that depends only on α and μ_t. Comparing choices for f and d, [7] conclude that assuming a model with both $d = 1$ and linear f provides a good fit to a patient's time series among the models they considered. This contrasts the original choice of a nonlinear model in [23], but agrees with [24] which finds that nonlinear models for f do not significantly improve forecasting over linear models. Thus, we will examine the linear model

$$f(X_t, X_{t-1}, \ldots, X_{t-d+1}) := aX_t + b. \tag{9}$$

Two-Dimensional Models

Mania and depression are often considered to be the two "poles" (as in bi-polar) of mood. For this reason, mood is often modeled as one-dimensional with high values for mania and low values for depression. Indeed, most models discussed here describe mood as one-dimensional, with exceptions being the biological rhythms model of Goldbeter [10] and DTMCs, where mood is categorical. However, a one-dimensional model cannot capture known features of BP: mixed states (states in which manic and depressive symptoms are present) and direct transitions between mania and depression without going through euthymia. Therefore, it may be necessary to model mood as two-dimensional.

We introduce two new models that model mood with two variables, a manic variable M_t and a depressive variable D_t. Our first model has multistability in the absence of noise. Each variable (marginalizing the other variable out) satisfies an stochastic differential equation known as a "double-well" model, which has been used to model particles that randomly fluctuate between two energy wells. To this end, we assume that each variable satisfies

$$dM_t = -a_M M_t (M_t - l_M)(M_t - h_M)dt + \sigma_M dW_t \tag{10a}$$
$$dD_t = -a_D D_t (D_t - l_D)(D_t - h_D)dt + \sigma_D dV_t \tag{10b}$$

where W_t, V_t are Wiener processes with correlation ρ. Constants $a_M, a_D, h_M, h_D, \sigma_M, \sigma_D$ are positive, and l_M and l_D are negative. In the absence of noise and provided $M_t, D_t \neq 0$, the manic variable M_t and the depressive variable D_t would each tend to either a low value (respectively, l_M and l_D) or a high value (h_M and h_D). Together, the patient would be inclined towards one of four stable states representing euthymia ($M_t = l_M; D_t = l_D$), mania ($M_t = h_M; D_t = l_D$), depression ($M_t = l_M; D_t = h_D$), or a mixed state ($M_t = h_M$; $D_t = h_D$). Correlated Wiener processes are used so that manic and depressive symptoms can either fluctuate together or in opposite directions.

Finally, we consider a two-dimensional model without multistability. The idea that severe symptoms can arise without multistability or biological rhythms was put forth in [25] for major depression. They describe depressive symptoms as the absolute value of an Ornstein-Uhlenbeck process and suppose that depression arises when depressive symptoms are above a certain threshold. To extend this model to BP, we suppose that mood is described by two variables M_t and D_t and suppose that $M_t := |X_t|$ and $D_t := |Y_t|$, where (X_t, Y_t) describe a two-dimensional Ornstein-Uhlenbeck process:

$$dX_t = -a_M X_t dt + \sigma_M dW_t \tag{11a}$$
$$dY_t = -a_D Y_t dt + \sigma_D dV_t \tag{11b}$$

where W_t, V_t are Wiener processes with correlation ρ and $a_M, a_D, \sigma_M, \sigma_D$ are positive constants.

Simulation

Each model was simulated in Matlab. All code has been made publicly available on https://sites.google.com/site/amylouisecochran/code. Parameters were chosen to agree with the associated reference when available. Parameter choices are documented in the companion Matlab code. Models were simulated for a total of 400 "simulated" years, and the first 200 years were removed, using the guideline for stochastic simulations of having a warm-up period equal to the period used for analysis. With the exception of the DTMC model where mood is categorical, daily samples of mood were normalized to have a mean of zero and sample variance of one, and the daily normalized mood scores were used for analysis. For these models, mania and depression were based on the quantiles of mood to be consistent across models. When mood was one-dimensional, we defined mania as values greater than 82nd quantile, depression as values less than 18th quantile, and euthymia as the remaining 64 % of the mood scores. For comparison, [26] found that BP patients spend a median of 62 % of their time asymptomatic. When mood was two-dimensional, we defined mania as a manic variable greater than its 0.8 quantile and depressive variable less than its 0.8 quantile, depression as a depressive variable greater than its 0.8 quantile and manic variable less than its 0.8 quantile, mixed state as manic and depressive variables greater than their respective 0.8 quantiles, and euthymia as manic and depressive variables less than their 0.8 quantiles. The value of 0.8 was chosen so that if mania and depression were independent, euthymia would constitute 64 % of the mood scores to agree with the one-dimensional models.

Results

The models of mood in BP discussed in section "Computational Methods" differ widely in their assumptions and equations (Table 1). So naturally, there is a question of what assumptions are supported by data. For each individual with BP, the only data available on mood is usually a set of mood scores from clinical surveys. Clinical surveys measure mood by scoring the presence and duration of specific symptoms (e.g. irritability, feeling of loss) which are associated with mania and depression. Symptoms may include other symptoms, such as poor sleep quality, that are present during episodes of mania and depression. In most observation studies of BP, clinical surveys are usually administered on a regular interval (daily, weekly, bi-monthly, or yearly) over an observation period that can last anywhere from several weeks to several years. Thus, it is reasonable to assume that each patient may have completed at most 100 surveys. The challenge is validating assumptions given such limited data.

Time Courses of Mood

For an example of limited data, we sampled the simulated mood data at weekly intervals and plotted these samples over a period of two years in Fig. 2. It is difficult to validate modeling assumptions by visually inspecting the time course of this simulated data alone. Consider, for instance, the assumption that mood is periodic. Mood is periodic in the biological rhythms model of Daugherty [8], but the oscillation period (over one year) is too long to clearly identify periodicity in the figure. Mood is also periodic in the biological rhythms model of Bonsall [7], but in this case the oscillation period (around eight weeks) is too short and noise is too strong to identify periodicity. At the same time, mood in this model appears similar to mood in the autoregressive model, which has no inherent periodicity. Thus, mood may even be mistaken as periodic in the autoregressive model, since it too constantly fluctuates between high and low values. Periodicity can be identified visually in the biological

Table 1 Models of mood in BP

Models	Features	Eq.	Ref.
Biological rhythms (Daugherty)	1D, continuous-time, periodic	(1)	[8]
Biological rhythms (Bonsall)	1D, continuous-time, periodic	(2, 3a and 3b)	[7]
Biological rhythms (Goldbeter)	2D, continuous-time, periodic	(4)	[10]
Behavioral activation	1D, continuous-time, multistable	(5a, 5b and 6)	[13]
Discrete-time Markov chain	Categorical, discrete-time	(6)	[22]
Autoregressive	1D, discrete-time	(7–9)	[23]
2D Well	2D, continuous-time, multistable	(10)	n/a
Modified 2D Ornstein-Uhlenbeck	2D, continuous-time	(11)	n/a

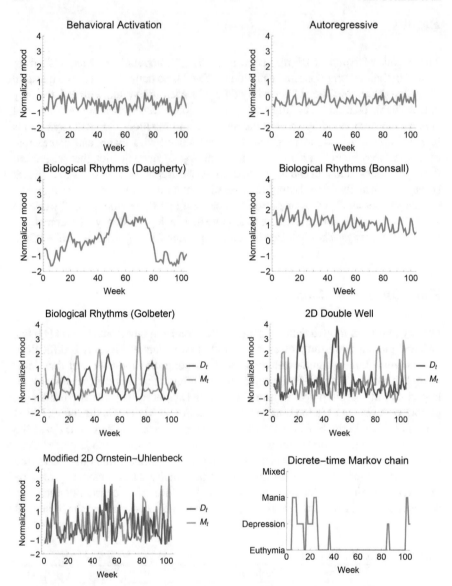

Fig. 2 Weekly samples of mood over 2 years from simulations of various models

rhythms model of Goldbeter [10], but here oscillation period is adequate (around 18 weeks) and noise is sufficiently weak to observe periodicity. In sum, it is very easy to mistake periodicity for random noise (and vice versa) when visually inspecting time courses of limited data.

Similar to periodicity, it is also difficult to surmise whether mood has multiple stable points from its time course. Both the 2D double well model and the behavioral activation model have multiple stable points in the absence of noise. In the 2D double well model, the sudden shift towards high values of a depressive variable (M_t) suggests the possibility of multistability. However, the 2D Ornstein-Uhlenbeck model appears to also have sudden shifts toward higher values of the depressive variable (D_t), and so, multistability remains ambiguous from the time courses.

Last, we mention that when measurements of mood are not sufficiently frequent, discrete-time models can recover similar patterns of mood as continuous-time models. As illustrated in Fig. 2, the discrete-time autoregressive models looks similar to the biological rhythms model of Bonsall [7] and the behavioral activation model.

Chronicity

If the time course of mood cannot verify modeling assumptions, then the data has to be examined in another way. One option would be use the data to evaluate symptom chronicity, that is, the relative amount of time that a patient spends with symptoms at a particular level. Clinicians and researchers already use chronicity to describe individuals with BP.

When mood is continuous, one way to measure chronicity is to estimate a probability density function (pdf) from the samples of mood. This approach ignores when mood is sampled and hence, correlation among samples; rather, it assumes (incorrectly) that the samples of mood are identically and independently distributed. The pdf approximates the relative probability that a sample of mood will be at a certain level, which in turn estimates the relative amount of time that mood is at a given level, i.e. symptom chronicity. Daily samples of mood were recovered from simulation of each model to estimate a pdf for mood (Fig. 3). The pdfs were estimated in Matlab using the ksdensity function, which uses a kernel-based approach.

For the one model where mood is categorical (i.e. the discrete-time Markov chain mood), mood has a probability mass function rather than a pdf. We calculated the probability mass function, which is shown in Fig. 3. However, parameters can always be adjusted to recover any probability mass function defined for the four states, and so, we cannot evaluate this model further based on chronicity alone.

For the remaining models, we can use qualitative differences in pdfs to start to distinguish between models, as Fig. 3 illustrates. For example, the pdfs for the biological rhythms models of Daugherty [8] and Bonsall [23] are less peaked and more symmetric than the others. These two biological rhythms models actually have similar shapes, even though mood oscillates more rapidly in the model of Bonsall [23], so symptom chronicity cannot differentiate between all models. Their symmetric pdf

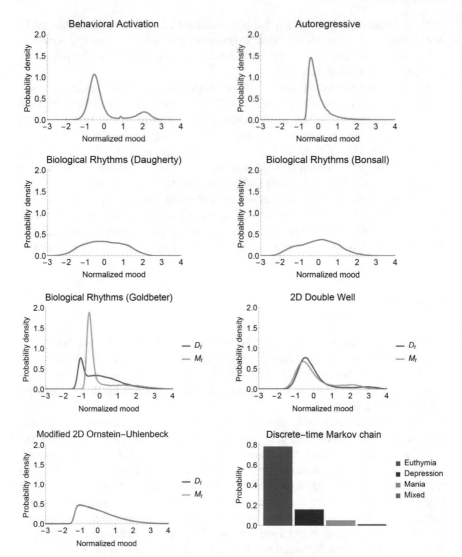

Fig. 3 Estimated probability density functions of mood for various models

suggests that an individual with BP spends relatively more time around their average mood level than any other mood level.

The remaining pdfs are positively skewed. A positively-skewed distribution would suggest that an individual with BP spends relatively more time around mood values that are lower than its average mood level. Bonsall et al. [23] find that this feature accurately reflects distribution of mood scores on a particular depressive survey known as the quick inventory for depressive symptoms (QIDS) [27]. Specifically, they found these mood scores are approximately Gamma-distributed, a positively-skewed distribution. Consequently, they proposed a Gamma transition density function for their autoregressive model. The modified 2D Ornstein-Uhlenbeck model also has a known pdf which is positively-skewed: the pdf of the absolute value of a bivariate Gaussian random variable. We note that positive random variables frequently have skewed distributions and both the autoregressive model and the modified 2D Ornstein-Uhlenbeck model mood as a non-negative variable.

The probability density functions also vary in their number of modes or peaks, a feature that is useful for detecting multistability. A mode is a value that is relatively more likely than nearby values. We expect that asymptotically stable points in the absence of noise would become modes once noise was added. This can be clearly seen in the behavioral activation model which has three modes in its respective pdf. Parameters were chosen so that the model had three asymptotically stable points in the absence of noise. Of the three stable points, the middle stable point has a much smaller region of mood values that would limit towards the stable point in the absence of noise. Similarly, the middle mode is the smallest of the three peaks of the density function.

Comparably, the 2D double well model also has stable points in the absence of noise and provided neither the manic or depressive variable is zero. Specifically, it has four stable points that occur exactly when the manic variable is one of two values (l_M and h_M) and the depressive variable is one of two values (l_D and h_D). There are two modes in the pdf for the manic variable M_t and two modes in the pdf for the depressive variable D_t. In each pdf, the mode with a smaller value corresponds to a higher peak. Model parameters were chosen to reflect this asymmetry, $|l_D| > |h_D|$ and $|l_M| > |h_M|$, so that patients spend more time in euthymia than in any other mood state.

Although both models with multistability have modes in their pdf that correspond with each stable point, modes are not sufficient for multistability. The biological rhythms model of Goldbeter [10], for instance, has two modes in the pdf for the manic variable and two modes in the pdf the depressive variable. Like a sinusoidal wave, mood oscillates in this model such that the rate of change in mood is slower near its two turning points than at other values. Hence, mood spends more time near its turning points relative to other values.

There are simple statistics to quantify differences between pdfs. One such statistic is skewness, which measures asymmetry. Another is kurtosis, which measures peakedness in a probability density function (Table 2). Skewness and kurtosis are the third and fourth moments, respectively, of a normalized random variable. For a baseline, a standard normal distribution has a skewness of zero and a kurtosis of three. Hartigan's dip test statistic provides yet another statistic; it measures deviation from unimodality [28]. Calculating skewness, we find that the density functions of the biological rhythms models of Daugherty [8] and Bonsall [23] have values of skewness closest to zero, whereas the density functions of the other models have a more positive skewness value. This agrees with our earlier observations. Also in agreement, the density functions of the biological rhythms models of Daugherty [8] and Bonsall [23] have a kurtosis less than three, so their pdfs are less peaked than a

Table 2 Statistics of probability density functions of mood for various models

Model	Skewness[a]	Kurtosis[a]	Dip value[a]
Biological rhythms (Daugherty)	0.015	2.25	0.0008
Biological rhythms (Bonsall)	0.045	2.48	0.0008
Biological rhythms (Goldbeter)	1.82, 0.86	5.54, 3.22	0.0012, 0.0032
Behavioral activation	1.38	3.60	0.0245
Autoregressive	23.94	1024.92	0.0006
2D double well	1.19, 1.87	3.64, 6.35	0.0023, 0.0034
Modified 2D Ornstein-Uhlenbeck	0.98, 0.94	3.82, 3.65	0.0008, 0.0007

[a]For the 2D models, values are listed for the manic variable and then the depressive variable

normal distribution. The remaining pdfs have a kurtosis larger than three and hence, are more peaked than a normal distribution.

Using the statistics, we also find that the behavioral activation model has the greatest deviation from unimodality with the largest dip value (Table 2), followed by the 2D Double well and the biological rhythms model of Goldbeter [10]. This result confirms the observation that the pdf for these models deviate more from unimodality than the others. Hartigan's dip test statistic is accompanied by a test for significance, from which a P-value can be calculated. The null distribution for the test statistic is generated by performing Hartigan's dip test on samples of a uniform distribution. To demonstrate this approach, we calculated the P-value for the dip test statistic calculated from a year of weekly samples of mood simulated with the biological rhythms model of Bonsall [7]. With a P-value of 0.35, unimodality cannot be rejected for this model's pdf at a significant level of 0.05.

Lastly, it is important to test whether mood scores are generated from a specific pdf whenever possible. To illustrate, consider the modified 2D Ornstein-Uhlenbeck model, which has a known pdf. One could test the null hypothesis that the data came from this pdf, which can help justify whether this model is appropriate for describing the data. In general, nonparametric tests in [29] can be performed to determine whether to reject a model based on goodness-of-fit tests between given distributions and modified survey scores. This approach consists of fitting a parametric density/distribution to the data using maximum likelihood estimation and then calculating a test statistic that measures the difference between a non-parametric and the parametric approximation to the relevant density/distribution. Non-parametric approximations can be based on kernel approximations which can be computed using Matlab's ksdensity function. After these two steps are completed, a P-value is calculated for the test statistic by approximating the null distribution via Monte Carlo simulation. For comparison, we tested whether we could reject the hypothesis that a year of weekly mood data from the biological rhythms models of Daugherty [8] or the autoregressive model [23] came from a normal distribution. With a P-value less than 0.0001, we can reject the assumption that the mood samples from the autoregressive model are normally-distributed. With a P-value of 0.52, we cannot reject the assumption that the mood samples from the biological rhythms model of Daugherty [8] are normally-distributed, which agrees with our earlier observation that this model yields a pdf that appear to be similar to a normal distribution.

Survival Functions

In addition to examining symptom chronicity, model differences can also be evaluated using survival functions. If T is the (random) time to a specific event, then the survival function for T is the function that maps each time t to the probability that the time-to-event T occurs after time t, which we denote by

$$S(t) = P(T > t). \tag{12}$$

In the context of BP, survival functions are calculated for the time until a patient in euthymia enters a mood episode, as well as the time until a patient in a mood episode switches to a new mood state. Survival functions are estimated from longitudinal measurements of mood and are frequently studied for BP. Examples of survival curves estimated for a population of BP patients can be found in [30]. Since population survival curves can be distinctly different than patient-specific survival curves and we are interested in knowing whether mood at the patient-level has a certain survival curve, we do not discuss the population survival curves further.

We calculated survival functions of mood states from daily samples of mood. In the discrete-time Markov chain model, mood states are the four mood categories: euthymia, depression, mania, and mixed. When mood is one-dimensional, three mood states were defined: euthymia, depression, and mania (see section "Computational Methods" for details). When mood is two-dimensional, four mood states were defined: euthymia, depression, mania, and mixed. Survival curves were estimated using Matlab's ksdensity function for each mood state after conditioning on the mood state lasting at least seven days. We conditioned this way since current criteria for diagnosing a mood episodes often requires that symptoms have lasted a week.

Survival curves were concave up for most of the models, with the exception of certain survival curves for the biological rhythms models of [10, 23] (Fig. 4). In fact, the survival curves of euthymia and mania in the biological rhythms model of [10] are sufficiently concave down that they are relatively flat in regions. Hence, this model predicts that it is unlikely that patients would leave euthymia or mania in a week given that symptoms have already lasted for a week. We also note that the survival curves of the biological rhythms models of [10, 23] cross one another. These types of qualitative observations can be used to distinguish between models.

With the survival functions, models can be differentiated based on whether one mood state lasts longer on average relative to another mood state. Longitudinal stud-

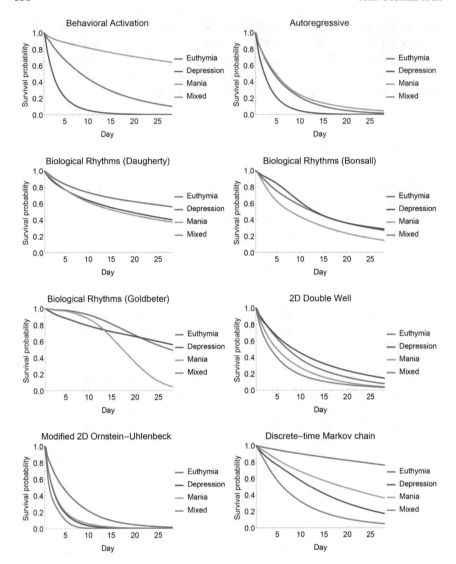

Fig. 4 Estimated survival functions for mood states after the mood episode has lasted 7 days

ies suggest that on average and for most BP patients, depression lasts longer than mania [30]. This is the case for the discrete-time Markov chain model where we see that the survival curve for euthymia is always above the other curves. This implies that the probability that euthymia lasts a given time is greater than the probability that a particular mood episode lasts the same time.

For the discrete-time Markov chain model, parameters for the discrete-time Markov chain can be specified to capture any ordering on the average duration of mood states. The survival function of this type of model also has a similar shape to survival curves of other models. Therefore, discrete-time Markov chains may be desirable when the main goal is to model duration of mood states. For other models, it may prove challenging to alter parameters to capture the relative duration of one state compared to other mood states. Consider, for instance, the behavioral activation model. Here, the probability that mania lasts a certain amount of time is always greater than the probability that another mood state lasts that same amount of time. It is not immediately clear how a parameter change could lead to (on average) euthymia lasting longer than depression and mania. In contrast, parameters l_M and l_D in the 2D double-well could be decreased to increase the time spent in euthymia. Therefore, while the behavioral activation model may describe the underlying biophysical process more accurately, the 2D double-well model may more useful when capturing duration of mood states.

In addition to survival functions, we can also examine the cumulative hazard function $C(t)$ which relates to the survival function:

$$C(t) = -\log(S(t)). \tag{13}$$

Unlike the survival function, the cumulative hazard function is less intuitive. When the cumulative hazard function is continuous, its derivative, called the hazard rate, measures the rate of an event conditioned on surviving up to a certain time. The cumulative hazard function also has a simple expression for exponential and Weibull distributions, which are common models for survival analysis. It is linear $C(t) = -\alpha t$ when the time-to-event has an exponential distribution with rate α and a power function $C(t) = (t/\alpha)^\beta$ when the time-to-event has a Weibull distribution with scale parameter α and shape parameter β. Therefore, it is easier to observe deviations from an exponential or a Weibull distribution by looking at the cumulative hazard function (Fig. 5).

As a final note on the survival functions, we point out that two-dimensional models may not lead to any mixed states. We have already noted that one-dimensional models do not allow for mixed states. The biological rhythms model of Goldbeter [10], for instance, is a two-dimensional model, but did not have mixed states that lasted over seven days throughout the simulation of this model. Accordingly, in Fig. 4, we observe that the biological rhythms model of Goldbeter [10] does not have a survival curve corresponding to mixed states.

Spectral Density

When examining time courses, we found that periodicity can be mistaken for random noise (and vice versa). This issue can be overcome using the power spectral density (assuming mood is stationary). Specifically, a periodic signal has a peak in its power spectral density, where the peak corresponds to the frequency of its cycles. For mood, the spectral density can be estimated from its time course. The simplest

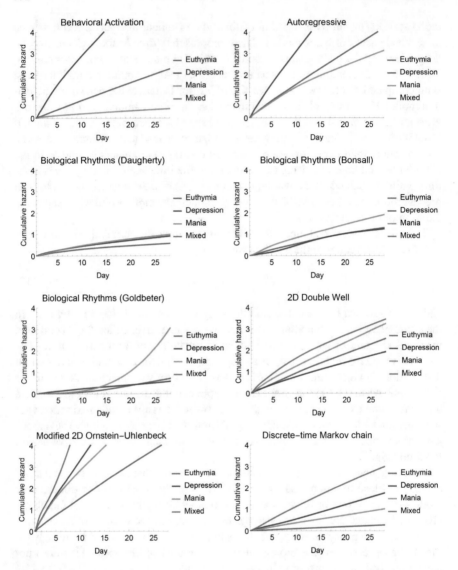

Fig. 5 Estimated cumulative hazard functions for mood states after the mood episode has lasted 7 days

way to estimate the spectral density is the periodogram (i.e. scaling the square of the discrete Fourier transform), but there are advantages to other techniques. One such technique is the multitaper method introduced by Thomson [31]. Thomson's multitaper method benefits from reduced bias and variance in the estimation of spectral density. It is also accompanied by a statistic test (Thomson's harmonic F test) to determine whether a particular frequency is statistically significant. This test is not available in Matlab, but is available in the multitaper package for R.

We applied Thomson's multitaper method to estimate spectral density functions for the simulated time courses for each model, with the exception of the discrete-time Markov chain model, since in that case, mood is categorical. The multitaper method was applied using Matlab's pmtm function. Figure 6 displays the spectral density functions for the various models.

The estimated spectral density functions can be used to distinguish between models with periodicity from models without periodicity. That is, each of the biological rhythms models has large peaks corresponding to frequencies at which they oscillate, whereas the other models do not have large peaks. The biological rhythms model of Daugherty [8] has the longest period with a sharp peak at 1 cycle per 489 days (0.75 cycles per year). The biological rhythms model of Bonsall [7] has the shortest period with two peaks, one at 1 cycle per 56 days (1 cycle per 7.9 weeks) and one at 1 cycle per 29 days (1 cycle per 4.2 weeks). Each peak captures periodicity of one of two (non-identical) oscillators that additively contributed to mood for this particular model. Lastly, the biological rhythms model of Goldbeter [10] has a peak at 1 cycle per 128 days (1 cycle per 18.3 weeks) for both the manic and depressive variable. The manic variable has a second peak around at 1 cycle per 67 days (1 cycle per 9.6 weeks). Except for initial conditions, parameters for the manic variable were identical to parameters for the depressive variable.

Manic and Depressive Variables

Of the models considered, only three (two of which were introduced here) model mood in BP as two-dimensional. The popularity of a one-dimensional model could stem from the common belief that mania and depression are the two "poles" in bipolar such that mania or depression can arise only in the absence of the other. Another possibility is that mood, an ambiguous term, refers to a narrow set of symptoms for which mood is one-dimensional by definition. In either case, the one-dimensional models are unable to describe mixed episodes, as well as direct transitions between mania and depression. The question is whether it is reasonable to ignore these events.

When the goal is to capture a broader set of symptoms relevant to BP, then two different surveys could be administered, one to measure severity of manic symptoms and another to measure the severity of depressive symptoms. Concurrent scores could be plotted together to determine whether a one-dimensional model describes the data. If a one-dimensional model is appropriate, we would expect that concurrent

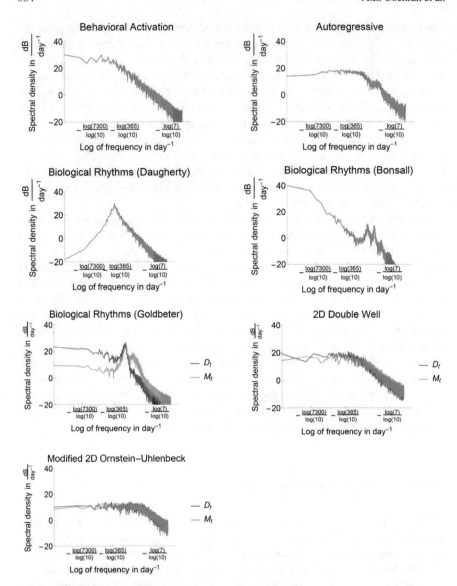

Fig. 6 Spectral density functions estimated using Thomson's multitaper method from the simulated time courses

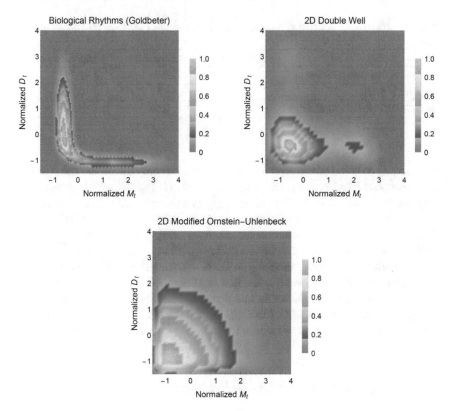

Fig. 7 Density plot of concurrent values of normalized manic (M_t) and depressive (D_t) variables

scores lie close to a one-dimensional line that is non-decreasing, so that an increase in a manic score is accompanied by a decrease in the depressive score.

In Fig. 7, we plot the density of concurrent values of mania and depressive variables for the two-dimensional models. These plots could be compared to a plot of concurrent manic and depressive survey scores. Mood from each model ranges in the prevalence of mixed states, i.e. how often manic and depressive variables are both at high levels. In the biological rhythms model of [10], for instance, mixed states never occur. Indeed, its plot appears similar to a plot needed to justify a one-dimensional model, where points lie close to a one-dimensional curve for which increases to the manic variable are accompanied by decreases in the depressive value. The biological rhythms model of Goldbeter [10] was designed so that mania and depression mutually-inhibit each other. Mixed states are more prevalent in the 2D double well, compared to the model of Goldbeter [10], and even more prevalent in the modified 2D Ornstein-Uhlenbeck model. For the latter, mania and depression appear to be entirely independent.

We can measure the degree to which mania and depression inhibit each other or cooperate using rank correlation. Correlation is more commonly measured using

Table 3 Measures of association between manic and depressive variables for two-dimensional models

Model	Pearson's ρ	Kendall's τ	Spearman's ρ
Biological Rhythms (Goldbeter)	−0.59	−0.54	−0.74
2D Double Well	−0.11	−0.11	−0.16
Modified 2D Ornstein-Uhlenbeck	0.06	0.02	0.04

Pearson's correlation coefficient, but this measure can only determine the degree to which two variables linearly correlated. As illustrated in Fig. 7, mania and depression may not be linearly-related, but still strongly inhibit each other. Rank correlations are alternative measures that depend only on the ranking of each variable; it has the advantage that it is invariant under nonlinear scaling. For the two-dimensional models, we measured rank correlation using two metrics, Spearman ρ and Kendall τ. Their values have been compared to Pearson correlation coefficient in Table 3. All three types of metrics were calculated in Matlab using the corr function. With a negative rank correlation, mania and depression inhibit each other more in the biological rhythms model of Goldbeter [10] than the other two models. Mania and depression are positively associated in the modified 2D Ornstein-Uhlenbeck.

Combining Inferences Across Individuals

So far, we have discussed how to determine whether a model is appropriate for an individual with BP. Data on an individual patient, however, may be too limited (in number of samples and/or frequency of samples) to provide enough statistical power to reject any particular model. To address this issue, we can combine statistical inferences across many patients to examine the collective evidence to support or oppose a hypothesis. The idea is to recover a P-value for each patient from each statistical test of interest and then aggregate the P-values into a single statistic. As described in [32], P-values can be combined across K patients into a single statistic of the form:

$$\phi = \sum_{k=1}^{K} \Psi^{-1}(P_k) \tag{14}$$

where P_k is the P-value for the kth patient and Ψ is a certain cumulative distribution function (e.g. a normal or uniform distribution).

The null distribution for the resulting statistic ϕ can be estimated through Monte Carlo simulation. The procedure consists of repeatedly sampling K independent random variables with distribution Ψ and calculating their sum $\phi^{(k)}$. Then, the samples $\phi^{(1)}, ..., \phi^{(k)}$ are then used to estimate the null distribution non-parametrically (which can be done with Matlab's ksdensity function). The cumulative distribution function Ψ can be chosen to place different emphasis between low, medium, and high P-values [32]. For example, a uniform distribution places more emphasis on high P-values than a normal distribution function.

Discussion

In this chapter, we evaluated existing models of mood in BP (Eqs. 1–9) and two new models we proposed here (Eqs. 10 and 11). Each model makes different assumptions about mood dynamics. Our objective was to differentiate between models using only time courses of mood. This addresses the question of how to identify an appropriate model of mood in BP when only longitudinal scores from clinical assessments are available. We focus on clinical assessments, because there are no validated alternatives to measuring mood. Albeit subjective, clinical assessments are ubiquitous to psychiatry, have been extensively studied, and can be administered by an experienced clinician. They also provide clinical standards for describing mood and for defining BP: through the presence, duration, and severity of a list of common symptoms associated with mania and depression.

Our results strongly encourage that models be compared using more than visual inspection of the time courses. Time courses for the models were sufficiently diverse and erratic to support any modeling assumption. Even a periodic assumption can be mistaken for a random model that has no inherent periodicity. This ambiguity calls into question whether previous assumptions about mood could have been proposed from visual inspection of time courses. These issues further stresses the need for empirical validation of models.

To address ambiguity in the time courses, we proposed different ways to analyze the time series. Below we highlight our conclusions:

Detecting Periodicity. Mood in BP is often believed to cycle, that is, to oscillate periodically [7–11]. Periodicity was best detected in a model using the spectral density function. Spectral density was estimated nonparametrically from the time courses of the simulated mood with a method known as Thomson's multitaper approach, which reduces estimation bias and variance over other popular methods, e.g. periodograms. From the estimated spectral density, we could clearly identify periodic models, as these models had sharp peaks in their spectral density. The sharp peak corresponded with the frequency or frequencies at which they oscillate. For example, we found two peaks in a model based on two oscillations.

A limitation of a spectral density approach is that it can only detect periodicities in a certain range. As such, we cannot determine if a period of oscillation lies outside this range. The frequencies that can be detected lie between half the sampling frequency and the fundamental frequency (i.e. one over the observation period). For example, if two years of mood were sampled weekly, then we could only detect if mood cycles at a frequency between one cycle every two years and one cycle every two weeks. Additionally, noise in the signal renders it difficult to determine whether a peak is significant. In this case, a Thomson's harmonic F test can be used to test whether a specific frequency is significant. Thomson's multitaper method assumes that the spectral density does not depend on time or varies slowly in time; extensions are discussed in [33] when these assumptions do not hold.

Detecting Multistability. Rather than a periodic assumption, it has been proposed that mania, depression, and euthymia can be considered stable states when ignoring external influences, such as random noise [10, 12, 13]. Our analysis indicates that models with multistability are best detected using the probability density function (pdf) of mood. The pdf of mood was estimated from the samples of mood using a non-parametric kernel approach. For the models with multistability, we found that each mode in the pdf corresponded with a point in the model that was stable in the absence of noise. Consequently, the pdf for these models was multimodal. We quantified deviation from unimodality (i.e. the degree to which a pdf is multimodal) using Hartigan's dip test statistic, which confirmed our observations.

In general, there are caveats to using modes to detect multistability. First, we note that our models with multistability had points that were not just stable, but asymptotically stable when ignoring noise, so our observations may not generalize for stable points that are not asymptotically stable. Second, the presence of multiple modes was not sufficient to conclude multistability. For one reason, we found that a model without multistability (a model with periodicity) can also have multiple modes. For another reason, a pdf estimated from samples of mood can be multimodal even when the samples are drawn from a unimodal distribution. To address this latter concern, Hartigan's dip test statistic is accompanied with test for significance, from which a P-value can be calculated. The null distribution for the test statistic is generated from samples of a uniform distribution. Intuitively, the null distribution would reflect the fact that an estimated pdf can be multimodal even when the actual pdf from which samples are drawn is not.

Deciding between a two-dimensional or one-dimensional model of mood. Even a simple question of whether to model mood as one-dimensional or two-dimensional remains unanswered. The answer may depend on what set of symptoms are used to defined mood, e.g. whether we are interested in a broad set of symptoms related to mania and depression or a narrow set of symptoms related to mood only. For BP, a one-dimensional model would not be able to capture mixed states or direct transitions between mania and depression.

If the model aims to capture mania and depression broadly, then survey scores can be gathered for both mania and depression separately and concurrent scores graphed in a 2D density plot. A one-dimensional model could be excluded on the basis that concurrent scores are not plotted closely to a one-dimensional curve where high values on the manic survey correspond with low values on the depressive survey and conversely, low values on the manic survey correspond with high values on the depressive survey. A one-dimensional model could also be excluded if mixed states are sufficiently prevalent.

For the two-dimensional models, we plotted concurrent values of the manic and depressive variable. We found that the prevalence of mixed states varies from model to model. On one extreme, one model had no mixed states and had plotted points that were relatively close to a one-dimensional curve. On the other, one model had many cases of mixed states, where the manic and depressive variables were simultaneously high. We proposed that rank correlation could measure the degree to which manic and depressive variables either cooperate or inhibit each other.

Qualitative differences in pdfs, survival functions, and cumulative hazard functions. We also found that the models could be differentiated based on qualitative differences in their pdfs, survival functions, and cumulative hazard functions, each of which can be estimated from times courses of mood. For example, certain pdfs were relatively more symmetric or more peaked than other distributions, observations that could be confirmed quantitatively by measuring kurtosis and skewness, respectively. We also found that certain models had survival functions for mood states that were concave down and crossed each other, whereas other models did not.

We recognize that changing parameters in a model will change the resulting pdfs, survival functions, and cumulative hazard functions. When available, parameters were matched to parameters in the original reference for each model. A thorough parametric study is needed to determine whether qualitative differences would remain under parameter changes. This type of analysis was beyond the scope of this chapter. However, we did find that certain models with the fixed set of parameters had more realistic properties, e.g. euthymia that lasts longer on average than depression, which lasts longer on average than mania.

Clinical implications. From a clinical standpoint, an accurate model of mood in BP at the patient-level may help to alleviate disease's burden through forecasting outcomes and subsequent clinical needs. Mood episodes, alone, are already debilitating. However, mood is often so erratic and unpredictable that uncertainty in one's future only adds to patient anxiety and stress [2]. The current classification system based in the Diagnostic and Statistical Manual 5 (reference for DSM 5 is: American Psychiatric Association, 2013) uses only a small portion of a patient's history of mood and typically transforms dimensional qualities into categories, thereby removing information. A proper modeling framework could be used to tailor models to an individual's historical illness course, thereby capturing more nuance to their illness description. Clinicians could then use this information to help manage the disease and provide personalized expectations for their future illness course.

At the same time, an accurate model of mood in BP may also help better understand the mechanisms that cause both BP and the occurrence of mood episodes. For example, knowledge that a periodic model is appropriate, and of its frequency of oscillation, would help to narrow a search for possible biological oscillations that could trigger mood changes. An accurate model also provides a theoretical model for clinicians to integrate their understanding of causal pathways in BP that span multiple levels (e.g. genes, molecular pathways, and neural systems).

Here, we examined direct (also referred to as, minimal or phenomenological) models of mood. As researchers narrow their focus on specific biophysical systems in BP, mathematicians can build models for these biophysical systems to help understand their connection to BP. Such a *bottom-up* approach was taken in [9], where they consider specific protein kinases and its possible role in producing periodic oscillation of mood in BP. Future research will create many more opportunities for modeling specific biophysical processes in BP.

References

1. Merikangas, K.R., Jin, R., He, J.P., Kessler, R.C., Lee, S., Sampson, N.A., Viana, M.C., Andrade, L.H., Hu, C., Karam, E.G., et al.: Prevalence and correlates of bipolar spectrum disorder in the world mental health survey initiative. Archives of general psychiatry **68**(3), 241–251 (2011)
2. Goodwin, F.K., Jamison, K.R.: Manic-depressive illness: bipolar disorders and recurrent depression. Oxford University Press (2007)
3. McGuffin, P., Rijsdijk, F., Andrew, M., Sham, P., Katz, R., Cardno, A.: The heritability of bipolar affective disorder and the genetic relationship to unipolar depression. Archives of General Psychiatry **60**(5), 497–502 (2003). doi:10.1001/archpsyc.60.5.497
4. Berrettini, W.: Evidence for shared susceptibility in bipolar disorder and schizophrenia **123**(1), 59–64 (2003)
5. Craddock, N., Sklar, P.: Genetics of bipolar disorder. The Lancet **381**(9878), 1654–1662 (2013)
6. Phiel, C.J., Klein, P.S.: Molecular targets of lithium action. Annual review of pharmacology and toxicology **41**(1), 789–813 (2001)
7. Bonsall, M.B., Geddes, J.R., Goodwin, G.M., Holmes, E.A.: Bipolar disorder dynamics: affective instabilities, relaxation oscillations and noise. Journal of The Royal Society Interface **12**(112), 20150,670 (2015)
8. Daugherty, D., Roque-Urrea, T., Urrea-Roque, J., Troyer, J., Wirkus, S., Porter, M.A.: Mathematical models of bipolar disorder. Communications in Nonlinear Science and Numerical Simulation **14**(7), 2897–2908 (2009)
9. Frank, T.: A limit cycle oscillator model for cycling mood variations of bipolar disorder patients derived from cellular biochemical reaction equations. Communications in Nonlinear Science and Numerical Simulation **18**(8), 2107–2119 (2013)
10. Goldbeter, A.: A model for the dynamics of bipolar disorders. Progress in biophysics and molecular biology **105**(1), 119–127 (2011)
11. Nana, L.: Bifurcation analysis of parametrically excited bipolar disorder model. Communications in Nonlinear Science and Numerical Simulation **14**(2), 351–360 (2009)
12. Bystritsky, A., Nierenberg, A., Feusner, J., Rabinovich, M.: Computational non-linear dynamical psychiatry: a new methodological paradigm for diagnosis and course of illness. Journal of psychiatric research **46**(4), 428–435 (2012)
13. Steinacher, A., Wright, K.A.: Relating the bipolar spectrum to dysregulation of behavioural activation: A perspective from dynamical modelling. PLoS ONE **8**(5), e63345 (2013). doi:10.1371/journal.pone.0063345
14. Bonnin, C., Sanchez-Moreno, J., Martinez-Aran, A., Solé, B., Reinares, M., Rosa, A., Goikolea, J., Benabarre, A., Ayuso-Mateos, J., Ferrer, M., et al.: Subthreshold symptoms in bipolar disorder: impact on neurocognition, quality of life and disability. Journal of affective disorders **136**(3), 650–659 (2012)
15. Hadaeghi, F., Golpayegani, M.R.H., Murray, G.: Towards a complex system understanding of bipolar disorder: A map based model of a complex winnerless competition. Journal of theoretical biology **376**, 74–81 (2015)
16. Kato, T.: The role of mitochondrial dysfunction in bipolar disorder. Drug News Perspect **19**(10), 597–602 (2006)
17. Murray, G., Harvey, A.: Circadian rhythms and sleep in bipolar disorder. Bipolar disorders **12**(5), 459–472 (2010)
18. Kaladchibachi, S.A., Doble, B., Anthopoulos, N., Woodgett, J.R., Manoukian, A.S.: Glycogen synthase kinase 3, circadian rhythms, and bipolar disorder: a molecular link in the therapeutic action of lithium. Journal of Circadian Rhythms **5**(1), 3 (2007)
19. Alloy, L.B., Abramson, L.Y.: The role of the behavioral approach system (bas) in bipolar spectrum disorders. Current Directions in Psychological Science **19**(3), 189–194 (2010)
20. Carver, C.S., White, T.L.: Behavioral inhibition, behavioral activation, and affective responses to impending reward and punishment: the bis/bas scales. Journal of personality and social psychology **67**(2), 319 (1994)

21. Fan, J.: On markov and hidden markov models with applications to trajectories. Ph.D. thesis, University of Pittsburgh (2015)
22. Lopez, A.: Markov models for longitudinal course of youth bipolar disorder. ProQuest (2008)
23. Bonsall, M.B., Wallace-Hadrill, S.M., Geddes, J.R., Goodwin, G.M., Holmes, E.A.: Nonlinear time-series approaches in characterizing mood stability and mood instability in bipolar disorder. Proceedings of the Royal Society of London B: Biological Sciences **279**(1730), 916–924 (2012)
24. Moore, P.J., Little, M.A., McSharry, P.E., Goodwin, G.M., Geddes, J.R.: Mood dynamics in bipolar disorder. International Journal of Bipolar Disorders **2**(1), 11 (2014)
25. van der Werf, S.Y., Kaptein, K.I., de Jonge, P., Spijker, J., de Graaf, R., Korf, J.: Major depressive episodes and random mood. Archives of general psychiatry **63**(5), 509–518 (2006)
26. LL, J., HS, A., PJ, S., et al: The long-term natural history of the weekly symptomatic status of bipolar i disorder. Archives of General Psychiatry **59**(6), 530–537 (2002). doi:10.1001/archpsyc.59.6.530
27. Trivedi, Madhukar H., et al. "The Inventory of Depressive Symptomatology, Clinician Rating (IDS-C) and Self-Report (IDS-SR), and the Quick Inventory of Depressive Symptomatology, Clinician Rating (QIDS-C) and Self-Report (QIDS-SR) in public sector patients with mood disorders: a psychometric evaluation." Psychological medicine 34.01 (2004): 73-82.
28. Hartigan, J.A., Hartigan, P.: The dip test of unimodality. The Annals of Statistics pp. 70–84 (1985)
29. Fan, J., et al.: A selective overview of nonparametric methods in financial econometrics. Statistical Science **20**(4), 317–337 (2005)
30. Solomon, D.A., Leon, A.C., Coryell, W.H., Endicott, J., Li, C., Fiedorowicz, J.G., Boyken, L., Keller, M.B.: Longitudinal course of bipolar i disorder: duration of mood episodes. Archives of general psychiatry **67**(4), 339–347 (2010)
31. Thomson, D.J.: Spectrum estimation and harmonic analysis. Proceedings of the IEEE **70**(9), 1055–1096 (1982)
32. Loughin, T.M.: A systematic comparison of methods for combining p-values from independent tests. Computational statistics & data analysis **47**(3), 467–485 (2004)
33. Thomson, D.J.: Multitaper analysis of nonstationary and nonlinear time series data. Nonlinear and nonstationary signal processing pp. 317–394 (2000)

Computational Neuroscience of Timing, Plasticity and Function in Cerebellum Microcircuits

Shyam Diwakar, Chaitanya Medini, Manjusha Nair,
Harilal Parasuram, Asha Vijayan and Bipin Nair

Introduction

Cerebellum is attributed to contain 50 % of brain's neurons although occupying only 10 % of its volume [1, 2]. Clinical implication of cerebellar functions were elucidated by Luciani in 1891 [3, 4] with his experimental and clinical observations that cerebellum excision in dogs and monkeys forming a triad of symptoms Atonia, Asthenia, Astasia. However, his work was related to acute results of cerebellar hemi-spherectomy in animals and hence missed many of the functions of the cerebellum including perception, action and cognition. During the first world war, Holmes [5] observed voluntary tremors and dyskinesia now categorized as ataxia. Camillo Golgi and Ramon y Cajal's histological analysis of brain tissues added on to the observations of cerebellum in motor function. Cerebellar roles in visceral functions, coordinating emotions has been proposed [6, 7]. Several studies have led to newer insights of cerebellar roles in perception, action and cognition based on studies on patients with cerebellar agenesis [8]. Although complete absence of the cerebellum are known to cause mild to moderate motor deficiency, dysarthria and ataxia, primary cerebellar agenesis has been shown to affect patients with mild mental retardation and cerebellar ataxia [9].

Neuronal networks in the cerebellum include feedforward and feedback connections contributing to specific functional roles related to timing [10, 11] and movement control [12–15] in addition to what was originally hypothesized due to orthogonal projections of parallel fibers on Purkinje neurons [16, 17]. Starting with

S. Diwakar (✉) · C. Medini · M. Nair · H. Parasuram · A. Vijayan · B. Nair
Amrita School of Biotechnology, Amrita Vishwa Vidyapeetham (Amrita University),
Amritapuri, Clappana P.O, Kollam 690525, Kerala, India
e-mail: shyam@amrita.edu

M. Nair
Amrita School of Engineering, Amrita Vishwa Vidyapeetham (Amrita University),
Amritapuri, Clappana P.O, Kollam 690525, Kerala, India

© Springer International Publishing AG 2017
P. Érdi et al. (eds.), *Computational Neurology and Psychiatry*,
Springer Series in Bio-/Neuroinformatics 6,
DOI 10.1007/978-3-319-49959-8_12

Camillo Golgi and Ramon y Cajal's studies on the organization and structure of the cerebellum [18] and the study of Eccles [19] on inhibitory synapses in the circuit, cerebellar information processing had gained interest. Structure of the cerebellum is known to have three different distinct layers i.e. granular, Purkinje and molecular layers. Optimal weight distributions of connections determine pattern recognition properties of neural networks [20]. Cerebellum granule neuron- Golgi neuron connections play a role as a feedback network that has been known to synchronize network activity exhibiting rhythmic behaviour [21–23].

Cerebellar networks have been known to play a crucial role in timing of motor actions more than planning/initiating movements [15, 24, 25]. Computationally feedback networks as well as feedforward circuits have been studied in the cerebellar microcircuits, where the latter results in faster information processing while the former would synchronize the circuit inhibiting the amount of information to be processed. Cerebellar network primarily receives input via pontine nuclei (PN) where several major tracts of information including those from upper cortical areas, brainstem, thalamus converge. Sub-thalamic nucleus (STN) of basal ganglia circuit known to form connection with PN supporting the role of cerebellum in parkinsonian disease condition (PD). Output of the cerebellar cortex was directed to the cortical areas via relay through thalamus where the basal ganglia has inhibitory connections to thalamus. Striatum in basal ganglia circuit known to be influenced by the disynaptic connection from deep cerebellar nuclei (DCN) of cerebellum via thalamus. Several hypotheses such as vestibulo-ocular reflex (VOR) where gain modulation and classical eye-blink conditioning reflexes were tested in cerebellar networks.

Additionally, modular architecture allowed cerebellar microzones have been known to perform common computational operations. Timing phenomena observed in cerebellar granular layer play a very important role in passage of time representation (POT), learning or adaptation to movements, modulation of information transfer to Purkinje cells (activation of granule cell subsets with respect to time). Some of the knockout [26, 27] and lesion studies showed that granular layer disruption leads to abnormal functioning of the cerebellar mossy fiber-granule cell relay, affects the learning-dependent timing of conditioning eyelid responses, loss of rapid spike processing in the cerebellum (Ataxia) which indirectly has effect on plasticity of parallel fiber (pf)-Purkinje cell (PC) synapses.

Cerebellum has also been known to act as a pattern classifier which resembles a perceptron [16] (see Fig. 1). Granular layer, the input stage of cerebellum is hypothesized in sparse signal encoding [28, 29] based on patterns and specialized in generalization and learning interference. Inspired from cerebellum's sparseness and pattern abstraction properties, computational models have been developed [12, 30] to classify real world data based on its architecture. Real-time control via spiking neural networks [31] and look-up methods allow kinematic design of robotic articulators [32]. Several encoding schemes (Gaussian convolution, filter based and look-up table) were proposed [33] to translate the real world data such as robotic arm datasets (motor angles or 3D coordinates) into known spike trains. Spike trains matching granule cell firing in vivo and in vitro allowed better encoding [34, 35] and pattern separation using standard machine learning methods.

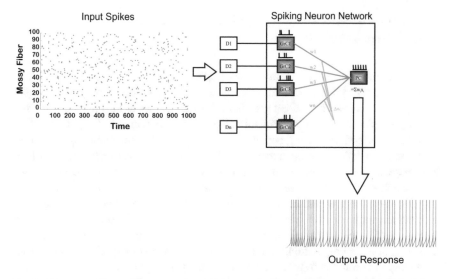

Fig. 1 Perceptron like spiking network with input spikes and output response. Many pattern capturing circuits like the cerebellum Purkinje layer act as spatial pattern capturing abstractions

Cerebellum was referred to as "head ganglion of the proprioceptive system" by Charles Sherrington depending on its anatomical position above the spinal cord [36]. Several studies [14, 24, 25, 37–40] have shown that cerebellum implicates both forward and inverse models to complete the learning of motor tasks. Forward model helps predict the next state from the present situation while inverse model uses mapping of environment to perform motor movement to reach target destination [15, 41, 42]. These kind of models help fine tune the motor movements so as to achieve desired target destination. Learning in these models performed via supervised or through error feedback mechanism which facilitates learning.

In this chapter, we focus on reconstructing cerebellar neuronal and circuit properties and an inference model for abstracting cerebellum pattern abstractions based on detailed and simple spiking models. It has been assumed that the randomization of connectivity and release probability of the synapses are the main sources of response variability in the granule neuron. The contribution of mossy fiber excitatory and Golgi cell inhibitory inputs on granule neuron firing was analyzed. We investigated the contribution of spike frequency and spike correlations in the information transmitted by the neurons. Applying communication analogs from electrical communication systems, we observed that the increase in spike frequency increased the information transmitted by the neurons while spike correlations had a negative impact on the declarative aspects of information. The behavior persisted during the induced plasticity states of the neurons meaning that spike correlations are not contributing much to enhance the learning and memory capabilities of neurons. We also employed an abstraction of the model for spike encoding of real world data and for pattern recognition in neurorobotics.

Background—On Cerebellum Granule Layer and Network Activity

The cerebellum receives and processes signals from various sensory modalities [43–45]. Sensory information fan into the cerebellum through mossy fibers. Spike trains patterns were processed at granular layer then passes to Purkinje neurons then to deep cerebellar nuclei for consolidation of motor memory [46]. Cerebellum granular layer activity during somatosensory function can be studied by recording the population activity during experimental conditions [47–49]. Granular layer local field potentials (LFP) and population activity have been estimated to be majorly contributed by the ionic currents generated in the densely packed granule neurons [50–52]. A direct assessment of population activity and underlying neuronal mechanism requires both single neuron based studies and correlation of neural dynamics [53–55]. Detailed network reconstruction as in [56, 57] from experimental data and computational modeling of population activity aids to understand how single neuronal activity [53] are represented in population code.

Granular layer population activity during network function have been studied by recording the LFP from Crus II circuits [49, 58, 59]. LFPs are population signals, recorded as low frequency components (<300 Hz) generated from complex spatio-temporal interaction of current sources during the network activity [60]. In this chapter, we employed models of granular layer circuit based on the functional and anatomical details [56] and the population behavior or ensemble activity was studied by modeling information.

A toolbox called LFPsim was developed to mathematically reconstruct local field potentials and a detailed mechanistic reconstruction from ionic currents are described elsewhere [61].

Methods

Spiking Models

Spiking models replicate neurophysiological firing properties with reliable accuracy in spike timing [57, 62, 63]. Several spiking models have been proposed to explain the complex neuronal firing behaviours observed in different cortical neurons. Using such models large-scale network dynamics employ a relatively lower computational overhead [64]. Spiking neuron models are chosen as a trade-off between the computational load and the complexity of the model, but with a (± 4 ms) tolerance window for their firing properties. These kind of models were known to maintain biological plausibility of Hodgkin-Huxley type dynamics while maintaining lower computational cost.

Spiking models include Hodgkin and Huxley (HH), leaky integrate and fire (LIF), quadratic integrate and fire (QIF) and Izhikevich, adaptive exponential integrate and fire (AdEx) [65]. Izhikevich models were widely used to study several neural networks which has the capability to reproduce known cortical firing behaviours [66]. Model complexity was decreased and also the computational load was reduced in our simulations as we moved from HH-type models to simple spiking models—employing two differential equations, one governing the membrane potential and the other governing the adaptation dynamics of the model. AdEx models have been shown to have reasonable accuracy with respect to spike timing when compared to conductance-based HH model as well as to real pyramidal neurons [67]. AdEx known to have coupled differential equations regulating membrane potential (V) and adaptation current (w).

$$C\frac{dV}{dt} = -g_L(V - E_L) + g_L * \Delta_T * e^{\left(\frac{V - V_T}{\Delta_T}\right)} + I - w \tag{1}$$

$$\tau_w \frac{dw}{dt} = a(V - E_L) - w \tag{2}$$

In the model, C is the membrane capacitance, g_L represents leak conductance, E_L denotes resting potential, Δ_T represents slope factor and V_T denotes threshold potential and 'a' represents the relevance of sub-threshold adaptation. Different neuronal firing patterns were reproduced by changing the values of the parameters [67].

Cerebellar firing patterns were reproduced and spiking network with validated population behaviour was reconstructed using AdEx model [68]. Primarily, there are 5 neuron types in cerebellar network: granule, Golgi, Purkinje, deep cerebellar nuclei (DCN) and inferior olive (IO). Stellate and basket cells function as inhibitory interneurons inhibiting Purkinje cell dendritic tree. Most of the cells are GABAergic (inhibitory) in nature while granule and deep cerebellar nuclei (DCN) are glutamatergic (excitatory) in nature. Spiking neurons are known to reproduce spike timing as well as spiking properties without reproducing ion channel dynamics and spatial geometry. In vitro and in vivo dynamics as well as electroresponsiveness properties of biophysical models were reproduced using AdEx models (see Fig. 2). To fit firing behaviour, we used particle swarm optimization (PSO) algorithm (Eq. 3) for comparing number of spikes and spike timing in each time window. N_{coin} is the number of coincidences, N_{exp} and N_{model} refer to number of spikes in experimental and model spike trains respectively and r_{exp} refer to average firing rate in experimental train [69].

$$\Gamma = \left(\frac{2}{1 - 2\delta r_{exp}}\right) \left(\frac{N_{coin} - 2\delta N_{exp} r_{exp}}{N_{exp} + N_{model}}\right) \tag{3}$$

This similarity measure compared the spike trains from the experimental data or biophysical models with the model data with a tolerance window of ±4 ms. Synaptic dynamics in the model was reproduced for α-amino-3-hydroxy-5-methyl-4-isoxazolepropionic acid (AMPA) receptor, N-methyl-D-aspartate

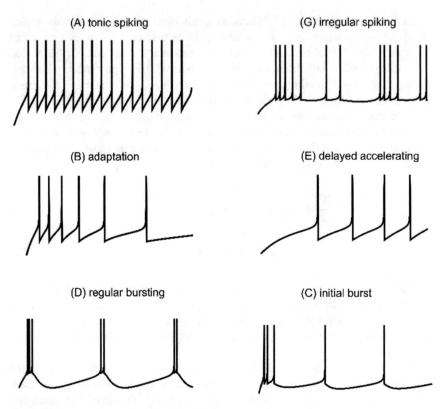

Fig. 2 Firing behaviours reproduced by AdEx model. Simulations were carried by injecting current and the parametric values taken from [67]

(NMDA) receptor and gamma aminobutyric acid (GABA) receptors. Synaptic equations were used to model the in vitro and in vivo behaviour of granule neurons in response to mossy fiber inputs. AMPA dynamics relate to fast spiking while NMDA relates to summing up of EPSPs and contributing to post synaptic potential (PSP) which is a slow current. GABA contributes negative polarization to the total PSP and reduces the excitatory effect. 'Mg' block in NMDA equation explains the gating dynamics of this channel. In our network, excitatory synapses (AMPA, NMDA) included mossy fiber (MF) connections to granule and MF to Golgi cells and granule to Purkinje cells and climbing fiber inputs to Purkinje cells. The other connections were inhibitory. Equations 4–10 show synaptic models for our simpler spiking network models.

$$g_{NMDA} = b_{Mg}(v_v) * (R_{on} + R_{off}) * g_{max} * f_{NMDA} \tag{4}$$

$$i_{NMDA} = g_{NMDA} * (v_v - 0) \tag{5}$$

$$b_{Mg} = \frac{1}{1 + e^{(0.062* - v_v)} * \left(\frac{C_{Mg}}{3.57}\right)} \tag{6}$$

$$g_{GABA} = B - A \tag{7}$$

$$B = -\frac{B}{\tau_2} \tag{8}$$

$$A = -\frac{A}{\tau_1} \tag{9}$$

$$i_{GABA} = g_{GABA} * (v_v + 70) \tag{10}$$

Representing synaptic dynamics, R_{on} refers to receptor on state and R_{off} refers to receptor off state. The variable g_{max} refers to maximal conductance and f_{NMDA} refers to scaling factor. For modulating magnesium block dynamics, b_{Mg} was used, which regulates gating of NMDA channels. Here, v_v refers to membrane potential of the spiking neuron model. Time constants τ_B and τ_A are used for modulating GABA dynamics by changing the behaviour of A and B variables over time. Conductance of different receptors (NMDA, GABA) was represented with g_{NMDA} and g_{GABA}. Synaptic currents for the computed conductance was represented with i_{NMDA} and i_{GABA}.

Cerebellar Network

Cerebellum granular layer forms the input stage which receives primary afferents from mossy fibers (MF). MF also provides excitatory input to Golgi cells and the inhibition from Golgi cells to granule cells arrives with a time delay of 4 ms (see Fig. 3). The granule cell axon extends into parallel fibers which provide excitatory input to Purkinje cells. The Purkinje cell (PC) also receives excitatory input from inferior olivary (IO) which carries proprioception information. PC serves as sole output of cerebellar cortex which inhibits deep cerebellar nuclei (DCN). Biophysical and simple spiking cerebellum networks were built as an attempt to understand the information processing, where the former was used to validate hypotheses such as coincidence detection, time-window and center-surround organization while the latter was used to test the reliability of spiking models to emulate similar behaviour as that of biophysical models so as to reduce computational load.

Using such network models, a previous study [57] had assessed the role of feed-forward inhibition as well as burst-burst transmission in the granule cells [70, 71]. The input was modelled as single spike for in vitro simulations while in vivo simulations modelled burst like input (see Fig. 4). For simulating intrinsic

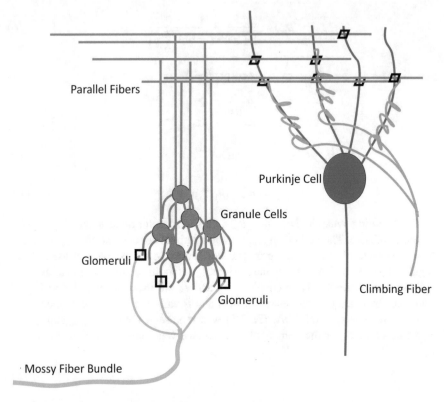

Fig. 3 Cartoon representation of cerebellum network architecture. 5 types of neurons are illustrated with their connectivity dynamics

excitability in biophysical networks, sodium channel on-off gating characteristics were modified to simulate similar behaviour. A limitation with simple spiking models was inability to intrinsically model plasticity and selective pharmacological effects in population code unlike with detailed biophysical models. Golgi cells play a critical role in modulating the information flow of granule cells through feed-forward and feed-back loops [72–74].

Modeling Cerebellar Granular Layer Evoked LFP

Modeling somatosensory pathway which activate Crus-IIa of cerebellum is crucial for understanding granule neuron computation in motor learning and other function. Cerebellar granular layer evoked local field potential (LFP) is characterized by trigeminal (T) and Cortical (C) waves; T wave corresponds to trigeminal afferents and C wave from cerebral cortex and pontine nuclei [49]. Extracellular electrodes

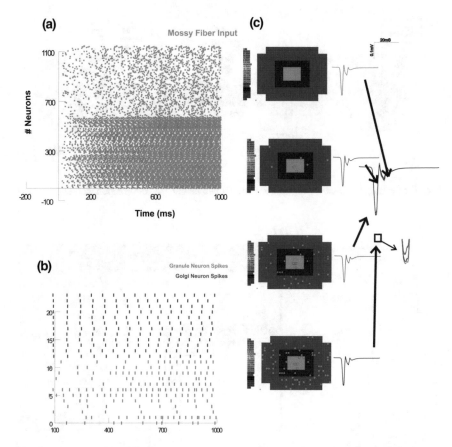

Fig. 4 Cerebellum Network Firing Properties (adapted from [57]). **a** Raster plot of the mossy fibers input to the cerebellum granular layer. **b** Raster plot of granule and Golgi cells firing for MF inputs. **c** Selective disabling of NMDA receptors reconstructing diseased (ataxia-like) condition

record field potential as an average response from the neurons within the visibility of the electrode (Figs. 5 and 6).

Cerebellar evoked LFP was modelled using point source approximation and line source approximation techniques to understand circuit function. Detailed biophysical model of rat Crus-IIa cerebellar granular layer was built to study central cerebellar function and dysfunction. The network model consisted of 730 multi-compartmental granule cells [75], 2 Golgi cell [76], 40 Mossy fibers (MF) and ~8,500 synapses to occupy 35 μm cubic slice of cerebellar cortex. The convergence and divergence ratios used to build the network model was adapted from earlier studies [77]. The LFP electrode was modeled based on point source approximation techniques to read signals from a population of neurons in the vicinity of the electrode see (Fig. 7a).

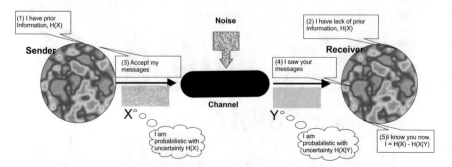

Fig. 5 Modeling information transmitted between two neurons via information theoretic approach

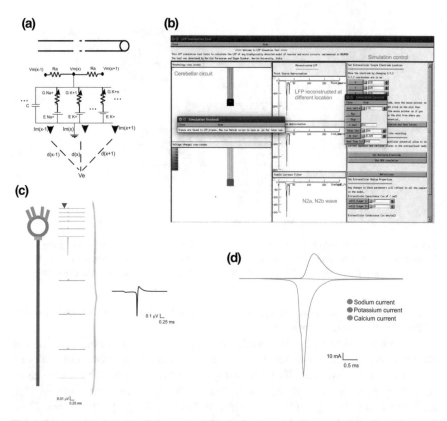

Fig. 6 Reconstructed extracellular potential of single cerebellar granule cell and circuit LFP. **a** Circuit model of intracellular and extracellular potentials. **b** Screenshot of LFPsim, a tool for reconstructing extracellular filed potential [61] and simulated cerebellar granular layer in vitro LFP, characterized by N_{2a} and N_{2b} wave [48]. **c** Single granule neuron LFP and compartmental contribution. **d** Compartmental individual ionic current contribution to single neuron LFP

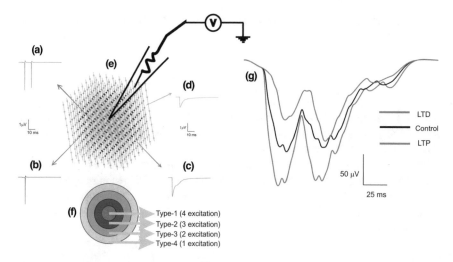

Fig. 7 Simulating in vivo LFP with center-surround excitation. Cerebellar network model was simulated with center-surround "spot" activation described in [51]. *Colored dots* in the figure (*red, blue, green, orange*) represent the soma of granule cells in the network (e). Type-1 cells receive 4 MF excitation which were located in the center of the network (*red*), Type-2 cells receive 3 MF excitation (blue), Type-3 cells receive 2 MF excitation (*green*) and Type-4 cells receive 1 MF excitation (*orange*). **a** doublet was observed from the cells which receive 4 MF inputs. **b, c** cells receive 3 and 2 excitation showed single spike. Cells with 1 MF input reproduced an EPSP (**d**). **f** Activation pattern applied to cells in the network **g** Simulating induced plasticity conditions in vivo

Modeling Extracellular Field Potential

Extracellular field potential was calculated from individual ionic currents using forward modeling schema. Point source approximation technique was used to compute LFP of cerebellar granular layer population. The electric field in extracellular space can be calculated by the Laplace equation, where $\emptyset$ is the extracellular potential at the boundary condition $(1/ \sigma)\emptyset = J_m$. J_m is the transmembrane current density and σ, the extracellular resistivity.

$$\nabla^2 \emptyset = 0$$

Extracellular potential generated from a neuronal compartment/segment can be approximated to a point source (see Fig. 6a). For a point source in the extracellular space, I represented current generated from the source, ρ denotes the conductivity of the medium and r denotes the distance from source to the point of measurement. Extracellular potential (φ) at a point r expressed as

$$\emptyset = \rho I / 4\pi r$$

Single neuron LFP can be reconstructed by linear summation of extracellular voltages from individual compartments [51]. The extracellular potential for each compartment was computed. Single neuron LFP (S_LFP) for a 'n' compartmental neuron was given by

$$S_{LFP}(x, y, z) = \sum_{i=0}^{n-1} \frac{\rho I}{4\pi r_i}$$

where, I represented the transmembrane current generated from neuronal compartments, ρ denotes the resistivity and r_i denotes the distance from calculating point to center of the neuronal compartment.

Population LFP can be computed as a linear summation of electric potential generated by single neurons in the network [53, 78]. In this technique, each active ionic dynamics of the neuronal compartment was modeled as individual current sources. Population LFP (P_{LFP}) generated from a network of 'k' neurons was calculated by

$$P_{LFP}(x, y, z) = \sum_{i=0}^{k-1} S_{LFP, r_i}$$

where S_{LFP, r_i} denotes the single neuron LFP of neuron i, r_i denotes the distance of individual current sources from point of calculation of population LFP. The models were implemented in NEURON Simulation environment. All simulations were performed on a workstation with 6-core Intel Xeon CPU running at 3.20 GHz processor with 8 GB of RAM.

LFPsim, an open-source software NEURON-based tool we had developed for mathematically modeling local field potentials using line source, point source approximations and passive resistance-conductance methods on multicompartmental models is freely available now on ModelDb with accession number = 190140 [79]. The study describes modeling local field potentials in cerebellum granular layer and extracellular action potentials in cortical neurons [61].

Neural Information Theory

An estimate of information processed could help quantify neural spiking activity and correlate local field potential responses with underlying neural circuit function. Through information theoretic methods, such methods allow to assess the role of integrative excitatory and inhibitory synaptic processes at level of neural population activity and possibly look into correlates with geometric symmetries observed in evoked local field potential. Since correlated local field potentials are representative to external world, stimuli-response estimates help decipher coding of firing information in terms of Crus II sensory-motor input-output fluctuations. We used neural information theory to estimate individual neuron roles in assessing network activity.

It has been observed that neurons elicit a variety of responses for the same input. This neuronal response variability inherent in the biological mechanisms of action potential generation limits the actual information that can be produced by the neuron. Noise introduces super-threshold and sub-threshold variations in the firing of neurons. With Shannon's information theory [80], declarative aspect of information transmitted between neurons have been proposed. Estimations of mutual information assume symbolic representation or codes for action potentials between communicating neurons. The average surprise contained in such symbols produced by neurons can be estimated efficiently with Shannon's Entropy,

$$H(R) = - \sum_R P_r log_2(P_r)$$

where R is the set of symbols and P_r is the probability of each such symbol in the entire response space. With an estimate of response space from recorded experiments, information capacity of the neuron, was maximum when occurrence probability of all the symbols are equally likely.

While calculating the maximum information capacity of a neuron, noise was considered to be involved in neuronal processing [53]. Noise increases the number of possible response symbols and the maximum information capacity of the neuron but information capacity was estimated by neuron's capability in selecting specific response patterns [81, 82]. Via information theory, estimates suggested responses of a diseased neuron regulated less or more information compared to control condition and inhibitory inputs or plasticity conditions affected the information carrying capacity of neurons.

Although entropy represents the information known to the system (neuron), it also represents lack of information the observer (receiver) neuron has before receiving the response. Information in neural communication was always measured as the decreased uncertainty of the receptor neuron after receiving the response. The statistical measure Mutual Information (MI) quantifies the average information the receiver neuron has about the source neuron after message reception (See Fig. 5).

As in other models, the mutual information between the set of stimulus $S = \{s1, s2,...\}$ and set of response $R = \{r1, r2,....\}$ was estimated as,

$$I(S,R) = H(R) - H(R|S)$$

where $H(R)$ quantifies the response variability and $H(R|S)$ quantifies the noise in the response which was calculated as,

$$H(R|S) = - \sum_S P(s) \sum_R P_{r|s} log_2(P_{r|s})$$

where (P_s) is the probability of the stimulus s from the stimulus set S and $P_{r|s}$ is the probability of observing a response r when the stimulus s is known. A variant of the

MI, called **Stimulus specific information (SSI)** is alternatively used and quantifies the input discrimination reliability of the neuron for each stimulus (input).

$$SSI(s) = H(R) - H(R|s) = -\sum_R P_r log_2(P_r) + \sum_R P_{r|s} log_2(P_{r|s})$$

Results

In this section, we will address the roles of biophysically detailed models and mathematical reconstructions of population responses, the information estimates of granular layer activity, the impact of spiking models in representing relevant circuit models and abstractions of sensory motor responses and behavior for modeling robotic articulation.

Reconstructed Extracellular Potential of Single Cerebellar Granule Cell and Circuit LFP

Single granular neuron LFP was simulated from detailed biophysical model of granule neuron (Fig. 6a) to study single neuron contributions to population code. Detailed granule neuron model was simulated with 3 mossy fiber excitatory synapses that generated a single intracellular action potential (Fig. 6c). Axon hillock compartments showed pronounced extracellular activity compared to other compartments, attributed to high density of sodium and potassium channels were concentrated in axon-hillock. The resultant linear summation of individual ionic currents yielded single neuron contribution of field potential to population activity (Fig. 6c). Total ionic current contribution of single granule neuron was estimated and inward sodium currents contributed a major part of negative potential (Fig. 6d). Granular layer in vitro LFP consisted of N_{2a} and N_{2b} wave, generated due to post synaptic activity [48]. The network reproduced in vitro behavior reported in the experiments [48].

Simulating STDP on Cerebellar Granular Layer LFP with Center-Surround Excitation

From the network model, LFP was successfully reproduced using point source approximation technique (see Fig. 7). Evoked cerebellum granular layer LFP during in vitro and in vivo conditions were modeled. Simulated in vivo LFP waveforms consisted of Trigeminal (T) and Cortical (C) wave attributed to trigeminal and cortical pathways [49]. Control condition (black), LTP (red) and

LTD (green) for in vivo like behavior were reproduced (Fig. 7g). While LTP caused increase in amplitude and width of evoked LFP responses, LTD showed reduced amplitude and width compare to control [53].

Spatial Attenuation of Single Granule Neuron Extracellular Field Potential

Ensemble responses replicate the spatio-temporal organization of underlying neural circuitry [83]. As observed in electrophysiological recordings [28, 49], the amplitude and width of the extracellular waveform exponentially decreased when the recording electrode point was moved away from the soma (see Fig. 8a).

To model the behavior of extracellular space, we varied the distance from which LFP was estimated and compared reconstructed LFP. Simulating the electrode recording point closest to soma of a granule neuron showed increased amplitude and width of the wave. It was observed that when the recording electrode was moved away from the soma, the extracellular wave amplitude decreased exponentially at an average rate of −0.011 mV (Fig. 8b).

Information Capacity of Cerebellar Granule Neurons for Sensory Inputs

The input layer of the cerebellum, the granular layer receives sensory information via mossy fibers through corticopontine, spinocerebellar and vestibulocerebellar tracks [84]. Computational models of Wistar rat granule neurons [75] were used to

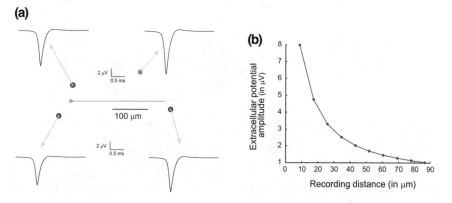

Fig. 8 Spatial attenuation of single granule neuron extracellular field potential. **a** Reconstructed extracellular potential around the granule neuron. **b** Attenuation of single neuron LFP over distance

quantify the information capacity of granule neurons for in vitro and in vivo like inputs coming through mossy fibers. Even though input to output mapping in the granule neurons for physiologically relevant conditions is not known, the tactile stimulation of the Crus IIa region of the rat via air puffs onto the whisker pad was reported to elicit short burst of spikes in the mossy- fiber [85].

Information capacity of granule neurons receiving electrical in vitro and tactile in vivo stimulations [49] were quantified [86] using information theoretic methods (Fig. 9a–d). It has been assumed that the neurons use different coding schemes to encode the information. Whether the neurons use the firing frequency, spike timings, inter-spike intervals or a combination of the above to encode information is still an area of research and debate [87]. Experimental evidence suggests neurons could be using selective strategies. Information capacity was estimated relative to

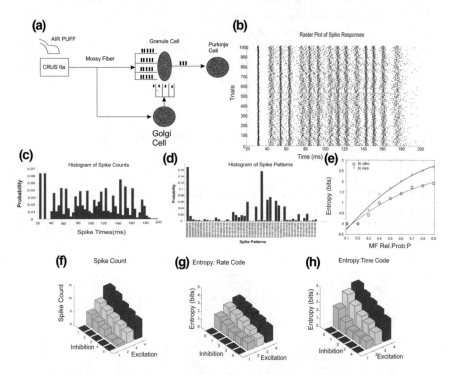

Fig. 9 Information capacity of granule neurons. **a** Tactile stimulation of the Crus IIa region of the rat by delivering air puffs to the whisker pad elicit short burst of spikes in the mossy- fiber. **b** Raster plot of spike responses recorded from 1000 neurons. **c, d** Histogram of spike count and spike patterns. **e** Entropy estimated for different release probabilities of the MF synapses for both in vitro like and in vivo like inputs: Entropy increased with the increase in release probability. **f** Spike count elicited from single neurons for each inhibitory—excitatory combination: Increase in excitatory inputs increased the spike frequency while increase in Inhibitory inputs reduced the spike frequency. **g** Entropy estimated from spike frequency code. **h** Entropy estimated from spike time code. The increase in entropy was in positive correlation with spike count both in the case of rate code and time code

the coding schemes used, like rate code and temporal code/pattern code. Absolute value of the capacity was found to be different under different coding schemes. In the temporal coding analysis of spike trains, we assumed that the time of occurrences of the spikes convey information while in rate code, the information was estimated from the firing frequency. Also for each type of inputs there was a theoretic maximum and minimum capacity different from the other type of input.

To maximize the information that a neuron can communicate about the inputs, it should maximize the entropy of the outputs. The codebook of the granule neurons under different release probabilities of the mossy fiber—granule cell synapses and under different synaptic combinations was measured (Fig. 9e–h). This code book was then used to estimate the theoretical maximum information that can be transmitted and the actual information that was transmitted through the neuron for the given input.

Stimulus Discrimination Capacity of Cerebellar Granule Neurons

Estimating the role of geometric organization through spiking information would help correlate local field potentials and information capacity in networks. However, to correlate spatio-temporal properties of individual neurons in network, input specific information signaled by granule neurons were estimated using the information theoretic quantities Mutual Information (MI) and Stimulus specific information (SSI). MI was used as an average measure to quantify the information content of a set of stimuli while SSI represented the information content of each stimulus. The change in mutual information while changing the connection geometry of the MF and Golgi cell synapses were estimated and it has been observed that the average stimulus discrimination capacity of the granule neurons was increased with the increase in MF excitatory inputs while increase in inhibition from the Golgi cells reduced this discrimination power (Fig. 10a–b).

NMDA receptor-dependent synaptic plasticity in the mossy-fiber granule cell relay was found to be depended on the excitatory-inhibitory balance of the mossy fiber excitatory and Golgi cell inhibitory connections to the granule neuron [48]. The effect of plasticity on the information capacity of the neuron was quantified using the entropic estimates reported in the previous section. It has been observed that plasticity changed the information content of the receptor neurons. Induced LTP (Long Term Potentiation) increased the input classification power of granule neurons while LTD (Long Term Depression) decreased the input classification power (Fig. 10c).

Stimulus Specific Information (SSI) estimated how each stimulus in the stimulus set was encoded by the granule neuron. A larger value of *SSI* signified the higher predictability of the response given the stimulus. Histogram plot of the stimulus

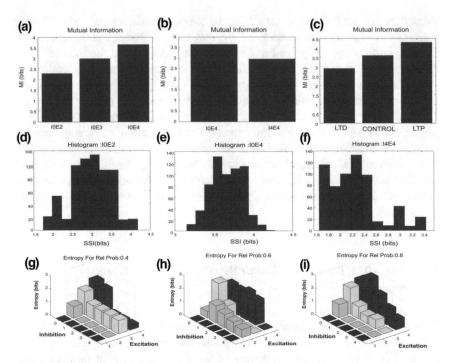

Fig. 10 Stimulus discrimination power and spike correlations. **a** MI was increased with the increase in MF excitatory inputs. I-Inhibition, E-Excitation. **b** MI was decreased with the increase in GoC inhibitory inputs. **c** MI increased during LTP from control, while MI decreased during LTD from control. **d–f**. Histogram of the SSI values for I0E2, I0E4 and I4E4 synaptic combinations respectively. Different stimuli are encoded in a similar way revealing the redundancy in neural code. **g–i**. Correlation on information transmission. Entropy was decreased in two cases —whenever overall neural activity was suppressed (which correlated with the increase in inhibition) and whenever the correlation between spike patterns was increased (with the increase in excitation)

specific information revealed that some subsets S_i of the stimulus set S were discriminated by the neurons in a similar way (Fig. 10d–f). This suggested that similarity in the stimulus features are encoded by different neurons in a similar way revealing the possibility of redundant information encoded by the granule neurons.

Since information change was observed both in rate coding and time coding schemes, it seems indicative that plasticity not only changes the rate of responses but also changes the correlation between responses. The role of correlation on the information capacity of the neurons was reconstructed and information capacity was noted to be reduced whenever the spike correlations were high (see Fig. 10g–i).

Robotics and Applications

Abstracting Models for Neuro-Robotics—Training Spiking Networks and Storage Capacity

We used the biophysics of the cerebellum (Fig. 3) to model training in neuro-robotics (see Fig. 1). For training the spiking neural networks, different learning algorithms have been proposed (see Multi-SpikeProp [88] and RProp [89]). In previous studies, high-threshold projection (HTP) and fast precision (FP) algorithms were used to train leaky integrate and fire (LIF) neurons [31]. In our study, a spiking network constructed used granule cells (600) and single Purkinje cell to train the synaptic weights which were assigned randomly resembling a simple perceptron model. One to one connection was used for the connectivity between mossy fibers and granule cells where the dynamics of mossy fiber was modeled as homogenous Poisson process with firing rate (r_{in}) and has a duration of T_{input}.

In the abstraction, granule cell and mossy fiber spike timings were almost similar as input from mossy fiber was multiplied with the respective assigned random weight. The output of all granule cells were integrated by a Purkinje cell using general Perceptron equation [90].

$$P = \sum_i w_i X_i$$

where w_i is the synaptic weight assigned to i^{th} afferent, i varies from 1 to N and X_i is 1, if there was a spike or 0 otherwise. The firing properties of granule and Purkinje cells were modeled using adaptive exponential integrate and fire structures and the parameters were same as in [68]. Purkinje cell with its auto rhythmic behavior has been reported to have a firing frequency of 30 Hz. The input current which was used to simulate the activity of granule cells is,

$$I = g * (V - V_0)$$

where g refers to conductance (4) and V, V_0 represent the membrane potential and reversal potential. The resultant spike trains were trained to obtain precisely timed spikes using FP algorithm [31]. Within each time window of $[t_d - \frac{\mu}{2}, t_d + \frac{\mu}{2}]$, the synaptic weights were updated to obtained desired spike times. The weights were updated using the following equation.

$$\Delta w_i = \mp \gamma X_i(t_{err})$$

An adapted version of the High Threshold Projection (HTP) algorithm [31] was developed as a learning rule to update weights for robotic control of a low-cost arm

[91] based on the large-scale spiking models. The algorithmic implementation of HTP in our encoding includes 3 conditions in order to have an output spike:

(1) $U(t_d) = U_{thr}$, where U_{thr} was the threshold potential (linear equality constraint)
(2) $\frac{dU}{dt(td)} > 0$ for all t_d (linear inequality constraint)
(3) $U(t) < U_{thr}$ for all t except t_d (linear inequality constraint)

Feed forward Spiking Algorithm (adaptation of HTP Algorithm [31]):
Begin

1. Initialize the weight ω with random numbers
2. For every pattern in the training set

   a. Present the pattern to the network and calculate the PSP using (9)
   b. For output neuron

      i. Create a set of error times, $\{t_{err}\}$ and find the error
      ii. Stop if no error found i.e. all equality and inequality constraints are satisfied.
      iii. Construct a set of input labeled patterns, $\{(x, y)\}$, where x = x(t_{err}), y = −1 if t_{err} is above the suprathreshold, else x = diff(x(t_{err}), y = +1 if $t_{err} = t_d$.
      iv. Add the constructed set to x.

   c. End

3. End

Storage Capacity Computation as a Pattern Recognition Circuit for Robotics

In our abstractions towards a bioinspired pattern separator, we used the large-scale model to assess storage capacity. Purkinje cell (PC) received a large number of parallel fiber (pf) inputs as patterns to produce an output [92]. At each of these synapses, some random input-output associations were stored. The capacity (input-output associations) of a Purkinje neuron per synapse was referred to as the maximal storage capacity (α_c).

Storage Capacity was defined as the number of desired spikes that a PF-PC synapse implemented. Several methods described in [20, 93] have estimated maximal storage capacity. We estimated storage capacity using methods described in [94], where stability constant K, total duration of spike train T and the mean rate of desired output spike times within the correlated time of post synaptic potential (PSP) $r_{out}\tau$ [31] were used. Stability constant K referred to as threshold stability constant [20] helped increasing the reliability of input patterns which lead to a spike, thereby acting as a resistance to noise. The constant ensured robustness of storage by avoiding erroneous threshold crossings [20]. Using this parameter taken from [94], we estimated the stability constant.

$$K = \frac{T}{\sqrt{\tau_s \tau_m}}$$

$$\alpha_c^p (b) = \frac{1}{b \int_x^\infty Dt(t-x)^2 + (1-b) \int_{-x}^\infty Dt(t+x)^2}$$

We were able to reproduce a spiking neural network and model the adaptive dynamics of Purkinje neuron (PC) and its pattern separation properties (Fig. 11). The network was simulated with auto rhythmic PC firing and the variations in the PC firing rate were shown when granule neuron (GrC) inputs were given as PF with the presence of parallel fiber and climbing fiber inputs (Figs. 11 and 12). To model storage unreliability at the PF-PC synapse, stability constant K, was employed.

In our network, storage capacity was estimated to be <1. Based on estimated values, we inferred ≥5 PF-PC synapse were required for the implementation of a desired spike in a pattern [35] (Fig. 13).

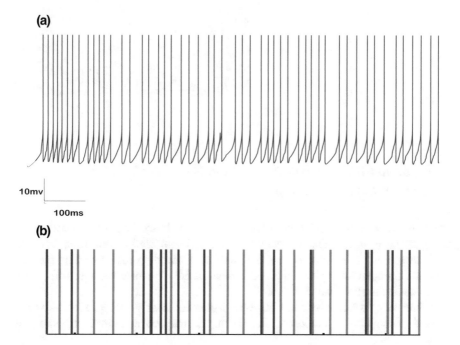

Fig. 11 PC learning after using FP algorithm. **a** Purkinje Cell (PC) response, after learning frequency of firing gets reduced. **b** PC response (*red*) and *blue line* indicates desired output sequence

Fig. 12 Mossy fiber, GrC
and PC Spike Patterns.
a Spike times generated from
Poisson distribution. **b** Firing
pattern of a single granule cell
when input as in A. **c** PC
spontaneous firing behavior
without synaptic inputs

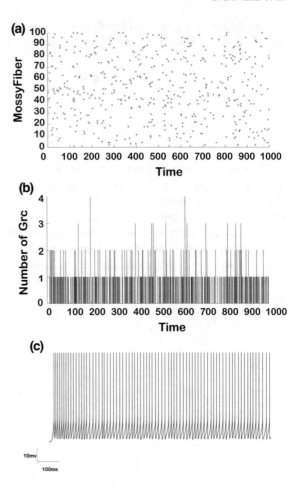

Abstraction of Real-World Data into Spike Trains Using Granular Layer Encoding

We compared a look-up based technique developed based on pattern classification of granular layer encoding of real-world data and compared it with other encoding schemes. Bens spike algorithm (BSA) used rate encoding for converting the real world data input into spike trains and contains a finite impulse response (FIR) filter with fixed threshold value of 0.86 [35].

Employing patterns based on large-scale granular layer encoding, a look-up table based encoding was proposed with the inputs mapped to known physiological

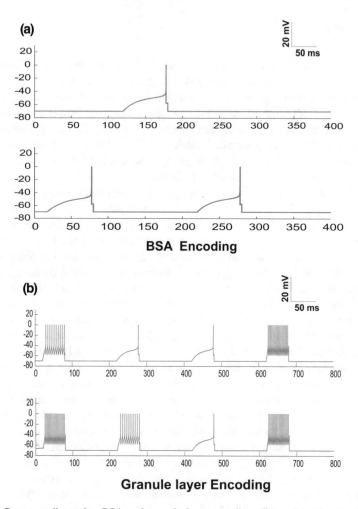

Fig. 13 Data encoding using BSA and granule layer encoding. Generation of spike trains as potential (*y-axis*) with time (*x-axis*). **a** Spike patterns using BSA encoding for two different input patterns. **b** Encoded spike patterns using granule layer encoding for the same two input patterns

conditions of granule neurons. In vitro and in vivo granule neuron firing properties [68] were matched using a look-up of firing behaviors mapped to a particular input. Machine learning based on such spiking datasets were performed to evaluate the usability as novel encoders.

In its role for spike-encoded classifiers, we found granular layer based encoding provided a higher classification accuracy [34].

Discussion

In this chapter, we have looked into how cerebellum networks may be involved in timing [57, 78], sparse recoding [53] and pattern abstraction [35, 68]. Using mathematical modelling, we focus on reconstruction of large-scale neuronal dynamics [57, 78], role of inhibition in modulating the intrinsic behaviour of neurons [53], translations of such behaviour as algorithms [35] as well as the study of network activity during diseased conditions [57]. Using AdEx spiking models, cerebellum granular layer and Purkinje layer was reconstructed. While spiking network models were significantly reliable for reconstructing spiking patterns [68] and passage of time [57], detailed biophysical models were crucial to reconstruct local field response [53] and to intrinsically model role of induced plasticity [78].

Via modeling, estimations of mutual information while changing the connection geometry of the MF and Golgi cell synapses were obtained and an increase in average stimulus discrimination capacity of the granule neurons was observed with the increase in MF excitatory inputs while increase in inhibition (I) from the Golgi cells reduced this discriminatory capacity. Increase in excitatory (E) connections alone did not necessarily guarantee the increase in information transfer capacity of the neuron at fixed release probability. Information capacity was maximized in certain cases for intermediate E-I balance. Estimated entropy was decreased when the overall neural activity was suppressed via increase in inhibition. E-I activity correlations may be used by neural populations to modulate the information flow.

Increased release probability of the MF synapses increased the information capacity of granule neurons. Also the capacity of the granule neuron to encode mossy fiber information was found to be higher with strong excitation from mossy fibers and low with strong inhibition from the Golgi neurons both in temporal coding and in rate coding schemes. There was no information transfer through the neuron for single synapse inputs. Presence of two or more excitatory synapses was a necessary condition for the information transfer through the synapses under all conditions suggesting combinatorial properties of granule neurons [95].

Assessments with MI and SSI information theoretic measures indicated that the geometry of synaptic connections and plasticity changed response characteristics of granule neurons. This was also quantified and reported in local field potential reconstructions although that LFP emphasized on spiking components in terms of lag and amplitude of generated waves. Spike frequency and spike correlations on information capacity and stimulus discrimination reliability suggest multiplexed information streams in LFPs have individual neural mechanisms. It was also observed that increase in frequency increased the information transfer while spike correlations negatively affected the information transmitted. The same behavior was found to be persisted in the plasticity states of the neurons as seen in model-based LFP reconstructions.

Oscillatory activity in awake and anesthetized rats in the granular layer have been reported in the same frequency ranges [54, 58]. Since rhythmic activity has been determined to be dependent on multi-unitary granule cell activity, through mathematically reconstructed LFP, we were able to look into plasticity roles in population response. The mathematical modeling also suggests the extended somatosensory processing circuitry in associating the role of clusters and independent activation patterns that may generate in vivo trigeminal and cortical waves in cerebellum granular layer evoked post-synaptic LFPs. This may suggest the spatio-temporal organization of granule layer oscillations and source reconstructions related to evoked LFPs in the range of 4–25 Hz. Although spike-time dependent plasticity did not affect spike amplitude or delay, population LFP showed induced LTP caused increased wave amplitude and decreased lag while LTD did the counter-effect [53]. The modulatory roles of generated local field potentials suggest a network role in optimized neural response in the context of sensorimotor-related stimuli. The attribution of changes reflected due to plasticity at the mechanism level and its upstream modifications in evoked LFPs has been shown crucial to link molecular mechanisms to their circuit functions and dysfunction especially in diseased condition. Ataxia like conditions where NR2A subunit cause NMDA receptor dysfunction in cerebellar granule cells were predicted by mathematical simulations to show decreased LFP waves although some of their spiking behavior remained unchanged [57].

Using large-scale spiking models, we also attempted to scale this problem to neurorobotics. With two algorithms [35], it was shown optimal encoding may use cerebellum granule cell firing representation and storage capacity of such networks rely on less than five neurons to capture unique patterns [34, 35]. Spike propagation and burst-burst transmission in granular layer may aid scaling responses for temporal signal encoders. A new study on robotic control where kinematic patterns are encoded as spikes in being explored (manuscript in preparation).

Conclusion

A preliminary study of a crucial cerebellar microcircuit and its applications in robotics have been initiated in this study. The circuits have important roles in neural dysfunctions and several neurological conditions. Understanding the functional role of such circuits also add to neuroinspired computing methods. Spiking networks for pattern abstraction, if extended with large number of clusters of cerebellum granule cells may help explore sparse recoding hypothesis. Although a precursor, this study could help understand movement-related pattern recognition and improve models for motor articulation control.

Acknowledgments This work derives direction and ideas from the Chancellor of Amrita University, Sri Mata Amritanandamayi Devi. Authors would like to acknowledge Egidio D'Angelo of University of Pavia, Giovanni Naldi of University of Milan, Sergio Solinas, Thierry Nieus of IIT Genova for their support towards work in this manuscript. This work is supported by Grants SR/CSI/49/2010, SR/CSI/60/2011, SR/CSRI/60/2013, SR/CSRI/61/2014 and Indo-Italy POC 2012-2014 from DST and BT/PR5142/MED/30/764/2012 from DBT, Government of India and partially by Embracing the World.

References

1. Lange W (1975) Cell number and cell density in the cerebellar cortex of man and some other mammals. Cell Tissue Res 157:115–24
2. Herculano-Houzel S (2009) The human brain in numbers: a linearly scaled-up primate brain. Front Hum Neurosci 3:31
3. Luciani L (1891) Il Cervelletto, nuovi studi di fisiologia normale e patologica. coi tipi dei successori Le Monnier, Firenze
4. Manni E, Petrosini L (1997) Luciani's work on the cerebellum a century later. Trends Neurosci 20:112–116
5. Holmes G (1917) The symptoms of acute cerebellar injuries due to gunshot injuries. Brain 40:461–535
6. Bower JM (1997) Is the cerebellum sensory for motor's sake, or motor for sensory's sake: the view from the whiskers of a rat? Prog Brain Res 114:463–96
7. Ivry RB, Baldo J V (1992) Is the cerebellum involved in learning and cognition? Curr Opin Neurobiol 2:212–6
8. Boyd CAR (2010) Cerebellar agenesis revisited. Brain 133:941–4
9. Yu F, Jiang Q, Sun X, Zhang R (2015) A new case of complete primary cerebellar agenesis: clinical and imaging findings in a living patient. Brain 138:e353
10. Medina JF, Garcia KS, Nores WL, Taylor NM, Mauk MD (2000) Timing mechanisms in the cerebellum: testing predictions of a large-scale computer simulation. J Neurosci 20:5516–5525
11. Vos BP, Volny-Luraghi A, Schutter E De (1999) Cerebellar Golgi cells in the rat: receptive fields and timing of responses to facial stimulation. Eur J Neurosci 11:2621–2634
12. Albus JS (1975) A New Approach to Manipulator Control: The Cerebellar Model Articulation Controller(CMAC). J. Dyn. Syst. Meas. Control
13. Tyrrell T, Willshaw D (1992) Cerebellar cortex: its simulation and the relevance of Marr's theory. Philos Trans R Soc Lond B Biol Sci 336:239–57
14. Eccles JC (1981) Physiology of motor control in man. Appl Neurophysiol 44:5–15
15. Ito M (2000) Mechanisms of motor learning in the cerebellum. Brain Res 886:237–245
16. Albus JS (1971) A theory of cerebellar function. Math Biosci 10:25–61
17. Marr D (1969) A theory of cerebellar cortex. J Physiol 202:437–470
18. Mazzarello P, Haines D, Manto M-U (2012) Camillo Golgi on Cerebellar Granule Cells. Cerebellum 11:5–24–7
19. Eccles JC, Llinás R, Sasaki K (1965) Inhibitory systems in the cerebellar cortex. Proc Aust Assoc Neurol 3:7–14
20. Brunel N, Hakim V, Isope P, Nadal JP, Barbour B (2004) Optimal information storage and the distribution of synaptic weights: Perceptron versus Purkinje cell. Neuron 43:745–757
21. D'Angelo E, Mazzarello P, Prestori F, Mapelli J, Solinas S, Lombardo P, Cesana E, Gandolfi D, Congi L (2011) The cerebellar network: from structure to function and dynamics. Brain Res Rev 66:5–15
22. D'Angelo E, De Zeeuw CI (2009) Timing and plasticity in the cerebellum: focus on the granular layer. Trends Neurosci 32:30–40
23. D'Angelo E, Koekkoek SKE, Lombardo P, Solinas S, Ros E, Garrido J, Schonewille M, De Zeeuw CI (2009) Timing in the cerebellum: oscillations and resonance in the granular layer. Neuroscience 162:805–15
24. Eccles JC (1982) The initiation of voluntary movements by the supplementary motor area. Arch Psychiatr Nervenkr 231:423–441
25. Horne MK, Butler EG (1995) The role of the cerebello-thalamo-cortical pathway in skilled movement. Prog Neurobiol 46:199–213
26. Prestori F, Rossi P, Bearzatto B, Lainé J, Necchi D, Diwakar S, Schiffmann SN, Axelrad H, D'Angelo E (2008) Altered neuron excitability and synaptic plasticity in the cerebellar granular layer of juvenile prion protein knock-out mice with impaired motor control. J Neurosci. doi:10.1523/JNEUROSCI.0409-08.2008

27. Goldfarb M, Schoorlemmer J, Williams A, et al (2007) Fibroblast growth factor homologous factors control neuronal excitability through modulation of voltage-gated sodium channels. Neuron 55:449–463
28. Bower JM, Woolston DC (1983) Congruence of spatial organization of tactile projections to granule cell and Purkinje cell layers of cerebellar hemispheres of the albino rat: vertical organization of cerebellar cortex. J Neurophysiol 49:745–66
29. Maex R, Vos B, Ã EDES, Volny-Luraghi a, Vosdagger B, De Schutter E (2002) Peripheral stimuli excite coronal beams of Golgi cells in rat cerebellar cortex. Neuroscience 113:363–73
30. Carrillo RR, Ros E, Boucheny C, Coenen OJ-MD (2008) A real-time spiking cerebellum model for learning robot control. Biosystems 94:18–27
31. Memmesheimer RM, Rubin R, Ölveczky B, Sompolinsky H (2014) Learning Precisely Timed Spikes. Neuron 82:925–938
32. Carrillo RR, Ros E, Tolu S, Nieus T, D'Angelo E (2008) Event-driven simulation of cerebellar granule cells. Biosystems 94:10–17
33. Gamez D, Fidjeland AK, Lazdins E (2012) iSpike: a spiking neural interface for the iCub robot. Bioinspir Biomim 7:25008
34. Medini C, Vijayan A, Zacharia RM, Rajagopal LP, Nair B, Diwakar S (2015) Spike Encoding for Pattern Recognition: Comparing Cerebellum Granular Layer Encoding and BSA algorithms. In: Adv. Comput. Commun. Informatics (ICACCI), 2015 Int. Conf. IEEE, Kochi, pp 1619–1625
35. Vijayan A, Medini C, Palolithazhe A, et al (2015) Modeling Pattern Abstraction in Cerebellum and Estimation of Optimal Storage Capacity. In: Fourth Int. Conf. Adv. Comput. Commun. Informatics. IEEE, Kochi, New York, USA, pp 335–347
36. Burke RE (2007) Sir Charles Sherrington's the integrative action of the nervous system: a centenary appreciation. Brain 130:887–94
37. Ghez C, Hening W, Gordon J (1991) Organization of voluntary movement. Curr Opin Neurobiol 1:664–671
38. Mehring C, Rickert J, Vaadia E, Cardosa de Oliveira S, Aertsen A, Rotter S (2003) Inference of hand movements from local field potentials in monkey motor cortex. Nat Neurosci 6:1253–4
39. Schaal S (2002) Arm and Hand Movement Control. 110–113
40. Hemminger S (2010) Linking Error, Passage of Time, the Cerebellum and the Primary Motor Cortex to the Multiple Timescales of Motor Memory By.
41. Kawato M (1999) Internal models for motor control and trajectory planning. Curr Opin Neurobiol 9:718–727
42. Gomi H, Kawato M (1996) Equilibrium-Point Control Hypothesis Examined by Measured Arm Stiffness During Multijoint Movement. Science (80-) 272:117–120
43. Snider RS, Stowell A (1944) Receiving Areas of the Tactile, Auditory, and Visual Systems in the Cerebellum. J Neurophysiol 7:331–357
44. Azizi SA, Woodward DJ (1990) Interactions of visual and auditory mossy fiber inputs in the paraflocculus of the rat: a gating action of multimodal inputs. Brain Res 533:255–62
45. Gao J-H, Parsons LM, Bower JM, Xiong J, Li J, Fox PT (1996) Cerebellum Implicated in Sensory Acquisition and Discrimination Rather Than Motor Control. Science (80-) 272:545–547
46. Eccles JC, Ito M, Szentágothai J (1967) The Cerebellum as a Neuronal Machine. doi:10.1007/978-3-662-13147-3
47. Morissette J, Bower JM (1996) Contribution of somatosensory cortex to responses in the rat cerebellar granule cell layer following peripheral tactile stimulation. Exp brain Res 109:240–250
48. Mapelli J, D'Angelo E (2007) The spatial organization of long-term synaptic plasticity at the input stage of cerebellum. J Neurosci 27:1285–96
49. Roggeri L, Rivieccio B, Rossi P, D'Angelo E (2008) Tactile stimulation evokes long-term synaptic plasticity in the granular layer of cerebellum. J Neurosci 28:6354–9

50. Diwakar S, Lombardo P, Solinas S, Naldi G, D'Angelo E (2011) Local field potential modeling predicts dense activation in cerebellar granule cells clusters under LTP and LTD control. PLoS One 6:e21928

51. Parasuram H, Nair B, Naldi G, D'Angelo E, Diwakar S (2015) Exploiting point source approximation on detailed neuronal models to reconstruct single neuron electric field and population LFP. In: 2015 Int. Jt. Conf. Neural Networks. IEEE, pp 1–7

52. Reimann MW, Anastassiou CA, Perin R, Hill SL, Markram H, Koch C (2013) A biophysically detailed model of neocortical local field potentials predicts the critical role of active membrane currents. Neuron 79:375–90

53. Diwakar S, Lombardo P, Solinas S, Naldi G, D'Angelo E (2011) Local field potential modeling predicts dense activation in cerebellar granule cells clusters under LTP and LTD control. PLoS One 6:e21928

54. Courtemanche R, Robinson JC, Aponte DI (2013) Linking oscillations in cerebellar circuits. Front Neural Circuits 7:125

55. Einevoll GT, Kayser C, Logothetis NK, Panzeri S (2013) Modelling and analysis of local field potentials for studying the function of cortical circuits. Nat Rev Neurosci 14:770–85

56. Solinas S, Nieus T, D'Angelo E (2010) A realistic large-scale model of the cerebellum granular layer predicts circuit spatio-temporal filtering properties. Front Cell Neurosci 4:12

57. Medini C, Nair B, D'Angelo E, Naldi G, Diwakar S (2012) Modeling spike-train processing in the cerebellum granular layer and changes in plasticity reveal single neuron effects in neural ensembles. Comput Intell Neurosci 2012:359529

58. Courtemanche R, Chabaud P, Lamarre Y (2009) Synchronization in primate cerebellar granule cell layer local field potentials: basic anisotropy and dynamic changes during active expectancy. Front Cell Neurosci 3:6

59. Bower JM, Woolston DC (1983) Congruence of spatial organization of tactile projections to granule cell and Purkinje cell layers of cerebellar hemispheres of the albino rat: vertical organization of cerebellar cortex. J Neurophysiol 49:745–766

60. Mitzdorf U (1985) Current source-density method and application in cat cerebral cortex: investigation of evoked potentials and EEG phenomena. Physiol Rev 65:37–100

61. Parasuram H, Nair B, D'Angelo E, Hines M, Naldi G, Diwakar S (2016) Computational Modeling of Single Neuron Extracellular Electric Potentials and Network Local Field Potentials using LFPsim. Front Comput Neurosci 10:65

62. Izhikevich EM, Edelman GM (2008) Large-scale model of mammalian thalamocortical systems. Proc Natl Acad Sci U S A 105:3593–3598

63. La Camera G, Rauch A, Lüscher H-R, Senn W, Fusi S (2004) Minimal models of adapted neuronal response to in vivo-like input currents. Neural Comput 16:2101–2124

64. Yoosef A, Rajendran AG, Nair B, Diwakar S (2014) Parallelization of Cerebellar Granular Layer Circuitry Model for Physiological Predictions. Proc. Int. Symp. Transl. Neurosci. {&} XXXII Annu. Conf. Indian Acad. Neurosci.

65. Brette R, Gerstner W (2005) Adaptive exponential integrate-and-fire model as an effective description of neuronal activity. J Neurophysiol 94:3637–3642

66. Izhikevich EM (2003) Simple model of spiking neurons. IEEE Trans Neural Netw 14:1569–1572

67. Naud R, Marcille N, Clopath C, Gerstner W (2008) Firing patterns in the adaptive exponential integrate-and-fire model. Biol Cybern 99:335–347

68. Medini C, Vijayan A, D'Angelo E, Nair B, Diwakar S (2014) Computationally Efficient Biorealistic Reconstructions of Cerebellar Neuron Spiking Patterns. Int Conf Interdiscip Adv Appl Comput - ICONIAAC '14 1–6

69. Rossant C, Goodman DFM, Fontaine B, Platkiewicz J, Magnusson AK, Brette R (2011) Fitting neuron models to spike trains. Front Neurosci 5:9

70. D'Angelo E, Nieus T, Maffei a, Armano S, Rossi P, Taglietti V, Fontana a, Naldi G (2001) Theta-frequency bursting and resonance in cerebellar granule cells: experimental evidence and modeling of a slow k + -dependent mechanism. J Neurosci 21:759–70

71. Rancz EA, Ishikawa T, Duguid I, Chadderton P, Mahon S, Häusser M (2007) High-fidelity transmission of sensory information by single cerebellar mossy fibre boutons. Nature 450:1245–8

72. Maex R, Schutter E De (1998) Synchronization of golgi and granule cell firing in a detailed network model of the cerebellar granule cell layer. J Neurophysiol 80:2521–2537

73. Vos BP, Maex R, Volny-Luraghi A, Schutter E De (1999) Parallel fibers synchronize spontaneous activity in cerebellar Golgi cells. J Neurosci 19:RC6

74. Prestori F, Person AL, D'Angelo E, Solinas S, Mapelli J, Gandolfi D, Mapelli L (2013) The cerebellar Golgi cell and spatiotemporal organization of granular layer activity. Front Neural Circuits 7:93

75. Diwakar S, Magistretti J, Goldfarb M, Naldi G, D'Angelo E (2009) Axonal Na + channels ensure fast spike activation and back-propagation in cerebellar granule cells. J Neurophysiol 101:519–532

76. Solinas S, Forti L, Cesana E, Mapelli J, Schutter E De, Angelo ED (2007) Computational reconstruction of pacemaking and intrinsic electroresponsiveness in cerebellar golgi cells. doi:10.3389/neuro.03/002.2007

77. Solinas S, Nieus T, D'Angelo E (2010) A realistic large-scale model of the cerebellum granular layer predicts circuit spatio-temporal filtering properties. Front Cell Neurosci 4:12

78. Parasuram H, Nair B, Naldi G, Angelo ED, Diwakar S, D'Angelo E (2011) A modeling based study on the origin and nature of evoked post-synaptic local field potentials in granular layer. J Physiol Paris 105:71–82

79. Hines ML, Morse T, Migliore M, Carnevale NT, Shepherd GM (2004) ModelDB: A Database to Support Computational Neuroscience. J Comput Neurosci 17:7–11

80. Shannon C (1948) A Mathematical Theory of Communication. Bell Syst Tech J 27:379–423

81. Brasselet R, Johansson RS, Arleo A (2011) Quantifying neurotransmission reliability through metrics-based information analysis. Neural Comput 23:852–81

82. Arleo A, Nieus T, Bezzi M, D'Errico A (2010) How synaptic release probability shapes neuronal transmission: Information-theoretic analysis in a cerebellar granule cell. Neural ...

83. Nicholson C, Llinas R (1971) Field potentials in the alligator cerebellum and theory of their relationship to Purkinje cell dendritic spikes. J Neurophysiol 34:509–531

84. D'Angelo E (2011) Neural circuits of the cerebellum: hypothesis for function. J Integr Neurosci 10:317–52

85. Chadderton P, Margrie TW, Häusser M (2004) Integration of quanta in cerebellar granule cells during sensory processing. Nature 428:856–60

86. Reinagel P, Reid RC (2000) Temporal coding of visual information in the thalamus. J Neurosci 20:5392–5400

87. Rieke F, Warland D, De Ruyter Van Steveninck R, Bialek W (1997) Spikes: Exploring the Neural Code. MIT Press 20:xvi, 395

88. Ghosh-Dastidar S, Adeli H (2007) Improved Spiking Neural Networks for EEG Classification and Epilepsy and Seizure Detection. Integr Comput Aided Eng 14:187–212

89. McKennoch S, Liu DLD, Bushnell LG (2006) Fast Modifications of the SpikeProp Algorithm. 2006 IEEE Int Jt Conf Neural Netw Proc 3970–3977

90. Rosenblatt F (1962) Principles of Neurodynamics.

91. Vijayan A, Nutakki C, Medini C, Singanamala H, Nair B (2013) Classifying Movement Articulation for Robotic Arms via Machine Learning. J Intell Comput 4:123–134

92. Hansel C, Linden DJ (2000) Long-Term Depression of the Cerebellar Climbing Fiber–Purkinje Neuron Synapse. Neuron 26:473–482

93. Clopath C, Nadal JP, Brunel N (2012) Storage of correlated patterns in standard and bistable Purkinje cell models. PLoS Comput Biol 8:1–10

94. Rubin R, Monasson R, Sompolinsky H (2010) Theory of spike timing based neural classifiers. 4

95. Mapelli J, Gandolfi D, D'Angelo E (2010) Combinatorial responses controlled by synaptic inhibition in cerebellum granular layer. J Neurophysiol 103:250–261

A Computational Model of Neural Synchronization in Striatum

Rahmi Elibol and Neslihan Serap Şengör

Introduction

Behavior is the output of the brain, and now is the first time in the history of humankind when science has the tools to understand how behavior emerges from the brain. Science is still at the very first step of modeling the brain, and the brain is conceptualized almost as a black box, creating feelings and thoughts from sensory stimuli, that is, creating all that makes us human. Thus, from systems point of view, the input of the black-box is the sensory stimuli and the output is the behavior as in Fig. 1. Tools to measure these outputs are either images obtained by different techniques like functional magnetic resonance imaging (fMRI) or waveforms obtained by field potential recordings, such as electroencephalogram (EEG), and scores of neuropsychological tests, such as the Iowa gambling task (IGT). Looking at these outputs along with the physiological tests, clinicians are trying to diagnose behavioral and neurological disorders and psychiatric problems. The subsequent models are usually black box models, since scientists are still trying to link molecular with phenomenological levels [1].

Even though a lot is known about what is happening at the cell level, understanding how these processes at the micro level give rise to behavior is far away from the knowledge of today's neuroscientists. Many treatment methods, e.g., deep brain stimulation (DBS), have been developed, and various drugs work well in many cases, but a clear explanation of why these help so many people with neurological disorders is missing. Once the gap between the molecular and phenomenological levels is better elucidated, then how the behavior emerges from the brain will be understood. This will enable the development of better diagnoses and treatment methods.

R. Elibol · N.S. Şengör (✉)
Istanbul Technical University, Maslak, Istanbul, Turkey
e-mail: sengorn@itu.edu.tr

R. Elibol
e-mail: rahmielibol@itu.edu.tr

© Springer International Publishing AG 2017
P. Érdi et al. (eds.), *Computational Neurology and Psychiatry*,
Springer Series in Bio-/Neuroinformatics 6,
DOI 10.1007/978-3-319-49959-8_13

Fig. 1 Tools such as fMRI, EEG, and neuropsychological tests provide us data and information which can be interpreted as the outputs of a *black-box* model of the brain. With computational models based on rigorous mathematical representations of the neural activity, we can build more descriptive models

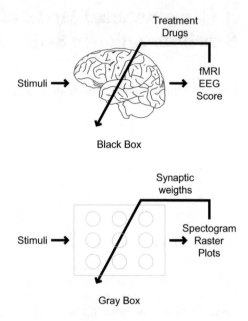

As pointed out in the paper of P.R. Montague and his colleagues [2], computational psychiatry aims to fulfill this gap for psychiatric conditions, since decision-making is the major cognitive process underlying many psychiatric problems. Here, we will consider the input structure of basal ganglia, the striatum, which has an important role in decision-making. The computational models are expected to constitute a means to understand the ongoing complex processes in the brain. As stated in [3], when based on carefully chosen assumptions, simulations obtained with computational models are not susceptible to statistical errors and deepen our systems-level understanding of complex processes. Thus, computational models would help to build gray boxes rather than black ones (Fig. 1). We will propose a mathematical model for striatum based on the biophysical properties of the striatum, especially focusing on the modulatory effect of dopamine. While concepts such as entropy and mutual information from information theory are considered to define measures for the complexity of whole-brain modeling [4, 5], temporal synchrony is considered a hallmark of activity within and between local neuronal groups, either for execution of behavioral tasks [6] or for malfunctioning of neural structures [7]. Thus, the synchronous behavior of the neuron populations can be considered as the computational basis of neural and cognitive processes. To relate model outputs to brain oscillations observed in EEG and local field potential (LFP) recordings [8–10], the simulation results are discussed focusing on spectrograms obtained with the proposed computational model and synchrony measure. Raster plots are also considered while evaluating the simulation results.

The role of basal ganglia in voluntary motor movements has been long studied [11], but their role in cognitive processes only began recently with the work of Alexander and his colleagues [12, 13]. Of particular interest is the relation between limbic and prefrontal basal ganglia circuits when considering reward related learning [14, 15] and decision-making [16, 17]. Impairment of basal ganglia circuits does not only manifest as deficits in motor actions observed in neurodegenerative diseases such as Parkinson's and Huntington's disease, but also as behavioral deficits observed in attention deficit hyperactivity disorder (ADHD), obsessive-compulsive disorder (OCD) and addiction [18–23]. These behavioral disorders, like motor movement disorders, are also treated by deep brain stimulation (DBS), a well-known treatment of Parkinson's disease [24, 25]. The role of basal ganglia in psychiatric disorders is considered more recently [1, 26], and their treatment by DBS makes basal ganglia a target for functional and restorative neurosurgery [27, 28].

Computational models of basal ganglia circuits mostly focus on action selection. While the work of Doya focused more on the computational aspects of cerebellum, basal ganglia and cortex [29], Taylor and his colleague focused on deriving a nonlinear dynamical system model of basal ganglia circuit for action selection. Their work gives an explanation of how action selection can be related to a nonlinear dynamical system behavior around stable equilibrium points [30]. With a series of publications, Gurney and his colleagues give a model of basal ganglia circuit based on a integrate and fire neuron model and show with simulations that the model is capable of producing functional properties of basal ganglia circuit for action selection [31–33]. Though all these models also consider the role of dopamine and mention the effect of dopamine on neurodegenerative diseases, the model proposed in [34] focuses on explaining cognitive deficits observed in Parkinson's disease by considering the striatum. In [35], the idea is to propose a computational model based on a realistic biophysical neuron model to investigate the role of different pathways of basal ganglia on action selection. In [36], the proposed action selection model of basal ganglia circuit is shown to be able to model the Stroop effect; the effect of dopamine on action selection is also discussed considering the time delays during the Stroop task.

All these models consider only the role of dorsal striatum during voluntary movement and action selection. There are also computational models where the effect of ventral tegmental area is considered for reward based learning. In particular, the well-known work of Schultz et al. [37], where reinforcement learning is considered as the computational basis of reward based learning, is a pioneer in modeling basal ganglia circuits. This paper declared a breakthrough idea about the computational aspects of basal ganglia circuits and a series of papers followed it [38–43]. In these papers, mostly actor-critic models developed in [44] are utilized. The relation between the ventral and dorsal parts of striatum had been anticipated in [45, 46] by considering a biologically realistic model of basal ganglia-thalamo-cortical pathways rather than actor-critic models of machine learning.

Based on the computational models of reward based learning, models of addiction have been proposed, where the role of dopamine from ventral tegmental area are considered to be important for the malfunctioning of the striatum and thus basal ganglia circuit [47–49]. These works, along with the computational models for DBS in basal ganglia circuits [50, 51], can be considered as a step towards computational psychiatry. The first three of these publications consider modeling neural structures and try to explain the phenomena behind a behavioral disorder through a computational model, while the last two focus on explaining why DBS works as a treatment procedure.

Neuropsychological tests and tasks are the means of measuring cognitive tasks and are also used as one of the measures to diagnose cognitive deficits. Working memory is considered to be a temporal storage unit of the brain, which is an important part of many cognitive tasks [52], and basal ganglia performs dynamic gating for working memory processes by the modulatory mechanism of disinhibition. There are papers focusing on modeling the basal ganglia circuits during working memory tasks [53–56]. In [54], it is discussed that the role of dopamine in modulating cognitive processes depends on basal ganglia. The relation of this modulatory effect to psychopharmacological studies is investigated with the proposed computational model. So, the computational model in [54] gives clues about the drug effects and shows that with computational psychiatry, it is possible to have a better understanding of drug effects.

EEG and LFP recordings provide information regarding synchronous activity of neuronal population and is thought as one way of coding/encoding the computational property of the brain [57]. Understanding the mechanism giving rise to the synchronous activity observed in EEG and LFP recordings has been considered in computational models [58]. Because the synchronous activity observed in EEG and LFP recordings are connected to cognitive tasks [8, 10] and also thought to be hallmarks of disorders [7, 59, 60], then if the outcomes of the computational model are related to oscillations, the model would be informative and provide an understanding of ongoing activity. The computational models of basal ganglia and striatum also considered these oscillations [33, 50, 51, 55, 61, 62] and showed that the model outcomes agree with the frequency bands observed in the experimental studies [22, 63].

One aspect of modeling is to understand the phenomenon of interest, but modeling also should boost the development of new tools. A model always gives a reductionist approach and its simplicity should help to develop computational tools that improve quality of life. So, the models of cognitive tasks and neuronal structures are also used to develop tools to ease the life of elderly and disabled, providing intelligent systems to help daily lifes [64]. The role of basal ganglia circuits in decision-making helps develop neuro-robots [65–68], providing an approach different than the rule based methods used in artificial intelligence.

Here, we will introduce a simple spiking neuron model of striatum with which the role of dopamine on synchronous activity can be captured and discuss the computational outcomes of the model with the spectrograms obtained from raster plots of the spiking neurons. The model is simple enough to be implemented for hardware applications [68], but still is informative and helps our understanding of decision-making which is thought to be key to psychiatric conditions.

The Scientific Problem: Modeling Striatum

All models lead to abstraction of the phenomenon considered. They give a partial explanation from the point of view of what is thought to be the most important aspect of the problem under investigation. Here, we will focus on investigating the role of dopamine on synchronous activity in striatum and show that even with a simple model, it is possible to build a connection between the observed data, such as LFP, and the physiological quantity, such as the neurotransmitter dopamine. There are a number of works showing the relation between neural oscillations in striatum and cognitive tasks and disorders [59, 69–74]. In order to show the oscillations of striatal neurons, we considered the properties of the medium spiny and interneurons and the connection properties within the striatum and to the striatum, but make some assumptions to keep the model simple while modeling all these properties.

Striatal Medium Spiny Neurons and Interneurons

Functionally, striatum coordinates simple body movements, voluntary motor actions and information processing in complex cognitive tasks [11, 13, 14, 75]. Even though dorsal and ventral striatum are similiar to each other in most respects, dorsal striatum (neostriatum—caudate and putamen) determines the actions in performing goal directed behaviors [18], whereas ventral striatum (nucleus accumbens) calculates the reward value of tasks and the error in expectation [15, 76, 77]. Striatum together with Subthalamic nucleus, Globus pallidus (internal, external and ventral pallidum) and substantia nigra (pars nigra and pars reticulata) [78] form the direct and indirect basal ganglia pathways. The cortical projections are processed in striatum and pass through direct and indirect pathways to the output nuclei of basal ganglia [13, 18]. Other afferents of striatum like the cortex are mostly excitatory and do not distinguish between dorsal and ventral striatum [76]. Striatum is mostly composed of GABAergic inhibitory medium spiny neurons (MSNs) which comprise 90–95 % [78] of striatum and spike rarely; the remainder of the striatum is mostly interneurons. Striatal interneurons are GABAergic and fast spiking and they form the main inhibitory input to MSNs and provide a winner-take-all mechanism [79].

Even though MSNs are structurally homogenous, they have different chemical properties and are classified according to their response to dopamine neurotransmitter [78]. The two most effective groups are MSNs with D1 type and D2 type receptors. While D1 type MSNs inhibit Globus Pallidus internal neurons and form a direct pathway, D2 type MSNs inhibit Globus Pallidus external neurons and form an indirect pathway. The direct pathway promotes the action initiation and selection; the indirect pathway prevents actions [13, 14]. MSNs with D1 type receptors form synaptic connections only with other D1 type while MSNs with D2 type receptors form synaptic connections with D2 and D1 types, where D2 types receive higher release probability for the excitatory synapses [75].

The role of dopamine on striatum behavior is vital: many neurological diseases and disorders are due to malfunctioning of dopamine neurons in striatum [34, 50, 51]. Also, striatal plasticity especially in reward based learning is due to dopamine modulation [18, 37, 80]. The role of striatal dopamine in schizophrenia has been known for long [81]. Thus modeling the role of dopamine on synchronization in striatum is important in understanding motor deficits, such as Parkinson's disease [7, 71] and neuropsychiatric disorders as ADHD [60]. Even though the neurode-generation is not obvious, as in Parkinson's disease, it is proposed that psychiatric diseases, such as obsessive compulsive disorder [1], major depressive disorder [73], schizophrenia [82], can be related to a cortico-basal ganglia circuit disorder.

Striatal Oscillations

The oscillations of basal ganglia structures have been studied for over a decade [7, 83]. While earlier recordings are mostly from rodents and primates, human data can now be collected using recordings from the implanted deep brain stimulation electrodes in human patients [71]. Though abnormally synchronized beta oscillations in basal ganglia are a hallmark of Parkinson's disease [8, 83, 84], beta oscillations in striatum are related to action selection network [69], whereas theta and gamma oscillations in striatum are observed for motor control [70]. Thus, even though MSNs rarely spike and striatal local field potential displays oscillations at low frequencies (delta band), still a rhythmic activity at a relatively high frequency is observed [51, 63, 71]. The difference between dorsal and ventral striatal oscillations are also considered [85], and the role of striatal oscillations in neuropsychiatric disorders are discussed [60, 73, 81]. In [84], the role of dopamine on striatal oscillations is discussed, where it is stated that the increases and decreases of dopamine level in basal ganglia push the dynamic state toward or away from beta oscillations.

Computational Methods: Point Neurons and Synaptic Connections

Since the aim is to propose a simple model that would help the investigation of dopamine on synchronous activity in striatum, the model will be composed of Izhike-vich point neurons and dynamic synaptic connections. The structure given in Fig. 2 is built to model the striatum as the input structure of basal ganglia circuits for cortical stimulations. This model consists of three groups of neuron populations: D1 type medium spiny neurons (MSND1), D2 type medium spiny neurons (MSND2) and interneurons (IN). MSND1 and MSND2 neuron populations consist of 100 neurons and IN group has 20 neurons.

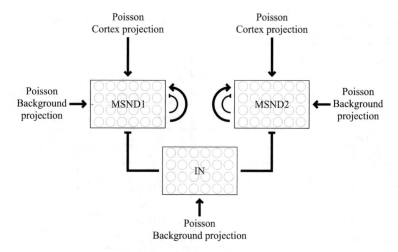

Fig. 2 A model of striatum which includes MSND1, MSND2 and IN. ⊣ is used for inhibitory and → for excitatory connections. Projections from the cortex and the background activity are modeled by poisson distributions

All three populations receive mostly excitatory afferents from different cortical and subcortical areas [76]. These are denoted as background oscillations in the model, but still some excitatory interconnections are implemented in the striatum, not only to model the effect of excitatory afferents further, but also the excitatory effect of neurotransmitters on MSNs. In order to represent background oscillations, these three different neuron groups are stimulated using Poisson distributions. The frequency of Poisson distributions is 20 spike/s for MSN groups, while this frequency is doubled for the IN group. These are denoted by arrows labeled Poisson in Fig. 2.

During the simulations, the sensory stimuli that cause activation in striatum from cortical excitation is modeled as a Poisson distribution with frequency 50 spike/s. As the sensory input will activate the basal ganglia circuit for a specific time interval, this sensory input is applied to the network for 1000 ms while the activation of the network is considered for 3000 ms.

To form the network of neurons, point neurons are considered and Izhikevich neuron model [86] is used. The equations governing the membrane potential and recovery variable are given in Eq. 1 and the reset condition is given in Eq. 2.

$$v' = 0.04v^2 + 5v + 140 - u + g_e - g_i$$
$$u' = a(bv - u) \tag{1}$$

$$if \ v > 30 \, mV, then \ v \leftarrow c \ and \ u \leftarrow u + d \tag{2}$$

MSNs are modeled as regular spiking and IN neurons are modeled as fast spiking neurons. The parameters for the neuron models are given in Table 1.

Table 1 Connection weights and Izhikevich model parameters. ⊣ is used for inhibitory and → for excitatory connections

Neuron parameters	MSN	IN
a	0.02/ms	0.1/ms
b	0.2/ms	0.2/ms
c	–65 * mV	–65 * mV
d	8 * mV/ms	2 * mV/ms
Connections	Weights	Probabilities
IN ⊣ MSND1	5 * mV/s	0.2
IN ⊣ MSND2	5 * mV/s	0.2
MSND1 → MSND1	α * 200 * mV/s	0.15
MSND1 ⊣ MSND1	10 * mV/s	0.25
MSND2 → MSND2	α * 200 * mV/s	0.15
MSND2 ⊣ MSND2	10 * mV/s	0.25
Poisson (Cortex projection) → MSND1	α * 10 * mV/s	One-to-one
Poisson (Cortex projection) → MSND2	α * 10 * mV/s	0.2
Poisson (Background activity) → MSND1	300 * mV/s	One-to-one
Poisson (Background activity) → MSND2	300 * mV/s	One-to-one
Poisson (Background activity) → IN	700 * mV/s	0.2

In all the Poisson groups that denote background activity in Fig. 2, it is assumed that there are 100 independent units and that the connections from Poisson units are one-to-one in the case of MSNs and all-to-all in the case of INs with probability of 20 %. All-to-all connections for a certain probability means that the connections are determined randomly and that only the specified percentage of possible connections exist. Thus, while each MSN gets one Poisson input, each IN gets approximately 20 Poisson inputs as background activation. The connections from cortex, representing the projection of cortical activity to striatum, are excitatory and assumed to have one-to-one connections for MSND1 population and all-to-all connections with probability of 20 % for the MSND2 population. The difference in connections between MSND1 and MSND2 populations is to model the higher release probability of the excitatory synapses in the indirect pathway.

Though these above mentioned connections are excitatory, there are also inhibitory connections in the network. The IN group has inhibitory connections to MSND1 and MSND2 populations, and these are assumed to be all-to-all connections with probability of 20 %. The connections within the MSND1 and MSND2 groups are all-to-all and excitatory with probability of 15 % and all-to-all and inhibitory with probability of 25 %. Though MSNs connect with each other through GABAergic synapses, we added excitatory connections to model the excitatory effect of neurotransmitters. This strategy of modeling helps to build the intrinsic oscillatory properties of MSNs, even with such a simple model. Otherwise the oscillations of MSNs in the model would just follow the oscillations of Poisson groups. All the changes in conductances (weights) and probabilities are given in Table 1.

All the connections to MSN groups and the IN group, from the IN group to MSN groups, and within groups are taken to be dynamical. Each MSND1 and MSND2 population has excitatory and inhibitory connections within its own group Fig. 2. The inhibitory connections within the MSN groups and from INs to MSN groups and the excitatory connections corresponding to background activity (the last three rows in the connections part of the Table 1) are modeled as in Eqs. 3 and 4. Synaptic activity occurs when presynaptic neurons fire; in this case, the conductance between pre- and post-synaptic neurons increases as given in Eq. 4.

$$\dot{g}_x = -\frac{g_x}{\tau_{syn}} \quad x \in \{e, i\}, \tag{3}$$

Here index e corresponds to excitatory, and i corresponds to inhibitory connections.

$$v^{(j)} > V_{thr} \quad then \quad g_x^{(k)} \rightarrow g_x^{(k)} + w_{x,j-k} \tag{4}$$

As the overall effect of dopamine is to promote activation, this effect is modeled using excitatory connections. That is why excitatory connections are added to the intraconnections of the striatum in the model. These excitatory connections and the connections for cortical projections are rendered to be modulatory due to the Spike Time Dependent Plasticity (STDP) rule [18, 78]. Modeling the effect of dopamine with STDP would also reveal its role in reward systems. In STDP, the weight of the synaptic connection between a presynaptic and postsynaptic neuron is modulated according to spike times. Whenever, presynaptic neuron fires before a post synaptic neuron, the connection weight is increased exponentially in proportion to how close together the spikes are. If the post synaptic neuron fires first, then the weight is decreased [87]. Synaptic time constants are also modulated to model the difference in how quickly dopamine affects D1 type versus D2 type MSNs, which is contrary due to the difference in the amount [78]. So, the connection dynamics are modeled to incorporate this difference as in Eqs. 5, 6 and 7, where $\alpha = DA$ for MSND1 neurons and $\alpha = 1/DA$ for MSND2 neurons. Here, DA denotes the amount of dopamine, and three different levels are considered throughout the simulations explained in section "Simulation Results". All parameters related to connections are given in Tables 1 and 2.

$$\dot{g}_x = -\frac{g_x}{\alpha \cdot \tau_{syn}} \quad x \in \{e\}, \tag{5}$$

$$v^{(j)} > V_{thr} \quad then \quad g_x^{(k)} \rightarrow g_x^{(k)} + w_{STDP,j-k} \tag{6}$$

$$w_{STDP,j-k} = \begin{cases} A_{pre}e^{-\tau/\tau_{pre}} & if \quad \tau > 0 \\ A_{post}e^{+\tau/\tau_{post}} & if \quad \tau < 0 \end{cases} \tag{7}$$

Table 2 Synaptic time constant and STDP parameters

Synaptic and STDP parameters	τ_{syn}	τ_{pre}	τ_{post}	g_{max}	A_{pre}	A_{post}
Values	10 ms	10 ms	10 ms	2	0.01	0.01

$$\tau = t_k^{post} - t_j^{pre} \qquad (8)$$

The network given in Fig. 2 and explained above is a simple model of striatum, when compared to other computational models using spiking neurons [88, 89]. The simulation results in the next section shows that still, this simple model is capable of providing necessary information about the oscillatory activity in the striatum.

Results: Simulations and Takehome Messages

The model of striatum introduced in the previous section is simulated in BRIAN environment [90] and the frequency analysis is done using [91] with three different DA levels to investigate the role of dopamine on synchronization of MSND1 and MSND2 groups. These three different levels correspond to normal dopamine level (DA = 1), depletion of dopamine (DA = 0.9) and excess of dopamine (DA = 1.1). Thus, the simulation results are given for the MSND1 and MSND2 populations. We especially considered the activity of two groups separately to investigate further the different role of two types of dopamine receptors on the overall activity of the basal ganglia circuits.

While running all simulations, initial values of point neurons are chosen to be randomly distributed around the resting values. Initial values for connection dynamics are all zero. The simulations are done 20 times for all the different cases, and spectrograms and histograms are obtained considering the mean of the spike rates for 20 trials. The synchronization measures are also obtained, considering 20 trials. For each case and for each population, the behaviour of a randomly choosen single neuron and synaptic connection is given, to give an idea about how a single neuron is behaving in a population. To show the activity of neuronal population also raster plots with spike rates are given.

Before begining to investigate the effect of dopamine on the activity of striatal medium spiny neurons, we first concentrated on the intrinsic activity of MSND1 and MSND2 population due to background stimulation. So, results for both MSN groups are first obtained without sensory stimuli for DA = 1 case. We considered this case as a resting state activity of MSNs and will benefit from these results to observe the effect of the sensory stimuli on striatum due to the cortical excitation. In order to show the activity of a single neuron for each population, membrane potential of one randomly chosen MSN from MSND1 group and MSND2 group are given in Fig. 3

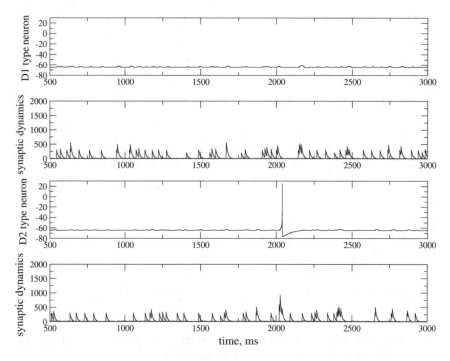

Fig. 3 Membrane potential and synaptic dynamics of a neuron, randomly selected in MSND1 and MSND2 groups, when DA = 1.0 and projections from cortex are removed. The *red lines* indicate the synaptic dynamics of the inhibitory connections

on the first and the third rows, respectively. The change in the synaptic behavior is also shown for a randomly chosen connection from each group in Fig. 3, on the second and the forth rows, respectively.

In order to follow the activity of the whole population for both groups, raster plots and spike rates are given in Fig. 4. As illustrated in these figures, there is almost no activity in both groups, which is expected as the MSNs spike rarely. Together, Figs. 3 and 4 show that the randomly choosen neuron in the MSND2 group spikes, but the neuron chosen from the MSND1 population does not. As these are just randomly choosen neurons, Fig. 4 gives a better view of overall activity and shows that the two groups behave almost similiar. The difference between MSND1 and MSND2 populations can be followed better from the synaptic dynamics in Fig. 3 and from spectrograms in Figs. 5 and 6. The difference between the synaptic dynamics of randomly choosen MSND1 and MSND2 indicates that MSND2 is fired more than MSND1. Even though this difference is not significant to effect the raster plots, the intrinsic behavior of MSND1 and MSND2 shows differences [92].

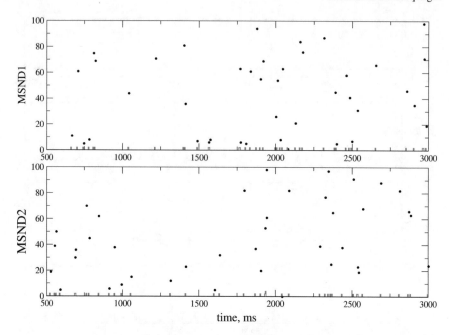

Fig. 4 Raster plot and spike rates of MSND1 and MSND2 groups when DA = 1.0 and projections from cortex are removed. The *red lines* indicate the spike rates

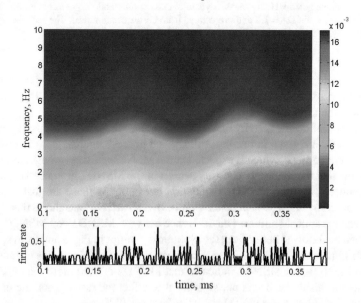

Fig. 5 The *above figure* is the time-frequency spectrogram and the *bottom figure* is the frequency spectrum of spike rates of MSND1 group when DA = 1.0 and projections from cortex are removed

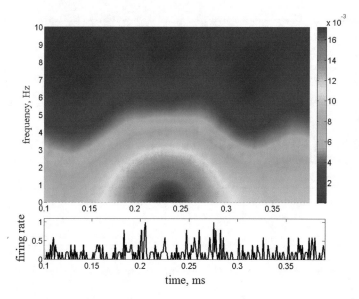

Fig. 6 The *above figure* is the time-frequency spectogram and the *bottom figure* is the frequency spectrum of spike rates of MSND2 group when DA = 1.0 and projections from cortex are removed

For this case, the frequency analysis for MSND1 and MSND2 are given in Figs. 5 and 6, respectively. The spectrograms obtained from the spike rates of each group, shows that there is delta band activity in resting as expected [51, 63]. This result verifies the validity of the model, as the oscillatory behavior of striatal neurons is observed rather than frequencies of the Poisson distributions applied for background oscillations. This is due to the modeling strategy explained in section "Computational Methods: Point Neurons and Synaptic Connections", where the behavioral properties are considered along with structural properties to obtain a simple but descriptive model.

The frequency range is cut at 10 Hz for both MSN groups, as the frequency histogram given in the inner figure in Fig. 21 denotes that there is no significant activity for higher frequencies.

Simulation Results

Now, we will proceed to obtain results for different DA levels, with the effect of the sensory stimuli causing activation in striatum due to cortical excitation. This effect is modeled is modeled as Poisson distribution with frequency 50 spike/s applied from 1000 to 2000 ms. Thus in the following figures in the first and last 1000 ms, there is no excitatory stimuli effecting the neuronal population in both MSN groups.

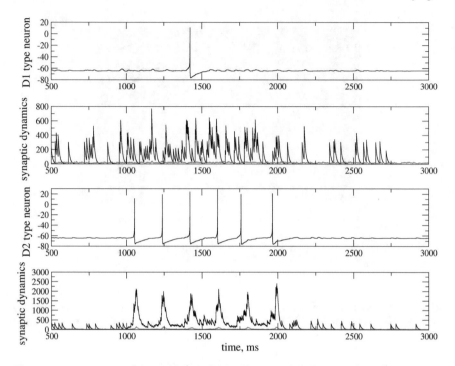

Fig. 7 Membrane potential and synaptic dynamics of a neuron, randomly selected in MSND1 and MSND2 groups, when DA = 0.9. The *red lines* indicate the synaptic dynamics of the inhibitory connections

First, we will look at the dopamine depletion case. Because the indirect pathway is active in the case of dopamine depletion, we expect to see activation in the MSND2 group, while the MSND1 group is quiescent [92, 93]. Looking at the membrane potential of a randomly chosen neurons from MSND1 and MSND2 groups in Fig. 7 on the first and the third rows, this case is very clearly obtained with the model. As it can be followed from Fig. 7 on the second and fourth rows, the activity in the D2 group renders active synaptic dynamics, while there is almost no change in the synaptic activity of the synaptic connection from the D1 group.

This activity of D2 group and quiescence of D1 group is also observed in raster plots of the populations as given in Fig. 8.

As it is followed from Figs. 5 and 6, only delta frequency band is observed in the resting case, but now in this case the delta band is the dominant frequency band for the MSND1 group (Fig. 9), whereas the beta band is also dominant for MSND2 group (Fig. 10). So, the simulation results obtained with the model gives a difference between D1 and D2 type dopamine receptors, while the over all activity of frequency bands are in agreement with the literature [8, 83, 84].

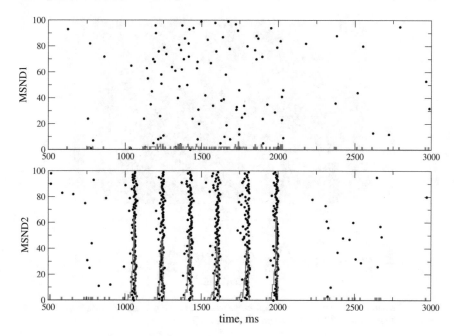

Fig. 8 Raster plot and spike rates of MSND1 and MSND2 groups when DA = 0.9. The *red lines* indicate the spike rates

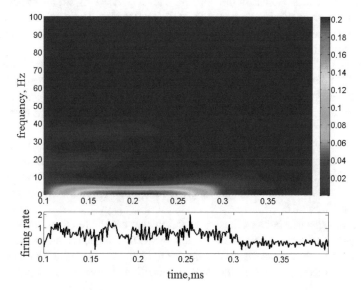

Fig. 9 The *above figure* is the time-frequency spectrogram and the *bottom figure* is the frequency spectrum of spike rates of MSND1 group when DA = 0.9

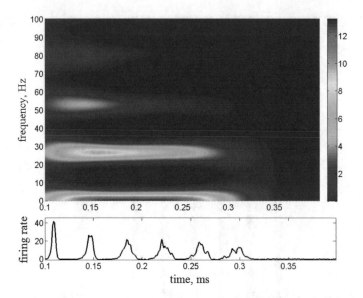

Fig. 10 The *above figure* is the time-frequency spectrogram and the *bottom figure* is the frequency spectrum of spike rates of MSND2 group when DA = 0.9

In case there is an excess in the dopamine level (DA = 1.1), we are expecting the reverse situation of what is observed in the DA = 0.9 case. The membrane potential and synaptic activity of randomly chosen single neuron are given in Fig. 11 and raster plots for both groups along with spike rates are given in Fig. 12.

As it can be followed from Figs. 11 and 12, the D1 population is more active than the D2 population, but when we look at the frequency analysis given in Figs. 13 and 14, the difference between delta and beta band activity is not great as in the D2 population for DA = 0.9 case. This is due to the connection difference between the two populations.

To see better the role of dopamine level on rhythmic activity, we will also look at the normal dopamine level, which is obtained by DA = 1 in the simulations. In this case, there is not much difference between activity of the MSND1 and MSND2 groups as it can be followed from single neuron behavior in Fig. 15 and neuronal populations in Fig. 16.

When the spectrograms of these populations are considered as given in Figs. 17 and 18 which correspond to MSND1 and MSND2 groups, respectively, then the difference in the frequency bands can be recognized. D2 population have more high frequency components, due to the dense connections between this population and cortex.

To have an overall understanding of the role of dopamine on different MSN groups, we will look at the power spectrograms and frequency histograms of the spike rates with different dopamine levels. In Fig. 19, it is clearly seen that as the dopamine level increases the activity of the MSND1 population increases. Mean-

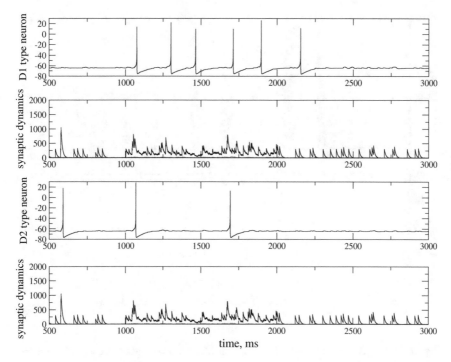

Fig. 11 Membrane potential and synaptic dynamics of a neuron, randomly selected in MSND1 and MSND2 groups, when DA = 1.1. The *red lines* indicate the synaptic dynamics of the inhibitory connections

while, only weak low frequency band delta are observed in the DA = 0.9 and DA = 1 cases, whereas delta oscillation is strengthened and beta oscillation emerges with the excess of dopamine.

In Fig. 20, the opposite of what is happening in the D1 population is occurring, but there is a difference, since the activity in the D2 population is greater when compared to the D1 population overall. Also, the strength of beta oscillations in the low dopamine case is more than the strength of delta oscillations, showing the role of D2 receptors in Parkinson's disease as mentioned in [51, 94].

Looking at the histograms for MSND1 and MSND2 groups with different dopamine levels (Fig. 21 gives a better understanding of high beta activity in MSND2 population).

To see the general effect of dopamine on the synchronization of MSND1 and MSND2 groups, the synchronization measure given by Eq. 9 [95] is used to obtain Fig. 22. The synchronization measure depends on the variance of neural activities of neurons in the considered time interval.

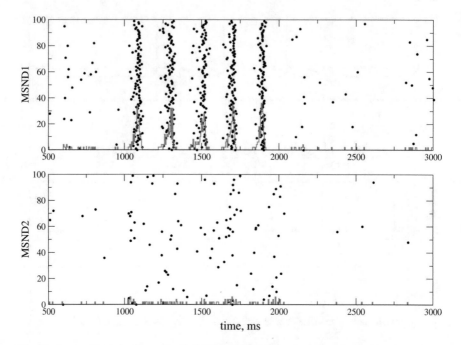

Fig. 12 Raster plot and spike rates of MSND1 and MSND2 groups when DA = 1.1. The *red lines* indicate the spike rates

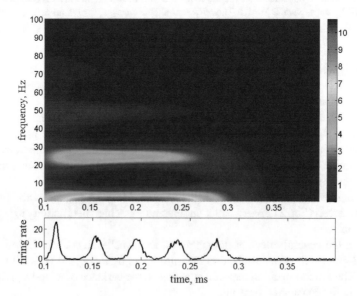

Fig. 13 The *above figure* is the time-frequency spectrogram and the *bottom figure* is the frequency spectrum of spike rates of MSND1 group when DA = 1.1

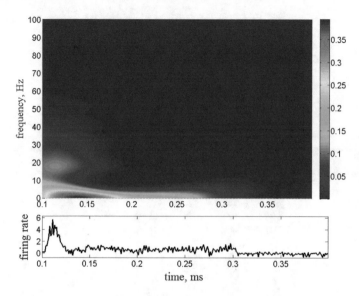

Fig. 14 The *above figure* is the time-frequency spectrogram and the *bottom figure* is the frequency spectrum of spike rates of MSND2 group when DA = 1.1

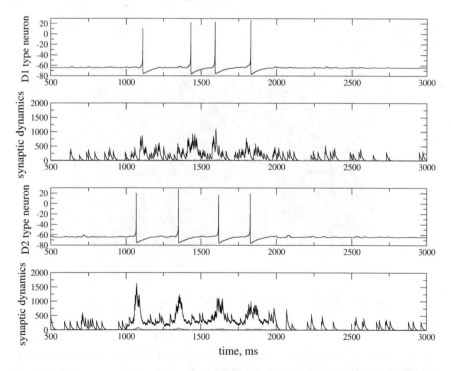

Fig. 15 Membrane potential and synaptic dynamics of a neuron, randomly selected in MSND1 and MSND2 groups, when DA = 1.0. The *red lines* indicate the synaptic dynamics of the inhibitory connections

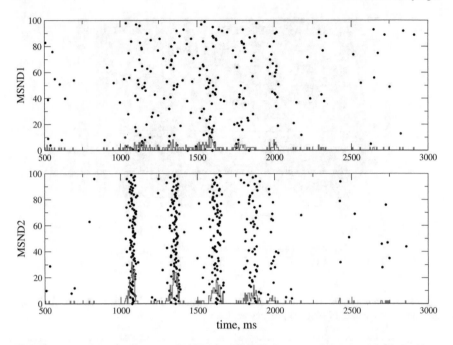

Fig. 16 Raster plot and spike rates of MSND1 and MSND2 groups when DA = 1.0. The *red lines* indicate the spike rates

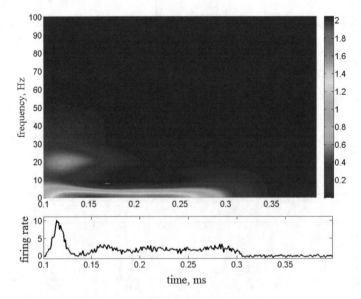

Fig. 17 The *above figure* is the time-frequency spectrogram and the *bottom figure* is the frequency spectrum of spike rates of MSND1 group when DA = 1.0

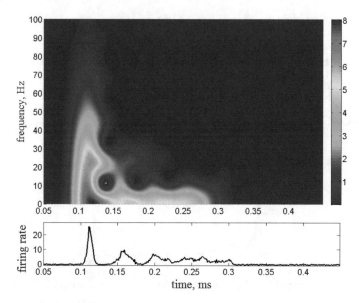

Fig. 18 The *above figure* is the time-frequency spectrogram and the *bottom figure* is the frequency spectrum of spike rates of MSND2 group when DA = 1.0

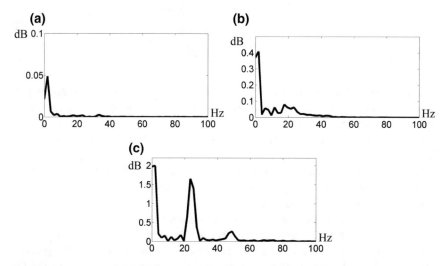

Fig. 19 Power spectrum of MSND1 groups for DA level is 0.9 (**a**), 1.0 (**b**) and 1.1 (**c**), respectively

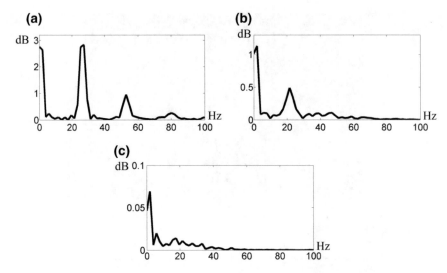

Fig. 20 Power spectrum of MSND2 groups for DA level is 0.9 (**a**), 1.0 (**b**) and 1.1 (**c**), respectively

$$
\rho = \frac{\langle \frac{1}{N} \sum_{j=1}^{N} v_j(t)^2 \rangle_t - \langle \frac{1}{N} \sum_{j=1}^{N} v_j(t) \rangle_t^2}{\frac{1}{N} \sum_{j=1}^{N} (\langle v_j(t)^2 \rangle_t - \langle v_j(t) \rangle_t^2)}
\tag{9}
$$

In the synchronization measure equation, N is the number of neurons and $v_j(t)$ is the membrane potential of the jth neuron. $\langle . \rangle_t$ denotes the mean over the time variable. Nominator corresponds to variance of $V(t)$ which is the mean over the neuron index variable and the denominator corresponds to the variance in membrane potentials.

In Fig. 22, the synchronization measures calculated for MSND1 and MSND2 groups are plotted with changing dopamine levels between 0.9 and 1.1 for 21 discrete values. For each value of dopamine level, 20 different simulations are carried out, and results of each are shown in Fig. 22 with stars and diamonds corresponding to MSND1 and MSND2, respectively. While the synchrony in MSND1 population increases as dopamine level increases, the reverse is true for the MSND2 population. This contrary effect of dopamine on synchrony is not symmetric for MSND1 and MSND2: D2 population are more sychronous in general.

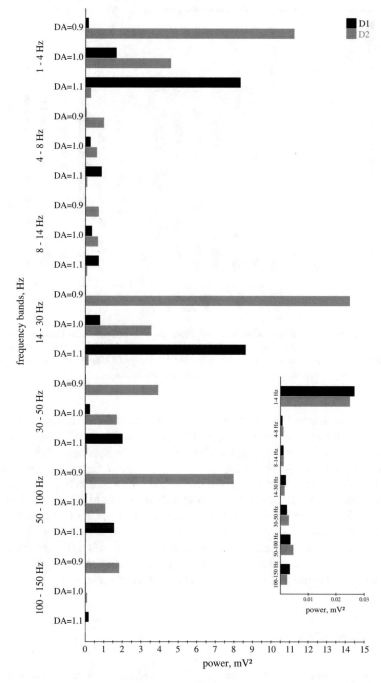

Fig. 21 Frequency histograms of spike rates of MSND1 and MSND2 groups when DA = 0.9, 1.0 and 1.1

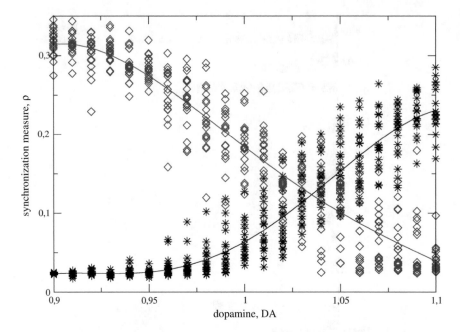

Fig. 22 DA versus synchronization measure. *Stars* show synchronization measure values for MSND1 and *diamonds* show synchronization measure values for MSND2 at DA levels which are changed from low DA level (0.9) to high DA level (1.1). *Black* and *blue lines* are obtained by 4th degree regression using the data for stars and diamonds, respectively

We also looked at the role of STDP on synchronization which can be can be followed from the raster plots for three different dopamine levels given in Fig. 23. The maximum value for the synaptic conductance and the amplitude of change in the conductance for postsynaptic neuron are increased here, compared to values given in Table 2. When the raster plots given in Fig. 23 is compared to Figs. 8, 12 and 16, it is obvious that STDP enhances synchronization in both MSN groups. It can be seen from Fig. 23 that the synchrony is preserved even after the stimuli is removed for MSND2 while DA = 0.9 and DA = 1 and for MSND1 when DA = 1.1. Though there are papers on STDP and DA's role in striatum, it is still not clear how STDP effects the coherent behavior of MSN [92, 96]. This result gives a clue that STDP's role could be to enhance the effect of dopamine on synchronization.

Take Home Messages

A simple neurocomputational model of striatal medium spiny neurons focusing on the role of dopamine is introduced and its simulation is carried out in the BRIAN environment. While the effect of dopamine on the synchronization is investigated,

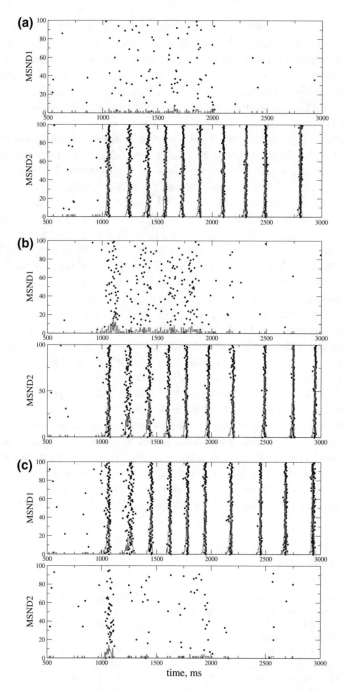

Fig. 23 Raster plot and spike rates of MSND1 and MSND2 groups when STDP weights are increased ($g_{max} = 5$, $A_{pre} = 0.1$ and $A_{post} = 0.1$) and DA = 0.9 (**a**), DA = 1.0 (**b**) and DA = 1.1 (**c**)

D1 and D2 types are considered separately and simulation results obtained show that dopamine level affects D1 and D2 type MSN differently. As the effect of dopamine is modeled similar to [88] where the role of STDP is considered, we also looked at the role of STDP on the synchronization for different levels of dopamine, which was not considered in [88]. It is observed that STDP enhances the synchronization of MSNs. The effect of STDP on striatal medium spiny neurons behavior has been mentioned in [92, 96] too.

The model employs Izhikevich point neurons and the synaptic connections are modeled with the simplest dynamic model. Even with such a simple model of neurons and synaptic connections, the intrinsic properties of striatal MSNs are captured. The role of dopamine is investigated at a single neuron level and also at population level and simulation results are displayed with membrane potential and synaptic connection dynamics, raster plots and frequency analysis. Also, a comprehensive analysis of synchronization is done using the synchronization measure given in [95].

The results show that even a simple computational model is capable of building a relation between data obtained by LFP and spiking neural network models. With further improvement of computational models in neuroscience based on rigorous mathematics, the community of neurologists and psychiatrists will benefit more from them when investigating their hypotheses [2, 3]. Even the simple model proposed here shows that computational models could be used as a step in building a more comprehensive understanding of cognitive processes and fulfill the gap between molecular and phenomenological levels. It can also provide tools to diagnose and treat psychiatric disorders besides neurodegenerative diseases.

While proposing computational models, it should be kept in mind that all models are abstractions and they give a partial explanation from the point of view of what is thought to be the most important aspect of the problem under investigation. So, the complexity level of the computational model should be determined based on the phenomenon to be modeled. In neuroscience, the computational model should be based on the functional and structural properties of the neuronal processes considered and if possible, the model results should be consistent with data observed at different levels, such as single neuron recordings, EEG/LFP and even fMRI.

References

1. Gunaydin, L. A., and Kreitzer, A. C., 2016. Cortico-basal ganglia circuit function in psychiatric disease. *Annual review of physiology* 78:327–350.
2. Montague, P.R., Dolan, R.J., Friston, K.J. and Dayan, P. 2012. Computational psychiatry. *Trends in cognitive sciences* 16(1):72–80.
3. Schroll, H. and Hamker, F., 2016. Basal Ganglia dysfunctions in movement disorders: What can be learned from computational simulations. *Movement Disorders*.
4. Deco, G., Jirsa, V. K., and McIntosh, A., 2011. Emerging concepts for the dynamical organization of resting-state activity in the brain. *Nature Reviews Neuroscience* 12(1):43–56.
5. Deco, G., Tononi, G., Boly, M., and Kringelbach, M., 2015. Rethinking segregation and integration: contributions of whole-brain modelling. *Nature Reviews Neuroscience* 16(7):430–439.

6. Sporns, O., Tononi, G., and Edelman, G. M., 2000. Connectivity and complexity: the relationship between neuroanatomy and brain dynamics. *Neural Networks* 13(8):909–922.

7. Hammond, C., Bergman, H., and Brown, P., 2007. Pathological synchronization in Parkinson's disease: networks, models and treatments. *Trends in neurosciences* 30(7):357–364.

8. Brown, P., and Williams, D., 2005. Basal ganglia local field potential activity: character and functional significance in the human, *Clinical neurophysiology*, 116(11):2510–2519.

9. Frederick, A., Bourget-Murray, J., Chapman, C. A., Amir, S., and Courtemanche, R. 2014. Diurnal influences on electrophysiological oscillations and coupling in the dorsal striatum and cerebellar cortex of the anesthetized rat, *Frontiers in Systems Neuroscience* 8(145):1–15.

10. Zhang, Y., Pan, X., Wang, R., and Sakagami, M., 2016. Functional connectivity between prefrontal cortex and striatum estimated by phase locking value. *Cognitive Neurodynamics* 10(3):245–254

11. Graybiel, A. M., Aosaki, T., Flaherty, A. W., and Kimura, M. 1994. The basal ganglia and adaptive motor control. *Science* 265(5180):1826–1831.

12. Alexander, G.E., Crutcher, M.D. and DeLong, M.R. 1989. Basal ganglia-thalamocortical circuits: parallel substrates for motor, oculomotor," prefrontal" and "limbic" functions. *Progress in brain research* 85:119–146.

13. Alexander, G.E. and Crutcher, M.D. 1990. Functional architecture of basal ganglia circuits: neural substrates of parallel processing. *Trends in neurosciences* 13(7):266–271.

14. Haber, S.N., Fudge, J.L. and McFarland, N.R. 2000. Striatonigrostriatal pathways in primates form an ascending spiral from the shell to the dorsolateral striatum. *The Journal of neuroscience* 20(6): 2369–2382.

15. Haber, S.N. and Knutson, B. 2010. The reward circuit: linking primate anatomy and human imaging. *Neuropsychopharmacology* 35(1):4–26.

16. Berns, G.S. and Sejnowski, T.J. 1996. How the basal ganglia make decisions. In *Neurobiology of decision-making*,eds. Damasio, A.R. Damasio, H., Christen, Y., 101–113. Berlin Heidelberg: Springer.

17. Balleine, B.W., Delgado, M.R. and Hikosaka, O. 2007. The role of the dorsal striatum in reward and decision-making. *The Journal of Neuroscience* 27(31):8161–8165.

18. DeLong, M.R. and Wichmann, T.2007. Circuits and circuit disorders of the basal ganglia. *Archives of neurology* 64(1):20–24.

19. Teicher, M.H., Anderson, C.M., Polcari, A., Glod, C.A., Maas, L.C. and Renshaw, P.F. 2000. Functional deficits in basal ganglia of children with attention-deficit/hyperactivity disorder shown with functional magnetic resonance imaging relaxometry. *Nature medicine* 6(4):470–473.

20. Chacko, R.C., Corbin, M.A. and Harper, R.G. 2000. Acquired obsessive-compulsive disorder associated with basal ganglia lesions. *The Journal of neuropsychiatry and clinical neurosciences*

21. Belin, D., Jonkman, S., Dickinson, A., Robbins, T.W. and Everitt, B.J. 2009. Parallel and interactive learning processes within the basal ganglia: relevance for the understanding of addiction. *Behavioural brain research* 199(1):89–102.

22. Murer, M.G., Tseng, K.Y., Kasanetz, F., Belluscio, M. and Riquelme, L.A. 2002. Brain oscillations, medium spiny neurons, and dopamine. *Cellular and molecular neurobiology* 22(5–6):611–632.

23. Heinz, A., Siessmeier, T., Wrase, J., Hermann, D., Klein, S. and others, 2004. Correlation between dopamine D2 receptors in the ventral striatum and central processing of alcohol cues and craving, *American Journal of Psychiatry*, 161:17831789.

24. Benabid, A.L. 2003. Deep brain stimulation for Parkinson disease. *Current opinion in neurobiology* 13(6):696–706.

25. Da Cunha, C., Boschen, S.L., Gmez-A, A., Ross, E.K., Gibson, W.S., Min, H.K., Lee, K.H. and Blaha, C.D. 2015. Toward sophisticated basal ganglia neuromodulation: review on basal ganglia deep brain stimulation. *Neuroscience & Biobehavioral Reviews* 58:186–210.

26. Sui, J., Pearlson, G.D., Du, Y., Yu, Q., Jones, T.R., Chen, J., Jiang, T., Bustillo, J. and Calhoun, V.D. 2015. In search of multimodal neuroimaging biomarkers of cognitive deficits in schizophrenia. *Biological psychiatry* 78(11):794–804.

27. Mayberg, H.S., Lozano, A.M., Voon, V., McNeely, H.E., Seminowicz, D., Hamani, C., Schwalb, J.M. and Kennedy, S.H. 2005. Deep brain stimulation for treatment-resistant depression. *Neuron* 45(5):651–660.

28. Kopell, B.H. and Greenberg, B.D. 2008. Anatomy and physiology of the basal ganglia: implications for DBS in psychiatry. *Neuroscience and Biobehavioral Reviews* 32(3):408–422.

29. Doya, K. 1999. What are the computations of the cerebellum, the basal ganglia and the cerebral cortex? *Neural networks* 12(7):961–974.

30. Taylor, J.G. and Taylor, N.R. 2000. Analysis of recurrent cortico-basal ganglia-thalamic loops for working memory. *Biological Cybernetics* 82(5):415–432.

31. Gurney, K., Prescott, T.J. and Redgrave, P. 2001. A computational model of action selection in the basal ganglia. I. A new functional anatomy. *Biological cybernetics* 84(6):401–410.

32. Gurney, K., Prescott, T.J. and Redgrave, P. 2001. A computational model of action selection in the basal ganglia. II. Analysis and simulation of behaviour. *Biological cybernetics* 84(6):411–423.

33. Humphries, M.D., Stewart, R.D. and Gurney, K.N. 2006. A physiologically plausible model of action selection and oscillatory activity in the basal ganglia. *The Journal of neuroscience* 26(50):12921–12942.

34. Guthrie, M., Myers, C.E. and Gluck, M.A., 2009. A neurocomputational model of tonic and phasic dopamine in action selection: a comparison with cognitive deficits in Parkinson's disease. *Behavioural brain research* 200(1):48–59.

35. Yucelgen, C., Denizdurduran, B., Metin, S., Elibol, R. and Sengor, N.S., 2012. A biophysical network model displaying the role of basal ganglia pathways in action selection. In *Artificial Neural Networks and Machine Learning-ICANN 2012* eds. A.E.P. Villa, W. Duch, P.Erdi, F. Masulli, G.Palm, 177–184. Berlin Heidelberg: Springer.

36. Sengor, N.S. and Karabacak, O. 2015. A computational model revealing the effect of dopamine on action selection. arXiv preprint arXiv:1512.05340.

37. Schultz, W., Dayan, P. and Montague, P.R. 1997. A neural substrate of prediction and reward. *Science* 275(5306):1593–1599.

38. Suri, R.E. and Schultz, W., 1998. Learning of sequential movements by neural network model with dopamine-like reinforcement signal. *Experimental Brain Research* 121(3):350–354.

39. Suri, R.E., Bargas, J. and Arbib, M.A., 2001. Modeling functions of striatal dopamine modulation in learning and planning. *Neuroscience* 103(1):65–85.

40. Dayan, P. and Balleine, B.W., 2002. Reward, motivation, and reinforcement learning. *Neuron* 36(2):285–298.

41. Joel, D., Niv, Y. and Ruppin, E., 2002. Actor-critic models of the basal ganglia: New anatomical and computational perspectives. *Neural networks* 15(4):535–547.

42. Montague, P.R., Hyman, S.E. and Cohen, J.D., 2004. Computational roles for dopamine in behavioural control. *Nature* 431(7010):60–767.

43. Haruno, M. and Kawato, M., 2006. Heterarchical reinforcement-learning model for integration of multiple cortico-striatal loops: fMRI examination in stimulus-action-reward association learning. *Neural Networks* 19(8):1242–1254.

44. Sutton, R.S., Barto, A.G. 1998. *Reinforcement Learning* (2nd printing), Cambridge: A Bradford Book. MIT Press.

45. Berns, G. and Sejnowski, T. 1994. A model of basal ganglia function unifying reinforcement learning and action selection. In *Joint Symposium on Neural Computation* 129–148.

46. Sengor, N.S., Karabacak, O. and Steinmetz, U. 2008. A computational model of cortico-striato-thalamic circuits in goal-directed behaviour. In *Artificial Neural Networks-ICANN 2008* eds.V. Kurkova, R.Neruda, J. Koutnik, 328–337. Berlin Heidelberg: Springer.

47. Gutkin, B.S., Dehaene, S. and Changeux, J.P., 2006. A neurocomputational hypothesis for nicotine addiction. *Proceedings of the National Academy of Sciences of the United States of America* 103(4):1106–1111.

48. Ahmed, S.H., Graupner, M. and Gutkin, B., 2009. Computational approaches to the neurobiology of drug addiction. *Pharmacopsychiatry* 42(1):S144.

49. Metin, S. and Sengor, N.S. 2012. From occasional choices to inevitable musts: A computational model of nicotine addiction. *Computational intelligence and neuroscience* 2012:18.
50. Terman, D., Rubin, J.E., Yew, A.C. and Wilson, C.J., 2002. Activity patterns in a model for the subthalamopallidal network of the basal ganglia. *The Journal of neuroscience* 22(7):2963–2976.
51. McCarthy, M.M., Moore-Kochlacs, C., Gu, X., Boyden, E.S., Han, X. and Kopell, N., 2011. Striatal origin of the pathologic beta oscillations in Parkinson's disease. *Proceedings of the National Academy of Sciences* 108(28):11620–11625.
52. Baddeley, A., 1974. Working memory. In *The Psychology of learning and motivation*, ed. G.H. Bower, 47–89. New York San Francisco London: Academic press.
53. O'Reilly, R.C. and Frank, M.J. 2006. Making working memory work: a computational model of learning in the prefrontal cortex and basal ganglia. *Neural computation* 18(2):283–328.
54. Frank, M.J. and O'Reilly, R.C. 2006. A mechanistic account of striatal dopamine function in human cognition: psychopharmacological studies with cabergoline and haloperidol. *Behavioral neuroscience* 120(3):497–517.
55. Celikok, U., Navarro-Lpez, E.M. and Sengor, N.S. 2016. A computational model describing the interplay of basal ganglia and subcortical background oscillations during working memory processes. *arXiv preprint* arXiv:1601.07740.
56. Schroll, H., and Hamker, F. H., 2013. Computational models of basal-ganglia pathway functions: focus on functional neuroanatomy, *Frontiers in systems neuroscience*, 7(122):1–18.
57. Varela, F., Lachaux, J.P., Rodriguez, E. and Martinerie, J. 2001. The brainweb: phase synchronization and large-scale integration. *Nature reviews neuroscience* 2(4):229–239.
58. Deco, G., Jirsa, V.K., Robinson, P.A., Breakspear, M. and Friston, K., 2008. The dynamic brain: from spiking neurons to neural masses and cortical fields. *PLoS Comput Biol* 4(8):e1000092.
59. Masimore, B., and Kakalios, J., and Redish, A.D., 2004. Measuring fundamental frequencies in local field potentials, *Journal of neuroscience methods*, 138(1):97–105.
60. Sukhodolsky, D. G., Leckman, J. F.,Rothenberger, A., and Scahill, L., 2007. The role of abnormal neural oscillations in the pathophysiology of co-occurring Tourette syndrome and attention-deficit/hyperactivity disorder, *European child & adolescent psychiatry*, 16(9):51–59.
61. Mandali, A., Rengaswamy, M., Chakravarthy, V.S. and Moustafa, A.A., 2015. A spiking Basal Ganglia model of synchrony, exploration and decision making. *Frontiers in neuroscience* 9.
62. Hélie, S., and Fleischer, P. J., 2016. Simulating the Effect of Reinforcement Learning on Neuronal Synchrony and Periodicity in the Striatum, *Frontiers in computational neuroscience*, 10.
63. Sharott, A., Moll, C.K., Engler, G., Denker, M., Grn, S. and Engel, A.K. 2009. Different subtypes of striatal neurons are selectively modulated by cortical oscillations. *The Journal of Neuroscience* 29(14):4571–4585.
64. Lenzi, T., De Rossi, S., Vitiello, N., Chiri, A., Roccella, S., Giovacchini, F., Vecchi, F. and Carrozza, M.C., 2009, September. The neuro-robotics paradigm: NEURARM, NEUROExos, HANDEXOS. In *Engineering in Medicine and Biology Society, 2009. EMBC 2009. Annual International Conference of the IEEE* 2430–2433. IEEE.
65. Prescott, T.J., Gonzlez, F.M.M., Gurney, K., Humphries, M.D. and Redgrave, P., 2006. A robot model of the basal ganglia: behavior and intrinsic processing. *Neural Networks* 19(1):31–61.
66. Denizdurduran, B. and Sengor, N.S., 2012. Learning how to select an action: A computational model. In *Artificial Neural Networks and Machine Learning ICANN 2012* 474–481. Berlin Heidelberg:Springer.
67. Fiore, V.G., Sperati, V., Mannella, F., Mirolli, M., Gurney, K., Friston, K., Dolan, R.J. and Baldassarre, G., 2014. Keep focussing: striatal dopamine multiple functions resolved in a single mechanism tested in a simulated humanoid robot. *Front. Psychol* 5(124):10–3389.
68. Ercelik, E. and Sengor, N.S., 2015, July. A neurocomputational model implemented on humanoid robot for learning action selection. In *Neural Networks (IJCNN), 2015 International Joint Conference* 1–6. IEEE.
69. Courtemanche, R., Fujii, N., and Graybiel, A.M., 2003. Synchronous, focally modulated β-band oscillations characterize local field potential activity in the striatum of awake behaving monkeys *The Journal of neuroscience*, 23(37):11741–11752.

70. Berke, J.D., Okatan, M., Skurski, J., and Eichenbaum, H.B., 2004. Oscillatory entrainment of striatal neurons in freely moving rats, *Neuron*, 43(6):883–896.

71. Gatev, P., Darbin, O., and Wichmann, T., 2006. Oscillations in the basal ganglia under normal conditions and in movement disorders, *Movement disorders*, 21(10):1566–1577.

72. Erbas, O., Oltulu, F., and Taskiran, D., 2013. Suppression of exaggerated neuronal oscillations by oxytocin in a rat model of Parkinson's disease, *Gen Physiol Biophys*, 32(4):517–525.

73. Northoff, G., 2015. Spatiotemporal psychopathology I: is depression a spatiotemporal disorder of the brains resting state, *Journal of Affective Disorder (in, revision)*

74. Belić, J.J., Halje, P., Richter, U., Petersson, P., and Kotaleski, J.H., 2016. Untangling cortico-striatal connectivity and cross-frequency coupling in L-DOPA-induced dyskinesia, *Frontiers in systems neuroscience*, 10.

75. Nambu, A., 2008. Seven problems on the basal ganglia. *Current opinion in neurobiology* 18(6):595–604.

76. Voorn, P., Vanderschuren, L.J.M.J., Groenewegen, H.J., Robbins, T.W. and Pennartz, C.M.A., 2004. Putting a spin on the dorsal–ventral divide of the striatum, *Trends in neurosciences*, 27(8):468–474.

77. Salamone, J.D., Correa, M., Farrar, A., and Mingote, S.M., 2007. Effort-related functions of nucleus accumbens dopamine and associated forebrain circuits, *Psychopharmacology*, 191(3):461–482.

78. Kandel, E.R., Schwartz, J.H. and Jessell, T.M. eds., 2000. *Principles of neural science* New York: McGraw-hill.

79. Koos, T. and Tepper, J.M., 1999. Inhibitory control of neostriatal projection neurons by GABAergic interneurons. *Nature neuroscience* 2(5):467–472.

80. Kreitzer, A.C. and Malenka, R.C., 2008. Striatal plasticity and basal ganglia circuit function. *Neuron* 60(4):543–554.

81. Abi-Dargham, A., Gil, R., Krystal, J., Baldwin, R.M., Seibyl, J.P. and others, 1998. Increased striatal dopamine transmission in schizophrenia: confirmation in a second cohort, *American Journal of Psychiatry*, 155:761767.

82. Graybiel, A.M., 2000. The basal ganglia, *Current Biology*, 10(14):509–511.

83. Bergman, H., Feingold, A., Nini, A., Raz, A., Slovin, H., and others, 1998. Physiological aspects of information processing in the basal ganglia of normal and parkinsonian primates, *Trends in neurosciences*, 21(1):32–38.

84. Leventhal, D.K., Gage, G.J., Schmidt, R., Pettibone, J.R., Case, A.C., and Berke, J.D., 2012. Basal ganglia beta oscillations accompany cue utilization, *Neuron*, 73(3):523–536.

85. Malhotra, S., Cross, R.W., Zhang, A., and Meer, M.A.A., 2015. Ventral striatal gamma oscillations are highly variable from trial to trial, and are dominated by behavioural state, and only weakly influenced by outcome value, *European Journal of Neuroscience*, 42(10):2818–2832.

86. Izhikevich, E.M., 2003. Simple model of spiking neurons. *IEEE Transactions on neural networks* 14(6):1569–1572.

87. Izhikevich, E.M., 2007. Solving the distal reward problem through linkage of STDP and dopamine signaling. *Cerebral cortex* 17(10):2443–2452.

88. Chersi, F., Mirolli, M., Pezzulo, G. and Baldassarre, G., 2013. A spiking neuron model of the cortico-basal ganglia circuits for goal-directed and habitual action learning. *Neural Networks* 41: 212–224.

89. Baladron, J. and Hamker, F.H., 2015. A spiking neural network based on the basal ganglia functional anatomy. *Neural Networks* 67:1–13.

90. Goodman, D.F. and Brette, R., 2008. The brian simulator. *Frontiers in neuroscience* 3, p.26.

91. Adam, G., Peter, R. 2006. LFP Analyser. *MATLAB Free Toolbox*.

92. Surmeier, D.J., Ding, J., Day, M., Wang, Z. and Shen, W., 2007. D1 and D2 dopamine-receptor modulation of striatal glutamatergic signaling in striatal medium spiny neurons. *Trends in neurosciences* 30(5):228–235.

93. Nicola, S.M., 2007. The nucleus accumbens as part of a basal ganglia action selection circuit, *Psychopharmacology*, 191(3):521–550.

94. Brooks, D.J., Ibanez, V., Sawle, G.V., Playford, E.D., Quinn, N., Mathias, C.J., Lees, A.J., Marsden, C.D., Bannister, R. and Frackowiak, R.S.J., 1992. Striatal D2 receptor status in patients with Parkinson's disease, striatonigral degeneration, and progressive supranuclear palsy, measured with 11C-raclopride and positron emission tomography. *Annals of neurology* 31(2):184–192.
95. Hansel, D., Mato, G., Meunier, C. and Neltner, L., 1998. On numerical simulations of integrate-and-fire neural networks. *Neural Computation* 10(2):467–483.
96. Surmeier, D.J., Plotkin, J. and Shen, W., 2009. Dopamine and synaptic plasticity in dorsal striatal circuits controlling action selection. *Current opinion in neurobiology* 19(6):621–628.

A Neural Mass Computational Framework to Study Synaptic Mechanisms Underlying Alpha and Theta Rhythms

Basabdatta Sen Bhattacharya and Simon J. Durrant

Introduction

Building biologically-inspired computational tools is gaining in popularity for advancing neuroscience research towards understanding and predicting neurological and psychiatric disorders [6, 16]. However, computational time and resources have been a major challenge towards such endeavours. Population-level representations (as opposed to networks of single neuronal models) that can simulate higher-level brain dynamics observed in Electroencephalogram (EEG) and Local Field Potentials (LFP) can address the computational constraints to a fair extent, for example neural mass models [21, 53, 54]. The term 'neural mass' was coined by Walter J. Freeman [32] to define the collective behaviour of a mesoscopic scale neuronal population ($\approx 10^4$–10^7 neurons) that are packed densely in a spatial area of 0.3–3 mm and may be assumed as a single entity [48]; Freeman's work comprised neuronal behaviour and dynamics in the olfactory pathway. Around the same time, Wilson and Cowan proposed the mathematical framework for modelling feed-forward and -back connections between excitatory and inhibitory 'point-neurons' (an ensemble representation of a neuronal population, along the lines of neural mass) that could mimic brain dynamics such as seen in EEG and LFP [75]. This mathematical framework forms the basis of a seminal work by daSilva [19] and Zetterberg et al. [76], where they introduced a block-diagram-like approach (from Control Engineering) to model a simple thalamocortical circuitry of the visual pathway for simulating alpha rhythms—oscillatory activity within 8–13 Hz seen in LFP and EEG record-

B. Sen Bhattacharya (✉)
University of Lincoln, INB 3225, Engineering Hub, Brayford Pool,
Lincoln LN6 7TS, UK
e-mail: basab.sen.b@gmail.com

S.J. Durrant
School of Psychology, University of Lincoln, Brayford Pool, Lincoln LN6 7TS, UK
e-mail: sidurrant@lincoln.ac.uk

© Springer International Publishing AG 2017
P. Érdi et al. (eds.), *Computational Neurology and Psychiatry*,
Springer Series in Bio-/Neuroinformatics 6,
DOI 10.1007/978-3-319-49959-8_14

ings. The interaction between thalamic and cortical networks are now well known to underlie brain oscillatory patterns, referred to commonly as 'brain rhythms' [14], corresponding to cognition, perception and sleep-wake transitions [45, 65]. Subsequently, Lopes da Silva's thalamocortical neural mass model mimicking alpha rhythm was further extended in [43, 66, 74]; these works set a 'trend' of adopting neural mass models of thalamocortical circuitry in clinical neuroscience research towards mimicking brain rhythms of both health and disease conditions [13, 37, 55, 63, 67, 70] (see [22] for a review).

Alpha rhythms are traditionally believed to represent an idling state of the brain and is most prominent in EEG from occipital scalp (the seat of visual cortex) when a subject is awake and resting with eyes closed; the rhythms subside when the eyes are opened. However, in current times and with advancing research, it has emerged that the alpha rhythms also play an integral role in various awake cognitive states [50]. Furthermore, alpha to theta (4–7 Hz) shift is an EEG marker of brain state transition from quiet wakefulness (preceding sleep) to a state of drowsiness (sleep stage-I). At the same time, anomalies of alpha rhythmic oscillations are indicators of several disease conditions, for example 'slowing' (reduced frequency of peak power) of the alpha rhythm is a hallmark of EEG in Alzheimer's disease [8, 44]. Similarly, thalamocortical dysrhythmia (TCD), a shift of peak frequency from alpha to theta, is an EEG marker of several disorders such as Tinnitus, Neurogenic Pain, Depression [41, 49, 60]. We have proposed a modification to the alpha rhythm model in [19] to show a significant effect of reduced synaptic connectivity from inhibitory cell populations in simulating AD related conditions in the model [8]. This is consistent with autopsy studies in AD showing impaired inhibitory pathways [31]. However, a major constraint in the classic neural mass computational models is the use of Ralls alpha function [56]; although the alpha function is a fair estimation of the synaptic transmission process [4], it falls short when investigating disease conditions, where attributes such as transmitter concentration and ion-channel states might play a significant role. More recent research have used parameters from an experimental study [36] on a thalamic slice from mammal to model receptor dynamics during synaptic transfer in a neural mass model [67]. Along these lines, another biologically plausible alternative for modelling ligand-gated and secondary-messenger-gated synaptic transmission is to implement kinetic models [24, 25], which takes into account the transmitter concentration in the synaptic cleft and subsequent state of channels involved in generating the post-synaptic potential. Discussing the future benefits of kinetic modelling of synaptic processes, the authors in [23] reflect thus: "A considerable amount of experimental data is available from measurements of the average activity of populations of brain cells: recordings of electroencephalogram, local field potentials, magnetoencephalograms, optical recordings, magnetic resonance images, etc. It would be interesting to attempt to establish a relationship between such global measurements and dynamics at the molecular level". Indeed, such an approach was already adopted by Aradi and Erdi [2] and Erdi et al. [30] in network models of hippocampal neurons to investigate neuropharmacological solutions to neuro-psychiatric disorders.

In prior works, a novel approach is adopted for classic neural mass models where Rall's alpha function is replaced by kinetic models of Glutamatergic and γ-amino-butyric-acid (GABA)-ergic synapses mediated by α-amino-3-hydroxy-5-methyl-4-isoxazolepropionic-acid (AMPA) and $GABA_{A,B}$ neuroreceptors respectively [5, 10]. The motivation for these works have been to take a step forward in building computational tools that can complement experimental research in understanding the underlying cellular mechanisms of anomalous EEG signals in neurological and psychiatric disorders. In addition, this novel approach reduces the computational time by an order of 10 compared to the classic neural mass model for similar thalamocortical structures. Implementing this approach for a thalamocortical model have demonstrated model sensitivity to neurotransmitter concentration, forward and reverse rates of reaction and leak conductance in the model in effecting change in time-series patterns as well as shifting the power spectrum [5].

In this work, we present a thalamocortical model representing the three neuronal population of the Lateral Geniculate Nucleus (LGN) viz. the thalamocortical relay (TCR) neurons that are the main carriers of sensory information to the cortex; the thalamic interneurons (IN); the thalamic reticular nucleus (TRN) that receive a 'copy' of all information transfer between the TCR and the cortex [61]. The synaptic layout of the model is based on data obtained from LGNd (dorsal) slices of cat and rat thalamus [40, 45, 62]. Each parameter corresponding to synaptic attributes in the model is a representation of the population average of the parameter value in the respective neural mass. The AMPA and $GABA_A$ synapses in the model are simulated by two-state kinetic models. In the visual pathway, simultaneous LFP and EEG recordings from the TCR cells (thalamus) and the cortex respectively show a high degree of correlation [17, 20]. Thus, the output from the TCR population in the model is assumed to be a simulation of LFP recordings from LGN. We assume a de-corticated LGN (thalamus), similar to an approach adopted in early experimental works [3, 46, 51, 52, 64] looking into independent thalamic cell behaviour, which allowed an in-depth understanding of the thalamus as a key player in generating and sustaining brain oscillations. These pioneering works on LGN slices in vitro showed that the thalamus is capable of displaying oscillations even in a de-corticated state, and that similar thalamic mechanisms underlie alpha and theta rhythms of EEG and LFP [42]. The results from our model conform to these experimental observations and identify the neurotransmitter concentration in the synaptic cleft as a crucial parameter that impact transition between alpha and theta oscillatory dynamics. In addition, our results indicate distinct inhibitory roles for the IN and the TRN population, the former acting as a 'balancing' element in the circuit, the latter taking a dominant role in effecting (spindle-like) 'waxing-and-waning' and limit cycle oscillations.

In section "A Kinetic Model Based Framework for Neural Mass Modelling of the Thalamic Circuit", we present an overview of kinetic modelling of AMPA and $GABA_A$ based neuroreceptors in context to their embedding in the neural mass models. This is followed by a description of the neural mass model of LGN presented in this work. The results of model simulation are demonstrated in section "Results and Discussion" and their implications are discussed. We conclude in section

"Conclusion" with a recapitulation of the salient observations made in section "Results and Discussion" and a discussion of ongoing and future work.

A Kinetic Model Based Framework for Neural Mass Modelling of the Thalamic Circuit

A simple mathematical model defining the chemical kinetics of ion-channel is proposed in [24] as a computationally efficient means of studying detailed synaptic attributes in neuronal models. The dynamics of each ion channel is represented by

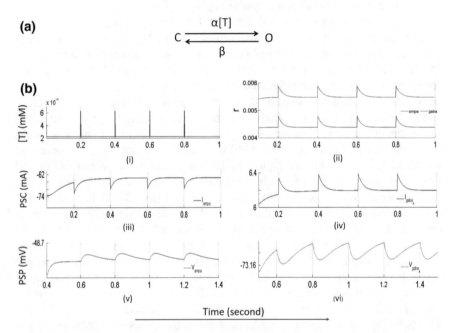

Fig. 1 **a** The state transition diagrams for AMPA and GABA$_A$ neuroreceptor dynamics defined in Eq. (1); α and β are rate of transitions between the two states; The desensitised state of the ion-channels are ignored in this work for brevity (see [25] for a detailed comparison of kinetic models simulating more than two-states). **b** The response function [T] corresponding to an impulse train of magnitude 2 mV and a base voltage of −65 mV applied at an arbitrarily selected rate of 4 Hz is shown in panel (i). (The readers may note that for the purposes of this work, all parameters in Eq. (2) are same for both AMPA and GABA$_A$ mediated synapses. Thus, any change in the equation parameters result in a change in [T] in all synaptic junctions of the model). Panel (ii) shows the corresponding change in proportion of open ion channels (r). Panels (iii) and (v) show the negative post synaptic membrane current (PSC) and the positive (excitatory) post synaptic membrane potential (PSP) corresponding to the AMPA neuroreceptor mediated synapses. Conversely, a positive PSC and a negative (inhibitory) PSP is shown for GABA$_A$ mediated synapses in panels (iv) and (vi) respectively

two states as shown in Fig. 1a: an unbound state referred to as 'C' denoting the closed state of the ion channels; a bound state referred to as 'O' denoting open states of the same. While the 'two-state' model is an abstracted representation of complex ion-channel dynamics, it is reported to be a fair approximation of the higher state models [25]. In the following sections, we present the two-state kinetic model of ion channels in the post-synaptic neuronal membrane responding to AMPA and $GABA_A$ neuroreceptor based signal transmission, which are then implemented in a neural mass model of the LGN to simulate LFP dynamics.

AMPA and GABA$_A$ Based Neurotransmission

The proportion of open ion channels on the post-synaptic cell membrane (Y) corresponding to the synapse mediated by the neurotransmitter receptor $\bar{\eta} \in \{AMPA, GABA_A\}$ is represented by $r_Y^{\bar{\eta}}$, and the two state dynamics are defined in Eq. (1):

$$\frac{dr_Y^{\bar{\eta}}(t)}{dt} = \alpha^{\bar{\eta}}[T]_\chi(1 - r_Y^{\bar{\eta}}(t)) - \beta^{\bar{\eta}}r_Y^{\bar{\eta}}(t) \tag{1}$$

where $\alpha^{\bar{\eta}}$ and $\beta^{\bar{\eta}}$ are the rate transitions from the open to the closed state and vice-versa respectively and corresponding to the synapses mediated by $\bar{\eta}$. Furthermore, $r_Y^{\bar{\eta}}$ is a function of the concentration of neurotransmitters in the synaptic cleft ($[T]_\chi$), which in turn is a function of the pre-synaptic cell (χ) membrane potential (V_χ) and is approximated as a sigmoid function shown in Eq. (2).

$$[T]_\chi(V_\chi(t)) = \frac{T_{max}}{1 + exp\left(-\frac{V_\chi(t) - V_{thr}}{\sigma}\right)} \tag{2}$$

T_{max} is the maximum neurotransmitter concentration and is well approximated by 1 mM (milliMole) [23]. The parameter V_{thr} represent the threshold at which $[T]_\chi = 0.5T_{max}$ while σ affects the steepness of the sigmoid. The resulting post-synaptic current (PSC) is defined in Eq. (3):

$$I_Y^{\bar{\eta}}(t) = g^{\bar{\eta}}r_Y^{\bar{\eta}}(t)(V_Y^{\bar{\eta}}(t) - E^{\bar{\eta}}) \tag{3}$$

where $g^{\bar{\eta}}$ and $E^{\bar{\eta}}$ are the maximum conductance and membrane reversal potential respectively of the post-synaptic cell corresponding to the $\bar{\eta}$ mediated synapse; $V_Y^{\bar{\eta}}$ is the post synaptic membrane potential (PSP) corresponding to the $\bar{\eta}$-mediated synapse, and is defined in Eq. (4):

$$V_Y^{\bar{\eta}}(t) = \frac{1}{\kappa_m} \int I_Y^{\bar{\eta}}(t)\, dt, \tag{4}$$

Table 1 The range of parameter values are referred from [9, 24, 36, 67, 71]. The exact parameter values in the model are set by trial simulations such that the model output time-series has a dominant frequency within the alpha band (8–13 Hz). (A) Data for the forward (α) and reverse (β) rates of synaptic transmission is according to the range mentioned in [24, 36]. Note that the units used in our model is at a different time scale (sec^{-1}), and thus absolute figures are different from these references. The data for maximal synaptic conductance $g^{\bar{\eta}}$ is in the range mentioned in [36, 71]; note that the unit for this parameter in our model is μS/cm^2. Data for $E^{\bar{\eta}}$ is as in [36, 71]. Specific data relating to the thalamic IN synapses are not mentioned in any of these sources, and are set as similar to those of TRN in this work. The 'RET' in the parameter superscripts refer to the retina as the source of input to the model. (B) The leakage current in the model cell populations are assumed to be due to Potassium (K) mainly. Thus, the leakage conductance and reverse potentials parameters in the model are in the range mentioned in [36, 71]. The resting state membrane potential for TCR and TRN are as in [71]; the resting state membrane potential for IN is set arbitrarily at a hyperpolarised value with respect to that of the TCR. The resting membrane potential for RET is set at -65 mV, and is simulated by a random white noise with mean -65 mV and standard deviation 2 mV^2. This signal represents the mean membrane voltage of the retina as an afferent to the TCR cell populations. Thus, there is no ODE corresponding to the RET in the model, and its leak conductance and leak reversal potentials are indicated with 'X'

(A) Neurotransmission parameters		
Parameters	Value	Synaptic pathway
$\alpha\ ((\text{mM})^{-1} \cdot (\text{s})^{-1})$	1000	AMPA, GABA$_A$
$\beta\ (\text{s}^{-1})$	50	AMPA
	40	GABA$_A$
$g^{\bar{\eta}}\ (\mu\text{S/cm}^2)$	300	AMPA (RET to TCR)
	100	AMPA (RET to IN) (TCR to TRN)
	100	GABA$_A$
$E^{\bar{\eta}}\ (\text{mV})$	0	AMPA
	-85	GABA$_A$ (TRN/IN to TCR)
	-75	GABA$_A$ (TRN (IN) to TRN (IN))

(B) Cell membrane parameters				
	RET	TCR	IN	TRN
$g^{leak}\ (\mu\text{S/cm}^2)$	X	10	10	10
$E^{leak}\ (\text{mV})$	X	-55	-72.5	-72.5
$V_{rest}\ (\text{mV})$	-65	-65	-75	-85

where κ_m is the post-synaptic membrane capacitance. The input V_{χ} is an impulse train of amplitude 2 mV at a base value of -65 mV and at an arbitrarily selected rate of 4 Hz. The resulting neurotransmitter concentration $[T]_{\chi}$, probability of opening of ion channel $r_{Y}^{\bar{\eta}}$, PSC ($I_{Y}^{\bar{\eta}}$) and PSP ($V_{Y}^{\bar{\eta}}$) are shown in Fig. 1b; all parameters used for generating Fig. 1b are as mentioned in Table 1(A).

In the following section, we present a neural mass model of the thalamic circuitry where the excitatory and inhibitory alpha functions are replaced with two-state kinetic models of AMPA and GABA$_A$ mediated synapses respectively as defined in Eqs. (1)–(4).

A Neural Mass Model of the Lateral Geniculate Nucleus Implementing Synaptic Kinetics

Experimental research on the Lateral Geniculate Nucleus (LGN) of mammals and rodents suggest two basic cell types: the thalamocortical relay (TCR) cells and the interneurons (IN). In addition, the thalamus is surrounded by a thin sheet of cells that receive copies of both afferent and efferent communication between thalamus and the cortex; this group of cells is considered as a part of the thalamus and is called the thalamic reticular nucleus (TRN). The neural mass model presented in this work is based on the synaptic layout of the LGN and is shown in Fig. 2. In addition to the fast excitatory (AMPA) and inhibitory ($GABA_A$) synapses in the LGN cell populations, the TRN also makes a slow inhibitory synapse on the TCR cells mediated by $GABA_B$ neuroreceptors [38]. However, in a previous study, the kinetic model of the $GABA_B$ synapse did not show any significant effect on the thalamocortical model output. Thus, this pathway is ignored in the current work (and is being investigated as a part of an ongoing work on the model). Also, while a feedback from the TCR to the IN cell

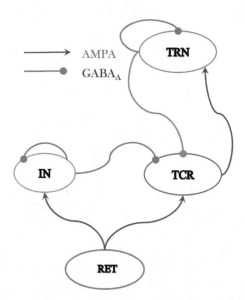

Fig. 2 The synaptic layout of the model is based on experimental data obtained from the dorsal Lateral Geniculate Nucleus (LGNd) of mammals and rodents. Both TCR and IN cell populations of the LGN receive excitatory Glutamatergic inputs from the retinal spiking neurons (RET) that are mediated by AMPA neuroreceptors. The IN cell populations make a $GABA_A$ receptor mediated inhibitory synapse on themselves as well as on the TCR population. Information on the synaptic pathway from the TCR to IN is ambiguous in literature and is ignored in the current work. The TCR population make AMPA mediated excitatory synapses on the TRN population, while the TRN population make $GABA_A$ mediated inhibitory synapses on the TCR population as well as on itself. All synaptic connectivity parameter values in the model are derived from experimental data presented in [40, 45] and are mentioned in Table 2

population is suggested based on extrapolations from EEG-based studies [18, 77], the biological plausibility of this pathway is yet to be confirmed from experimental studies to the best of our knowledge. Thus, this pathway is ignored in the current model (as in our previous works [9, 11, 16]).

The underlying mathematical framework for the neural mass model implementing synaptic kinetics (Fig. 1a) is defined in Eqs. (5)–(9). The input to the model is assumed to be the ensemble membrane potential of pre-synaptic retinal cell populations (V_{ret}) in a resting state with no sensory input and is simulated using a Gaussian white noise [20] with a biologically plausible mean value of −65 mV. The standard deviation of the noise is set by trial and error to 2 mV^2 to reduce stiffness of the differential equations. All variables and parameters defined in the equations below are assumed to be the 'ensemble representation' corresponding to a neural mass; this is similar to the concept of a 'point neuron' representation of a localised neuronal population acting in synchrony. The output of the TCR population, hereafter referred to as the 'model output', is considered as the simulation of LFP dynamics recorded from LGN in mammals and rodents (for example from dogs in [19]).

$$[T]_{\bar{\chi}}(V_{\bar{\chi}}(t)) = \frac{T_{max}}{1 + exp(-\frac{V_{\bar{\chi}}(t) - V_{thr}}{\sigma})} \tag{5}$$

$$\frac{dr_{\bar{Y}}^{\bar{\eta}}(t)}{dt} = \alpha^{\bar{\eta}} \cdot [T]_{\bar{\chi}}(V_{\bar{\chi}}(t)) \cdot (1 - r_{\bar{Y}}^{\bar{\eta}}(t)) - \beta^{\bar{\eta}} \cdot r_{\bar{Y}}^{\bar{\eta}}(t) \tag{6}$$

$$I_{\bar{Y}}^{\bar{\eta}}(t) = g^{\bar{\eta}} \cdot r_{\bar{Y}}^{\bar{\eta}}(t) \cdot (V_{\bar{Y}}(t) - E^{\bar{\eta}}) \cdot C_{conn} \tag{7}$$

$$\kappa_m \frac{dV_{\bar{Y}}(t)}{dt} = - \sum_{\bar{Y} \in \{TCR,IN,TRN\}} (I_{\bar{Y}}^{\bar{\eta}}(t) + I_{\bar{Y}}^{leak}(t)), \tag{8}$$

$$I_{\bar{Y}}^{leak}(t) = g_{\bar{Y}}^{leak}(V_{\bar{Y}}(t) - E^{leak}), \tag{9}$$

where $\bar{\chi} \in \{RET, TCR, IN, TRN\}$ represent the pre-synaptic cell populations; $\bar{Y} \in \{TCR, IN, TRN\}$ represent the post-synaptic cell populations. The variables and parameters in Eqs. (5)–(8) are similar to those defined in Eqs. (1)–(4) and mentioned in Table 1(A). The normalised synaptic connectivity parameter C_{conn} in Eq. (7) represent the collective 'fan-in' of synapses from a pre-synaptic population on to the post-synaptic population dendrites. Specific connectivity parameters in each synaptic pathway of the model shown in Fig. 2 is denoted as $C_{\bar{u}\bar{v}\bar{w}}$; see the legend of Table 2 for further details. All synaptic connectivity parameter values are obtained from literature on experimental data, for example in the TCR population, experimental studies on a sample population of cells indicate that approximately 7.1 % of the total number of incoming synapses are from the retinal ganglion cells, while 30.9 % of the synapses are from inhibitory cell populations. However, to the best of our knowledge, a quantitative distinction between the inhibitory inputs from the TRN and IN populations on to the TCR is not yet available. Thus, we have set the inhibitory connectivity parameter values for TCR afferents to an arbitrarily selected proportion of

Table 2 Base values of the synaptic connectivity parameters C_{conn} in Eq. (7) and derived from experimental data on LGNd of mammals and rodents [40, 45, 62] (as in our previous works [9, 16]). The nomenclature for the specific connectivity parameter in each synaptic pathway is $C_{\tilde{u}\tilde{v}\tilde{w}}$: each parameter value in the table is a normalised figure that represents the percentage of the synaptic contacts made on the post-synaptic cell population u by the pre-synaptic cell population v, and w represents the nature of the synapse i.e. excitatory (e) or inhibitory (i). The model neuronal populations are represented by the letters t for TCR, n for TRN, i for IN and r for retina. For synaptic contacts by a cell population on itself, v is represented by s, which stands for a connection from 'self'. All 'X' indicate a lack of biological evidence for any synaptic connectivity in the specific pathway

Efferents → Afferents ↓	TCR	IN	TRN	Retinal
TCR	X	C_{tii} $\frac{5}{8}$ of 30.9	C_{tni}^a $\frac{3}{8}$ of 30.9	C_{tre} 7.1
IN	X	C_{isi} 23.6	X	C_{ire} 47.4
TRN	C_{nte} 35	X	C_{nsi} 20	X

62.5(IN) : 37.5(TRN) in the model.[1] A similar abstraction is followed for parameterising the synaptic connectivity in the remaining model pathways and are mentioned in Table 2. While there is variation in reported data for the connectivity in literature on experimental studies, we follow the data specified in [40, 45]. The parameter $I_{\tilde{Y}}^{leak}$ in Eq. (8) is the ensemble leak current of the post-synaptic population membrane defined in Eq. (9), where $g_{\tilde{Y}}^{leak}$ and E^{leak} are the maximum leak conductance and leak reversal potential respectively of the post-synaptic cell population $\tilde{Y}$. The leak parameters as well as the resting membrane potentials for the model cell populations are mentioned in Table 1(B).

Empirical Methods

The ODEs in Eqs. (5)–(9) are solved using the 4th/5th order Runge-Kutta-Fehlberg method (RKF45) in Matlab for a total duration of 40 s at a resolution of 1 ms. The output voltage time-series is averaged over 20 simulations, where each simulation

---

[1] In previous works with GABA$_B$ pathway from the TRN to the TCR, we have maintained an equal proportion of fan-in on the TCR from the IN and TRN. Thus, total synaptic contact from TRN to TCR has been $\frac{1}{2}$ of 30.9 %. Now, TRN makes both GABA$_A$ and GABA$_B$ contact on the TCR; thus the proportionality of GABA$_A$: GABA$_B$ was maintained at 3 : 1, as the GABA$_B$ pathway showed minimal effect on the model. Here, we have ignored the GABA$_B$ pathway and diverted the proportion of connectivities in this pathway i.e. $\frac{1}{8}$ of 30.9 % to the GABA$_A$ pathway from the IN to the TCR. A combinatorial study on the possible proportionates in the GABA$_A$ pathway in the model remains to be explored in ongoing and future works.

runs with a different seed for the noisy input. For frequency analysis, a 30 s epoch (from 9–39 s) of the output signal from each of the 20 simulations is bandpass filtered between 1–100 Hz with a Butterworth filter of order 10. A 4-point Fast Fourier Transform (FFT) at a sampling frequency of 1000 Hz is applied to each of these filtered signals. The power spectral density (psd) is derived using the Welch periodogram with a Hamming window of segment length spanning 500 data points (half the size as that of the sampling frequency) and with overlap of 50 %. All of the 20 psd thus obtained are then averaged for further analysis. The bar plots show the mean power within the frequency bands theta (4–7 Hz) and alpha (8–13 Hz).

Results and Discussion

The objective is to mimic the EEG corresponding to the state of 'quiet wakefulness' (i.e. when a subject is in an awake resting state with eyes closed, for example just before transition to a state of sleep) in the model output with a dominant frequency within the alpha band. The model input is simulated with a random white noise, representing the pre-synaptic mean membrane potential $V_{\bar{\chi}}$ producing low-level background firing in the retinal spiking neurons in a state of quiet wakefulness and no sensory input. This state of the model is taken as the 'base' state, and the corresponding set of model parameter values are referred to as 'base values'. All parameter variations in the model are carried out with respect to these base values to study the synaptic correlates of EEG band power alterations in both healthy (e.g. sleep-wake transition) and disease states. The results are presented in the following sections along with discussion on their implications in context to alpha and theta rhythms in the LGN.

The Causality of Neurotransmitter Concentration

Presynaptic membranes are rich in a diverse range of potassium channels that are likely relevant to the fine-tuning and regulation of neurotransmitter release [28]. While membrane-derived lipids such as arachidonic acid can act to inhibit presynaptic potassium channels [15], it is reported that Na+/K+-ATPase is involved in the maintenance of the synaptic vesicles filled with transmitters to be released [68]. On the other hand, release of neurotransmitters is well known to be mediated by calcium ion-channel dynamics in the pre-synaptic membrane [17, 39]. However, and to the best of our knowledge, there is a lack of experimental data establishing a correlation between brain states and neurotransmitter concentration and/or rate of release in the synaptic cleft. Here, we use the neural mass computational framework to look into this aspect.

We speculate that the exact amount of neurotransmitter released in clefts is bound to be varying in time corresponding to varying brain states. We simulate this in the model by varying specific parameters, at the same time identifying a set of base parameter values defining the neurotransmitter concentration levels such that the dominant power in the model output is within the alpha frequency band.

Selecting Base Parameter Values

In Eq. (5), an increase in the steepness parameter σ leads to a decrease in gradient of the sigmoid representing the amount of transmitter released for a given change in the pre-synaptic population mean membrane potential V_χ. On the other hand, a decrease in the threshold voltage V_{thr} effect an increase in the release of neurotransmitter into the synaptic cleft. A simultaneous variation of V_{thr} and σ is done to observe the correlation between neurotransmitter concentration $[T]$ and the region of alpha band dominance in the power spectra of the post-synaptic population average membrane potential $V_{\bar{Y}}$.

The readers may note that in the model, the parameters V_{thr} and σ are set as equal for both AMPA or GABA$_A$ mediated synapses. Thus, any variation of these parameters will result in a variation of neurotransmitter concentration in all the synaptic clefts in the model. This may be thought to be analogous to an 'overall system slow down' during reduced cognitive states such as falling asleep. A separate study on neurotransmitter concentration levels for AMPA and GABA mediated synapses will be carried out in future versions of the model. The results are shown in Fig. 3a–d.

Figure 3a and b show the inverse relation between $[T]$ and the model output V_{TCR}, indicating a dominant GABA-ergic influence on the TCR from both IN and TRN cell populations. The corresponding peak power plot in Fig. 3d shows a higher power content for depolarised output values in Fig. 3b and is mainly due to the increasing power of the dc component corresponding to progressive depolarisation in the TCR population. However, Fig. 3d show that the dominant alpha rhythmic region correspond to a mean output voltage of ≈ -70 mV. The contour plots showing a distinct alpha peak in the region where $-32 \geqslant V_{thr} \geqslant -33$ and $3.7 \leqslant \sigma \leqslant 3.8$ (indicated by an arrow in Fig. 3c). Based on this observation, we set the base values for V_{thr} and σ to -32 mV and 3.7 mV respectively and study the alpha and theta rhythmic content in the model output.

Alpha and Theta Band Power Variations

Figure 4 shows that the maximum power content with lower values of neurotransmitter concentration and higher mean membrane potential is primarily due to a high corresponding theta band power. Furthermore, the bar-plots show a left skew in the theta band power for progressively increasing values of σ, in contrast to a right-skew for the corresponding alpha band power. With progressively decreasing values of V_{thr}, the alpha band power shifts until for $-35 \geqslant V_{thr} \geqslant -38$, the maximum powers in both alpha and theta bands decrease exponentially with increasing values of σ. For $V_{thr} \leqslant -39$, the power in theta band falls significantly, and there is minimal effect on both bands for varying values of σ.

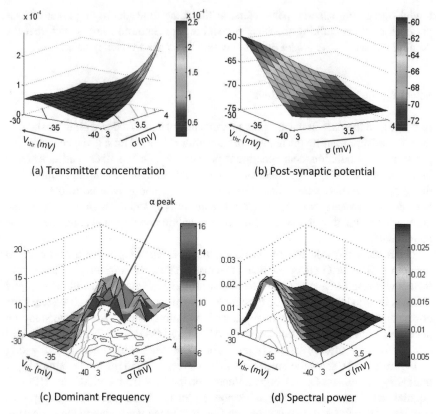

(a) Transmitter concentration   (b) Post-synaptic potential

(c) Dominant Frequency   (d) Spectral power

Fig. 3 **a** The neurotransmitter concentration $[T]$ with simultaneous variation of V_{thr} (from -30 to -40 in steps of -1) and σ (from 3 to 4 in steps of 0.1), and **b** the corresponding effect on the average population post-synaptic membrane potential V_{TCR}. The corresponding **c** dominant frequency of oscillation show specific regions of alpha and theta band dominance, while **d** the power in the frequency spectra indicate a high power within the theta band for lower values of the steepness parameter σ. It is worth reminding the readers at this point (see Fig. 1 legend) that the parameters V_{thr} and σ are set as equal for both AMPA or $GABA_A$ mediated synapses. Thus, any variation of these parameters will result in a variation of neurotransmitter concentration in all the synaptic clefts in the model, thus simulating conditions of an overall reduction in synaptic activity in the model

Role of the Thalamic Interneurons

Neural mass models of the thalamocortical circuitry has been simulated traditionally with two neural populations viz. the TCR and the TRN [12, 57]. These models were intended to simulate brain rhythm alterations and abnormal oscillations corresponding to disease conditions. Several of these studies have demonstrated that the feed-forward and -back connections between the TCR and the TRN can well mimic the dynamics of several disease conditions for example bifurcation of EEG time-series seen in epilepsy [13, 35, 58, 67, 69, 72, 73], EEG anomalies in Alzheimer's

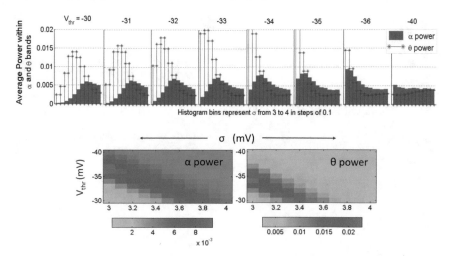

Fig. 4 (*Top*) Bar plots showing the relative distribution of alpha and theta band power in the model output with varying values of V_{thr} and σ. (*Bottom*) A 2-d representation to demonstrate the complementary pattern of dominance between the alpha and theta bands

disease [7, 16, 27], sleep-wake transition [1]. However, experimental studies indicate the presence of the IN cells in the LGN across mammals and rodents, comprising 20–25 % of the total cell population in the LGN and receive around 47 % of their total synaptic inputs from the retinal spiking neurons. Along these lines, further experimental investigation into the specific role of IN cells in the LGN may be suggested [17].

In section "The Causality of Neurotransmitter Concentration", we have tuned the neurotransmitter concentration parameters so that the model output oscillates with a dominant frequency within the alpha band. The time-series and power spectral density of all three cell populations are shown in Fig. 5 panels (a) and (c) respectively. The TCR and IN peak-to-peak oscillation is ≈1 mV; the same for TRN, however, is suppressed and is in the range of a few μV. The mean membrane potential of both the inhibitory populations viz. IN and TRN are greater than that of the TCR population. The power spectra indicates dominant power within the alpha band for both IN and TCR cell populations, while the dominant frequency of oscillation for the TRN is within the theta band.

Next, we remove the IN from the circuit to observe the output behaviour when all parameters are maintained at their respective base values. The time-series and power spectral density are shown in Fig. 5 panels (b) and (d) respectively. The TCR and TRN outputs show synchronised spindle oscillations with a dominant frequency at ≈11 Hz and within the alpha band. The amplitude of peak-to-peak oscillations is also increased in both TCR and TRN at ≈1.5 mV.

These observations may indicate a vital role of the IN in waking state EEG dynamics. Alpha rhythmic waxing-and-waning high amplitude oscillations are EEG markers of quiet wakefulness with eyes closed i.e. absence of sensory inputs, while alpha

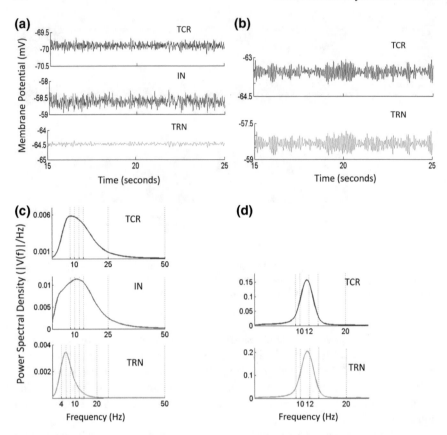

Fig. 5 **a** Time-series output of the thalamic cell populations in the model when all parameters are set at their base values as indicated in Tables 1 and 2. The corresponding power spectra in (**c**) indicate a dominant alpha rhythmic oscillation in the TCR and IN population output, while the TRN oscillates with a dominant frequency within the theta band. **b** The time-series output of TCR and TRN when the IN is disconnected from the circuit by setting the synaptic connectivity parameter $C_{tii} = 0$ (refer to Table 2). High amplitude synchronised oscillations is observed in both TCR and TRN populations. The corresponding power spectra in (**d**) indicate a sharp alpha peak at around 11.5 Hz for both TCR and TRN

rhythmic noisy oscillations are now known to be important indicators of several cognitive brain states. Thus, the model predicts specific role for each inhibitory cell population in the thalamic circuitry: the TRN assumes prominence during wake to sleep transition and in sleep states, while the IN dominates the inhibitory influence on TCR during the waking state. In the context of neurological disorders, the IN may play a role in maintaining homeostasis, while any anomaly in this circuitry may trigger an onset of abnormal high amplitude synchronous oscillations in the TCR and TRN.

Effects of the Leak Conductance

Anaesthetics are known to decrease excitability in muscles by increasing membrane leakage conductance. Furthermore, literature review of both experimental and modelling studies show evidence of the role of potassium leak channels in depolarisation/hyperpolarisation of population membrane potentials [34]. In addition, acetylcholine is known to cause hyperpolarisation in IN cells by increasing membrane potassium conductance. These results motivated us to investigate the effects of membrane leakage conductance in the model on its output.

To test the effect of increased leak conductance in our model, g^{leak} for any one cell population is increased to 100 progressively, while the values of the same for the other two cell populations remain at their base values of 10.

When $g^{leak} = 100$ for the TCR population, both TCR and TRN population are depolarised, and removing the IN did not show any drastic change in the output characteristics.

When $g^{leak} = 100$ for the IN population, their mean membrane potential is hyperpolarised, causing a reduced effect on TCR and TRN populations, both of which show a depolarisation. This is similar to the case in section "Role of the Thalamic Interneurons" when the IN population is removed from the circuit.

When $g^{leak} = 100$ for the TRN population, their mean membrane potential is hyperpolarised and the TCR population is depolarised as shown in Fig. 6b (compare with Fig. 6a corresponding to base parameter values). However, if the IN is removed

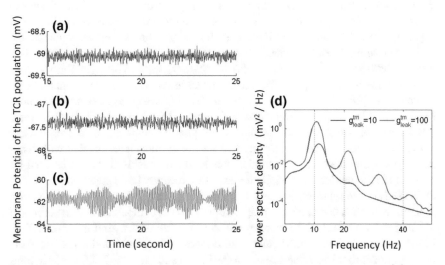

Fig. 6 The model output time-series corresponding to **a** base parameter values of g^{leak} for all cell populations in the model; **b** $g^{leak} = 100$ for TRN population and $g^{leak} = 10$ for TCR and IN; **c** $g^{leak} = 100$ for TRN and IN is removed from the circuit, while $g^{leak} = 10$ for TCR. **d** The power spectral density of the output obtained in (**c**) show a peak frequency within the alpha with harmonics (*violet*). A comparison is made with the power spectra for when $g^{leak} = 10$ (*blue*)

from the circuit, increasing g^{leak} for the TRN population cause a depolarisation in both TRN and TCR populations. The time-series of both cell populations show a bifurcation to high amplitude oscillations shown in Fig. 6c, while the power spectral plot in Fig. 6d shows harmonics of the dominant frequencies within the alpha band. A comparison with the power spectra when all parameters are at base values, i.e. $g^{leak} = 10$ for all populations, and IN is removed from the circuit, is also shown for the convenience of readers.

In summary, these results show that for a model of the LGN tuned to oscillate within the alpha band, an increase of leak conductance in the inhibitory cell populations, IN or TRN, will lead to a depolarisation of the excitatory TCR cell population for normal neurotransmitter concentration levels, thus conforming to experimental evidence [34]. However, if the influence of IN is reduced while leak conductance of TRN is increased, both TCR and TRN are depolarised leading to bifurcation of the time-series to a high amplitude limit cycle oscillations with harmonics in the power spectra. Once again, the results imply a role of the IN in maintaining an overall system stability.

In the following section, we summarise the results and outline future research directions.

Conclusion

We have presented a novel neural mass modelling approach to link the attributes of neuronal chemical synapses to higher level brain dynamics observed in Electroencephalogram (EEG) and Local Field Potential (LFP); this is done by replacing the traditional 'alpha function', which are used to model neuronal synaptic information transfer, with kinetic models of AMPA (excitatory) and $GABA_A$ (inhibitory) neuroreceptor mediated synapses. The paradigm was introduced to neuro-computational modelling in [26]. Subsequently, kinetic model of $GABA_A$-ergic synapse has been implemented in a single neuronal network model of the hippocampal circuit towards novel neuro-pharmacological paradigms and possible application in drug discovery [30]. In prior works, we have used neural mass models embedded with 'two-state' (open and closed states of ion channels; the desensitised state is ignored for brevity) kinetic models of synaptic transmission to study brain state transitions and the underlying cellular mechanisms [5]. Our study showed that replacing the alpha function with kinetic models of synapses improved the computational time of model simulation by a factor of 10 in comparison to the 'classic' neural mass modelling framework. In this chapter, we have built on these prior works and investigated the underlying neuronal correlates of alpha (8–13 Hz) and theta (4–7 Hz) EEG rhythms that are useful biomarkers in several neurological and psychiatric disorders.

Experimental evidence suggest that similar thalamic mechanisms underline alpha and theta rhythms, and that the thalamus is a key player in thalamocortical generation of these rhythms. This is not surprising as thalamocortical dysrhythmia (TCD) is a known feature in neurological disorders such as Depressive disorders,

Neurogenic Pain, Parkinson's disease, Tinnitus, Alzheimer's disease. On the other hand, transition from alpha to theta rhythm in a healthy, adult brain corresponds to brain state transition from quiet wakefulness (dominant alpha rhythm) to that of drowsiness (dominant theta rhythm). Thus, a computational model-based study to underpin the synaptic correlates of alpha and theta rhythms seems appropriate and useful in current times.

The model in this work emulates the neuronal populations of the Lateral Genicu- late Nucleus (LGN), the thalamic nucleus in the visual pathway of mammals and rodents, and the intra-population synaptic connectivity mediated by AMPA and $GABA_A$ neuroreceptors. Please refer to Fig. 2 in the text for a schematic of the model consisting of the thalamocortical relay (TCR) cells that are the main carriers of sen- sory information from the retina to the visual cortex, the inhibitory interneurons (IN) that constitute around 20–25 % of the cells in the LGN, and the thalamic reticular nucleus (TRN), which is a thin sheet of inhibitory cell populations that are consid- ered a part of the thalamus and receives 'copies' of all efferent and afferent commu- nications between the thalamus and the cortex. The model input is a random noise with a white power spectrum, and may be thought to emulate the background fir- ing activity of retinal ganglion cells under conditions of eyes closed i.e. no sensory input, and when the subject is in an awake but resting state, often referred to as a state of 'quiet wakefulness'. Both EEG and LFP recordings in quiet wakefulness, for example in the stage preceding sleep, indicate a strong alpha rhythmic content. Furthermore, simultaneous EEG recordings from the occipital scalp electrode (the seat of visual cortex) and LFP recordings from the TCR cells show a high coher- ence in their time-series. Thus, the output of the TCR cell population in the model is considered as the 'model output'.

Traditionally, neural mass models of the thalamocortical circuitry have explored the rich dynamics of the feed-forward and -back loop between the TCR and TRN that can emulate time-series and frequency domain behaviour of several neurological and psychiatric disorders. Thus, the IN cells have largely been ignored, in spite receiv- ing around 47 % of their afferents from the retinal spiking neurons. This may be due to lack of experimental data on the IN characteristics, which in turn is attributed to insufficient technological advances. In this scenario, a computational model seems to be an apt tool to make investigations into the IN and its role in the thalamocortical dynamics. Towards this, a simple test is adopted by disconnecting the IN population from the circuitry. Indeed the time-series of both TRN and TCR are synchronised with a high magnitude of oscillation. Furthermore, waxing-and-waning patterns are observed and the power spectra indicates the dominant alpha rhythmic content with a peak at around 11.5 Hz. This result is consistent with several other prior work on the classic neural mass model of the thalamocortical circuitry consisting of just the TCR and TRN cell populations. Next, the IN is re-connected to the circuitry. The time- series output of the TCR changes significantly with a noisy pattern and low average amplitude reflecting the noisy input to the model, and a peak frequency within the alpha band; the time-series of the IN has a similar characteristic, albeit with a broader power spectral density peak within the alpha frequency band. The TRN output under- goes a remarkable change and appears to be suppressed by the dominant inhibitory influence of the IN on the TCR, and its peak frequency of oscillation is ≈6 Hz within the theta frequency band. Thus, the model predicts a significant influence of the IN on the TRN in spite an absence of direct synaptic contact between the two populations. The results raise the speculation that in a healthy adult brain, the IN plays a dominant role in modulating TCR response in awake cognitive states; the diminished activity of IN in a quiet resting brain state establishes the dominant inhibitory influence of TRN on the TCR, leading to high amplitude synchronous oscillations. These observations call for further experimental research correlating the role of IN in higher level brain dynamics.

Another attribute that has received relatively reduced attention in computational modelling of neurological disorders, albeit with a few exceptions e.g. [29, 30], is the neurotransmitter concentration in synaptic clefts. Once again, the model presented here facilitates such a research direction. The neurotransmitter concentration in the synaptic cleft (refer to Eq. 2 in section "AMPA and GABA$_A$ Based Neurotransmission") is simulated with a sigmoid function that is sensitive to (a) the threshold voltage at which the neurotransmitter concentration is half the maximum value, as well as to (b) the steepness parameter of the sigmoid that indicates the proportionality of the amount of transmitter released to the pre-synaptic membrane potential. Our results show that an increase in the neurotransmitter concentration in the synaptic cleft is effected by a decrease in the threshold pre-synaptic membrane voltage, which agrees to an intuitive understanding of the phenomenon. However, the neurotransmitter concentration is also increased by a decrease in the steepness of the sigmoid. From a systems perspective, the phenomenon can be explained thus: for a steep sigmoid, the neurotransmitter concentration reaches saturation ($\approx T_{max}$) or 'cut off' (≈ 0) for low-range fluctuation of the pre-synaptic voltage about the threshold voltage. Thus, the neurotransmitter concentration follows a spike train-like all-or-none pattern, which effectively reduces the average value of the parameter. On the other hand, with lesser steepness, the 'operating region' of the sigmoid is much larger prior to the concentration reaching either saturation or cut-off, thus effecting higher average concentration levels in the synaptic cleft.

In terms of frequency domain response with varying neurotransmitter concentrations, we note an interesting complementary behaviour in the alpha and theta band power content. With lower neurotransmitter concentration levels in the circuit, the theta band dominates and with an overall higher magnitude of power in the spectra. Peak alpha band power is observed when neurotransmitter concentration is at a mid-range value. This may also be interpreted as a shift from alpha to theta band power for raised threshold pre-synaptic membrane potential and steepness values for the neurotransmitter release/concentration, and may reflect abnormalities of chemical synaptic attributes corresponding to TCD. It may be noted that in the current work, we have considered simultaneous changes in the neurotransmitter concentration in all synaptic clefts in the model; the objective has been to study a more holistic effect of fluctuations in synaptic activity in the system. For example during transition from wakefulness to sleep, it may be speculated, intuitively, that there is a change in overall system behaviour leading to high-amplitude low-frequency synchronous oscillations across larger brain areas. However, this needs further investigation with distinct neurotransmitter concentration for each synaptic cleft in the model. Furthermore, a dynamic adaptability in this attribute may be desirable when investigating specific disease conditions; these directions are planned as future work on the model.

A preliminary observation on the effects of leakage conductance in the model re-iterates the prominent role of IN over TCR in normal brain states, maintaining an overall homeostasis in the circuit. Recent research proposes a possible role of potassium leakage currents in epileptic seizures [33], while potassium leak currents are known to affect cell response by causing de-(hyper-)polarisation in cell membrane potentials. Also, the effects of anaesthetics in decreasing muscle excitability is facilitated by increasing population leakage conductance. In the model, increasing the leakage conductance of both IN and TRN populations cause a depolarisation of the TCR cells. However, when the IN is disconnected from the circuit, an increase in leakage conductance in the TRN cells causes a bifurcation in the TCR output leading to high amplitude limit-cycle like oscillations with slight waxing and waning envelope modulation. Frequency analysis shows alpha band dominance. It may be noted that theta band limit cycles are seen with decreased neurotransmitter concentrations under these conditions. Once again, the results indicate an overall homeostatic role of the IN in the LGN circuitry and conforming to a healthy, awake and cognitive brain state. Disruption in factors affecting homeostasis in the brain is implicated in several disease conditions. The model-based observations in this work implicates disruption of the IN circuitry as a possible underlying factor for certain disease related homeostatic abnormalities that reflect in higher level brain dynamics recorded via EEG and LFP.

It is worth mentioning here that in a similar work on the classic neural mass models, we have looked into synaptic connectivity parameters that effect a 'slowing' (left-shift of peak frequency of oscillation) of the alpha rhythm, a definite biomarker of Alzheimer's disease [8]. However, investigation into further synaptic attributes has not been possible due to model limitations on detailed synaptic attributes; in comparison, the modified neural mass modelling approach presented in this work have alleviated this constraint to a fair extent. However, several levels of abstraction are adopted in the model for brevity—first, the $GABA_B$ pathway from the TRN to the TCR is ignored; second, the kinetic models of the synapses are two-state models i.e. the desensitised state is not considered here; third, feedback from the TCR to the IN is ambiguous in literature, and thus are not explored in this work; fourth, the neurotransmitter concentration in synaptic clefts has similar parameters for all cell populations (afore-mentioned in this section); fifth, while the present model aims to underpin the rhythmic behaviour of the LGN when de-corticated, i.e. disconnected from the visual cortex, however, cortico-thalamic feedback is an integral factor in the generation and sustenance of brain rhythms observed in EEG. All of these abstractions will be looked into in a future work.

Ongoing research is looking into implementing a neural mass cortical circuitry that will then be linked to the LGN model presented in this work. While such a model has already been explored in a prior work, the novelty will be the introduction of synaptic kinetics in such a thalamo-cortico-thalamic neural mass framework. Furthermore, a recent research has used the classic alpha rhythm neural mass model to emulate EEG signals corresponding to trains of flickering visual stimuli, commonly termed as steady-state-visually-evoked-potentials; the model results were validated with experimental data [47]. Using computational models to emulate steady-state-visually-evoked-potentials was initiated in [59]; the potential of the neural mass framework presented herewith will be tested along these lines.

Overall, the study presented herewith have contributed in justifying the ongoing endeavours to build biologically-inspired computational paradigms that are computationally efficient and can contribute to progressing the diagnosis, prognosis and prediction of neurological and psychiatric disorders. The observations made in this work call for further experimental data for the purposes of model validation and continued advancement of research in computational neurology and neuropsychiatry.

Acknowledgements The authors would like to acknowledge the contribution of Vincent N. Martin, second year undergraduate student of Mechanical Engineering at the University of Lincoln, for his contributions to Fig. 3 in this work as a part of his coursework. We thank the reviewers for very useful suggestions towards improving the manuscript. BSB is grateful to Piotr Suffczynski for useful discussions on the model during his visit to the University of Lincoln (July-September 2015) supported by DVF1415/2/35 grant, awarded by the Royal Academy of Engineering, UK.

References

1. Abeysuriya, R., Rennie, C., Robinson, P.: Prediction and verification of nonlinear sleep spindle. Journal of Theoretical Biology **344**, 70–77 (2014)
2. Aradi, I., Érdi, P.: Computational neuropharmacology: dynamical approachers in drug discovery. TRENDS in Pharmacological Sciences **27**(5), 240–243 (2006)
3. Bal, T., von Krosigk, M., McCormick, D.A.: Synaptic and membrane mechanisms underlying synchronized oscillations in the ferret lateral geniculate nucleus *in vitro*. Journal of Physiology **483**, 641–663 (1995)
4. Bernard, C., Ge, Y., Stockley, E., Willis, J., Wheal, H.: Synaptic integration of NMDA and non-NMDA receptors in large neuronal network models solved by means of differential equations. Biological Cybernetics **70**, 267–273 (1994)
5. Bhattacharya, B.S.: Implementing the cellular mechanisms of synaptic transmission in a neural mass model of the thalamo-cortico circuitry. Frontiers in Computational Neuroscience **81**, 1–11 (2013)
6. Bhattacharya, B.S., Chowdhury, F.N. (eds.): Validating neuro-computational models of neurological and psychiatric disorders, *Springer series in Computational Neuroscience*, vol. 14. Springer (2015)
7. Bhattacharya, B.S., Coyle, D., Maguire, L.P.: Thalamocortical circuitry and alpha rhythm slowing: an empirical study based on a classic computational model. In: Proceedings of the International Joint Conference on Neural Networks (IJCNN), pp. 3912–3918. Barcelona, Spain (2010)
8. Bhattacharya, B.S., Coyle, D., Maguire, L.P.: Alpha and theta rhythm abnormality in Alzheimer's disease: a study using a computational model. In: C. Hernandez, J. Gomez, R. Sanz, I. Alexander, L. Smith, A. Hussain, A. Chella (eds.) Advances in Experimental Medicine and Biology, Volume 718, pp. 57–73. Springer New York (2011)
9. Bhattacharya, B.S., Coyle, D., Maguire, L.P.: A thalamo-cortico-thalamic neural mass model to study alpha rhythms in Alzheimer's disease. Neural Networks **24**, 631–645 (2011)
10. Bhattacharya, B.S., Coyle, D., Maguire, L.P., Stewart, J.: Kinetic modelling of synaptic functions in the alpha rhythm neural mass model. In: A.V. et al (ed.) ICANN 2012 Part I, Lecture Notes in Computer Science 7552, pp. 645–652. Springer Verlag Berlin Heidelberg (2012)

11. Bond, T., Durrant, S., O'Hare, L., Turner, D., Bhattacharya, B.S.: Studying the effects of thalamic interneurons in a thalamocortical neural mass model. In: BMC Neuroscience (Suppl 1), vol. 15, p. P219 (2014)

12. Breakspear, M., Roberts, J., Terry, J., Rodrigues, S., Mahant, N., Robinson, P.A.: A unifying explanation of primary generalized seizures through nonlinear brain modelling and bifurcation analysis. Cerebral Cortex **16**, 1296–1313 (2006)

13. Breakspear, M., Terry, J.R., Friston, K.J.: Modulation of excitatory synaptic coupling facilitates synchronization and complex dynamics in a nonlinear model of neuronal dynamics. Neurocomputing **52**, 151–158 (2003)

14. Buzsáki, G.: Rhythms of the Brain, first edn. Oxford University Press, New York (2006)

15. Carta, M., Lanore, F., Rebola, N., Szabo, Z., da Silva, S.V., Lourenco, J., Verraes, A., Nadler, A., Schultz, C., Blanchet, C., Mulle, C.: Membrane lipids tune synaptic transmission by direct modulation of presynaptic potassium channels. Neuron **81**(4), 787–799 (2014)

16. Coyle, D., Bhattacharya, B.S., Zou, X., Wong-Lin, K., Abuhassan, K., Maguire, L.: Neural circuit models and neuropathological oscillations. In: N. Kasabov (ed.) Handbook of Bio-Neuro-Informatics, pp. 673–702. Springer (2014)

17. Crunelli, V., Cope, D.W., Hughes, S.W.: Thalamic t-type calcium channels and nrem sleep. Cell Calcium **40**, 175–190 (2006)

18. Crunelli, V., Haby, M., Jassik-Gerschenfeld, D., Leresche, N., Pirchio, M.: Cl⁻-and k⁺-dependent inhibitory postsynaptic potentials evoked by interneurons of the rat lateral geniculate nucleus. Journal of Physiology **399**, 153–176 (1988)

19. daSilva, F.H.L., Hoeks, A., Smits, H., Zetterberg, L.H.: Model of brain rhythmic activity. Kybernetic **15**, 27–37 (1974)

20. daSilva, F.H.L., van Lierop, T.H., Schrijer, C.F., van Leeuwen, W.S.: Organisation of thalamic and cortical alpha rhythms: spectra and coherences. Electroencephalography and Clinical Neurophysiology **35**, 627–639 (1973)

21. David, O., Friston, K.J.: A neural mass model for MEG/EEG: coupling and neuronal dynamics. NeuroImage **20**, 1743–1755 (2003)

22. Deco, G., Jirsa, V.K., Robinson, P.A., Breakspear, M., Friston, K.: The dynamic brain: from spiking neurons to neural masses and cortical fields. PLOS Computational Biology **4**(8), e1000,092 (2008)

23. Destexhe, A.: Synthesis of models for excitable membranes, synaptic transmission and neuromodulation using a common kinetic formalism. Journal of Computational Neuroscience **1**, 195–230 (1994)

24. Destexhe, A., Mainen, Z., Sejnowski, T.: An efficient method for computing synaptic conductances based on a kinetic model of receptor binding. Neural Computation **6**, 14–18 (1994)

25. Destexhe, A., Mainen, Z., Sejnowski, T.: Kinetic models of synaptic transmission. In: C. Koch, I. Segev (eds.) Methods in neuronal modelling, pp. 1–25. MIT Press, Cambridge, MA (1998)

26. Destexhe, A., Mainen, Z., Sejnowski, T.: Kinetic models for synaptic interactions. In: M. Arbib (ed.) The handbook of brain theory and neural networks, pp. 1126–1130. MIT Press, Cambridge, MA (2002)

27. de Haan, W., Mott, K., van Straaten, E.C.W., Scheltens, P., Stam, C.J.: Activity dependent degeneration explains hub vulnerability in Alzheimer's disease. PLOS Computational Biology **8**(8), e100,252 (2012)

28. Dodson, P.D., Forsythe, I.D.: Presynaptic k+ channels: electrifying regulators of synaptic terminal excitability. Trends in Neurosciences **27**(4), 210–217 (2004)

29. Érdi, P., John, T., Kiss, T., Lever, C.: Discovery and validation of biomarkers based on computational models of normal and pathological hippocampal rhythms. In: B.S. Bhattacharya, F.N. Chowdhury (eds.) Validating neuro-computational models of neurological and psychiatric disorders, pp. 15–42. Springer (2015)

30. Érdi, P., Kiss, T., Tóth, J., Ujfalussy, B., Zalányi, L.: From systems biology to dynamical neuropharmacology: proposal for a new methodology. IEEE Proceedings of Systems Biology **153**(4), 299–308 (2006)

31. Francis, P.T., Palmer, A.M., Snape, M., Wilcock, G.K.: The cholinergic hypothesis of alzheimer's disease: a review of progress. Journal of Neurology and Neurosurgical Psychiatry **66**(2), 137–147 (1999)
32. Freeman, W.J.: Mass action in the nervous system, first edn. Academic Press, New York (1975)
33. Gentiletti, D., Gnatkovsky, V., de Curtis, M., Suffczyński, P.: Changes of ionic concentrations during seizure transitions - a modeling study. Under Review (2016)
34. Goldstein, S., adn Ita O'Kelly, D.B., Zilberberg, N.: Potassium leak channels and the kcnk family of two-p-domain subunits. Nature Reviews Neuroscience **2**, 175–184 (2001)
35. Golomb, D., Wang, X.J., Rinzel, J.: Synchronization properties of spindle oscillations in a thalamic reticular nucleus model. Journal of Neurophysiology **72**(3), 1109–1126 (1994)
36. Golomb, D., Wang, X.J., Rinzel, J.: Propagation of spindle waves in a thalamic slice model. Journal of Neurophysiology **75**, 750–769 (1996)
37. Grimbert, F., Faugeras, O.: Bifurcation analysis of Jansen's neural mass model. Neural Computation **18**, 3052–3068 (2006)
38. Guillery, R.W., Sherman, S.M.: The thalamus as a monitor of motor outputs. Philosophical Transactions of the Royal Society of London B Biological Science **357**(1428), 1809–1821 (2002)
39. Harris, K.P., Littleton, J.T.: Transmission, development and plasticity of synapses. Genetics **210**(2), 345–375 (2015)
40. Horn, S.C.V., Erisir, A., Sherman, S.M.: Relative distribution of synapses in the A-laminae of the lateral geniculate nucleus of the cat. The Journal of Comparative Neurology **416**, 509–520 (2000)
41. Hughes, S.W., Crunelli, V.: Thalamic mechanisms of eeg alpha rhythms and their pathological implications. The Neuroscientist **11**(4), 357–372 (2005)
42. Hughes, S.W., Lorincz, M., Cope, D.W., Blethyn, K.L., Kekesi, K.A., Parri, H.R., Juhasz, G., Crunelli, V.: Synchronised oscillations at α and θ frequencies in the lateral geniculate nucleus. Neuron **42**, 253–268 (2004)
43. Jansen, B.H., Rit, V.G.: Electroencephalogram and visual evoked potential generation in a mathematical model of coupled cortical columns. Biological Cybernetics **73**, 357–366 (1995)
44. Jeong, J.: Eeg dynamics in patients with alzheimer's disease. Clinical Neurophysiology **115**, 1490–1505 (2004)
45. Jones, E.G.: The Thalamus, Vol. I and II, first edn. Cambridge University Press, Cambridge, UK (2007)
46. von Krosigk, M., Bal, T., McCormick, D.A.: Cellular mechanisms of a synchronised oscillation in the thalamus. Science **261**, 361–364 (1993)
47. Labecki M, Kus R, Brzozowska A, Stacewicz T, Bhattacharya BS and Suffczynski P (2016). Nonlinear origin of SSVEP spectra–a combined experimental and modeling study. Front. Comput. Neurosci. 10:129. doi:10.3389/fncom.2016.00129
48. Liljenstrom, H.: Mesoscopic brain dynamics. Scholarpedia **7**(9), 4601 (2012)
49. Llinas, R., Urbano, F.J., Leznik, E., Ramirez, R.R., van Marle, H.J.: Rhythmic and dysrhythmic thalamocortical dynamics: Gaba systems and the edge effect. Trends in Neuroscience **28**(6), 325–333 (2005)
50. Lörincz, M.L., Crunelli, V., Hughes, S.W.: Cellular dynamics of cholinergically induced α (8–13 hz) rhythms in sensory thalamic nuclei *In Vitro*. The Journal of Neuroscience **628**(3), 660–671 (2008)
51. McCormick, D.A., Pape, H.C.: Properties of a hyperpolarization-activated cation current and its role in rhythmic oscillation in thalamic relay neurones. Journal of Physiology **431**, 291–318 (1990)
52. McCormick, D.A., Prince, D.A.: Actions of acetylcholine in the guinea-pig and cat medial and lateral geniculate nuclei, *In Vitro*. Journal of Physiology **392**, 147–165 (1987)
53. Modolo, J., Thomas, A., Legros, A.: Neural mass modelling of power-line magnetic fields effects on brain activity. Frontiers in computational neuroscience **7**, 34 (2013)
54. Moran, R., Pinotsis, D.A., Friston, K.: Neural masses and fields in dynamic causal modeling. Frontiers in Computational Neuroscience **7**, 1–12 (2013)

55. Pons, A.J., Cantero, J.L., Atienza, M., Garcia-Ojalvo, J.: Relating structural and functional anomalous connectivity in the ageing brain via neural mass modelling. NeuroImage **52**(3), 848–861 (2010)
56. Rall, W.: Distinguishing theoretical synaptic potentials computed for different soma-dendritic distributions of synaptic inputs. Journal of Neurophysiology **30**, 1138–1168 (1967)
57. Robinson, P., Phillips, A., Fulcher, B., Puckeridge, M., Roberts, J.: Quantitative modelling of sleep dynamics. Philosophical Transactions of the Royal Society A **369**, 3840–3854 (2011)
58. Robinson, P., Rennie, C., Rowe, D.: Dynamics of large-scale brain activity in normal arousal states and epileptic seizures. Physical Review E **65**, 041,924 (2002)
59. Robinson, P.A., Postnova, S., abeysuriya, R.G., Kim, J.W., Roberts, J.A., McKenzie-Sell, L., Karanjai, A., Kerr, C.C., fung, F., Anderson, R., Breakspear, M.J., Drysdale, P.M., Fulcher, B.D., Phillips, A.J.K., Rennie, C.J., Yin, G.: A multiscale "working brain" model. In: B.S. Bhattacharya, F.N. Chowdhury (eds.) Validating neuro-computational models of neurological and psychiatric disorders, pp. 107–140. Springer (2015)
60. Sarnthein, J., Morel, A., von Stein, A., Jeanmonod, D.: Thalamic theta field potentials and eeg: high thalamocortical coherence in patients with neurogenic pain, epilepsy and movement disorders. Thalamus and related systems **2**, 231–238 (2003)
61. Sherman, S.M.: Thalamus. Scholarpedia **1**(9), 1583 (2006)
62. Sherman, S.M., Guillery, R.W.: Exploring the thalamus, first edn. Academic Press, New York (2001)
63. Sotero, R.C., Tujillo-Barreto, N.J., Iturria-Medina, Y.: Realistically coupled neural mass models can generate EEG rhythms. Neural Computation **19**, 479–512 (2007)
64. Steriade, M., McCormick, D.A., Sejnowski, T.J.: Thalamocortical oscillations in the sleeping and aroused brain. Science **262**(5134), 679–685 (1993)
65. Steriade, M.M., McCarley, R.: Brain control of wakefulness and sleep, second edn. Kluwer Academic/Plenum Publishers, New York (2005)
66. Suffczyński, P.: Neural dynamics underlying brain thalamic oscillations investigated with computational models. Ph.D. thesis, Institute of experimental physics, University of Warsaw (2000)
67. Suffczyński, P., Kalitzin, S., Silva, F.L.D.: Dynamics of non-convulsive epileptic phenomena modelled by a bistable neuronal network. Neuroscience **126**, 467–484 (2004)
68. Taruno, A., Ohmori, H., Kuba, H.: Inhibition of pre-synaptic na(+)/k(+)-atpase reduces readily releasable pool size at the avian end-bulb of held synapse. Neuroscience Research **72**(2), 117–128 (2012)
69. Taylor, P.N., Wang, Y., Goodfellow, M., Dauwels, J., Moeller, F., Stephani, U., Baier, G.: A computational study of stimulus driven epileptic seizure abatement. PLOS one pp. 1–26 (2014)
70. Ursino, M., Cona, F., Zavaglia, M.: The generation of rhythms within a cortical region: Analysis of a neural mass model. NeuroImage **52**(3), 1080–1094 (2010)
71. Wang, X.J., Golomb, D., Rinzel, J.: Emergent spindle oscillations and intermittent burst firing in a thalamic model: specific neuronal mechanisms. Proceedings of the National Academy of Sciences **92**, 5577–5581 (1995)
72. Wang, X.J., Rinzel, J.: Alternating and synchronous rhythms in reciprocally inhibitory model neurons. Neural Computation **4**, 84–97 (1992)
73. Wang, Y., Goodfellow, M., Taylor, P.N., Baier, G.: Dynamic mechanisms of neocortical focal seizure onset. PLOS Computational Biology **10**(8), e1003,787 (2014)
74. Wendling, F., Bartolomei, F., Bellanger, J.J., Chauvel, P.: Epileptic fast activity can be explained by a model of impaired GABAergic dendritic inhibition. European Journal of Neuroscience **15**, 1499–1508 (2002)
75. Wilson, H.R., Cowan, J.D.: A mathematical theory of the functional dynamics of cortical and thalamic nervous tissue. Kybernetik **13**, 55–80 (1973)
76. Zetterberg, L.H., Kristiansson, L., Mossberg, K.: Performance of a model for a local neuron population. Biological Cybernetics **31**, 15–26 (1978)
77. Zhu, J.J., Lytton, W.W., Xue, J.T., Uhlrich, D.J.: An intrinsic oscillation in interneurons of the rat lateral geniculate nucleus. Journal of Neurophysiology **81**, 702–711 (1999)

The Role of Simulations
in Neuropharmacology

Jean-Marie C. Bouteiller and Theodore W. Berger

Introduction

Experimental techniques have improved remarkably in the past decades, allowing a deeper understanding of the processes that take place in the central nervous system at different spatial (spanning from biomolecular mechanisms and synapses, to neurons and networks), and temporal scales.

These notable improvements have led to an exponential increase in the amount of data acquired. They have yielded a more quantitative view of the mechanisms underlying the central nervous systems' functions and dysfunctions, and the effects of drugs. Availability of such data enables the development of computational models that simulate the brain and its changes in response to the application of exogenous compounds. This chapter gathers examples of biosimulation efforts aimed at facilitating the generation of new working hypotheses in a structured and efficient manner, and translating the gained quantitative understanding of the brain, its normal and pathological hallmarks into the discovery of more efficient therapies.

J.-M.C. Bouteiller (✉) · T.W. Berger
University of Southern California, 1042 Downey Way,
Los Angeles, CA 90089, USA
e-mail: jbouteil@usc.edu

T.W. Berger
e-mail: berger@usc.edu

© Springer International Publishing AG 2017
P. Érdi et al. (eds.), *Computational Neurology and Psychiatry*,
Springer Series in Bio-/Neuroinformatics 6,
DOI 10.1007/978-3-319-49959-8_15

Scientific Problem

The Nervous System and Its Complexity

The nervous system is arguably one of the most complex organs of the body. Despite decades of relentless efforts, much remains to be learnt on how it performs its wide range of tasks and to this day, many questions remain unanswered. Our fascination for this complex organ feeds the headlines of health and science journals, drawing the attention of the neuroscientific community but also society at large.

The nervous system is afflicted by a variety of dysfunctions, with pathologies that may appear from a young age (e.g. Tourette syndrome, autism) to aged adulthood (Alzheimer's disease, Parkinson's disease). Understanding these dysfunctions have proven to be challenging for many reasons, especially due to the nervous system's highly multi-temporal and multi-hierarchical nature. An additional complexity stems from the multifactorial nature of the disease process. Indeed, even for diseases such as Huntington's disease in which the well-characterized mutation affects a single gene (which in the case of Huntington's disease consists of a trinucleotide repeat disorder caused by the length of a repeated section of a gene that exceeds normal range), this single mutation causes a variety of changes that result in the disease's pathological hallmarks (Fig. 1).

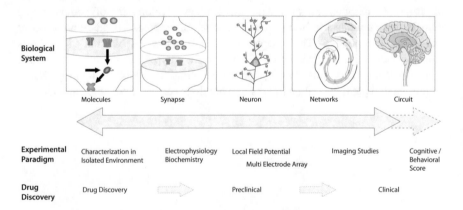

Fig. 1 Illustration of the different biological scales that comprise the nervous system, the corresponding experimental paradigms used to characterize their functions and dysfunctions, and the different phases in the drug discovery and development pipeline

Therapeutics Development

Drug discovery and development (DD&D) for disorders of the central nervous system (CNS) have been plagued with a high and continuously rising attrition rate. DD&D is a risky business, especially for the nervous system where the probability that an agent entering clinical development reaches the marketplace is below 7 %. This number is much lower than the industry average of 15 % across other therapeutic areas [1, 2]. Similarly, Development and regulatory approval for cardiovascular and gastrointestinal indications took an average of 6.3 and 7.5 years, versus 12 years for CNS indications [3]—and was reported to reach 18 years from laboratory bench to market in 2011 [4].

These difficulties result in higher costs for CNS DD&D and undoubtedly explain why since 2011, GSK, AstraZeneca, Novartis, Pfizer, Sanofi, Janssen and Merck have initiated a significant downsize in their CNS operations.

The poor success rates outline the prevalent dichotomy often observed in the DD&D process: drugs that have a potent effect in experimental protocols (e.g. strong affinity to the desired target, resulting in significant changes in synaptic and neuronal function or metabolism, etc.) end up having a modest or nonexistent effect, or prohibitive side effects at the macroscopic level. This is due to several factors including the CNS drugs' propensity to cause CNS-mediated side effects (e.g. nausea, dizziness and seizures), and the additional pharmacokinetic hurdle of the brain-blood barrier that therapeutic agents must face. Contrary to the development of a new antibiotic where the outcome is relatively simple (the bacterium is killed—or not—in a given and oftentimes relatively short treatment window), CNS compounds lead to a wide range of effects at different time scales. Recent examples of failure include suspicions of suicidal thoughts induced by anti-obesity or smoking cessation drugs. This led the Food and Drug Administration (FDA) to announce a change in policy in 2008 to mandate drug manufacturers to study the potential for suicidal tendencies during clinical trials. Compounds may even generate adverse effects. An example of adverse effect was reported by some patients taking antidepressants consisting of selective serotonin reuptake inhibitors (SSRIs) such as Prozac (fluoxetine), Paxil (paroxetine) or Zoloft (sertraline): they experienced suicidal thoughts during the initial phase of the treatment.

Necessity for a Quantitative Understanding of Mechanisms Underlying Pathology

These problems outline our lack of quantitative understanding of CNS dynamics. Unlike antibiotics, a CNS molecule may have a very small therapeutic window to induce a positive outcome without generating an army of harmful or undesired side effects. They also underscore our limited understanding of the mechanisms underlying pathologies. Finally, the DD&D pipeline is fragmented in multiple phases. From compound identification, to preclinical to clinical, preparations are inherently different and yield readouts that are arguably disconnected from the biological system, whose behavior we ultimately try to alter. Additionally, the use of animal models may contribute to a poor translatability of the observed effects of drug candidates to human patients population [5].

These compounded factors inherently result in a limited success rate, plaguing both the pharmaceutical industry and academic laboratories. They lead to increasingly higher costs, slow down the discovery process and delay the availability of potent drugs for patients.

A FDA report published in 2004 [6] outlined the need for innovative solutions to the healthcare challenges. It outlined how critical it has become to (i) integrate the data obtained in a uniform and standardized manner, while (ii) taking into account the dynamical properties inherent to biological systems. Together, these two measures will facilitate the generation of new working hypotheses in a structured and efficient manner, and translate our deepened understanding into more efficacious therapies. To this end, computational models constitute an innovative approach that may integrate up-to-date knowledge on the biological system; they may span multiple biological and temporal levels, encompassing mechanisms at the microscopic level as well as the resulting observations at the macroscopic level in a dynamic and integrated manner. They can replicate observables of function and dysfunctions, the effects of drugs on their respective target(s), and their subsequent effects on neuronal function, network function and ultimately on macroscopic observables, such as those obtained with functional neuroimaging.

Computational Methods

Unifying Computational Neuroscience

Computational neuroscience has witnessed tremendous growth in the past decade. However, given the complexity of the system studied, no unified methodology or tool exists that is able to span all hierarchical biological scales. Indeed, modeling methodologies are numerous, and differ quite significantly depending on the scale of the system under investigation. We refer the reader to other readings that provide an overview of the different modeling techniques as a function of the system investigated (see [7], Chap. 9). One notable point lays in the conceptual difference that separates computational neuroscience to computational neurology. Bridging this gap will deepen our understanding of CNS function and pave the path to individualized medicine.

Computational Neurology: Linking Observed Dysfunctions, Underlying Mechanisms and Individualized Treatments

In the context of pathologies, the computational methods described above often focus on modeling the mechanisms that underlie functions and dysfunctions. Whether at the biomolecular level (e.g. downstream effects of amyloid beta accumulation observed in Alzheimer's disease patients), neuron or network level (changes in spiking frequency and/or network-level activity), or systems level (models of brain function and its biophysics [8]), each of these computational models focuses on the pathology, and more precisely the mechanisms underlying the pathology and the consequences on the system (and its scale) of interest.

On the other hand, computational neurology and psychiatry focus primarily on the patient and its diagnosis to suggest efficacious therapies. Consequently, computational neurology has historically involved primarily a top-down approach that is oftentimes disconnected from the actual mechanisms underlying the pathology. Instead, it often relies on inference modeling (through database analysis, mathematical modeling, clinical algorithms) and statistical intelligence (Fig. 2a bottom).

From a methodological standpoint, the constraints of computational neurology imply unifying top-down and bottom-up approaches to link macroscopic observables with nanoscopic mechanisms. The necessary steps are summarized in Fig. 2b: (i) Starting from the patient, generate measurements; (ii) these measurements provide insights on the amplitude of dysfunctions observed at the macroscopic level, and allow calibration of the parameters of a simulation platform (i.e. virtual in silico patient), enabling quantification of biomolecular and neuron/network levels changes and characterization of the pathogenic features; (iii) the parameters of the computational platform are calibrated to replicate the measurements from (ii) (-generation in silico of the same 'virtual' observables), leading to (iv) a precise diagnosis; (v) different therapies may be tested on the virtual model—leading to the identification of the optimal treatment.

Computational neuropharmacology may greatly benefit from such centralized and integrated approach, removing gaps in the different phases of drug discovery and development. This will reduce error-prone interpolations and increase success rates.

The computational methods used in the results presented here use open tools and software. They comprise NEURON [9], EONS [10–12] and Libroadrunner [13]. Systems Biology Markup Language [14] is the preferred format in which all models are implemented and stored.

Fig. 2 a Illustration of the
current states of
computational neuroscience
and computational neurology.
b The future of computational
neurology: a patient-centered
discipline that is a superset of
computational neuroscience
and includes observables of
mechanistic (micro) and
behavioral (macro) nature

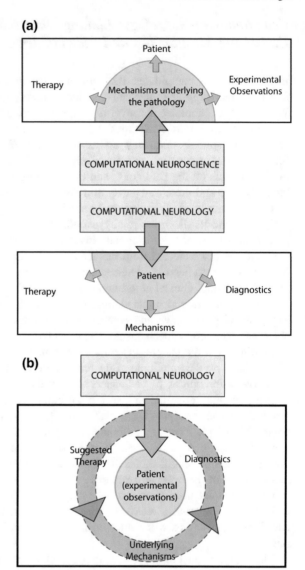

Results

NMDA Receptor Antagonism: Friend or Foe

We herein present results we obtained through Modeling and Simulation that
yielded a better comprehension of the effects of compounds on the NMDA receptor
and its associated channel, with direct significance on neuropharmacology of the
excitatory synaptic transmission. The N-methyl-D-aspartate receptor (also known

as NMDA receptor or NMDAR), is a glutamate receptor and ion channel found on excitatory postsynaptic spines. Its activation requires the binding of glutamate and glycine (or D-serine), and leads to the opening of its ion channel that is nonselective to cations with a reversal potential close to 0 mV. While the opening and closing of the ion channel is primarily gated by ligand binding, the flow of ions through the channel is voltage-dependent, due to the blockade of the channel by extracellular magnesium and zinc ions. Blockade removal allows the flow of sodium Na+ and small amounts of calcium Ca2+ ions into the cell, and potassium K+ out of the cell.

NMDA receptors are thought to play a critical role in the central nervous system. Interestingly, while competitive antagonists such as D-2-amino-5-phosphopentanoic acid (AP5) impair learning and memory, memantine, a non-competitive receptor antagonist has been reported to be paradoxically beneficial to patients with mild to moderate Alzheimer's disease (AD). In this study, we use a Markov kinetic model and look at the differences in the receptor dynamics and its associated channel current in response to changes in the presence of either molecule.

The kinetic schema used was proposed by Schorge [15] and presented in Fig. 3. The receptor model takes into account the two binding sites for glutamate, the two binding sites for the co-agonist glycine, and two open states O1 and O2; the non-linear voltage dependency is taken into account and depends on surrounding concentration of magnesium. Additional details and parameter values of the model may be found in [16]. Modifications of the kinetic model allow analysis of the effects of AP5 on receptor-channel current as it binds to the NMDAR in a competitive manner to the glutamate binding site. Association and dissociation rate constants (kon and koff) for AP5 were set at $0.38 \text{ mM}^{-1} \text{ ms}^{-1}$ and 0.02 ms^{-1}

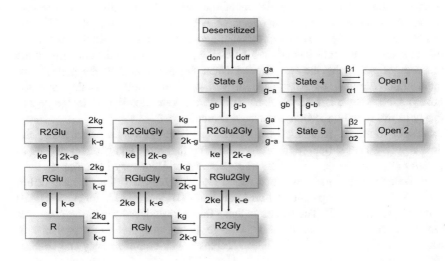

Fig. 3 The 15 states kinetic model of the NMDA receptor

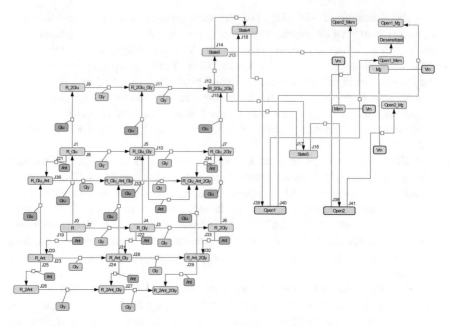

Fig. 4 NMDA receptor model modified to account for AP5 (competitive) and memantine (non-competitive) binding

respectively, based on published experimental results [17]. On the other hand, memantine has been shown to bind to the NMDA receptor in a voltage-dependent manner [18] (Fig. 4).

Learning and memory is associated on a cellular basis with changes in synaptic strengths due to repetitive stimulation of the synaptic connection. We therefore proposed to study the inhibitive effects of AP5 on the response of the NMDA receptor to trains of pulses of neurotransmitters at different frequencies; we quantified the cumulated inhibition on a 5 s window, and compared the response in the presence of AP5 with the one elicited when the receptor is in the presence of memantine. Simulations take place in voltage-clamp mode, meaning that the postsynaptic voltage is held constant. This removes the non-linear voltage dependency of the NMDA receptor associated channel.

Results outlined in Fig. 5 show that for both AP5 and memantine, the dose-responses are shifted to the right, indicating that more antagonist is needed to obtain the same level of inhibition when stimulation frequency increases.

In voltage-clamp mode, we then varied the postsynaptic voltage from -120 to $+20$ mV in low and high stimulation frequencies (10 and 100 Hz for AP5, and 10 and 200 Hz for MEM). The results indicate that AP5-induced inhibition decreases with frequency, but increases with voltage (when magnesium is present). In contrary, memantine-induced inhibition decreases with both frequency and voltage. This result is more clearly outlined in Fig. 6 which plots variations of the IC50

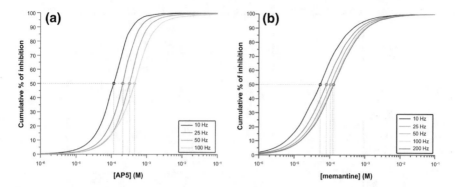

Fig. 5 Cumulative inhibition of the NMDA receptor current in the presence of AP5 (**a**) and memantine (**b**), in response to glutamate pulses applied at different frequencies (postsynaptic voltage is held constant at −60 mV). The inhibitory effect of both the competitive and non-competitive antagonists decreases as the stimulation frequency increases indicated by a shift to the right of the IC50 as pulses frequency increases [16]

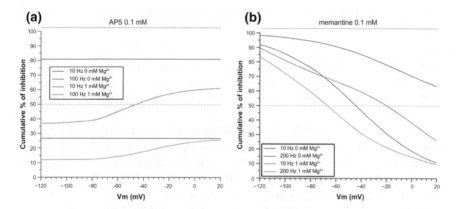

Fig. 6 Variations of cumulative inhibition of AP5 and memantine as a function of voltage at different glutamate application frequencies [16]

(concentration at which the amplitude of the output reaches 50 % of its maximum value) as a function of voltage at different frequencies (Fig. 7).

Our results indicate clear differences in the effects produced by AP5 and memantine on their target receptor and the resulting dynamics of its associated channel. AP5-induced inhibition is characterized by a weak voltage dependence, and an increase in IC50 values with heightened glutamate stimulation frequencies, indicating reduced inhibition. This suggests that in the presence of a large quantity of glutamate, AP5 loses its competitive advantage on the receptor's binding site, rendering it less potent. This may account for the failure of competitive antagonists reported in clinical trials, especially with respect to stroke. Indeed, stroke may be

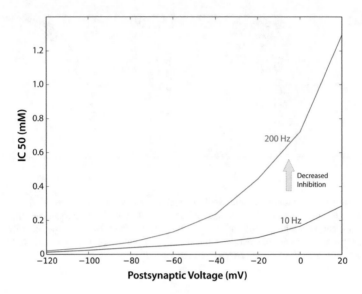

Fig. 7 Evolution of memantine IC50 as a function of voltage for 10 and 200 Hz glutamate pulse frequencies

characterized by an increase in neuronal firing frequency in the affected area due to excess glutamate. Maintaining AP5-induced inhibition would require several fold increases in AP5 concentration, which would then lead to serious and potentially toxic side effects.

On the contrary, our results indicate that memantine has a strong dependence on both voltage and stimulation frequency, which supports the notion that memantine provides a tonic blockade of the receptor in basal conditions. This could account for its neuroprotective attribute. However, this inhibition is lifted when stimulation frequency increases and postsynaptic membrane depolarizes (conditions presumably associated with learning new information), which would explain why memantine does not negatively impact learning.

Neuropharmacology of Combinations: Modeling from Biomolecular Mechanisms to Neuronal Spiking

One of the fundamental characteristics of the brain is its hierarchical and temporal organization; both space and time must be considered to fully grasp the impact of the system's underlying mechanisms on brain function. Complex interactions taking place at the molecular level regulate neuronal activity, which further modifies the function and structure of millions of neurons connected by trillions of synapses. This ultimately gives rise to phenotypic function and behavior at the system level. This spatial complexity is also accompanied by a complex temporal integration of events taking place at the microsecond scale, leading to slower changes occurring at the second, minute and hour scales. Simulation of mechanisms spanning multiple scales makes modeling a challenging task both from an implementation and numerical standpoint. Yet, these integrations are necessary for studying the effects of combinations of therapeutic agents that target very distinct systems and determining how these drugs interact to shape function at more integrated levels of complexity.

To illustrate proposed solutions to this challenge, we combine the NMDA-R competitive antagonist AP5 described earlier with a molecule known to act on the GABA A receptor. The GABA A receptor is an ionotropic receptor and ligand-gated ion channel that is found in inhibitory synapses; its endogenous ligand is GABA (γ aminobutyric acid), the major inhibitory neurotransmitter in the brain. Activation of this receptor leads to opening of its associated channel pore which selectively conducts chlorine ions inside the cell, resulting in hyperpolarization of the postsynaptic neuron. We propose to study the effect of bicuculline, a competitive antagonist of GABA A receptors, not only on the target's function, but also on the resulting spiking pattern when these receptors are placed on a CA1 pyramidal neuron [19].

The modeling framework consists of a co-simulation comprising multiple instances of the EONS simulator linked to NEURON through the message-passing interface MPJ Express [20] distributed on a high-performance computer cluster. The stimulation protocol consists in presenting a train of action potentials with random inter-pulse intervals at a mean frequency of 10 Hz (within the range of physiological frequencies reported in the hippocampus) as presynaptic inputs to both excitatory (i.e. glutamatergic) and inhibitory (i.e. GABAergic) synapses of a CA1 pyramidal neuron. The neuron model used is the pyramidal cell described in [21], which uses digitally reconstructed dendritic morphology described in [22] in which synaptic currents are integrated along dendritic branches (112 excitatory glutamatergic synapses located in the stratum radiatum area and 14 inhibitory synapses located close to the soma). The kinetic model for the GABA A receptor is the one presented in [23] (Fig. 8).

The model allows for readouts of molecular, synaptic (postsynaptic current and voltage) and neuronal nature (somatic potential, and spiking activity). Four conditions were simulated: a control condition (i.e. no modulator), with IC50 concentration of AP5 (established at 100 μM in the section "NMDA Receptor Antagonism: Friend or Foe"), with IC50 concentration of bicuculline, and with both antagonists combined at IC50 concentrations. The somatic potentials resulting from the 10 Hz random interval train is presented in Fig. 9 for all four conditions.

The somatic potentials obtained in the four conditions outline the high levels of non-linearity that arise at different levels of neuronal integration. Decreases in NMDA receptor current (50 % of the peak amplitude) at the molecular level result in a dramatic reduction in somatic spiking (77 % reduction) once placed in synapses and integrated along the dendritic tree of the pyramidal cell. Adversely, antagonizing the GABA A receptor results in over 50 % increase in the number of action potentials generated. When both modulators are applied, the number of spikes

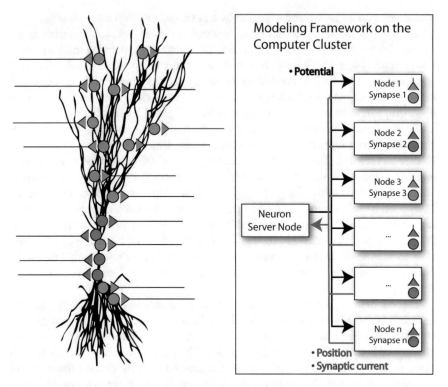

Fig. 8 *Right* Illustration of excitatory (*green*) and inhibitory (*red*) synapses distributed on a CA1 pyramidal neuron. *Left* Diagram illustrating the distribution of processes and communication flow on the high performance computer cluster

returns to a value close to the control condition (8 versus 9 action potentials), but with a very different (regularized) spiking pattern—likely to result in a very different outcome at the network level.

From Mechanistic to Non-mechanistic Modeling

The previous section outlined the notion that phenomena taking place at a specific scale in the nervous system may often interact in a non-linear manner and thereby yield emerging properties at other (often overlooked) scales. Indeed, exogenous compounds, modulators of excitatory and inhibitory synaptic receptors function were shown to interact and modulate the spike-timing patterns of a CA1 pyramidal neuron. The previous example, computed on several nodes of a high performance computer cluster, also outlines that integrating multiple levels of complexity (whether temporal or hierarchical) may result in a prohibitively large computational burden. Each synapse may de facto represent thousands of differential equations,

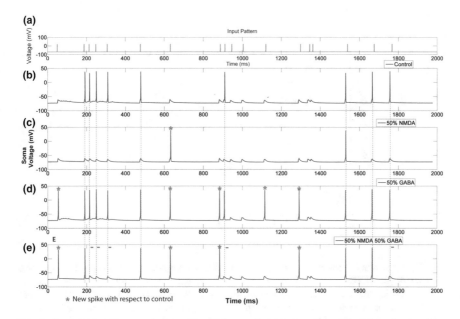

Fig. 9 From [19]. **a** Presynaptic input stimulation applied (random inter-pulse interval train with a Poisson distribution and a 10 Hz mean firing frequency). **b–e** CA1 hippocampal neuron somatic potential in response to stimulation **a**. **b** with no modulation. **c** Response with 100 μM AP5 (IC50 concentration). **d** Response with a 50 % decrease in GABA A current. **e** Response with both glutamatergic and gabaergic modulations combined

yet neurons can comprise thousands of synapses, and brain structures are composed of different neurons populations, each potentially containing millions of cells. Simulating such computational load requires (i) increased computational muscle (e.g. IBM BlueGene) and/or (ii) better management of the complexity of the models simulated. This section proposes to focus on the latter—suggesting the use of non-mechanistic modeling methodologies capable of capturing the non-linear dynamics of the system of interest, while significantly reducing the computational load.

The biochemical mechanisms underlying synaptic function have been shown to display a high level of non-linearities—both on the presynaptic and postsynaptic sides [24–27]. These non-linearities are critical and most likely play a significant role in shaping the functions of synapses and neurons, giving them the ability to learn and generate long term changes used to encode memories. Yet these mechanisms, if they are to be modeled in their mechanistic dynamical complexity yield a large number of differential equations, thereby resulting in substantial (and potentially prohibitively large) computational complexity.

An alternative approach is to consider the system of interest as a black box, focusing the computational complexity on replicating the functional dynamics, i.e. the outputs the system generates in response to a series of inputs, rather than the internal mechanisms comprised in the system. This approach was reported to be

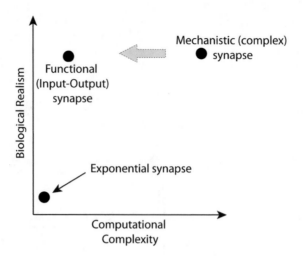

Fig. 10 Conceptual representation of different models of synapses in a realism—computational complexity plane. Exponential synapses are a good example of a relatively simple model of synapse that is often used in large scale simulations to save computational complexity. However, they lack in representing the non-linear dynamics of synaptic function (e.g. facilitation/depression, short term plasticity, etc.). Mechanistic models of synapses on the contrary may comprise a large number of mechanisms, thereby providing a high level of realism—but consequently impose more computationally heavy calculations which may become prohibitive in large scale neuronal network models. Examples of synaptic models spanning different levels of complexity may be found in [32]; example of very complex models are also described in [11, 33–35]

successfully applied with the use of Volterra kernels [28] on the dynamics of neuronal populations [29, 30], yielding highly predictive models with minimal computational complexity. We proposed to adapt this methodology to model electrical properties of glutamatergic synapses [31] and evaluate it both in terms of predictability and computational complexity (Figs. 10 and 11).

The properties modeled are the changes in conductance values of AMPA and NMDA receptors elicited by the response to presynaptic release of neurotransmitter. Estimation of parameters values in non-mechanistic models (interchangeably labeled input-output models in this context) is a crucial step that requires training the model with respect to reference input-output sequences obtained with the mechanistic model. Using long sequences captures a large number of nonlinear behavior thereby minimizing prediction errors by improving parameters estimation. The model structure comprises two sets of kernels with a slow and a fast time constant for each receptor model; we estimated the parameters of the model using a train of 1000 pulses at a 2 Hz random interval train (i.e. using a 500 s long simulation) (Fig. 12).

Having established that the response of the input-output model is very close to the response obtained with the mechanistic synapse model in the dynamical range of behaviors, we can now focus our attention on determining the computational speed gain obtained by replacing the mechanistic synapse model with an IO model.

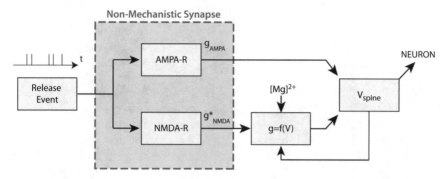

Fig. 11 The non-mechanistic model replaces the kinetic models of AMPA and NMDA receptors, corresponding to 16 and 15-states kinetic Markov models respectively. gAMPA is the conductance of the AMPA receptor channel; g* NMDA corresponds to the voltage-independent conductance value of the receptor

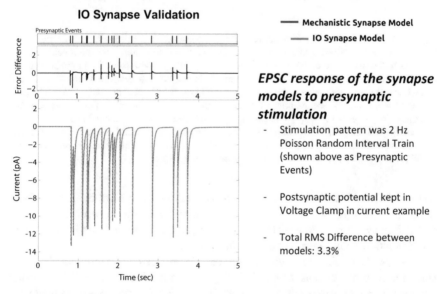

EPSC response of the synapse models to presynaptic stimulation

-   Stimulation pattern was 2 Hz Poisson Random Interval Train (shown above as Presynaptic Events)

-   Postsynaptic potential kept in Voltage Clamp in current example

-   Total RMS Difference between models: 3.3%

Fig. 12 5 s sample of dynamic response of the mechanistic model (used to calibrate the parameters values of the non-mechanistic model) and the non-mechanistic (IO) model. Visual inspection yields virtually identical response; the total root mean square (RMS) error calculated on a 500 s simulation with novel presynaptic events yield a 3.3 % error

To benchmark the models, multiple instances of the IO synapse model were integrated in a neuron model and simulation duration was compared to the one obtained with the original mechanistic models.

We chose the hippocampal pyramidal neuron model proposed by Jarsky [36]. To minimize the computational load necessary to perform neuron-related calculations, we assigned the weight of each synapse with a zero value—thereby ensuring that

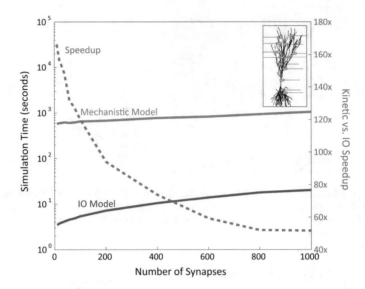

Fig. 13 Adapted from [31]. Simulation Time (represented in logarithmic scale) varies as a function of the number of synapse instances. Computation time required for the kinetic synapse model is within the range of 10–20 min, while the computation time required for the IO synapse model is consistently at least an order of magnitude lower, ranging between 3 and 30 s. Dashed green line represents the speedup obtained with the IO synapse model compared to the mechanistic model as a function of the number of synapses modeled. The insert on the top right corner illustrates the process of having glutamatergic synapses added onto a pyramidal neuron model. At low number of synapses, the speedup of the IO synapse model is highest (around 150× faster than the computation time required for the kinetic synapse model). The speedup decreases, but stabilizes at around 50× speedup for larger number of synapses

(i) the time needed to calculate the neuron model remains constant independently of the number of synapses, and (ii) the largest extent of computational time is spent calculating synapses outputs. The number of synapses was varied and we recorded the simulation times. Results are presented in Fig. 13.

The results in Fig. 13 indicate that the IO model consistently yields faster computation times even as the number of synapses modeled increases. The speedup decreases and seems to reach a plateau at around 50× (this value was verified with higher number of synapses—up to 10,000—not represented in the figure to focus on the non-asymptotic range).

Appropriately trained non-mechanistic models constitute a viable replacement for detailed mechanistic models in large scale models, yielding a significant gain in computational speed, while maintaining high predictability levels—thereby allowing larger scale models to be simulated while preserving biologically relevant subtleties in dynamics and non-linearities.

The non-mechanistic modeling methodology presented is generalizable and fully applicable, not only to other types of synapses (i.e. inhibitory or modulatory), but also other processes with identified inputs-outputs—yielding a natural solution for hierarchical large scale multiscale modeling challenges.

Messages for Neurologists and Computer Scientists

Experimental techniques have grown tremendously in the past decades, leading to a deeper understanding of the mechanisms that take place in the nervous system at multiple levels—ranging from genes and biomolecular mechanisms to brain level activity obtained through imaging methodologies. These notable improvements lead to an exponential increase in the amount of data acquired. They also lead to a more quantitative understanding of the roles of the mechanisms underlying the central nervous system's functions and dysfunctions. Finally, they yield a better understanding of the effects of perturbations that can occur within these processes, as well as those of exogenous compounds. This deepened understanding enables the construction of predictive computational models capable of simulating the nervous system (in its control and pathological states), and the changes that can take place in response to exogenous compounds.

One of the advantages of multiscale computational models is their inherent ability to integrate, within the same model, experimental observations obtained using different (often incompatible) experimental modalities (or experimental paradigms as defined in Fig. 1), leading to the creation of a single modeled entity which characteristics reproduce all (or at least most) observations. This inherent ability can lead to the creation of virtual patients—patients with normal neuronal function, or with pathological dysfunctions. These virtual patients comprise within the same simulation framework the biomolecular mechanisms that have been demonstrated to take place in the normal and pathological cases, along with the consequences at higher levels of hierarchical (i.e. neuronal, structural and behavioral) and temporal (e.g. neurodegeneration) aggregated scales. The creation of this integrated model will constitute a major accomplishment of the emerging field of computational neurology, the patient-centered successor of computational neuroscience.

This also constitutes a tremendous opportunity for the pharmaceutical industry—leading to an integrated model on which multiple steps of the drug pipeline may be performed in silico in a unified manner—encompassing a large number of processes from target identification to efficacy, side effects and toxicity evaluations. This can lead to much needed defragmentation of the drug discovery process, consequently reducing the attrition rate that plagues drug discovery and development.

The 'dream' described above, and more broadly discussed in this book is met by several roadblocks, amongst which are computational power: despite our computers seemingly ever increasing processing power, creating large integrated models spanning several hierarchical and temporal levels constitutes a real challenge. This challenge may be faced using hybrid models that consist of a combination of mechanistic and non-mechanistic models interacting in a seamless manner—

leading to efficient simulation of complex non-linear dynamics with a high level of functional realism. Another roadblock is constituted by our limited knowledge of many of the intricate mechanisms that underlie the complex biology of the nervous system. This challenge will ultimately be tackled as the field matures: using open standards and good computational modeling practices will ensure reproducible, iterative and collaborative modeling, consequently allowing iterative incorporation of additional findings while tuning and optimizing the rest of the model. This maturation will also see the development of frameworks compatible with the large number of methodologies and standards currently in use for modeling the multiple spatial and temporal scales of the nervous system.

Finally, a major shift will place the patient at the center. This shift has already started with the emergence of personalized medicine. It will expand to computational neurology to generate in silico models with personalized parameters values for utmost personalized medical prognosis and diagnosis.

Acknowledgments The authors would like to acknowledge all collaborators from Rhenovia Pharma, Serge Bischoff, Michel Baudry, Saliha Moussaoui, Florence Keller, Arnaud Legendre, Nicolas Ambert, Renaud Greget, Merdan Sarmis, Fabien Pernot, Mathieu Bedez and Florent Laloue, and the Center for Neural Engineering at University of Southern California, Dong Song, Eric Hu, Adam Mergenthal, Mike Huang, and Sushmita Allam for fruitful work and collaborations; Sophia Chen for proofreading this chapter. Finally, we thank the reviewers for their constructive comments and suggestions towards improving the manuscript.

References

1. "Pace of CNS drug development and FDA approvals lags other drug classes," *Tufts Cent. Study Drug Dev.*, vol. 14, no. 2, 2012.
2. "Longer clinical times are extending time to market for new drugs in US - Tufts CSDD Impact Report 7," 2005.
3. I. Kola and J. Landis, "Opinion: Can the pharmaceutical industry reduce attrition rates?," *Nat. Rev. Drug Discov.*, vol. 3, no. 8, pp. 711–716, Aug. 2004.
4. K. I. Kaitin and C. P. Milne, "A Dearth of New Meds," *Sci. Am.*, vol. 305, no. 2, pp. 16–16, Jul. 2011.
5. "Evaluation of Current Animal Models," Washington (DC). National Academies Press (US), 2013.
6. FDA, "Innovation or Stagnation: Challenge and Opportunity on the Critical Path to New Medical Products," 2004.
7. K. Gurney and M. Humphries, "Methodological Issues in Modelling at Multiple Levels of Description," in *Computational Systems Neurobiology*, N. Le Novere, Ed. Springer, 2012, pp. 259–281.
8. K. J. Friston and R. J. Dolan, "Computational and dynamic models in neuroimaging," *Neuroimage*, vol. 52, no. 3, pp. 752–765, Sep. 2010.
9. N. T. Carnevale and M. L. Hines, *The NEURON Book*. Cambridge: Cambridge University Press, 2006.
10. J.-M. C. Bouteiller, Y. Qiu, M. B. Ziane, M. Baudry, and T. W. Berger, "EONS: an online synaptic modeling platform," *Conf. Proc. Int. Conf. IEEE Eng. Med. Biol. Soc.*, vol. 1, pp. 4155–4158, 2006.

11. J.-M. C. Bouteiller, M. Baudry, S. L. Allam, R. J. Greget, S. Bischoff, and T. W. Berger, "Modeling Glutamatergic Synapses: Insights Into Mechanisms Regulating Synaptic Efficacy," *J. Integr. Neurosci.*, vol. 7, no. 2, pp. 185–197, Jun. 2008.

12. J.-M. C. Bouteiller, S. L. Allam, E. Y. Hu, R. Greget, N. Ambert, A. F. Keller, F. Pernot, S. Bischoff, M. Baudry, and T. W. Berger, *Modeling of the nervous system: From molecular dynamics and synaptic modulation to neuron spiking activity*, vol. 2011. IEEE, 2011, pp. 445–448.

13. E. T. Somogyi, J.-M. Bouteiller, J. A. Glazier, M. König, J. K. Medley, M. H. Swat, and H. M. Sauro, "libRoadRunner: a high performance SBML simulation and analysis library: Table 1," *Bioinformatics*, vol. 31, no. 20, pp. 3315–3321, Oct. 2015.

14. M. Hucka, A. Finney, H. M. Sauro, H. Bolouri, J. C. Doyle, H. Kitano, and the rest of the SBML Forum: A. P. Arkin, B. J. Bornstein, D. Bray, A. Cornish-Bowden, A. A. Cuellar, S. Dronov, E. D. Gilles, M. Ginkel, V. Gor, I. I. Goryanin, W. J. Hedley, T. C. Hodgman, J.-H. Hofmeyr, P. J. Hunter, N. S. Juty, J. L. Kasberger, A. Kremling, U. Kummer, N. Le Novere, L. M. Loew, D. Lucio, P. Mendes, E. Minch, E. D. Mjolsness, Y. Nakayama, M. R. Nelson, P. F. Nielsen, T. Sakurada, J. C. Schaff, B. E. Shapiro, T. S. Shimizu, H. D. Spence, J. Stelling, K. Takahashi, M. Tomita, J. Wagner, and J. Wang, "The systems biology markup language (SBML): a medium for representation and exchange of biochemical network models," *Bioinformatics*, vol. 19, no. 4, pp. 524–531, Mar. 2003.

15. S. Schorge, S. Elenes, and D. Colquhoun, "Maximum likelihood fitting of single channel NMDA activity with a mechanism composed of independent dimers of subunits," *J. Physiol.*, vol. 569, no. Pt 2, pp. 395–418, Dec. 2005.

16. N. Ambert, R. Greget, O. Haeberlé, S. Bischoff, T. W. Berger, J.-M. Bouteiller, and M. Baudry, "Computational studies of NMDA receptors: differential effects of neuronal activity on efficacy of competitive and non-competitive antagonists.," *Open Access Bioinformatics*, vol. 2, pp. 113–125, 2010.

17. M. Benveniste, J. M. Mienville, E. Sernagor, and M. L. Mayer, "Concentration-jump experiments with NMDA antagonists in mouse cultured hippocampal neurons," *J. Neurophysiol.*, vol. 63, no. 6, pp. 1373–84, Jun. 1990.

18. S. E. Kotermanski and J. W. Johnson, "Mg2 + Imparts NMDA Receptor Subtype Selectivity to the Alzheimer's Drug Memantine," *J. Neurosci.*, vol. 29, no. 9, pp. 2774–2779, Mar. 2009.

19. J.-M. C. Bouteiller, A. Legendre, S. L. Allam, N. Ambert, E. Y. Hu, R. Greget, A. F. Keller, F. Pernot, S. Bischoff, M. Baudry, and T. W. Berger, "Modeling of the nervous system: From modulation of glutamatergic and gabaergic molecular dynamics to neuron spiking activity," in *2012 Annual International Conference of the IEEE Engineering in Medicine and Biology Society*, 2012, pp. 6612–6615.

20. M. Baker, B. Carpenter, and A. Shafi, "MPJ Express: Towards Thread Safe Java HPC," in *2006 IEEE International Conference on Cluster Computing*, 2006, pp. 1–10.

21. M. Ferrante, K. T. Blackwell, M. Migliore, and G. A. Ascoli, "Computational models of neuronal biophysics and the characterization of potential neuropharmacological targets," *Curr. Med. Chem.*, vol. 15, no. 24, pp. 2456–71, 2008.

22. G. A. Ascoli, D. E. Donohue, and M. Halavi, "NeuroMorpho.Org: a central resource for neuronal morphologies," *J. Neurosci.*, vol. 27, no. 35, pp. 9247–51, Aug. 2007.

23. J. R. Pugh and I. M. Raman, "GABAA receptor kinetics in the cerebellar nuclei: evidence for detection of transmitter from distant release sites," *Biophys. J.*, vol. 88, no. 3, pp. 1740–54, Mar. 2005.

24. J. S. Dittman and W. G. Regehr, "Calcium dependence and recovery kinetics of presynaptic depression at the climbing fiber to Purkinje cell synapse," *J. Neurosci.*, vol. 18, no. 16, pp. 6147–62, Aug. 1998.

25. W. G. Regehr, "Short-term presynaptic plasticity," *Cold Spring Harb. Perspect. Biol.*, vol. 4, no. 7, p. a005702, Jul. 2012.

26. S. L. Allam, J.-M. C. Bouteiller, E. Y. Hu, N. Ambert, R. Greget, S. Bischoff, M. Baudry, and T. W. Berger, "Synaptic Efficacy as a Function of Ionotropic Receptor Distribution: A Computational Study," *PLoS One*, vol. 10, no. 10, p. e0140333, Oct. 2015.

27. J.-M. C. Bouteiller, S. L. Allam, R. Greget, N. Ambert, E. Y. Hu, S. Bischoff, M. Baudry, and T. W. Berger, "Paired-pulse stimulation at glutamatergic synapses - pre- and postsynaptic components," *Eng. Med. Biol. Soc. EMBC 2010 Annu. Int. Conf. IEEE*, vol. 2010, pp. 787–790, 2010.

28. V. Volterra, *Theory of functionals and of integral and integro-differential equations*. New York: Dover, 1959.

29. D. Song, Z. Wang, V. Z. Marmarelis, and T. W. Berger, "Parametric and non-parametric modeling of short-term synaptic plasticity. Part II: Experimental study," *J. Comput. Neurosci.*, vol. 26, no. 1, pp. 21–37, Feb. 2009.

30. V. Z. Marmarelis, *Nonlinear dynamic modeling of physiological systems*. John Wiley & Sons, 2004.

31. E. Y. Hu, J.-M. C. Bouteiller, D. Song, M. Baudry, and T. W. Berger, "Volterra representation enables modeling of complex synaptic nonlinear dynamics in large-scale simulations," *Front. Comput. Neurosci.*, vol. 9, Sep. 2015.

32. A. Roth and M. C. W. van Rossum, "Modeling Synapses," in *Computational Modeling Methods for Neuroscientists*, The MIT Press, 2009, pp. 139–160.

33. S. Nadkarni, T. M. Bartol, T. J. Sejnowski, and H. Levine, "Modelling vesicular release at hippocampal synapses," *PLoS Comput. Biol.*, vol. 6, no. 11, p. e1000983, Jan. 2010.

34. S. L. Allam, V. S. Ghaderi, J.-M. C. Bouteiller, A. Legendre, N. Ambert, R. Greget, S. Bischoff, M. Baudry, and T. W. Berger, "A computational model to investigate astrocytic glutamate uptake influence on synaptic transmission and neuronal spiking," *Front. Comput. Neurosci.*, vol. 6, p. 70, Jan. 2012.

35. R. Greget, F. Pernot, J.-M. C. Bouteiller, V. Ghaderi, S. Allam, A. F. Keller, N. Ambert, A. Legendre, M. Sarmis, O. Haeberle, M. Faupel, S. Bischoff, T. W. Berger, and M. Baudry, "Simulation of Postsynaptic Glutamate Receptors Reveals Critical Features of Glutamatergic Transmission," *PLoS One*, vol. 6, no. 12, p. e28380, Dec. 2011.

36. T. Jarsky, A. Roxin, W. L. Kath, and N. Spruston, "Conditional dendritic spike propagation following distal synaptic activation of hippocampal CA1 pyramidal neurons," *Nat. Neurosci.*, vol. 8, no. 12, pp. 1667–1676, Dec. 2005.

Printed in the United States
By Bookmasters